Phase Change Materials

Simone Raoux • Matthias Wuttig
Editors

Phase Change Materials

Science and Applications

Springer

Editors
Simone Raoux
IBM Almaden Research Center
650 Harry Road
San Jose, CA 95120
USA

Matthias Wuttig
1. Physikalisches Institut (1A)
RWTH Aachen University
52056 Aachen
Germany

ISBN 978-0-387-84873-0 e-ISBN 978-0-387-84874-7
DOI 10.1007/978-0-387-84874-7

Library of Congress Control Number: 2008935619

Printed on acid-free paper

springer.com

Foreword

It is a pleasure to write a foreword to this much needed book on "Phase Change Materials: Science and Applications." The book is a measure of the growing importance of the field. Phase change materials cover almost every aspect of material science from materials in the amorphous state to intermediate range order and to nano-crystalline and micro-crystalline states. Nanotechnology is considered to be a new science, yet we have utilized nano-materials in phase change memories since 1960. The richness of material science involved cannot be overstated. We are dealing with truly atomically engineered materials. Three examples are the Ovonic threshold switch, the multi-state Ovonic Universal Memory (OUM), and the Ovonic cognitive device which emulates the biological neurons with its plasticity and synaptic activity.

The field of amorphous and disordered materials created not only a basic new area of science, but also important new technologies. It should be kept in mind that to do so we had to invent the materials, products and manufacturing technology.

Amorphicity and disorder opened up new degrees of freedom that by-pass the constraints of crystalline periodicity. This freedom enabled me to synthesize materials with new physical, chemical, electronic and structural properties of unique functionality.

Referring to the Ovonic electrical phase change memories, Intel's chief technology officer in flash memory, Ed Doller, said at a meeting in 2007, "The phase change memory gets pretty close to Nirvana." Ironically, Bob Noyce and Gordon Moore came out to visit Iris and me before they set up Intel because there existed no solid state memory except our Ovonic Phase Change Memory. At that time, all computer memories were wire wound magnetic core devices.

The background of my work goes back to 1955, where I was determined to lay the basis of cognitive computing by creating an inorganic analog of biological neurons and their synapses. Fortunately, my approach attracted great interest from Dr. Ernest Gardner, head of the Department of Anatomy of Wayne State Medical School of Detroit, a highly respected neurophysiologist, who invited me to join his team to continue work on my ideas.

Such an analog neuronal device had to respond to various thresholds, exhibit plasticity and non-volatility and, of course, reversibility between a conducting and a non-conducting state. Since there was no prior literature to guide me, I had to create new materials and initiate a field of atomically designed amorphous and disordered materials.

The phase change materials that I developed had to have plasticity and a reversible structural change. I applied polymer science to inorganic polymeric materials such as Te and Se. Te/ Se are lone pair, polymeric, divalent materials with chain structures. They can be designed to bond with cross-linking elements of different bond strengths such as Ge and Sb. Such atomically designed, cross-linked Te/Se alloys have a huge number of non-bonded lone-pair electrons which could easily be excited by optical and electrical fields. When the amorphous state cannot contain the excitation energy, a phase change to a crystalline state occurs.

It is important to understand that such polymeric lone-pair chain structures are vibronic in nature wherein electronic transitions are made possible by vibrational motion of the chains. These simultaneous vibrational and electronic transitions are important to understanding the mechanism of both the optical and the electrical Ovonic phase change memories.

Lone-pair electrons are crucial for both the non-volatile phase change memory and the electronic Ovonic threshold switch which is volatile. The crucial difference between them is the amount of cross links and their bonding energies. Strong and numerous cross-linking atoms such as Si, As and Ge provide an amorphous material that is structurally not crystallizable, but electronically switchable when an electrical field beyond a certain threshold value is coupled to the large number of non-bonding lone pairs in the material. This results in a highly conducting filament, which expands and contracts in a servo manner to a load. These conducting filaments have a plasma nature which can sustain about 50 times the current density encountered in a typical transistor; an important implication for future devices.

The Ovonic phase change memory is a non-volatile device, which is commonly used for binary operations. But the phase change memory is capable of much more. We have demonstrated that it can operate reproducibly in many resistance states which correspond to different amorphous to crystalline fractions in the material. The plasticity of phase change memory and its ability to have many other states in both the amorphous and crystalline portions of the resistance vs. current curve (the U curve) enabled me to develop the cognitive computing device that resembles the biological neurons and synaptic activity.

Now, what is the future of our field? The Ovonic universal phase change memory is called universal because it can replace flash memory, DRAM and SRAM. These are not only basic computer memory devices but also are becoming the driving force for the ongoing revolutionary growth of cell phones and other mobile devices, which are in desperate need of memory providing higher density, faster speed and lower power consumption. As we write, the OUM has penetrated those areas already and it will excel with its multi-state operations. It is also important to keep in mind the point that I've been making since the very beginning that information is encoded energy.

What else will be coming along in the future? The phase change memories hold great promise for the following reasons: our devices are scalable, the smaller we make them, the better they are; they can be reduced in size to the quantum limit. The crystalline Si based transistors might be replaced by Ovonic three-terminal devices. The various multi-states Ovonic devices will be able to replace the logic and memory functions. In other words, logic and memory will become one. Huge parallelism which is not available in conventional computing will become standard, more and more re-configurable and interactive circuits will be integrated with the new parallelism. The Ovonic cognitive devices will make cognitive computing possible. Recall that Bill Gates said, "If you invent a breakthrough in artificial intelligence, so machines can learn, that is worth 10 Microsofts."

The work described by the authors in this book illustrates the range of contributions that are being made throughout the world. The future of our field continues to be expanded by ongoing creative work. The editors deserve much credit for assembling the important work of noted authors in this book. The collegiality and collaboration I have enjoyed is remarkable.

Bloomfield Hills, MI
March 2008

Stanford R. Ovshinsky

Preface

Phase change materials are characterized by a unique property portfolio. They possess a pronounced difference of optical and electronic properties depending upon their atomic arrangement, i.e. whether they are amorphous or crystalline. At the same time they can be rapidly switched between the amorphous and crystalline state. This combination of properties is attractive for applications and also provides a unique opportunity to test our understanding of the relationship between bonding and the atomic arrangement in solids and their correlation to solid state properties.

In rewriteable optical data storage employing phase change materials a short pulse of a focused, high intensity laser beams locally heats the phase change material above its melting temperature. Rapidly cooling the phase change alloy with rates higher than 10^9 K/s quenches the liquid-like state into a disordered, amorphous phase. This amorphous state has different optical properties than the surrounding crystalline state. Hence detecting regions with amorphous structure is straightforward employing a laser beam of low intensity. To erase the stored information a laser pulse with intermediate power is utilized. The laser locally heats the phase change film above the crystallization temperature. At elevated temperatures above the glass transition temperature the atoms become increasingly mobile and can revert to the energetically favourable crystalline state, erasing the recorded information. In electronic memory applications of phase change materials a short and relatively high current pulse is used to locally melt the crystalline material. The resulting amorphous state has a high resistance which exceeds the resistance of the crystalline state by several orders of magnitude. An intermediate power pulse is used to heat the material above its crystallization temperature to switch it back to the crystalline, low resistance state, while a low power pulse is used to determine the resistance of the phase change material.

While the simplicity of the storage concept has attracted the interest of industry a number of questions will decide how far reaching the potential of phase change materials is. For rewriteable optical data storage the most prominent questions are linked to a strong optical contrast also at short laser wavelengths as required for next generation optical storage media. Another important aspect is the quest for very high speed materials that enable crystallization on the nanosecond time scale. Finally mechanisms to bypass the diffraction limit are of paramount interest. For electronic memories again the ultimate data transfer rate is of utmost importance. Since crystallization of phase change materials is the time limiting step in storage applications a detailed understanding of ultra-fast crystallization processes is a prerequisite for the design of even faster materials and the identification of intrinsic material limits. It would be ideal if it was possible to correlate the atomic arrangement in the amorphous and crystalline state with the speed of the transformation process. Up to now this goal has not yet been reached. Nevertheless a detailed understanding of the atomic arrangement in the amorphous and crystalline state should facilitate an in-depth understanding of the bonding and the physical properties of phase change materials. To fully exploit the potential of phase change materials in electronic memory applications a detailed understanding of electronic switching phenomena is a prerequisite.

Material scientists are faced with formidable challenges if phase change solid state memory is considered for new applications. For example, in automotive applications rather high operation temperatures of about 150 °C are required at which none of the currently applied phase change materials is stable in the amorphous phase. In case of a possible DRAM replacement extremely high switching speed and very high numbers of switching cycles (10^{16}-10^{18}) are necessary. The quest for the ultimate scaling limits will decide if this is a viable technology that can be developed for several next generations of lithographies following the roadmap of the semiconductor industry.

This book covers our current understanding of the science and the status of the applications of phase change materials. In the foreword, written by the founding father of the field, Stanford Ovshinsky, the past, the present status and future perspectives of phase change materials are discussed. Chapter 1 provides a discussion of the history of phase change materials. The theoretical understanding of phase change materials and the nature of glasses are discussed in Chapters 2 and 3, while the structure of amorphous Ge-Sb-Te alloys is presented in Chapter 4. Selection strategies for phase change materials are provided in Chapters 5 and 10, with the latter chapter focusing on the development of materials for third generation rewriteable optical data storage. The crystallization kinetics which are important to maximize data transfer rates are the subject of chapter 7. Chapters 8 and 9 discuss the short and long-range order and the resulting optical and electrical properties with an emphasis on the prototype material $Ge_2Sb_2Te_5$. Chapter 6 discusses the scaling behavior of phase change materials, while novel deposition methods are presented in Chapter 11. The following chapters focus on applications of phase change materials. In Chapter 12 the first three generations of optical memories are discussed, while 4th generation optical memories employing near-field effects are presented in Chapter 13. The last five chapters focus on applications utilizing electronic switching phenomena for storage and logic. Chapter 14 provides an in-depth discussion of phase change memory device modeling, while advanced prototype devices and their scaling are presented in Chapter 15. Phase change memory cell concepts are introduced in Chapter 16 and integration aspects are provided in Chapter 17. Finally applications in reconfigurable logic are the focus of Chapter 18.

Santa Clara, CA
July 2008

Simone Raoux

Aachen, Germany
July 2008

Matthias Wuttig

Contents

Part II: Applications: Optical, Solid State Memory and Reconfigurable Logic

Abbreviations

ABC - Atanasoff-Berry Computer
ADC - analog-to-digital converter
AFM - atomic force microscope
AIMD - *ab initio* molecular-dynamics
AIST - Ag- and In-doped Sb_2Te
ALD - atomic layer deposition
ANN - artificial neural network
ASIC - application specific integrated circuit
AWSAM - adjusting write sequenced amplitude margin
BD - blu-ray disc
BEC - bottom electric contact
BEOL - back-end-of-line
BER - bit error rate
BJT - bipolar junction transistor
BL - bit line
BOE - buffered oxide etch
CB - conduction band
CCT - complete crystallizing time
CD - compact disc
CD - critical dimension
CD-ROM - compact disc read-only memory
CD-RW - compact disc re-writable
CET - complete erasure times
CLB - configurable logic block
CLK - clock
CMOS - metal-oxide-semiconductor
CMP - chemical mechanical polishing
CNR - carrier-to-noise ratio
COHP - chemical orbital Hamilton population
CP - Car-Parrinello
CPMD - Car-Parrinello molecular dynamics
CVD - chemical vapor deposition
DAC - digital-to-analog converter
DC - direct current
DFT - density functional theory
DOS - density of states

dpm - defect-per-million
DRAM - dynamic random access memory
DSC - differential scanning calorimetry
DTA - differential thermal analyzer
DUT - device under testing
DVD - digital versatile disc
DVD-RAM - digital versatile disc random access memory
DVD+RW - digital versatile disc re-writable
EC-ALE - electrochemical atomic layer epitaxy
EDX - energy-dispersive x-ray spectroscopy
EMA - effective medium approximation
E.M.F. - electromotive force
EXAFS - extended x-ray absorption fine structure
fcc - face-centered cubic
FDM - finite difference method
FET - field-effect transistor
FPGA - field programmable gate array
FSDP - first sharp diffraction peak
FSM - finite state machine
FTDT - finite-differential time-domain
GD - growth dominated
GDM - growth-dominated material
GFT - glass formation tendency
GGA - general gradient approximation
GND - ground
GST - Ge-Sb-Te
HD - high definition
HDTV - high definition television
IC - integrated circuit
ILD - inter layer dielectric
I/O - input/output
IP - intermediate phase
IRO - intermediate-range order
ITRS - International Technology Roadmap for Semiconductors
JDOS - joint density of states
KS - Kohn-Sham
LDA - local density approximation
LOCOS - local oxidation of silicon
LSV - linear sweep voltammetry
LUT - look-up table
m-DSC - modulated differential scanning calorimetry
MD - molecular-dynamics
MEMS - micro-electro-mechanical system
MLC - multi-level-cell

MOCVD - metal-organic chemical vapor deposition
MOS - metal oxide semiconductor
MOSFET - metal oxide semiconductor field effect transistor
MRO - medium-range order
MSRD - mean-square relative displacement
NA - numerical aperture
ND - neutron diffraction
NDM - nucleation-dominated material
NDR - negative differential resistance
N/G - nucleation and growth
NMR - nuclear magnetic resonance
nn - nearest-neighbor
nnn - next-nearest-neighbor
NSPS - nanoscale phase separate
NVM - non-volatile memory
OTP - one-time-programmable
PC - phase change
PCE - phase change element
PCM - phase change material
PCM - phase-change memory
PCRAM - phase change random access memory
PDE - partial differential equation
PDF - pair distribution function
PDS - photo-thermal deflection spectroscopy
PECVD - plasma-enhanced chemical vapor deposition
PEL - potential energy landscape
PF - Poole-Frenkel
PMRL - partial-response maximum likelihood
ppb - parts-per-billion
ppm - parts-per-million
PVD - physical vapor deposition
RAM - random access memory
RDF - radial distribution function
RENS - resolution near-field structure
RF - radio frequency
RIE - reactive ion etching
RMC - reverse Monte Carlo
ROM - read-only memory
SA - self-aligned
SAM - sequenced amplitude margin
SAT - stochastic agglomeration theory
SEM - scanning electron microscope
SLL - superlattice-like
SRAM - static random access memory

SRO - short-range order
STD - standard
STI - shallow trench isolation
STS - sub-threshold slope
SU - somewhat useless
TDDFT - time-dependent density functional theory
TEM - transmission electron microscopy
TLUT - thresholded look-up table
TTL - transistor-transistor logic
UHD - ultra-high definition
UOM - Ovonic unified memory
UPD - under potential deposition
UPS - ultraviolet photoemission spectroscopy
VAP - valence alternation pairs
VB - valence band
VEC - valence electron concentration
VLS - vapor-liquid-solid
VLSI - very-large-scale integration
WL - word line
XAFS - x-ray absorption fine structure
XANES - x-ray absorption near-edge structure
XAS - x-ray absorption spectroscopy
XPS - x-ray photoemission spectroscopy

1. History of Phase Change Memories

Chung H. Lam

Abstract This chapter reviews the history of phase change materials particularly in the applications of information storage. The chapter starts with the discovery of a one way resistance transformation phenomenon in a chalcogenide, namely molybdenite (MoS_2). Then the evolution of the understanding of the underlying physics governing the phase change characteristics by various investigators is reviewed along with the applications of the phase change characteristics in information storage applications. The chapter ends with a table summarizing critical events in the phase change memory developments.

1.1 The Discovery of Phase Change Materials

While Stanford Ovshinsky is generally credited as the inventor of phase change materials for information storage, the discovery of phase changing electrical characteristics dates back to the early 1900s in the little known and seldom cited pioneering work of Alan Tower Waterman of Yale University. While studying thermionic emission of certain hot salts [1.1], Waterman noted some peculiarities in the conductivity of MoS_2. This observation led him to further investigation of the conductivity of MoS_2. He observed a large negative coefficient of resistance of the MoS_2 with respect to temperature inasmuch Ohm's law is not obeyed. This large negative coefficient of resistivity is typical of semiconducting chalcogenide materials. More significantly, he also observed a breakdown characteristic when the device under test was heated by means of the electronic current as shown in Fig. 1.1 reprinted from his published findings on the conductivity of MoS_2 [1.2]. Essentially, the breakdown is characteristic of the MoS_2 when heated by means of electric current. In Fig. 1.1, he demonstrated that the conductivity of MoS_2 can be changed progressively until the heating was carried beyond the breakdown voltage threshold whence the increase in conductivity was permanent. He pointed out that MoS_2 may exist in two forms, α of high resistance and β of low resistance. The breakdown phenomenon and progressive conductivity changes are prominent of the phase change behavior in chalcogenide materials. In his study, however, without high speed melt-quench technique, the phase change or conductivity change is either permanent or semi-permanent.

S. Raoux, M. Wuttig (eds.), *Phase Change Materials*,
DOI 10.1007/978-0-387-84874-7_1, © Springer Science+Business Media, LLC 2009

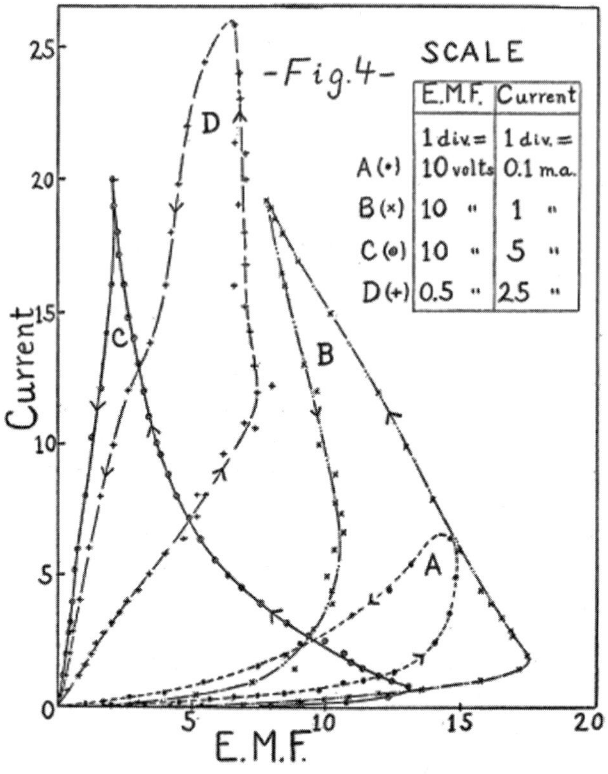

Fig. 1.1. Progressive change in conductivity of a single strip MoS_2 under current heating (reprinted from [1.1]). E. M. F. is the electromotive force.

Waterman pointed out that the transition from α form to β form was initiated by heat, electric field or light. Without the use of modern physical analysis tools such as transmission electron microscopy and x-ray crystallography, Waterman was not able to observe any physical changes of MoS_2 along with the conductivity changes. He did notice an increase in hardness accompanying increased conductivity.

1.2 Early Electronic Computers and Memory Systems

Waterman's finding on the peculiar conductivity characteristic of MoS_2 drew little attention since no practical application could have been envisioned with this characteristic before the first electronic computer was built. The basic principle of the electronic computer in which both operating data and operating instructions of the computing machine are to be stored in the memory of the computing machine was introduced by Alan Turing in his abstract "Universal Computing Machine" of 1936 at Cambridge University. In 1939, John Vincent Atanasoff of then Iowa

State College (now University) with his student Clifford Berry built the first electronic computer which has come to be known as the Atanasoff-Berry Computer, or ABC, using approximately 300 vacuum tubes. While the electronic part of the ABC worked successfully, the mechanical memory card-reader often introduced fatal errors rendering the computer unreliable. Punched cards, originally invented by Herman Hollerith, were used before computers in intensive statistics tabulation. It was a natural choice to use punched cards as the main memory system in early electronic computers. Although the reliability of electromechanical punched-card machines had greatly improved with the advance of electronic computers, thanks to the effort of Thomas Watson of International Business Machines, the mechanical punched card memory system was obviously the bottle neck for electronic computer performance.

In Turing's lecture to the London Mathematical Society in 1947 while outlining the ideas of the automatic computing engine, Turing emphasized the need of a practical form of memory for storing instructions and data aiming at replacing the paper punched cards. The development of magnetic core memory [1.3] in the 1950's provided a fast and reliable main memory system for early commercial electronic computers. Magnetic core memories were labor intensive hand strung wire cores. A solid state memory was more desirable particularly with the rapid advance of the transistor electronics since the invention of the junction transistor by John Bardeen, Walter Brattain and William Shockley unveiled by Bell Labs in 1948. The invention of the integrated circuit by Jack Kilby in 1958 at Texas Instruments enabled dense transistor-transistor logic circuits and spurred an explosive growth in the computer industry. In the late 1960's, the non-volatile magnetic core memory was replaced by much higher density integrated Random Access Memory or RAM such as Static RAM [1.4] or SRAM and Dynamic RAM [1.5] or DRAM. Both SRAM and DRAM are volatile. An alternative high density integrated non-volatile solid state memory was very much desired for the replacement of the non-volatile magnetic core memory.

Although integrated solid state non-volatile memories based on the Metal Oxide Semiconductor Field Effect Transistor or MOSFET such as floating-gate MOSFET [1.6] and metal nitride oxide semiconductor [1.7] memory cells were invented shortly after the demonstration of the MOSFET, these devices were complex requiring four terminals and high voltages. In theory, the simplest memory cell structure would be a two-terminal device with a minimal aerial cell size of four minimum feature squares or $4F^2$. Since the 1950's, two-terminal resistive memory devices attracted much attention from many research scientists and engineers. Two groups of materials received most investigation and development: transition/complex metal oxide or semiconductor oxides and chalcogenides. Chalcogenides are compounds or alloys which contain at least one element from Column VI of the periodic table. Generally, only sulfur (S), selenium (Se) and tellurium (Te) are commonly associated with chalcogenides used in phase change memory. These investigations and developments persist to present days. In this book, we are only concerned with the latter group of materials or chalcogenides.

1.3 Pioneers in Phase Change Memory

In 1962, in studying the chemical, physical, and electrical properties of some unusual inorganic glasses, A. David Pearson et al. [1.8] of Bell Laboratories observed the extraordinary current-voltage characteristics when a point contact is set down on a slice of the As-Te-I glass with a broad-area low-resistance contact on the back. Two regions of stable conductivity were revealed as shown in the reproduction of their current-voltage plot in Fig. 1.2.

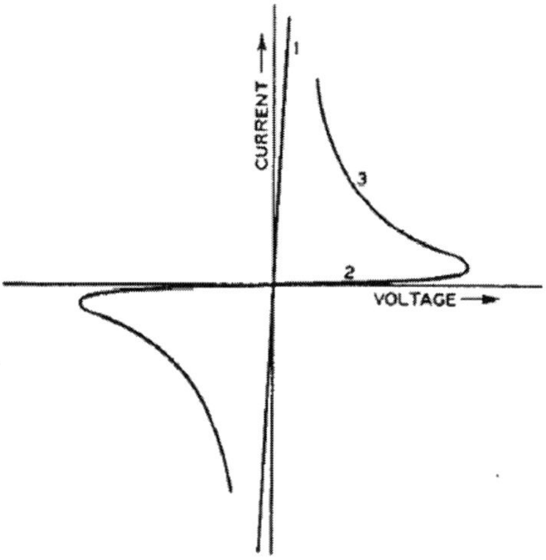

Fig. 1.2. Current-voltage of As-Te-I glass with two stable reversible regions (reproduced from [1.8] with kind permission of Springer Science and Business Media).

Region 1 is of high conductivity and region 2 is of low conductivity. Transitions from one region to another were reversible with appropriate electrical pulses. In their paper, the current-voltage characteristics were merely reported as they were observed. There was no attempt to explain the mechanism which might have caused such extraordinary characteristics. Nor was there an attempt to exploit the characteristics for any practical applications. Pearson et al. presented the novel switching, memory, and negative resistance effects they observed with the ternary As-Te-Br or -I semiconducting glasses at the Electronics Division Semiconductor Symposia in the Electrochemical Society Meeting in May 1962. However, only the abstract of their presentation [1.9] is available today.

Since 1960, Ovshinsky has been working with amorphous chalcogenides. He developed both electrically controlled threshold and memory switching devices and first reported his findings in a paper published in Physical Review Letters in 1968 [1.10]. In this paper, Ovshinsky detailed the operations of reversible switching in memory devices composed of 48 at. % tellurium, 30 at. % arsenic, 12 at. % silicon, and 10 at. % germanium. The most significant contribution of Ovshinsky's work is that the practicality of the switching phenomenon was demonstrated in continuous successful switching operations of multiple devices over periods of many months. Interest in phase change memory was practically initiated by this ground-breaking paper by Ovshinsky which remains the most cited literature in the field of phase change memory.

The significance of the reversible switching characteristics of chalcogenides as solid state computer memory was obvious in the applications of patents in the early 1960's before any literatures on the subject were published. In November 1961, Dewald, Northover and Pearson filed their first patent on a "Multiple Resistance Semiconductor Elements." They were granted the patent in 1966 [1.11]. Claim #1 in their patent, reproduced below, precisely described the operations of a memory component which is true to present day phase change memory devices:

"1. An electrical component capable of operating in two resistance states comprising a glass body having an electronic resistivity within the range 10^2 to 10^8 ohm-cm and having electrical means for impressing an electric signal on said body, said glass comprising means responsive to voltage signals of a minimum value for changing the resistance state of said body from a first discrete high resistance state to a second discrete low resistance state and said means being further responsive to the rate of decrease of the value of the voltage signals to revert said body to the high resistance state when the rate of decrease of the voltage signal exceeds a given rate and to maintain said body in the low resistance state when the rate of decrease of the voltage signal is less than the given rate."

In this claim, the "set" operation requires the applied voltage to exceed a "threshold" value across the memory component and the "reset" is a melt-quench process which requires quick removal of applied voltage across the memory component. Dewald et al. [1.11] also disclosed the basic structural variations associated with the low and high resistance states of the memory elements depicted in Fig. 1.3 reproduced from the first page of the patent.

Along with the "Multiple Resistance Semiconductor Elements" disclosure, Northover and Pearson also filed another disclosure at the same time to cover the materials used in the resistance switching device. Northover and Pearson were granted the patent in 1964 [1.12], two years ahead of the device patent. This patent basically claims the compositions of the specific ternary semiconductor glasses which exhibit the resistance switching characteristics. Figure 1.4 reproduces the ternary phase diagram of As-Te-I depicted in the patent wherein the shaded region represents the operable resistance switching compositions.

As a very successful entrepreneur, Ovshinsky also submitted a series of disclosures to protect his discoveries and findings of resistance switching or current controlling devices and materials. The most notable patent on the subject by Ovshinsky is US Patent 3,271,591 [1.13] in which he disclosed the symmetrical current controlling device. The patent contains 33 claims. In contrast to the "Multiple resistance semiconducting elements" patent by Dewald et al. [1.11], the claims in Ovshinsky's patent are more specific yet covering broad possible device configurations and operations. Ovshinsky is a master inventor. It is typical of his patents to contain broad yet extensive specific claims.

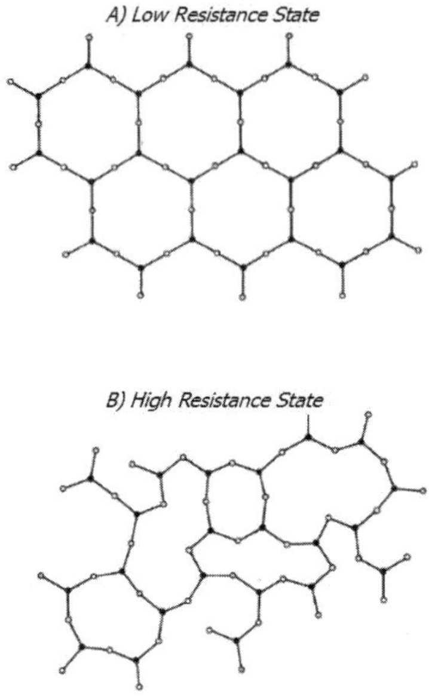

Fig. 1.3. A) Crystalline phase for low resistance state, and B) Amorphous phase for high resistance state (reproduced by permission from [1.11]).

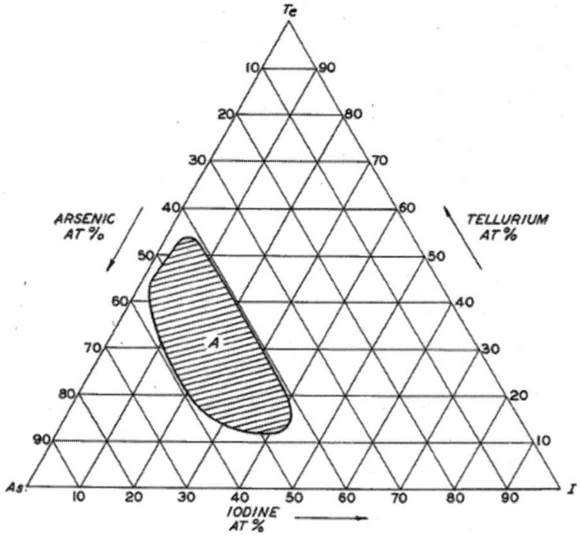

Fig. 1.4. Phase diagram of As-Te-I where the shaded region A represents resistance switching compositions (reproduced by permission from [1.12]).

One can learn a lot by studying Ovshinsky's patents. For comparison with Dewald et al. [1.11], claim 1 of this patent is reprinted here:

"1. A symmetrical current controlling device for an electrical circuit including a mechanism type semiconductor material means and electrodes in non-rectifying contact therewith for connecting the same in series in said electrical circuit, said semiconductor material means being of one conductivity type, said semiconductor material means including means for providing a first condition of relatively high resistance for substantially blocking current therethrough between the electrodes substantially equally in each direction, said semiconductor material means including means responsive to a voltage of at least a threshold value in either direction or alternately in both directions applied to said electrodes for altering said first condition of relatively high resistance of said semiconductor material means for substantially instantaneously providing at least one path through said semi-conductor material means between the electrodes having a second condition of relatively low resistance for conducting current therethrough between the electrodes substantially equally in each direction, said semiconductor material means including means for maintaining said at least one path of said semiconductor material means in its said second relatively low resistance conducting condition and providing a substantially constant ratio of voltage change to current change for conducting current at a substantially constant voltage therethrough between the electrodes which voltage is the same for increase and decrease in current above a minimum current holding value, and providing a voltage drop across said at least one path in its said second relatively low resistance conducting condition which is a minor fraction of the voltage drop across said semiconductor material means in its said first relatively high resistance blocking condition near said threshold voltage value, and said semiconductor material means including means responsive to a decrease in current, through said at least one path in its said second relatively low resistance conducting condition, to a value below said minimum current holding value for immediately causing realtering of said second relatively low resistance conducting condition of said at least one path to said first relatively high resistance blocking condition."

Essentially, claim 1 of Ovshinsky's patent disclosed a similar device as claim 1 of the Dewald et al. patent [1.11]. However, Ovshinsky also exhausted the possible embodiments of the current controlling device as depicted on the first page of the patent, reproduced in Fig. 1.5. While Dewald et al. and Ovshinsky had invented practically the same device, the main differentiation between them is the persistent and tireless efforts of Ovshinsky to pioneering phase change materials in the optical storage and semiconductor memory industries. In this chapter, we are mainly concerned with the development of solid state phase change memory. A recount of Ovshinsky's contributions in optical storage with phase change technology can be found in an article on phase change technology written by Ovshinsky himself in 1994 [1.14].

Another pioneer in phase change memory is Daniel J. Shanefield. While at International Telephone & Telegraph (ITT) in the 1960s, Shanefield was assigned the task to develop the phase change memory device ITT licensed from Ovshinsky. Along with co-workers at ITT, he perfected the alloy [1.15] to build the first phase change memory device. To assure reliable operations of the phase change memory they had developed, Shanefield also devised the circuit for controlling and adapting to inevitable fluctuations of the threshold voltage of the device in the

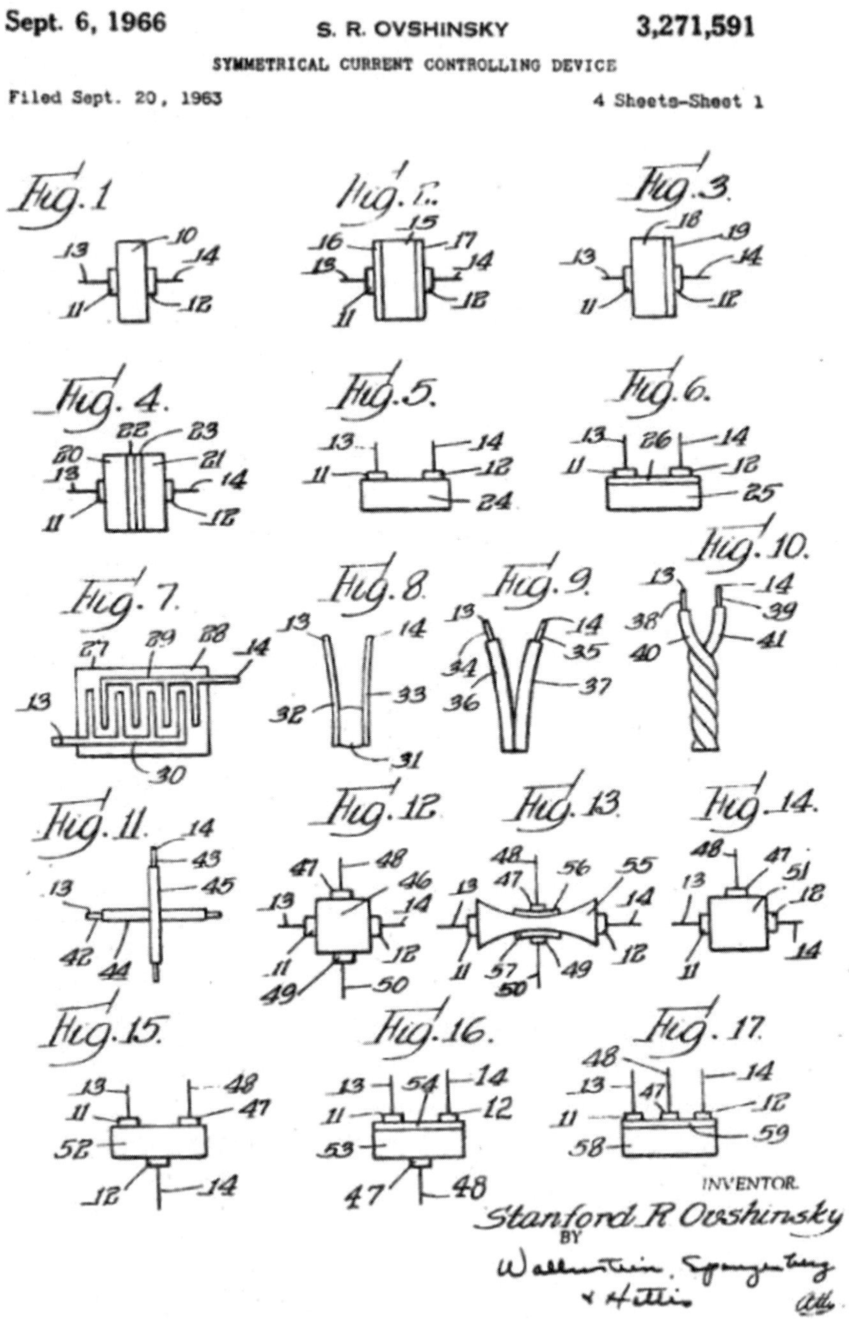

Fig. 1.5. Various device configurations in US Patent 3,271,591 [1.13], reprinted with kind permission of the inventor.

"off" state due to device geometry variations in manufacturing as well as variations from cycled on/off operations [1.16]. Although the telephone switching system which the device was intended to be used in was never commercialized, it represented the first proof of the phase change memory concept.

1.4 Early Attempts with Phase Change Memory

In the early 1970's, phase change memory drew a lot of interest in the industry and academia with the rapid expansion of applications of computers. The most notable work was the development of a 256-bit array comprised of a 16x16 matrix of phase change memory cells by R. G. Neale and D. L. Nelson of Energy Conversion Devices along with Gordon E. Moore of Intel [1.17]. Their memory cell consisted of a storage element, coined as the Ovonic amorphous semiconductor switching device, in series with a silicon p-n junction diode. The Ovonic memory element comprised of a thin film of phase change material sandwiched between two molybdenum electrodes was a nonvolatile bi-stable resistor. The high-to-low resistance ratio was about 10^3 providing ample data signal. The chip size, 122-by-131-mil or 3.099-by-3.327-mm, was as dense as any comparable bipolar or metal oxide semiconductor transistors memory array at the time. Back in 1970, the minimum feature size which could be printed on integrated circuit was about 10 μm. The write or programming operation of phase change memory from low resistance to high resistance, known as the reset operation, involves heating the phase change memory element to the melting point and cooling it quickly (quenching) so the material remains in the amorphous state. The programming operation of phase change memory from high to low resistance, known as the set operation, involves thermally annealing the disordered amorphous phase change memory element to the ordered crystalline state. The melting points of commonly used phase change materials are about 600 °C. The annealing temperatures are normally above 300 °C. With 10-μm geometry, the memory cells in Neale et al.'s memory array required 5 μs, 200 mA at 25 V bias for the reset operation and 10 ms, 5 mA at 25 V bias for the set operation. The volume of phase change material needed to be switched for the operation of the device was too large thus resulting in these enormous power consumptions during the programming operations. Compared to competitive electrical programmable read-only memories in the same period [1.18-20], phase change memory was inferior in power efficiency. The 256-bit phase change memory array was never commercialized.

The peculiar switching characteristics of phase change semiconductor devices along with other devices with similar characters such as binary metal or semiconductor oxide and solid state electrolytic devices continued to stir up interest in industry and academia. In 1971, while at Purdue University, Leon Chua classified these devices, along with resistor, inductor and capacitor, as the fourth of basic two-terminal elements he called memristor [1.21]. The fourth element, memristor, is a contraction of memory and resistor. Chua argued that this new element is necessary to complete the symmetric relationships connecting two of the four fundamental circuit variables: the current, the voltage, the charge and the flux-linkage. The amorphous Ovonic threshold switching device was used in Chua's paper to demonstrate the validity of the memristor. Due to the lack of commercial interest, the attention in the peculiar switching characteristics of the Ovonic phase change memory device and memristor subsided. However, the search for better phase change materials and the studies of the fundamental switching mechanisms persisted while attention of early investigators was drawn towards the development of phase change materials in optical storage systems. An organized attempt to stir up bewildering interest in semiconducting phase change devices was the special issue on amorphous semiconductor devices of the IEEE Transactions on Electron Devices published in February 1973 [1.22].

Before the phase change memory device went into hibernation, another attempt to build a phase change memory device was disclosed by Roy R. Shanks and Craig Davis of the Burroughs Corporation. They published their result of a 1024-bit phase change memory in 1978 [1.23]. The memory cell consisted of a phase change memory element in series with a Schottky diode which occupied an area of 4.3 mil^2 or 0.05267 mm^2. The reset operation was achieved using one hundred 25 mA, 2 μsec. pulses. The set operation was achieved using a single 6 mA, 10 ms pulse with a trailing edge of 5 ms. Both programming operations had the word line of the memory cell biased at 12 V. The programming voltages and currents in this chip were lower than Neale et al.'s earlier attempt but were still a far cry from being competitive with conventional metal oxide semiconductor memories. Aside from a micrograph of the chip, reproduced in Fig. 1.6, there were no electrical hardware data. The authors admitted that there were limited reliability data of the memory device. Again, this 1024-bit memory chip never made it to the manufacturing line.

1.5 Rebirth of Phase Change Memory

Since Gordon Moore published his first outlook on the trend of the number of components per integrated circuit silicon chip in 1965 [1.24], the semiconductor integrated circuit industry has been faithfully following the trend which has been since known as Moore's Law.

The validity of Moore's Law is made possible by the rapid advance of photolithography used to define minimum critical feature sizes on silicon chips. By the dawn of the new millennium, photolithographic engineers were able to print minimum features on silicon chips as small as 180 nm. Tyler Lowrey and Ward Parkinson, armed with more than 40 years of experiences in semiconductor memory business, envisioned that it was about time to bring phase change memory

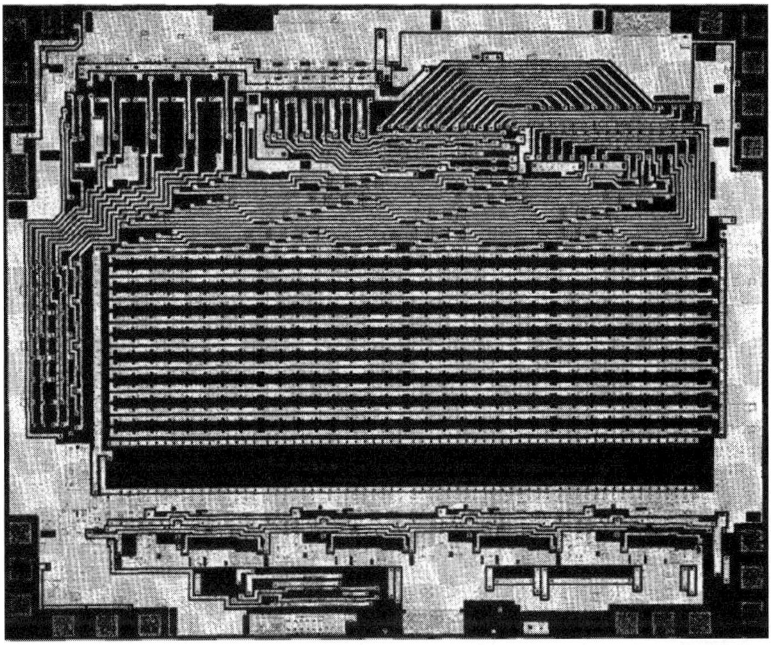

Fig. 1.6. Photograph of 1024-bit phase change memory chip ([1.23], © (1978) IEEE).

back to life. Recall that one major flaw in the practicality of phase change memory was the enormous power required to program the phase change memory.

The amount of energy required to switch the phase change memory is directly proportional to the volume of phase change material. With the minimum feature size shrinking from 10 μm to 180 nm or 55.56 times, the volume which shrunk in cubic progression was reduced by as much as 171,468 times. The power required to program phase change memory devices made with the 180 nm technology seemed reasonably competitive. In 1999, Tyler and Parkinson formed a joint venture called Ovonyx with Energy Conversion Devices to commercialize phase change memory as the Ovonic Unified Memory. Ovshinsky founded Energy Conversion Devices with his wife, Iris, in 1960 which was originally called Energy Conversion Laboratories.

Immediately after the formation of Ovonyx in 1999, BAE Systems, formerly Lockheed Martin, licensed the phase change memory technology for space applications. By virtue of the phase change material, phase change memory devices are radiation hard and suitable for space applications. With C-RAM as trade mark, BAE introduced the first commercially available, radiation-hardened phase change memory device in August 2006. It is a 3.3V, 512-kbit by 8 for a 4-Mbit memory device fabricated in 0.25 μm bulk complimentary metal-oxide-semiconductor (CMOS) process.

Perhaps, the most celebrated and important win for Ovonyx is the license agreement with Intel in February 2000. In the agreement, Intel also included a substantial investment in Ovonyx from Intel Capital. Effectively Intel was made a joint venture partner of Ovonyx, thus started a parade of semiconductor memory developers to license the phase change memory from Ovonyx. Before the end of the year in December 2000, Ovonyx signed an agreement with ST Microelectronics, their third licensee. In 2001, Intel and Ovonyx re-introduced the phase change memory to the industry as Ovonic Unified Memory at the International Electron Devices Meeting [1.25]. Industry's and academia's interest in phase change memory exploded. Since 2000, the number of patent applications and papers related to phase change memory materials and devices has increased rapidly. Fig. 1.7 depicts the number of US patents awarded since 1966 when the first three patents disclosing the original phase change memory devices were issued.

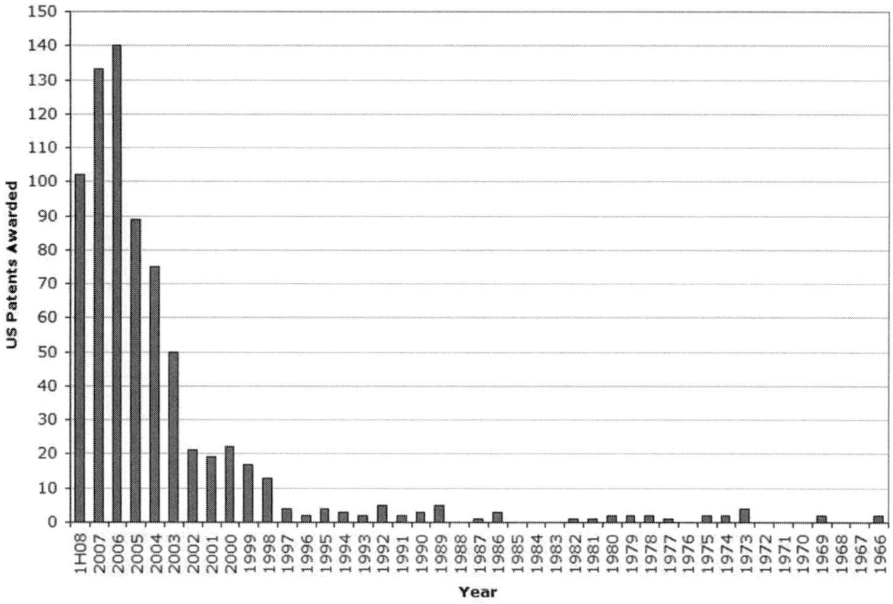

Fig. 1.7. Number of US Patents related to phase change memories and associated materials awarded since 1966.

At present, summer 2008, the only commercially available phase change memory device is the 4Mb C-RAM produced by BAE Systems. C-RAM is produced for specific military and space applications, thus relatively expensive compared to conventional semiconductor memories. While mass production of general purpose phase change memory products for computers and consumer applications are still years away, the industry is definitely busy with activities in the development of production worthy high density phase change memory devices. Finally, the critical events in phase change memory development are summarized in Table 1.1.

Table 1.1. Critical events in phase change memory by years since 1962.

Year	Events
1962	A. D. Pearson et al. reported switching phenomena in AsTeI
1968	S. R. Ovshinsky published phase change threshold switching
1970	R. G. Neale et al. demonstrated a 256-bit Phase Change Memory
1972	J. Feinleib et al. demonstrated Reversible Optical Memory
1973	IEEE Electron Devices February Special Issue on Amorphous Semiconductor Devices
1978	R. R. Shanks et al. demonstrated 1024-bits bipolar memory
1990	Panasonic introduced R/W phase change optical disc drive
1999	Ovonyx formed BAE licensed Ovonic Unified Memory (OUM) from Ovonyx 11/4/1999
2000	Intel Capital invested in Ovonyx and licensed OUM from Ovonyx 2/8/2000 STMicro licensed OUM from Ovonyx and announce Joint Development Project on 12/21/2000
2003	STMicroelectronics and Ovonyx expanded license agreement and extended Joint Development Project on 2/4/2003 Samsung published first paper on phase change RAM at VLSI 2003 Hitachi published first paper on phase change RAM in IEDM 2003
2004	Samsung announced full-scale phase change RAM production in 2006 on 8/23/2004 Nanochip licensed OUM from Ovonyx for Micro-Electro-Mechanical Systems (MEMS) based storage on 8/31/2004
2005	Elpida licensed OUM from Ovonyx on 2/3/2005 Philips Research published first paper on phase change random access memory in Nature Materials April issue IBM, Infineon and Macronix announced Joint Research Initiative on phase change memory on 5/23/2005 Samsung announced 512Mb phase change RAM availability in 2008 replacement on 12/28/2005
2006	ITRI of Taiwan and local memory vendors - PowerChip, Nanya, ProMOS and Winbond formed PCM Alliance on 9/26/2006
2007	Ovonyx and Qimonda sign technology licensing agreement for phase change memory on 1/16/2007 Samsung ships 90 nm 512Mb phase change RAM assessment sample to mobile phone manufacturers on 2/26/2007 Intel to sample 90 nm 128Mb phase change memory in first half of 2007 3/6/2007 Hynix licenses Ovonyx's phase change memory on 10/01/2007

References

[1.1] Waterman, A. T.: Positive ionisation of certain hot salts, together with some observations on the electrical properties of molybdenite at high temperatures. Phil. Mag. **33**, 225 (1917)

[1.2] Waterman, A. T.: The electrical conductivity of molydenite, Phys. Rev. **21**, 540-549 (1923)

[1.3] Wang, A.: Pulse transfer controlling devices. US Patent 2,708,722 (1955)

[1.4] Burke, H. K. and Michon, G. J.: Integrated bistable memory cell. US Patent 3,389,383 (1968)

[1.5] Dennard, R. H.: Field-effect transistor memory. US Patent 3,387,286 (1968)

[1.6] Kahng, D.: Field effect semiconductor apparatus with memory involving entrapment of charge carriers. US Patent 3,500,142 (1970)

[1.7] Wegener, H. A. R.: Electrically alterable non-destructive readout field effect transistor memory. US Patent 3,508,211 (1970)

[1.8] Pearson, A. D., Northover, W. R., Dewald, J. F. and Peck Jr., W. F.: Chemical, physical, and electrical properties of some unusual inorganic glasses. Adv. in Glass Tech., pp. 357-365, Plenum Press, New York (1962)

[1.9] Dewald, J. F., Pearson, A. D., Northover, W. R. and Peck Jr., W. F.: Semiconducting glasses. J. of Electrochem. Soc., p. 243C, (1962)

[1.10] Ovshinsky, S. R.: Reversible electrical switching phenomena in disordered structures. Phys. Rev. Lett. **22**, 1450-1453 (1968)

[1.11] Dewald, J. F., Northover, W. R. and Pearson, A. D.: Multiple resistance semiconductor elements. US Patent 3,241,009 (1966)

[1.12] Northover, W. R. and Pearson, A. D.: Glass composition. US Patent 3,117,013 (1964)

[1.13] Ovshinsky, S. R.: Symmetrical current controlling device. US Patent 3,271,591 (1966)

[1.14] Ovshinsky, S. R.: A history of the phase change technology. Memoires Optiques et Systemes, (1994), http://ovonic.com/PDFs/Optical_Memory_Research_Report/

[1.15] Shanefield, D. J. and Lighty, P. E.: Solid state element comprising semiconductive glass composition exhibiting negative incremental resistance. US Patent 3,448,425 (1969)

[1.16] Shanefield, D. J.: Operating circuit for phase change memory devices. US Patent 3,448,302.

[1.17] Neale, R. G., Nelson, D. L. and Moore, G. E.: Nonvolatile and reprogrammable, the read mostly memory is here. Electronics, pp. 56-60, Sept. 30 (1970)

[1.18] Frohman-Bentchkowsky, D: An integrated Metal-Nitride-Oxide-Silicon (MNOS) memory. Proc. IEEE Lett. **57** 1190-1192 (1969)

[1.19] Frohman-Bentchkowsky, D: A fully decoded 2048-bit electrically programmable FAMOS read-only memory. IEEE J. of Solid-State Circuits, **SC-6,** pp. 301-306 (1971)

[1.20] Iizuka, H., Sato, T., Masuoka, F., Ohuchi, K., Hara, H. and Takeishi, Y.: A fully-decoded 2048-bit avalanche-injection type, electrical alterable ROM. IEEE Int.. Electron Devices Meeting, Washington, D.C. (1972)

[1.21] Chua, L. O.: Memristor – the missing circuit element. IEEE Trans. on Circuit Theory, **CT-18**, pp. 507-519 (1971)

[1.22] Special Issue on Amorphous Semiconductor Devices. IEEE Trans. on Electron Devices, **ED-20**, February (1973)

[1.23] Shanks, R. R. and Davis, C.: A 1024-bit nonvolatile 15ns bipolar read-write memory. ISSCC Digest of Technical Papers, pp. 112-113, February (1978)

[1.24] Moore, G. E.: Cramming more components onto integrated. Circuits. Electronics, **38**, pp. 114–117, April 19 (1965)

[1.25] Lai, S. and Lowrey, T.: OUM – A 180 nm nonvolatile memory cell element technology for stand alone and embedded applications. IEDM Digest of Technical Papers, pp. 803-806, December (2001)

Part I:
Material Science: Theory and Experiment

2. Density Functional Theory Calculations for Phase Change Materials

Wojciech Wełnic

Abstract In this chapter the fundamentals of Density Functional Theory (DFT) are introduced. After presenting the Hohenberg-Kohn theorem, the most common concepts and approximations in DFT-calculations, namely the Kohn-Sham equations and the Local Density Approximation (LDA), are summarized. They allow mapping the many-body problem on a single particle system which ultimately enables numerical calculations of the electronic ground state of a solid.

Furthermore, applications of DFT in the field of phase change materials are presented. DFT proves to be a powerful tool to reveal structural properties of the crystalline, liquid and amorphous phases as well as to explain the nature of the electronic ground state and bonding properties.

2.1 Introduction

Computational methods are becoming more and more important in the field of solid state physics and material science. Often properties of a solid are not directly accessible from experiments for various reasons. A material might oxidize easily, which changes the electronic properties at the surface and makes the interpretation of surface-sensitive experiments such as photo emission difficult. Furthermore, the sample size or mass might not be sufficient for certain experiments. The beam diameter of a synchrotron for example is typically in the range of several micrometers, hence a single nanotube cannot be easily studied with this powerful light source. Finally one might be interested in studying fictitious systems e.g. in order to predict the stability or the properties of a composition which has not (yet) been produced. All three cases do actually occur for phase change materials. For these tasks numerical simulations provide a valuable tool. In order to be able to predict as well as reproduce experimental results, these simulations should be based simply on first principles of quantum mechanics, i.e. independent of any empirical (experimental) parameters[1]. An exact mathematical *ab initio* treatment of the Schrödinger equation of a many-atom-system composed of electrons and

[1] In the following these methods will be called *ab initio*, which is the Latin expression for "from the beginning".

S. Raoux, M. Wuttig (eds.), *Phase Change Materials*,
DOI 10.1007/978-0-387-84874-7_2,

nuclei is highly complex and virtually impossible. Nevertheless suitable methods to derive a wide range of physical properties of condensed matter systems are available in fields such as material science, mechanical engineering, chemistry and physics. Within this section Density Functional Theory (DFT) is introduced, a computational method which has been highly successful in calculating the electronic ground state properties of many-atom systems and is becoming a valuable tool in the field of phase change materials.

In order to study such systems with moderate computational effort a range of approximations is introduced. As the change in the electronic states occurs rapidly compared to the nuclear motion the electrons can be considered to be in their ground state for every ionic configuration. This approximation which is called Born-Oppenheimer (or adiabatic) approximation [2.1], allows one to completely separate the calculation of the electronic structure from the ionic motion. Usually the many-body-problem of interacting electrons is then reduced to a single-particle problem. A typical example of such a single-particle method is the Hartree-Fock approximation [2.2, 2.3]. This method provides an approximate solution to the many-body problem and uses a single Slater determinant to express the many electron wave function. The so-called Hartree-Fock equation which has to be solved includes a Coulomb term (Hartree term) and an exchange term (Fock term), which stems from the antisymmetry of the wave function with respect to two-particle permutation, due to Pauli's exclusion principle. The effects of electron correlation beyond that of exchange energy are completely neglected in this method. A more appropriate approach - though still based on the single particle approximation - to correctly describe the electronic system is provided by Density Functional Theory. In contrast to Hartree-Fock, DFT provides in principle an exact solution of the many-body problem in the electronic ground state. However for practical calculations a series of approximations is employed. The following section will provide a description of the basic principles of DFT, its strengths and limitations, introduce computational methods which go beyond DFT and the single-particle approximation and finally show several examples how DFT can be used to reveal fundamental electronic and structural properties of phase change materials.

2.2 The Theorem of Hohenberg and Kohn

In DFT all ground state properties such as the total energy are expressed as a functional of the charge density distribution. The fundamentals of this concept were derived by Hohenberg and Kohn [2.4]. They proved that the following theorem holds exactly:

a) There is a universal functional $F[\rho(\mathbf{r})]$ *of the spatial electron density distribution* $\rho(\mathbf{r})$, *that defines the total energy of the electronic system by*

$$E = \int v(\mathbf{r})\rho(\mathbf{r})d\mathbf{r} + F[\rho(\mathbf{r})] \tag{2.1}$$

*b) The total energy E has a minimum E_0 when the charge density ρ (**r**) coincides with the true charge density ρ_0 (**r**) in the external potential v (**r**). This theorem is exact if the ground state has no degeneracy.*

Thus the Hohenberg-Kohn theorem states that the ground state energy E is a universal functional of the charge density and that the ground state charge density can be obtained by applying the variation principle to the energy. Levy [2.5] found an alternative formulation of the Hohenberg-Kohn theorem where the functional of the ground state energy is given by

$$E[\rho(\mathbf{r})] = \min_{[\rho(\mathbf{r})]} \langle \Psi | H | \Psi \rangle \tag{2.2}$$

under the constraint, that the wave functions Ψ used for variation reproduce the charge density distribution $\rho(\mathbf{r})$

$$\rho(\mathbf{r}) = \langle \Psi | \sum_i \delta(\mathbf{r} - \mathbf{r}_i) | \Psi \rangle . \tag{2.3}$$

Here H is the Hamiltonian and i the index of the spatial coordinate. This leads to the relation

$$E[\rho(\mathbf{r})] \geq E[\rho_0(\mathbf{r})] = E_0 . \tag{2.4}$$

At this point the exact form of the total energy functional remains still undetermined, however the scheme used to derive electronic ground state properties can already be outlined here. First of all one has to find a suitable approximation of $E[\rho_0(\mathbf{r})]$ and then apply the variation principle to the total energy functional

$$\delta E[\rho(\mathbf{r})] = 0 \tag{2.5}$$

under the constraint, that the number of particles N is conserved

$$\int \rho(\mathbf{r}) d\mathbf{r} = N . \tag{2.6}$$

This constraint is expressed via the Lagrange parameter[2] μ

$$\delta\left\{E\left[\rho(\mathbf{r})\right]-\mu\left[\int d\mathbf{r}\,\rho(\mathbf{r})-N\right]\right\}=0\,. \tag{2.7}$$

A strategy how to calculate the electronic states of complex systems in the ground state according to DFT has been first presented by Kohn and Sham [2.6] and will be introduced in the following two sections.

2.3 The Kohn-Sham Equation

The main idea of Kohn and Sham [2.6] was to map the system of N interacting electrons onto an auxiliary system of N non-interacting electrons with the same density as the interacting system. In this case the charge density can be expressed in terms of single particle orbitals Φ_i

$$\rho(\mathbf{r})=\sum_{i=1}^{N}\left|\Phi_i(\mathbf{r})\right|^2\,. \tag{2.8}$$

With the kinetic energy T of the non-interacting particle given by[3]

$$T\left[\rho(\mathbf{r})\right]=-\sum_{i=1}^{N}\int d\mathbf{r}\,\Phi_i^*(\mathbf{r})\frac{1}{2}\nabla^2\Phi_i(\mathbf{r})\,. \tag{2.9}$$

Kohn and Sham rewrote the total energy functional in Eq. (2.1) in the following way:

$$E\left[\rho(\mathbf{r})\right]=T\left[\rho(\mathbf{r})\right]+E_{tot}\left[\rho(\mathbf{r})\right] \tag{2.10}$$

$$=T\,[\rho(\mathbf{r})]+\frac{1}{2}\int d\mathbf{r}d\mathbf{r}'\,\frac{\rho(\mathbf{r})\rho(\mathbf{r}')}{\left|\mathbf{r}-\mathbf{r}'\right|}+\int d\mathbf{r}\,v(\mathbf{r})\rho(\mathbf{r})+E_{xc}[\rho(\mathbf{r})]\,. \tag{2.11}$$

---

[2] In this case the Lagrange parameter is equivalent to the chemical potential since Eq. (2.7) can be written as $\frac{\delta E[\rho(\mathbf{r})]}{\delta\rho(\mathbf{r})}=\mu$.

[3] In the following atomic units will be used: $\hbar = 1$, $e = 1$, $m_e = 1$.

Besides T given by the first term, Eq. 2.10 contains a second term E_{tot} consisting of the Hartree energy, which describes the electrostatic energy of the electronic system (second term in Eq. 2.11), the interaction energy with the external field (third term in Eq. 2.11) and the exchange-correlation energy E_{xc}, which accounts for the exchange and correlation effects of the electrons. This term is unknown and - in contrast to all other terms on the right side of the equation - cannot be determined exactly for a complex, inhomogeneous system. Therefore finding a suitable description for E_{xc} poses the basic challenge in DFT calculations.

Within this scheme, the system of interacting particles in an external potential v is now replaced by a system of non-interacting particles in an effective potential V_{tot}. The single particle wave functions in Eq. 2.8 are orthogonal to each other and fulfill the norm

$$\langle \Phi_i(\mathbf{r}) | \Phi_i(\mathbf{r}) \rangle = 1 . \tag{2.12}$$

If the variational principle for the total energy E is applied with respect to Φ_i, the norm becomes the constraint instead of the particle conservation and one obtains

$$\delta \left\{ E\left[\Phi_i, \Phi_i^* \right] - \sum_{i=1}^{N} \varepsilon_i \left(\langle \Phi_i(\mathbf{r}) | \Phi_i(\mathbf{r}) \rangle - 1 \right) \right\} = 0 . \tag{2.13}$$

The solution of Eq. 2.13 can be obtained by solving the effective single-particle Schrödinger equation for Φ_i for N particles

$$\left\{ -\frac{1}{2}\nabla^2 + v(\mathbf{r}) + \frac{1}{2}\int d\mathbf{r}' \frac{\rho(\mathbf{r}')}{|\mathbf{r}\text{-}\mathbf{r}'|} + \frac{\delta E_{xc}\rho(\mathbf{r})}{\delta\rho(\mathbf{r})} \right\} \Phi_i(\mathbf{r}) = \varepsilon_i \Phi_i(\mathbf{r}) . \tag{2.14}$$

This equation is called Kohn-Sham (KS) equation and the effective one-electron Hamiltonian associated with Eq. 2.14 is called the KS Hamiltonian. It is worth mentioning that - although providing a suitable description e.g. for the electronic band structure or the charge distribution within a solid - its eigenvalues ε_i and the single particle wave functions Φ_i have no physical meaning. In fact the single particle wave function Φ_i is simply an object which is related to the true charge density ρ and will be called KS wave function in the following. Thus the Slater determinant constructed from Φ_i is not the true many-electron wave function either. Equation 2.14 rather represents a single-electron approximation which contains the many-body interactions in an effective potential V_{tot}

$$V_{tot}(\mathbf{r}) = V_{ext}(\mathbf{r}) + V_H(\mathbf{r}) + V_{xc}(\mathbf{r}) \tag{2.15}$$

where V_{ext} represents the external potential, V_H the Hartree potential and V_{xc} the exchange-correlation potential. As the effective potential is constructed from the charge density and in turn the charge density is constructed from the wave functions, one obtains a set of equations which has to be solved self-consistently.

2.4 The Local Density Approximation

Approximations enter density functional theory due to the fact that the exchange-correlation energy is unknown for inhomogeneous systems. The simplest approximation proposed by Kohn and Sham is the Local Density Approximation (LDA) [2.6]. Their idea is to use the exchange-correlation energy of the homogeneous electron gas, which can be calculated using Quantum Monte Carlo simulations [2.7]. Here one puts $\rho = \rho(\mathbf{r})$ and neglects the spatial dependence of ρ. Then the exchange-correlation energy at $\mathbf{r}$ is replaced by the corresponding energy of the homogeneous electron gas with density ρ. Mathematically this is described in the following way: The exchange-correlation energy of the free, homogeneous electron gas is given by

$$E_{xc}^{hom} = \rho \varepsilon_{xc}^{hom}, \tag{2.16}$$

where ε_{xc}^{hom} is the spatially constant exchange-correlation density of the homogeneous gas. For the inhomogeneous system one obtains

$$\varepsilon_{xc}^{LDA}(\mathbf{r}) = \varepsilon_{xc}^{hom}[\rho(\mathbf{r})]. \tag{2.17}$$

Thus

$$E_{xc}^{LDA}[\rho(\mathbf{r})] = \int d\mathbf{r}\, \rho(\mathbf{r})\, \varepsilon_{xc}^{LDA}(\mathbf{r}) \tag{2.18}$$

and

$$V_{xc}^{LDA}([\rho(\mathbf{r})]) = \frac{\delta E_{xc}^{LDA}[\rho(\mathbf{r})]}{\delta\rho(\mathbf{r})} = \frac{\delta\left[\rho(\mathbf{r})\varepsilon_{xc}^{LDA}(\mathbf{r})\right]}{\delta\rho(\mathbf{r})}. \tag{2.19}$$

The simplest form of the LDA to the exchange-correlation potential has been proposed by Slater et al. [2.8], where V_{xc} becomes

$$V_{xc}([\rho]) = -\left(\frac{3\alpha}{2}\right)\left(\frac{3\rho}{\pi}\right)^{1/3}. \tag{2.20}$$

This choice for the exchange-correlation functional is usually called $X\alpha$ method. The coefficient α can be determined from known results for the electron gas. For example it is known that the electron gas gives the same form for the exchange term. Thus if only the exchange term evaluated in the electron gas is considered one obtains $\alpha = 2/3$ [2.9]. Slater et al. originally obtained $\alpha = 1$ by averaging the exchange term of the Hartree-Fock equation.

Other now commonly used parametrizations based on the results by Ceperley and Adler [2.7] have been proposed by Perdew and Zunger [2.10], or Goedecker et al. [2.11]. Calculations using these functionals yield good results for the electronic ground state for a wide range of materials. Gunnarson et al. [2.12] pointed out that this is due to the fact that only the spherical average of the exchange-correlation hole (the hole near the origin of the pair distribution function) influences the exchange-correlation potential and that this spherical average is reproduced well by the LDA.

An extension of the LDA is the General Gradient Approximation (GGA) which describes the exchange-correlation energy not only as a function of the local density but also of its variation

$$E_{xc}^{GGA}\left[\rho(\mathbf{r})\right] = \int d\mathbf{r}\ \rho(\mathbf{r})\ \varepsilon_{xc}^{GGA}\left(\rho(\mathbf{r}), \nabla\rho(\mathbf{r})\right). \tag{2.21}$$

In common parametrizations ε_{xc} is an analytical function designed to satisfy certain conditions, e.g. that the exchange hole is negative everywhere and represents a deficit of one electron [2.13]. Among these GGAs one finds those of Perdew and Wang [2.13] or of Perdew, Burke and Ernzerhof [2.14].

2.5 Beyond Density Functional Theory

Density Functional Theory and the various methods to solve the KS equations presented here is nowadays widely and very successfully used to calculate electronic ground state properties of solids, atoms and molecules, clusters, surfaces, nanotubes or even biological systems. In solids structural properties such as the lattice constant are reproduced within an error of ± 2%; the bulk modulus is reproduced with an error of about ± 10%. Calculations of the total energy serve to determine the ground state structure, which is given by the structure with the lowest energy. In liquids DFT calculations combined with molecular dynamics are used to determine the structure factor and the pair correlation function, which is usually also determined within an error of a few percent. Molecular binding energies and band structures (for the occupied states) are obtained with good agreement to experimental data. DFT is even applied to calculate magnetic properties or phonon frequencies. However, as it is strictly a theory describing the electronic ground state of an independent-particle system it fails in the calculation of excited state properties such as the determination of the electronic band gap, the calculation of the optical response or of the transport properties. Therefore we have to go beyond ground state DFT to describe these properties accurately with computational methods. As the optical response and transport properties are of considerable importance in the field of phase change materials, these methods such as time dependent density functional theory or many-body perturbation theory (see e.g. [2.15]) shall become of great importance in the field in the near future.

2.6 Application of DFT in the Field of Phase Change Materials

The first numerical simulations for materials which are used as phase change materials today were reported by O'Reilly et al. [2.16] and Rabe et al. [2.17]. O'Reilly et al. used empirical potentials for their simulations, which are therefore not from first principles. Nevertheless both studies beautifully exemplify the opportunities provided by calculations of the electronic structure. O'Reilly et al. calculated the density of states for different structural models of amorphous GeTe, a basic phase change material, compared their results with experimental data from x-ray photoemission spectroscopy/ultraviolet photoemission spectroscopy (XPS/UPS) measurements and finally identified a model for the local structure in the amorphous phase from this comparison. They concluded that the structure is mainly chemically ordered, i.e. most bonds are heteropolar with Ge atoms surrounded by four and Te atoms surrounded by two neighbors. Up to today the local structure of an amorphous solid is difficult to access experimentally. Only very recently EXAFS (Extended X-ray Absorption Fine Structure) measurements have been used to reveal the local structure of phase change materials. EXAFS data for GeTe agree well with the findings of [2.16]. In the experiments a mixture of 4(Ge):2(Te) and 3(Ge):3(Te)-coordinated structural units [2.18] is found. Hence this is an early example how numerical simulation can be used to predict material properties that are not accessible with experimental techniques.

Rabe et al. [2.17] studied the structural transition from the trigonal to the rocksalt phase in GeTe which is accompanied by a ferroelectric phase transition. They predicted a transition temperature of 657 ± 100 K which is in good agreement with experimental data. This nicely shows that *ab initio* ground state calculations can be successfully applied even to calculate finite-temperature properties. Today two principal applications for numerical simulations of phase change materials, which are now mainly based on DFT, can be identified: Most often DFT is used to predict structural properties, either by comparing the total energies of different structural models or by performing *ab initio* molecular dynamics simulations in particular for the liquid or the amorphous phase. Furthermore calculations are often employed to compute electronic properties of phase change materials in order to characterize the bonding and correlate the electronic with the structural properties of the material. The following chapter will provide an overview of the various activities in this field.

2.6.1 Structure Determination

The application of phase change materials is based on the phase transition between the crystalline and the amorphous state. Hence a precise knowledge of the structure in these phases - and in the liquid phase, as the amorphization process involves melting - is crucial for a detailed microscopic understanding of phase change materials. In particular for the amorphous and the liquid state the determination of the structure is very cumbersome. A general characteristic attribute of these phases is the lack of long range order and periodicity. Therefore neither a unit cell as given in the crystalline state can be defined, nor the atomic positions within such a unit cell. Thus this lack of long range order significantly complicates a precise identification of the structural properties of the amorphous phase. Nonetheless, average, statistical properties like the structure factor or correlation functions are successfully employed to determine and classify the structure of amorphous solids by analyzing their short and medium range structural order. Besides experimental methods like neutron diffraction or EXAFS (Extended X-ray Absorption Fine Structure) spectroscopy, molecular dynamics and recently *ab initio* molecular dynamics (MD) have been employed in order to study these properties in the amorphous and liquid phase of a wide range of elements and alloys. In order to perform an *ab initio* MD simulation a random structure[4] is initialized within a unit cell. The number of atoms in such a structure should be sufficiently large to allow for statistical analysis (typically in *ab initio* MD simulations this number ranges from 100 to 1000). By coupling the system to a fictitious heat bath, usually using a Nosé thermostat [2.19, 2.20], the system is then heated to a high temperature to randomize the atomic distribution. Afterwards the system is cooled to the desired temperature and thermalized in order to obtain an ensemble of configurations which can be analyzed afterwards [2.21]. The trajectories of the atoms are computed according to Langevin dynamics and hence follow

$$M_i \frac{d\mathbf{v}_i}{dt} = -M_i \gamma \mathbf{v}_i + \mathbf{G}_i(\gamma, T) + \mathbf{F}_i \tag{2.22}$$

where M_i is the mass of the ith particle, $\mathbf{v}_i$ its velocity, $\mathbf{F}_i$ the interatomic force acting on the particle, γ the viscosity and $\mathbf{G}_i$ a random temperature- and viscosity-dependent force coupling with the heat bath [2.22-2.25].

In this spirit the microstructure and the dynamical properties of several elements and alloys including Si, Ge, GaAs and GeTe in the liquid phase [2.21, 2.26-2.28] have been studied. From the calculations the structure factor and pair correlation functions at different temperatures are obtained and compared with experimental data from neutron diffraction measurements. The agreement with the experiment is very good (see Fig. 2.1) and allows for further analysis, e.g. the extrac-

---

[4] However the structure is subject to constraints, e. g. there should be no unphysically short bonds.

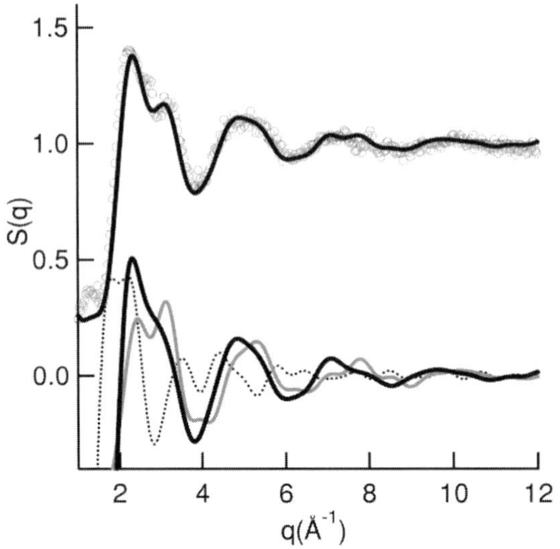

Fig. 2.1. Top: Calculated structure factor for GeTe at 1000K (solid line) compared with neutron diffraction data (symbols). Bottom: Calculated partial structure factors (shifted by -1): S_{GeTe} thick line, S_{GeGe} gray line, S_{TeTe} dots. Reprinted from [2.27], Copyright (2000), with permission from Elsevier.

tion of partial structure factors and pair correlation functions, which are very difficult to access experimentally.

Concerning the local structure a remarkable temperature-dependent behavior in the liquid phase of the phase change material GeTe is found. At low temperature just above the melting point the structure is described to be driven by a reentrant Peierls distortion [2.29], exhibiting short and long bonds, similar to the trigonal crystalline ground state of GeTe. This distortion slowly disappears with higher temperatures accompanying a semiconductor to metal transition. For liquid $Ge_{15}Te_{85}$ a Peierls transition has been observed as well, using neutron scattering, EXAFS and *ab initio* molecular dynamics [2.30]. No such studies are available in the literature for the liquid phase of ternary phase change materials such as $Ge_2Sb_2Te_5$ or $Ge_1Sb_2Te_4$.

Structural properties of glassy binary IV-VI alloys have also been studied [2.31, 2.32], however molecular-dynamics (MD) simulations of amorphous GeTe or ternary GeSbTe-alloys have become feasible only very recently. Alternatively simple structural models based on experimental data have been employed in DFT calculations. As DFT yields very good values for total energy differences of various structures, the stability of structural models can be conveniently studied. Inspired by EXAFS measurements [2.33] a structural model for $Ge_1Sb_2Te_4$ has been proposed, whose local geometry resembles a spinel structure with Ge atoms occupying tetrahedral positions while Sb and Te atoms remain in the octahedral positions already present in the crystalline phase [2.34]. Good agreement is found with experimental data for the structural parameters, i.e. bond length and density change compared to the crystalline phase. It is furthermore revealed that such an arrangement is energetically stable. The energy difference compared to the calculation of the crystalline phase has been determined to 30 meV per atom. This corresponds well to differential scanning calorimetry measurements of the crystallization energy[5] yielding values of 28 - 42 meV per atom [2.35]. Going beyond simple structural models *ab initio* molecular dynamics simulations of the amorphous phase of GeTe and $Ge_2Sb_2Te_5$ have become feasible recently thanks to massively improved computer performance [2.36, 2.37]. The analysis of these calculations results in a very detailed model of the local structure of amorphous $Ge_2Sb_2Te_5$ [2.36]. After verifying that the structure factor of the calculated system agrees well with experimental data a detailed account of the local bonding coordination for each atom type can be provided. It is found that Ge and Sb atoms have four nearest neighbors while Te exhibits three nearest neighbors. The bond angles have maxima mainly at 90 and 180 degrees recalling the distorted rocksalt environment found in the metastable crystalline phase. However 38 % of the Ge atoms are tetrahedrally coordinated in particular if homopolar bonds are present. This can be analyzed using a local order parameter q [2.38]

$$q = 1 - \frac{3}{8}\sum_{i>k}\frac{1}{3} + \cos\left(\theta_{ijk}\right), \tag{2.23}$$

where the sum runs over the couples of atoms bonded to a central atom j (see Fig. 2.2) and where θ_{ijk} is the angle between these atoms again with j being the central atom. Further information is then obtained from an analysis of the ring structures in amorphous GeTe and $Ge_2Sb_2Te_5$ [2.37]. Fourfold rings are found to be the dominant ring structure in $Ge_2Sb_2Te_5$. Most of these rings are of ABAB alternation with A: Ge and/or Sb, B: Te. This is also true for GeTe, however in this case no clear maximum for any n-fold ring configuration can be identified. Furthermore the MD calculations reveal the vacancy structure in the amorphous alloys. It is found that cavities or voids occupy a total volume of 11.8 % in $Ge_2Sb_2Te_5$ and 6.4 % in GeTe, respectively. The vacancies which tend to be surrounded by Te atoms repel each other. Hence only few di- or multivacancies are identified. These findings allow for an interpretation of the phase transition process. The study suggests that it can be viewed as a vacancy-supported reorientation of ABAB-rings. As there are more ABAB-rings and less homopolar bonds in $Ge_2Sb_2Te_5$ compared to GeTe the crystallization process is facilitated in the ternary alloy.

One of the largest difficulties in the determination of the structural properties of crystalline phase change materials is the complexity of the structures. The most commonly used materials lie on the $(GeTe)_x(Sb_2Te_3)_y$ pseudobinary line and exhibit a stable hexagonal phase composed of GeTe and Sb_2Te_3 building blocks and a metastable rocksalt phase with Te atoms occupying one sublattice and Ge, Sb atoms as well as vacancies occupying the other sublattice. Besides the random occupation EXAFS measurements [2.33] as well as DFT calculations [2.34] have proven that the rocksalt lattice is not perfectly symmetrical.

---

[5] which is an upper limit for the total energy difference between the two phases

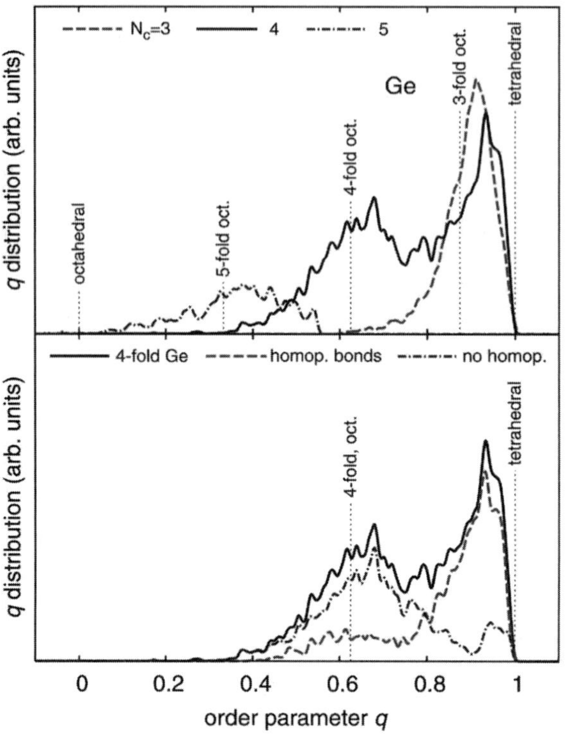

Fig. 2.2. Distribution of the local order parameter q for Ge with different coordination numbers N_c. q=1 for the ideal tetrahedral geometry, q=0 for the six-coordinated octahedral site, and q=5/8 for a four-coordinated defective octahedral site. Top panel: q-distribution resolved for Ge with different coordination number. Bottom panel: q-distribution for four-coordinated Ge further resolved for Ge with at least one homopolar bond with Ge or Sb or bonding with Te only no homopolar bonds Reprinted with permission from [2.36]. Copyright (2007), American Institute of Physics.

The DFT calculations show that the perfect rocksalt lattice of $Ge_1Sb_2Te_4$ is unstable against local distortions and relaxes into a rocksalt-like lattice with short and long bonds. Such a configuration exhibits a lower total energy and smaller interatomic forces.

Furthermore, with experimental methods it is very challenging to deduce the correct stacking order of the building blocks in the hexagonal phase and even more complicated to reveal the local order on the Ge/Sb sublattice in the rocksalt phase. Hence DFT calculations are employed to facilitate the structure determination of crystalline phase change materials. Studying the structural properties of $Ge_2Sb_2Te_5$ in the hexagonal phase it has been suggested that the lowest total energy, i.e. the most stable phase is found for a stacking order of Te-Ge-Te-Sb-Te-Te-Sb-Te-Ge- in the [0001]-direction [2.39]. This result agrees with a model based on electron microscopy measurements [2.40]. Furthermore the same authors studied several different structural models for the metastable crystalline phase and came to the conclusion that the site occupation on the Ge/Sb-sublattice is not random as has been assumed so far but follows a certain stacking sequence in the [111]-direction, which corresponds to the [0001]-direction in the hexagonal phase.

The sequence Te-Ge-Te-Sb-Te-v-Te-Sb-Te-Ge- is found to exhibit the lowest total energy. The “v” represents a vacancy layer or a vacancy position. This arrangement coincides with the most stable hexagonal configuration described above. However, it should be considered that the phases which have been identified to be the most stable ones only differ by 2 meV in the hexagonal phase and 7 meV in the rocksalt phase from the next-stable configuration. Hence entropic contributions could play an important role for the ground state at room temperature. Furthermore obviously only a limited number of different configurations can be studied, indicating that configurations with even lower energy can be found. Eom et al. [2.41] for instance found by studying different configurations that the vacancies in $Ge_2Sb_2Te_5$ are randomly distributed in the metastable crystalline phase, in contrast to the results of Sun et al. [2.39]. Hence the local structure of the crystalline phase still remains to some extent in dispute. A promising method to solve this problem by means of computational physics is to combine density functional theory calculations with data mining methods employed to predict crystal structures (see e.g. [2.42]).

2.6.2 Electronic Properties

Besides the information on the structure, *ab initio* calculations provide insight into the electronic properties of materials. In fact, the charge density computed in DFT for the independent particle system corresponds to the density of the full many body system. As structural and electronic properties are often correlated, a typical application of DFT is to identify the nature and energetical level of structural defects within a material. Phase change materials contain a large amount of vacancies in the crystalline state, e.g. in $Ge_1Sb_2Te_4$ 12.5 % of the lattice sites are empty. Hence they represent an interesting class of materials for electronic structure calculations of defect states. For GeTe these defect states have been studied with *ab initio* methods for the trigonal ground state and the metastable rocksalt phase (see [2.43]). By calculating a point defect within a supercell - i.e. using a basic structural unit consisting of several Wigner cells - the formation energies as well as the electronic levels of the defects can be derived. Formation energies are calculated by

$$\Delta E_{form}(q) = E_d(q) - E_p(q=0) - \sum_i^n \left(N_i \mu_i\right) - q\left(\varepsilon_F\right). \tag{2.24}$$

Here $E_d\,(q)$ is the total energy of the defective supercell in the charge state q, $E_p\,(q = 0)$ is the energy of the neutral perfect supercell, ε_F is the Fermi energy, measured from the top of the valence band, i is the chemical potential of the ith constituent atom, and N_i is the number of atoms of that type that has been either removed or added. The electrical level indicated by $(n-1)/n$ is the Fermi level at which the defect occupation changes from n to n - 1, i.e. one electron is added. It is calculated by

$$E\left((n-1)/n\right)=\left[E_d(n-1)+E_p(n)\right]-\left[E_d(n)+E_p(n-1)\right] \quad (2.25)$$

E_p and E_d refer to the energy of the perfect and the defective supercell.

The DFT calculations for both trigonal and rocksalt GeTe show that the Ge-vacancy is the most stable defect for all values of the Fermi energy in this system. In fact at Fermi energies above ≈ 0.25 eV Ge vacancies are predicted to form spontaneously (see Fig. 2.3).

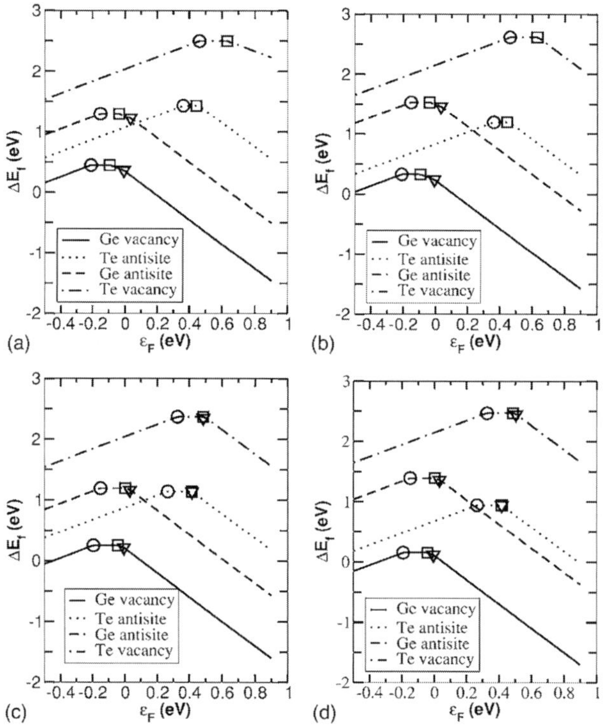

Fig. 2.3. Formation energy for the vacancy and antisite defects in rhombohedral (a) and (b) and fcc (c) and (d) germanium telluride for two growth environments: germanium rich (a) and (c) and tellurium rich (b) and (d). The symbols indicate electrical levels: ○ ↔ 0/+1; □ ↔ -1/0; ∇ ↔ -2/ -1. The zero of energy for the Fermi-level is the valence band edge. Reprinted with permission from [2.43]. Copyright (2006) by the American Physical Society.

Furthermore the 0/+1-level is below the valence band edge. As for reasonable values of the entropy of reaction, the Fermi level will be at or below the valence band edge at room temperature or above. This has clear implications for the transport properties. A large number of holes will be formed at the valence band edge, resulting in p-type conductivity. This furthermore explains the metallic Drude peak found in the optical spectra of crystalline GeTe.

Ternary GeSbTe alloys intrinsically contain a large amount of vacancies in the rocksalt phase. The electronic origin of these high vacancy concentrations can also be explained with the help of DFT calculations. By calculating fictitious systems along the stochiometric line $Ge_1Sb_2Te_4$ - $Ge_2Sb_2Te_4$ in the rocksalt phase, the influence of differences in the vacancy concentration on the electronic properties can be easily studied numerically [2.44]. $Ge_1Sb_2Te_4$ contains 25 % of vacancies on the Ge/Sb sublattice while in $Ge_2Sb_2Te_4$ all vacant lattice sites have been filled with Ge atoms. An interesting tool to study the chemical bonding is provided by the so-called Chemical Orbital Hamilton Population (COHP) analysis which allows rewriting the band structure energy in terms of atomic and bonding contributions [2.45, 2.46]. It is given by the following expression

$$\mathrm{COHP}(\varepsilon) = \sum_{R,L} \sum_{R',L'} \left\langle \chi_{R,L} \left| H \right| \chi_{R',L'} \right\rangle \underbrace{\sum_j f_j u^*_{R,L,j}\, u_{R',L',j}\, \delta(\varepsilon_j - \varepsilon)}_{DOS-matrix} \tag{2.26}$$

where $\chi_{R',L'}$ are local orbitals of which the single particle wave functions are constructed together with the eigenvectors $u^*_{R,L,j}$. R and L are the indices for the spatial coordinates and the momentum quantum numbers l and m, respectively, f_j is the occupation number, ε is the energy and ε_j the eigenvalue for the eigenstate j. Hence the COHP is the density of states multiplied with the matrix element of the Hamiltonian which corresponds to the overlap of two orbitals $\chi_{R,L}$ and $\chi_{R',L'}$. Figure 2.4 shows the COHP of the covalent Ge-Te and Sb-Te interactions in the three compounds $Ge_1Sb_2Te_4$, $Ge_{1.5}Sb_2Te_4$ and $Ge_2Sb_2Te_4$ both in the unrelaxed rocksalt lattice (top) as well as after the structural relaxation (bottom) which leads to local distortions of the lattice. The plot shows, that the composition $Ge_2Sb_2Te_4$ exhibits a large degree of antibonding Ge-Te and Sb-Te interactions in the highest occupied bands close to the Fermi level. Thus, the composition $Ge_2Sb_2Te_4$ stands for a too high valence-electron concentration (VEC) which must be lowered to achieve better stability. This is achieved by removing atoms from the system. In these compounds tellurium is the anionic atom with a high absolute electronegativity whose orbital contributions are dominant in the lower-lying valence bands. Hence, expelling Te atoms immediately weakens low-lying bonding states and is energetically unfavorable. Thus Ge and Sb whose orbitals both contribute to the states close to the valence band edge are energetically more favorable to remove. The comparison with the COHPs of the compositions $Ge_{1.5}Sb_2Te_4$ and $Ge_1Sb_2Te_4$ shows that these antibonding interactions are getting smaller and eventually vanish for an increasing number of vacancies. On the other hand such a consecutive removal of Ge-Te bonds also decreases the degree of bonding states in the system. Thus the most stable configuration with the lowest total energy is found for $Ge_{1.5}Sb_2Te_4$. The stability of this compound has been predicted by these calculations before it has been synthesized experimentally [2.44].

Furthermore, upon structural relaxation (Fig. 2.4, bottom) the band gap between the valence and conduction band increases significantly. Such a behavior, where a structural distortion is accompanied by a decrease in the electronic energy and the opening of a band gap, is called a Peierls instability (see also above).

Hence in this case DFT is employed to explain physical properties of phase change materials as well as to predict new materials with possibly superior properties.

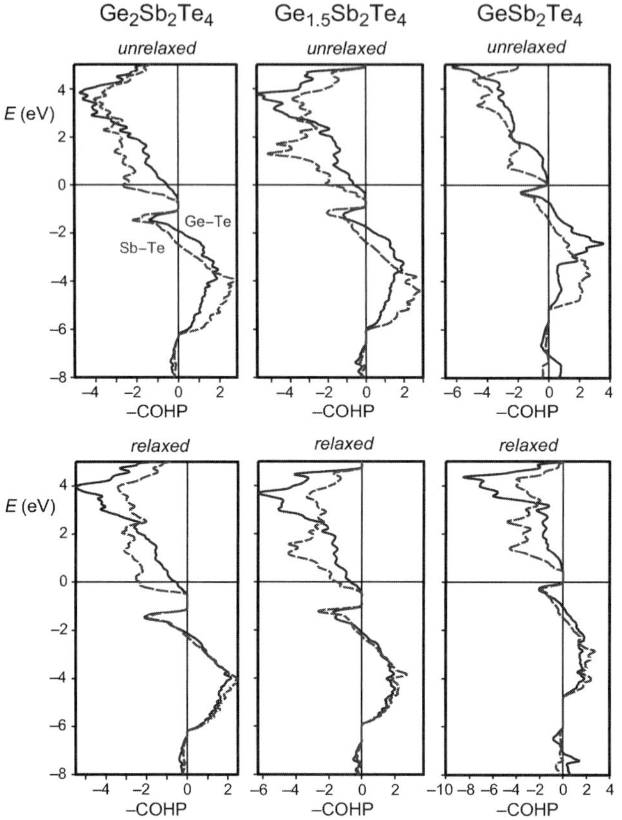

Fig. 2.4. Crystal-Orbital Hamilton Population (COHP) bonding analysis: The plot displays the Ge-Te (solid line) and Sb-Te (dashed line) interactions in $Ge_2Sb_2Te_4$ (left), $Ge_{1.5}Sb_2Te_4$ (middle) and $Ge_1Sb_2Te_4$ (right); the upper panel shows the unrelaxed NaCl-lattice whereas the lower panel corresponds to the structure with full atomic relaxation. The Fermi levels are set to zero. Bonding interactions to the right, antibonding interactions to the left. Reprinted by permission from Macmillan Publishers Ltd: Nature Materials [2.44], copyright (2007).

The predictive power of DFT can be also applied to a wider stochiometric range going beyond GeSbTe alloys. Calculating the structural stability and bonding properties of $XInTe_2$ and $XSbTe_2$ with X= Cu, Ag, Au it has been shown that only the alloys containing antimony exhibit the p-type bonding which is found in phase change materials and leads to the rocksalt-type crystalline phase. The In-containing alloys on the other side exhibit sp^3 hybridization resulting in a chalcopyrite structure with tetrahedral bonds. Chalcopyrites have been shown to have insufficient optical contrast and too long re-crystallization times which makes them unsuitable for phase change applications [2.47]. By calculating the structural stability it can be shown that the alloys favoring the rocksalt structure have a number of valence s- and p-electrons higher than 4.1 while those favoring the chalcopyrite structure contain a number of s- and p-electrons below this threshold [2.48].

As pointed out above it is much more challenging to study the amorphous phase with *ab initio* methods as the structure of this phase is not well known. Hence simple models have been used to calculate electronic properties of amorphous phase change materials, however with considerable success. Using a spinel structure for amorphous $Ge_1Sb_2Te_4$ with Ge in fourfold coordinated positions it was shown that the band gap of the amorphous phase widens compared to the crystalline state [2.34]. This result is important for the application of phase change materials in electronic data storage as the conductivity in the amorphous state is determined by localized trap states [2.49]. The existence of such states requires a sufficiently large band gap. However it is shown that - following this model - the trap states are not made of valence alternation pairs (VAP) as commonly found e.g. in As_2Se_3 [2.50]. VAPs require twofold coordinated Te with their lone pairs. However in the simple structural model for the amorphous phase as well as in the more detailed models derived from MD-simulations Te is mainly threefold coordinated in the amorphous phase (see [2.51]). A thorough analysis of the electronic structure calculations explains how the changes in the local arrangement are correlated to the change of electronic properties and the opening of the gap. Even though the Ge atoms experience a change in their local coordination from octahedral in the crystalline state to tetrahedral in the amorphous state, the change in the electronic properties is most pronounced for the Te and Sb atoms. In particular Te-p states which are dominant at the valence band edge in the crystalline phase decrease in energy which finally results in the opening of the electronic band gap. This is illustrated in Fig. 2.5, showing the differences in the calculated electronic density of states between the crystalline and the amorphous state.

A positive amplitude represents a higher weight in the rocksalt phase while a negative amplitude represents a higher weight in the amorphous phase. A large weight in the crystalline phase is found directly below the Fermi level (0 eV). The amorphous phase displays a large weight between - 4.5 eV and - 3 eV. This is mainly due to stronger hybridization between s and p-states in the spinel phase, which serves as a model for the local order in the amorphous state. The electronic states of the Ge atom on the other hand do not change considerably. The predictions of this simple structural model are confirmed by the analysis of MD-simulations for GeTe and $Ge_2Sb_2Te_5$. The structural model derived from these calculations is much more detailed and precise (see above and [2.36, 2.37]). The changes in the electronic properties however are very similar. The density of states has a stronger weight in the crystalline phase directly at the Fermi level, which is mainly due to contributions from Te-p and Sb-s, p and d states. In total the band below the Fermi level is narrower in the amorphous phase compared to the crystalline phase. Also in these calculations this is explained by a change from sp^3-bonding to p-bonding upon crystallization [2.37].

Optical properties as well as transport properties are important within the field of phase change materials. They define if the optical reflectivity or the electrical conductivity differs sufficiently between the amorphous and the crystalline phase and hence whether a material is suitable for application in phase change technology. However, DFT is a theory of the electronic ground state while optical and transport properties represent electronic excitations. Hence it is necessary to go beyond density functional theory in order to treat them computationally.

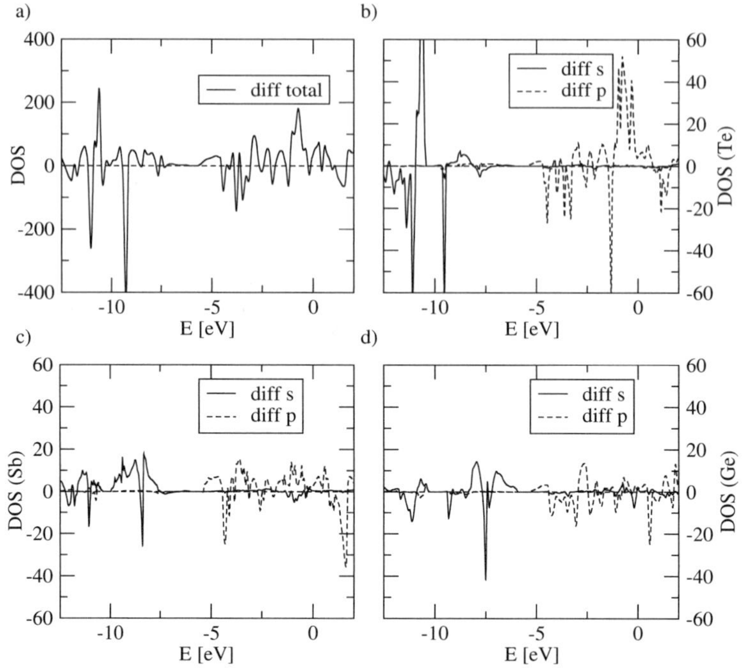

Fig. 2.5. Difference of the density of states (DOS) between the crystalline rocksalt and amorphous structures for all atoms and for the s-and p-bands for each element. Positive values indicate larger weight in the crystalline phase, negative values larger weight in the amorphous phase. Positive values are found just below the Fermi-level from 0 eV (= E_F) down to -2 eV. The difference in the density of states can be attributed mostly to the p-orbitals of Te, as Ge shows significant changes only in the s-band. The changes in the Sb s-and p-band are in between the extremes of Ge and Te. Reprinted by permission from Macmillan Publishers Ltd: Nature Materials [2.34], copyright (2006).

In particular time-dependent density functional theory (TDDFT) and many-body perturbation theory have been successfully applied to calculate excited electronic states in many solids (see [2.15]). To calculate optical properties of phase change materials and reveal the origin of their difference between the amorphous and the crystalline phase TDDFT within the so-called random-phase approximation has been employed. Using simple structural models for the amorphous phase of GeTe and $Ge_1Sb_2Te_4$ where Ge atoms partially occupy fourfold coordinated positions, the "optical contrast", i.e. the difference in the optical properties upon amorphization observed experimentally, has been reproduced [2.52]. The calculations then allow for a more detailed analysis of the data. The imaginary part of the dielectric function ϵ_M is calculated using Fermi's golden rule

$$Im(\epsilon_M(\omega)) = 2\pi \lim_{\mathbf{q}\to\mathbf{0}} v_0(\mathbf{q}) \sum_{vc} \left|\langle c | e^{-i\mathbf{q}\cdot\mathbf{r}} | v \rangle\right|^2 \delta(\omega - (\varepsilon_c - \varepsilon_v)) \tag{2.27}$$

where c, v and ε_c, ε_v are eigenstates and eigenenergies respectively, while ω is the frequency of the excitation. Hence it can be decomposed in the part which originates from the band structure, i.e. the so-called joint density of states (JDOS)

$$\mathrm{JDOS}/\omega^2 \equiv \frac{1}{N_{\mathbf{k}}\omega^2}\sum_{v,c,\mathbf{k}} \delta\left(\varepsilon_{c\mathbf{k}} - \varepsilon_{v\mathbf{k}} - \omega\right) \tag{2.28}$$

and by the velocity ($\mathbf{v}$) matrix elements of the optical transitions $\mathbf{m}_{v,c,\mathbf{k}} \equiv \langle c|\mathbf{v}|v\rangle$. Usually it is assumed for a semiconductor that the JDOS smears out upon amorphization and therefore generates a small change in the optical absorption, while the matrix elements remain constant. However in GeTe and $Ge_1Sb_2Te_4$ changes in the JDOS cannot explain the decrease in the optical absorption in the amorphous phase as demonstrated in Figs. 2.6a and 2.6c.

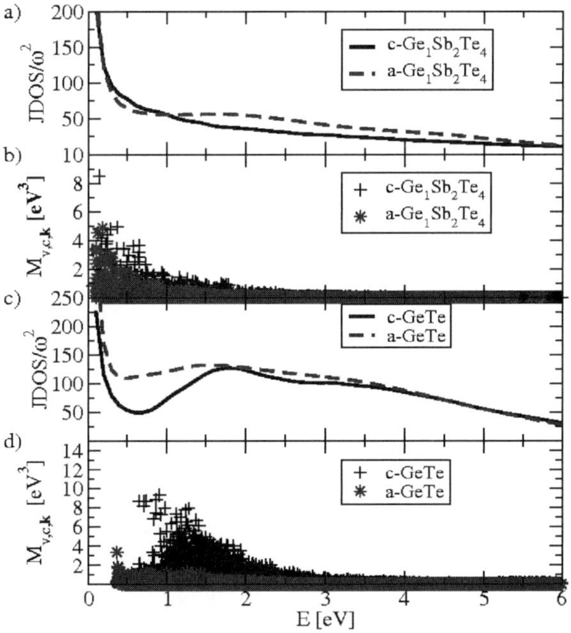

Fig. 2.6. a) JDOS/ω^2 of $Ge_1Sb_2Te_4$ in the crystalline and the amorphous state displayed as the number of transitions/eV^3 per cell and k-point. Up to 1.4 eV the JDOS in the crystalline phase is stronger than in the amorphous state. For GeTe (c) up to 1.7 eV more transitions are found in the amorphous state compared to the crystalline phase. For both materials the decrease in the absorption upon amorphization can only be explained taking into account the velocity matrix elements shown in (b) ($Ge_1Sb_2Te_4$) and (d) (GeTe) with $M_{v,c,\mathbf{k}} = 2\frac{4\pi^2}{\Omega}\lim_{\mathbf{q}\to 0}\frac{1}{\mathbf{q}^2}|m_{v,c,\mathbf{k}}|^2$. Reprinted with permission from [2.52]. Copyright 2007 by the American Physical Society.

For both alloys the JDOS is larger in the amorphous phase for a large part of the studied spectral range. Thus the optical contrast can only be understood if the contribution arising from changes in the oscillator strength of the optical transitions is taken into account. This contribution is shown in Figs. 2.6b ($Ge_1Sb_2Te_4$) and 2.6d (GeTe). For $Ge_1Sb_2Te_4$ up to about 2 eV stronger matrix elements are found in the crystalline phase, for crystalline GeTe they are significantly stronger in the entire spectral range. Hence the calculations show that the unusual change in the local order in phase change materials upon amorphization induces another remarkable effect, namely a significant decrease of the oscillator strength of the optical transitions which finally results in the large optical contrast employed in optical data storage.

A very challenging field for computational physics is the correct modeling of transport properties with *ab initio* calculations. The importance for phase change materials is evident. As they are emerging as a promising candidate for electronic data storage a precise microscopic understanding of their transport properties is of great importance. However, despite important progress has been made to describe quantum transport e.g. with TDDFT and nonequilibrium Green's functions (see e.g. [2.53]), no well established theory is available at the moment.

The development and numerical implementation of a method to calculate transport properties from first principles shall provide new interesting opportunities to employ computational methods in the field of phase change materials in the future.

References

[2.1] Born, M., Oppenheimer, J. M.: Zur Quantentheorie der Molekeln. Ann. Phys. **84**, 457-484 (1927)

[2.2] Hartree, D. R.: Proc. Cambridge Philos. Soc. **24**, 89-110 (1928)

[2.3] Fock, V.:. Z. Physik **61**, 126-148 (1930)

[2.4] Hohenberg, P., Kohn, W.: Inhomogeneous electron gas. Phys. Rev. 136, B864-871 (1964)

[2.5] Levy, M.: Universal variational functionals of electron densities, first-order density matrices,, natural spin-orbitals and solution of the nu-representability problem. Proc. Nat. Acad. Sci. **76**, 6062-6065 (1979)

[2.6] Kohn, W., Sham, L. J.: Self-consistent equations including exchange and correlation effects. Phys. Rev. **140**, A 1133-A1138 (1965)

[2.7] Ceperley, D. M., Alder, B. J.: Ground state of the electron gas by a stochastic method. Phys. Rev. Lett. 45, 566-569 (1980)

[2.8] Slater, J. C., Wilson, T. M., Wood, J. H.: Comparison of several exchange potentials for electrons in the Cu^+ ion. Phys. Rev. **179**, 28-38 (1969)

[2.9] Ohno, K, Esfarjani, K., Kawazoe, Y.: Computational Materials Science. Springer, Berlin (1999)

[2.10] Perdew, J. P., Zunger, A.: Self-interaction correction to density-functional approximations for many-electron systems. Phys. Rev. B **23**, 5048-5079 (1981)

[2.11] Goedecker, S., Teter, M., Hutter, J.: Separable dual-space Gaussian pseudopotentials. Phys. Rev. B **54**, 1703-1710 (1996)

[2.12] Gunnarsson, O., Jonson, M., Lundqvist, B. I.: Descriptions of exchange and correlation effects in inhomogeneous electron. Phys. Rev. B **20**, 3136-3164 (1979)

[2.13] Perdew, J. P., Wang, Y.: Accurate and simple density functional for the electronic exchange energy: Generalized gradient approximation. Phys. Rev. B 33, 8800-8802 (1986)

[2.14] Perdew, J. P., Burke, K., Ernzerhof, M.: Generalized gradient approximation made simple. Phys. Rev. Lett. **77**, 3865-3868 (1996)

[2.15] Onida, G., Reining, L., Rubio, A.: Electronic excitations: density-functional versus many-body Green's-function approaches. Rev. Mod. Phys. **74**, 601-659 (2002)

[2.16] O'Reilly, E. P., Robertson, J., Kelly, J. M.: The structure of amorphous GeSe and GeTe. Solid State Comm. **38**, 565-568 (1981)

[2.17] Rabe, K. M., Joannopoulos, J. D.: Theory of the structural phase transition of GeTe. Phys. Rev. B **36**, 6631-6639 (1987)

[2.18] Kolobov A. V., Fons, P., Tominaga, J., Ankudinov, A. L., Yannopoulos, S. N., Andrikopoulos, K. S.: Crystallization-induced short-range order changes in amorphous GeTe. J. Phys.: Condens. Matter 16, S5103-S5108 (2004)

[2.19] Nosé, S.: A molecular-dynamics method for simulations in the canonical ensemble. Mol. Phys. **52**, 255-268 (1984)

[2.20] Nosé, S.: A unified formulation of the constant temperature molecular dynamics methods. J. Chem. Phys. **81**, 511-519 (1984)

[2.21] Chelikowsky, J. R., Derby, J. J., Godlevsky, V. V., Jain, M., Raty, J. Y.: Ab initio simulations of liquid semiconductors using the pseudopotential-density functional method. J. Phys.: Condens. Matter. **13**, R817-R854 (2001)

[2.22] Kubo, R.: Fluctuation-Dissipation Theorem. Rep. Prog. Phys. **29**, 255-284 (1966)

[2.23] Stratonovitch, R. L.: Topics in the Theory of Random Noise. New York: Gordon and Breach (1963)

[2.24] van Kampen, N. G.: Stochastic Processes in Physics and Chemistry. North-Holland, Amsterdam (1981)

[2.25] Risken, H.: The Fokker-Planck Equation. Springer, Berlin (1984)

[2.26] Raty, J. Y., Godlevsky, V. V., Ghosez, P., Bichara, C., Gaspard, J. P., Chelikowsky, J. R.: Evidence of a reentrant Peierls distortion in liquid GeTe. Phys. Rev. Lett. **85**, 1950-1953 (2000)

[2.27] Raty, J. Y., Gaspard, J. P., Bichara, C., Bergman, C., Bellissent, R., Ceolin, R.: Re-entrant Peierls distortion in IV-VI compounds. Physica B 276, 473-474 (2000)

[2.28] Raty, J. Y., Godlevsky, V. V., Gaspard, J. P., Bichara, C., Bionducci, M., Bellissent, R., Céolin, R., Chelikowsky, J. R., Ghosez, P.: Local structure of liquid GeTe via neutron scattering and ab initio simulations. Phys. Rev. B **65**, 115205 (2001)

[2.29] Peierls, R. E.: Quantum Theory of Solids. Oxford University Press, Oxford (1956)

[2.30] Bichara, C., Johnson, M., Raty, J. Y.: Temperature-induced density anomaly in Te-rich liquid germanium tellurides: p versus sp^3 bonding? Phys. Rev. Lett. 95, 267801 (2005)

[2.31] Blaineau, S., Jund, P., Drabold, D.: Physical properties of a GeS_2 glass using approximate ab initio molecular dynamics. Phys. Rev. B 67, 094204 (2003)

[2.32] Tafen, D. N., Drabold, D. A.: Models and modeling schemes for binary IV-VI glasses. Phys. Rev. B 71, 054206 (2005)

[2.33] Kolobov, A. V., Fons, P., Frenkel, A. I., Ankudinov, A. L., Tominaga,, Uruga, J. T.: Understanding the phase-change mechanism of rewritable optical media. Nature Materials **3**, 703-708 (2004)

[2.34] Wełnic, W., Pamungkas, A., Detemple, R., Steimer, C., Blügel, S., Wuttig, M.: Unraveling the interplay of local structure and physical properties in phase-change materials. Nature Materials **5**, 56-62 (2006)

[2.35] Kalb, J., Spaepen, F., Wuttig, M.: Calorimetric measurements of phase transformations in thin films of amorphous Te alloys used for optical data storage. J. Appl. Phys. **93**, 2389-2393 (2003)

[2.36] Caravati, S., Bernasconi, M., Kühne, T. D., Krack, M., Parrinello, M.: Coexistence of tetrahedral- and octahedral-like sites in amorphous phase change materials. Appl. Phys. Lett. **91**, 171906 (2007)

[2.37] Akola, J., Jones, R. O.: Structural phase transitions on the nanoscale: The crucial pattern in the phase-change materials $Ge_2Sb_2Te_5$ and GeTe. Phys. Rev. B **76**, 235201 (2007)
[2.38] Errington, J. R., Debenedetti, P. G.: Relationship between structural order and the anomalies of liquid water. Nature **409**, 318-321 (2001)
[2.39] Sun, Z., Zhou, J., Ahuja, R.: Structure of phase change materials for data storage. Phys. Rev. Lett. **96**, 055507 (2006)
[2.40] Kooi, B. J., Groot, W. M. G., de Hosson, J. T. M.:In situ transmission electron microscopy study of the crystallization of $Ge_2Sb_2Te_5$. J. Appl. Phys. 95, 924-932 (2004)
[2.41] Eom, J.-H., Yoon, Y.-G., Park, C., Lee, H., Im, J., Suh, D.-S., Noh, J.-S., Khang, Y., Ihm, J.: Global and local structures of the Ge-Sb-Te ternary alloy system for a phase-change memory device. Phys. Rev. B **73**, 214202 (2007)
[2.42] Fischer, C., Tibbetts, K., Morgan, D., Ceder, G.: Predicting crystal structure: merging data mining with quantum mechanics. Nature Materials **5**, 641-646 (2006)
[2.43] Edwards, A. H., Pineda, A. C., Schultz, P. A., Martin, M. G., Thompson, A. P., Hjalmarson, H. P., Umrigar, C. J.: Electronic structure of intrinsic defects in crystalline germanium telluride. Phys. Rev. B **73**, 045210 (2006)
[2.44] Wuttig, M., Lüsebrink, D., Wamwangi, D., Wełnic, W., Gilleßen, M., Dronskowski, R.: The role of vacancies and local distortions in the design of new phase change. Nature Materials **6**, 122-128 (2007)
[2.45] Dronskowski, R., Blöchl, P. E.: Crystal orbital Hamilton populations (COHP): energy-resolved visualization of chemical bonding in solids based on density-functional calculations. J. Phys. Chem. **97**, 8617-8624 (1993)
[2.46] Glassey, W. V., Papoian, G. A., Hoffmann, R.: Total energy partitioning within a one-electron formalism: A Hamilton population study of surface-co interaction in the c(2 x 2)-CO/Ni(100) chemisorption system. J. Chem. Phys. **111**, 893-910 (1999)
[2.47] Detemple, R., Wamwangi, D., Wuttig, M., Bihlmayer, G.: Identification of Te alloys with suitable phase change characteristics. Appl. Phys. Lett. **83**, 2572-2574 (2003)
[2.48] Luo, M. B., Wuttig, M.: The dependence of crystal structure of Te-based phase-change materials on the number of valence electrons. Adv. Mat. **16**, 439-443 (2004)
[2.49] Pirovano, A., Lacaita, A. L., Benvenuti, A., Pellizzer, F., Bez, R.: Electronic switching in phase-change memories. IEEE Trans. Electron Devices **51**, 452-459 (2004)
[2.50] Kastner, M., Adler, D., Fritzsche, H.: Valence-alternation model for localized gap states in lone-pair semiconductors. Phys. Rev. Lett. **37**, 1504-1507 (1976)
[2.51] Robertson, J., Xiong, K., Peacock, P. W.: Electronic and atomic structure of $Ge_2Sb_2Te_5$ phase change memory material. Thin Solid Films **515**, 7538-7541 (2007)
[2.52] Wełnic, W., Botti, S., Reining, L., Wuttig, M.: Origin of the optical contrast in phase-change materials. Phys. Rev. Lett. **98**, 236403 (2007)
[2.53] Stefanucci, G., Almbladh, C.-O.: An exact ab initio theory of quantum transport using TDDFT and nonequilibrium Green's functions J. Phys. Conf. Ser. **35**, 17-24 (2006)

3. Nature of Glasses

Punit Boolchand, Matthieu Micoulaut, and Ping Chen

Abstract Glasses exist in *three* generic elastic phases: *flexible*, *intermediate* and *stressed-rigid,* which are determined by the connectivity of their backbones. Measurements of glass transition temperatures (T_gs) using modulated-differential scanning calorimetry permits distinguishing these phases by their characteristic non-reversing enthalpies (ΔH_{nr}) at T_gs. In Raman scattering, characteristic elastic power-laws are observed in *intermediate* and *stressed-rigid* phases. Liquid fragilities are found to correlate with ΔH_{nr} terms in covalent networks but not in modified oxide or H-bonded networks. In the latter systems weak network links exist, which cease to constrain networks as the temperature $T > T_g$ and viscosities plummet. *Intermediate phase glasses* are composed of rigid but unstressed networks that are in a state of *quasi-equilibrium* and age minimally. Such glasses usually form space filling networks and are structurally *self-organized.*

3.1 Introduction

Most liquids upon cooling crystallize when their temperature is lowered below their melting temperature (T_m). But some liquids can bypass crystallization and become solid at a lower temperature T_g (glass transition temperature) to form a *glass*. Basic challenges about glasses have confronted materials scientists, physicists, chemists, geologists and engineers alike for over fifty years - such as: what is special about glass-forming liquids and what determines the glass transition temperatures? Are glass network structures usually random and fully polymerized [3.1]? In this chapter we review the remarkable progress which has taken place on the subject in the past few years both in theory and experiment. Although glass formation occurs over a very small part of phase (i.e., composition) space, it is remarkable that the phenomenon spans liquids possessing a wide range of chemical bonding. Covalent, ionic, superionic, metallic, organic polymers and biopolymers can each exist in a glassy state. Once formed, some glasses appear to be generally quite robust and are used today in several mature technologies which require long-term stability. These observations are all the more puzzling given that glasses are intrinsically *non-equilibrium solids* (supercooled liquids), and most tend to continuously evolve with time, i.e., they age. In microelectronics, a

S. Raoux, M. Wuttig (eds.), *Phase Change Materials*,
DOI 10.1007/978-0-387-84874-7_3, © Springer Science+Business Media, LLC 2009

3-terminal MOSFET device forms the heart of all solid state chips in circuits and systems. The gate oxide is a thin overlayer of *glassy* SiO_2 on an active crystalline Si medium. These devices routinely operate for over 10^9 cycles without degradation of the gate oxide insulator. How can that come about? The very special role of the Si-SiO_2 interface in microelectronics has come to the fore in recent years as attempts to replace the silica gate material with high-K (dielectric constant) oxides (HfO_2, ZrO_2) in aggressively scaled devices have met with utter failure. We have recently come to realize that the Si-SiO_2 interface is rather special in that it forms a *self-organized stress-free* structure [3.2]. The recognition has opened a new avenue to engineer high dielectric constant (high-K) gate dielectrics on the self-organized Si-monoxide layer with remarkable success [3.3]. Another example of a functional glass is window glass which is an empirically engineered alloy of silica, lime and soda. In an early form, glass [3.4] has shaped human civilization since the times of the Phoenicians and it continues to do so at present as well. It has found applications in windows of houses and churches, in arts and crafts, in the evolution of experimental science, and variants of window glass are now used in state of the art fiber optics, lenses, microscopes, cars and flat-panel displays. Window glass is a remarkably robust material and our recent understanding reveals that this is not an accident. By tuning the soda to lime ratio, one can cooperatively drive the ternary alloy to *self-organize* [3.5]. A number of other network systems adapt and reconfigure their structure to *self-organize*; for example, covalent network glasses [3.6, 3.7], computational phase transitions [3.8], high temperature superconductors [3.9], H-bonded alcohols and saccharides [3.10], and protein folding [3.11].

Just now we are beginning to understand how disordered networks *self-organize* [3.12, 3.13] to form systems that exist in a state of supercooled *quasi-equilibrium*.

The present volume is dedicated to phase change materials that are used in rewritable DVDs or CDs (Chapters 12 and 13), Phase Change Random Access Memory (PCRAM) devices (Chapters 14–17), as well as in potential new applications such as reconfigurable logic (Chapter 18). The technology utilizes [3.14] amorphous thin-films of alloys of the group IV and group V elements with tellurium. Films of interest include $Ge_2Sb_2Te_5$ [3.15], which broadly form covalently bonded networks and are poor glass formers. Their networks may even be intrinsically nanoscale phase separated or demixed, i.e., composed of more than one type of a backbone structure. The interaction of light and independently heat permits films to switch locally between their amorphous and crystalline state as a group of atoms shuffle reversibly and repeatedly between the ordered and disordered state. The underlying structure changes have attracted recent interest [3.16] because of the commercial implications of the technology. It is thus appropriate that ideas on glass formation and structure, and particularly the nature of glass transition in covalently bonded solids be reviewed in this context. Many of the ideas on glass structure developed on Se- and S-based systems in this review will be relevant to Te-based systems as well. However, one must remember that Te-rich melts are usually semi-metallic with Te taking on a coordination number of greater than 2, which has the consequence that it is usually not feasible to obtain melt-quench glasses in Te-rich melts. Vapor deposition is the preferred method to obtain amorphous thin-films of the tellurides. The propensity of Te cations to easily switch their coordination number between 2- to 3- or even quasi 6-fold from the amorphous to the crystalline state by a heat pulse or light flux is the functionality of interest in their applications as phase-change materials [3.15].

3.2 Thermodynamics of the Glass Transition

Cooling of a liquid leading to crystallization is schematically illustrated in Fig. 3. 1, in which thermodynamic variables such as enthalpy (*H*), entropy (*S*) and molar volumes (*V*) undergo a first order change near T_m. Alternatively, a glass forming liquid can be supercooled below T_m and displays a higher order transition to a glass state near T_g (~2/3 T_m). Many years ago, Kauzmann [3.17] visualized the glass transition as an event that circumvents the entropy crisis that would result if the liquid entropy S_{liquid} continued to decrease linearly as T is lowered to the Kauzmann temperature T_K (Fig. 3.1) to become less than the crystal entropy $S_{crystal}$. Trajectories of two glass forming liquids cooled at different rates are shown in Fig. 3.1 as well; glass a (glass transition T_{ga}) is realized when a liquid is cooled at a high rate, and glass c (glass transition T_{gc}) is realized when it is cooled at a low rate. A higher cooling rate kinetically shifts the glass transition to higher temperatures, i.e., $T_{ga} > T_{gc}$. We also show the effect of aging glass a below its T_{ga}; its enthalpy H decreases (trajectory b) with time in relation to the as quenched (trajectory a) glass. Aging generally leads networks to compact as molar volumes, enthalpies and entropies decrease. The aging process involves relaxation from local metastable configurations produced by the particular quenching history. This will lower the entropy and allow a more efficient use of space.

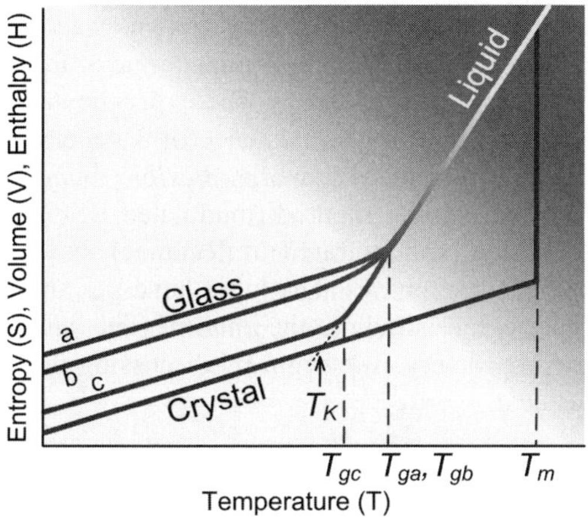

Fig. 3.1. Normal liquids crystallize at T_m, and glass forming liquids form a glass. Trajectory a and c show quench rate effects, while a and b aging effects. T_K represents the Kauzman temperature. Reprinted by permission from Macmillan Publishers Ltd: Nature [3.18]], copyright (2001).

Historically, glass transitions are discussed within theories that emphasize the kinetic aspect [3.19] of the underlying process. Good glasses form even when liquids are cooled very slowly (1°C/min). Aspects of network structure [3.20, 3.21], particularly connectivity, determine the magnitude of T_g as well as the nature of glass transitions [3.22].

The large reduction in T_g of SiO_2 glass from 1200°C to nearly 600°C upon alloying a few mole percent of soda was known as early as 1500 BC. These large changes in T_g brought about by chemical alloying far exceed those brought about by quench rates. Currently, we do not have a theory of the glass transition to account for all these effects. Not surprisingly, Anderson [3.23] noted several years ago that, "the deepest and most interesting unsolved problem in solid state theory is probably the theory of the nature of the glass transition; this could be the next breakthrough in the coming decade….".

Changes in the *nature* of T_g brought about by chemical alloying can produce far more spectacular effects than hitherto recognized. The motional arrest which takes place during the glass formation and in the reverse process during melting can be thought of as involving two components: the first, which we will call *non-ergodic* (or *T*-irreversible) involves the progressive restriction of allowed structures and loss of configurational entropy, and the second, the continuous change in vibrational specific heat at each configuration, which we will call *ergodic* (or *T*-reversible). Glass formation has its roots in non-ergodic behavior. Up until recently we have not had a clear experimental handle on separating these two types of processes that underlie the glass transition. With the advent of modulated Differential Scanning Calorimetry (m-DSC), one is now in a position [3.22] to separate these two fundamentally different behaviors. These experiments have led to the exciting discovery of compositional *windows of almost reversible* glass formation. Taken together with other remarkable features presented in this chapter of glasses in these composition windows, they represent a *new* phase of network solids, which has come to be known as the Intermediate Phase (IP).The sharpness of *reversibility windows* in composition space, and the quasi-equilibrium nature of network structures in these windows justifies the use of the term IP. These structures, neither random nor crystalline, are thought to be of a variety known as *self-organized.* The IP is a novel phase of condensed matter, and is probably one of the select few to be discovered since polymers in the early 1920s. The experimental landscape supported by numerical simulations reveals the existence of 3 generic classes of glasses based on the elastic response of their networks, *flexible*, *intermediate* and *stressed-rigid.* In addition to the thermal method (modulated DSC), vibrational spectroscopies (Raman scattering and infrared reflectance) have proved to be rather insightful in elucidating these phases. The latter have permitted characteristic elastic power-laws to be measured in the intermediate and stressed-rigid phases, which are supported by theory. We begin by discussing the nature of glass transitions in these phases next.

3.3 Glass Transition from Dynamics

Insights into the glass transition have also come from the dynamic response of glass-forming liquids [3.24]. In supercooled liquids changes of structure with temperature are small but those in dynamical properties are large. Viscosity (η) through the Maxwell relation, $\eta = G_\infty \tau$, where G_∞ is the high frequency shear modulus and τ the shear-stress relaxation time, is a probe of liquid dynamics. At high T (> T_g), glass-forming liquids have low viscosities (η) and short relaxation times (τ) of the order of atomic vibrations (10^{-12} sec). Upon cooling to $T \sim T_g$, liquid viscosities increase by orders of magnitude (Fig. 3.2).

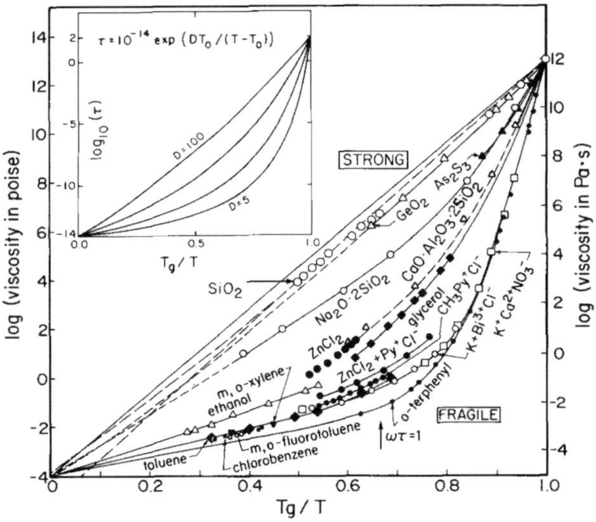

Fig. 3.2. Log of the viscosity as a function of the inverse temperature for different glass-forming liquids. The slope at T=T_g serves to define the fragility index m. Reprinted from [3.25]. Copyright (1988), with permission from Elsevier.

In covalent systems (such as silica or chalcogenides), changes in $\eta(T)$ are moderate and follow an Arrhenius-like behavior

$$\eta = \eta_0 \exp[\Delta E_A / k_B T] \tag{3.1}$$

Here ΔE_A represents the activation energy for viscosity, k_B is the Boltzman constant, and η_0 the high temperature limit of η. In liquids possessing softer interatomic interactions (ionic, van der Waals), $\eta(T)$ increases by several decades over a small T- range, a behavior first described by Vogel, Tamman, Hesse and Fulcher [3.26] in the form:

$$\eta = \eta_0 \exp[\Delta E_A / k_B (T - T_0)] \tag{3.2}$$

Here T_0 is the temperature at which η diverges and is also identified with the Kauzmann temperature T_K (Fig. 3. 1). By convention, the "dynamic" T_g is defined as the T when η acquires a value of 10^{12} Poise. This usually corresponds to the T at which the time needed to equilibrate the system becomes of the order of the timescale of an experiment (100 s), and correlates fairly well with "calorimetric" T_g. For $T < T_g$, relaxation times increase astronomically and liquids freeze into glassy solids. At $T \sim T_g$ a glass has the viscosity of a solid, but has a measurable diffusion constant which allows for a moderate relaxation on a reasonable experimental time scale. As a consequence, aging is manifested in thermal and dynamical properties, usually measured as a function of waiting time [3.27] after samples are synthesized by either a melt-quench or evaporation. In Fig. 3.2, liquids such as SiO_2 or GeO_2 , which display an Arrhenius T-variation of viscosity are recognized as *strong*, while liquids such as o-terphenyl, glycerol and toluene, which display a bowed T-variation of viscosity leading to a high slope or activation energy near T_g are termed as *fragile*. Glassy behavior is also realized in colloids or soft sphere liquids by jamming [3.28]. The latter occurs when a system develops extremely long stress relaxation times in a disordered state.

To summarize, glassy behavior is realized either by increasing density of a liquid (jamming effect), by shearing [3.29], or by super-cooling a liquid. The nature of atomic interactions usually determine relevant parameters upon vitrification: kinetic factors in soft sphere or organic supercooled liquids, structural factors in covalent glass-forming liquids, and increase of η results when chemical bonds form between atoms.

3.4 Glass Forming Tendency

From the point of view of the configuration space of a system of N atoms, crystalline structures occupy a very small region because, on average, the atoms in one unit cell define the entire structure. This simplification begins to break down near, say, a melting transition where there are a significant number of thermally activated defects. Insight into glasses and the glass transition is challenged by the need to consider much larger regions of the configuration space. The *potential* energy of all possible configurations, defined over the 3 (or 6 including rotations) N dimensional hyperspace, is called the potential energy landscape (PEL) and is a highly inhomogeneous distribution of peaks and valleys – for which a statistical approach is needed to extract predictions [3.30]. Physically reasonable but speculative statistical assumptions about the distribution of metastable regions permit a qualitative treatment of the dynamics on the PEL and have yielded useful insights on features like melting (T_m) and diffusion.

An important feature that is not directly derivable from the generic PEL is the particular molecular structures – not even average structures. For these, more specific methods are required. The experiments and theory in the present chapter are primarily concerned with elucidating local structures. The emerging picture of network glass structures as falling into three distinct mechanically elastic types (or phases), namely, floppy, isostatically rigid (the IP) and stressed rigid are as yet not describable with landscape arguments, particularly, the IP which, like an ordered phase, is not readily located on the PEL. It is hoped that future conceptual progress will enable the PEL picture to delineate (and predict) these mechanical phases.

A new beginning in understanding glasses at a basic level emerged in 1979 when Phillips [3.31] anticipated that a subtle balance between atomic degrees of freedom and mechanical constraints can arise in covalent networks when they possess optimal connectivity and lead liquids to undergo a glass transition with minimal enthalpic change, i.e., the glass formation tendency (GFT) is optimized. For liquids constrained only by bond-bending and bond-stretching interactions, the number of mechanical constraints per atom (n_c) can be easily computed, and when these match the available degrees of freedom (3) per atom in 3D, one can expect a glass transition with low enthalpic changes to occur. Thus the condition $n_c = 3$, when glasses are optimally constrained, has served as an interesting basis for the search of optimal glass forming compositions.

To see examples of optimally constrained networks leading to facile glass formation, one can examine as a function of composition the minimal liquid cooling rate needed to form a glass or avoid crystallization. In Table 3.1 we show examples of chalcogenides [3.32] where critical cooling rates show a global minimum in select systems.

Table 3.1. Predicted optimally constrained network compositions and observed minimal critical cooling rate compositions for glass formation.

Glass composition	Ge_xSe_{1-x}	As_xSe_{1-x}	$(1-x)SiO_2$-xNa_2O	$(1-x)SiO_2$-xK_2O
Predicted optimally constrained ($n_c = 3$) network composition ‘x’	20%	40%	20%	17%
Observed minimal critical cooling rate composition ‘x’	16%	~35%	24%	22%

For the case of the two covalent systems the predicted optimal glass composition ($n_c = 3$) correlates rather well with the location of the minimal cooling rate composition. Phillips’ predictions [3.31] are widely supported by experiments on covalent systems.

Once constraints associated with terminal atoms could be enumerated [3.33], it became feasible to extend constraint theory to networks containing dangling ends such as halogens in chalcohalide glass systems. The correlation between glass forming tendency and regions of compositions where networks become optimally constrained (n_c = 3) in the chalcohalides is found to be remarkably impressive [3.34]. Finally, the interplay between high T_gs and broken bond-bending constraints came to the fore as one began to understand the microscopic origin of the pronounced glass forming tendency of SiO_2 [3.35]. The hierarchical nature of constraints with strong ones (due to bond-stretching forces) usually intact while weaker ones (due to bond-bending forces) possibly breaking in systems with high T_gs has opened new avenues to extending constraint theory from covalent systems to ionic glasses.

Another factor promoting glass formation, particularly in metallic systems and even modified oxides comes from the variation of η with composition at a fixed T or at the liquidus temperature. Mysen and Richet have suggested [3.36] that the GFT is increased in systems where a melt viscosity increase can be shifted to low temperatures, such as at eutectics. In the silicates (Table 3.1) optimal glass compositions are close to the prediction of constraint counting algorithms and also to the eutectic compositions (respectively 23% and 19% in modifier composition for sodium and potassium silicates). This suggests that both mechanisms probably contribute to the GFT in these cases.

3.4.1 Compositional Trends of the Glass Transition Temperature

It is well known that an increase of network connectivity (or degree of cross-linking) leads to an increase of T_g. Indeed, liquids with soft interactions such as organic systems are able to more easily equilibrate during quench than liquids with strong interactions such as covalent systems. Energy barriers are small in the former while they are large in the later resulting in low glass transition temperatures in the former but high ones in the latter (T_g = 268 K for sorbitol, and T_g = 1450 K for silica). Beyond this very general view, it appears that a detailed study of the behavior of T_g with composition, when measured under standard conditions at a fixed cooling rate, provides insight into the structure of glasses through their network connectivity. In binary and ternary alloy glasses, the observation of an Arrhenius-like increase of viscosity as T approaches T_g is related to covalent bond creation, which can be encoded in local structural units stochastically agglomerating. In the stochastic agglomeration theory (SAT) one seeks a solution of a dynamical master equation [3.21] in which coordination numbers of the local structural units (such as pyramids, tetrahedral, etc.) and their concentrations are specified, and the solution yields parameter-free predictions of T_g variation with modifier composition. Thus, for example, in binary alloys of the form B_xA_{1-x} (e.g. Ge_xSe_{1-x}), the slope of T_g with x at low modifier concentration (x) becomes [3.20]:

$$\left[\frac{dT_g}{dx}\right]_{x=0} = \frac{T_g(x=0)}{\ln\left[r_B / r_A\right]} \tag{3.3}$$

where r_B and r_A are the coordination units of the agglomerating species. In Ge_xSe_{1-x}, this slope is for instance T_g ($x = 0$)/ln 2. These predictions are in remarkable agreement with experimental data on a large class of binary and ternary glass-forming systems including oxides and chalcogenides (see below). Noteworthy is the fact that these predictions fail when the glasses enter the IP regime as the mean coordination number increases to $<r> \sim 2.4$ (see below).

An alternative view [3.37] to understanding T_g variation with chemical composition is provided by combining rigidity theory (see below) with the Lindemann's melting criterion [3.38]. Crystals are thought to melt when the thermal mean square displacement $< u^2(T) >$ of atoms acquires 10% of the nearest-neighbor bond length. If the same criterion is applied [3.39] to highly viscous liquids at T_g, one can obtain the $< u^2(T) >$ by using

$$f = 2 - \frac{5}{6} <r> \tag{3.4}$$

the floppy mode fraction (f) modified vibrational density of states. The approach provides a statistical physics basis for the empirical Gibbs-DiMarzio [3.40] equation:

$$T_g = \frac{T_g(<r> = 2)}{1 + B(<r> - 2)} \tag{3.5}$$

where the parameter B is related to the average floppy mode frequency of the system. This parameter can be also computed from the topology of the network [3.41]. Measurements of the Lamb-Mössbauer mean square displacements [3.42] in binary Ge-Se glasses have permitted predicting variations in T_g as a function of network connectivity (see below).

In experimental investigations of glasses variations in physical properties as a function of composition in multi-component systems have proved to be far more incisive than merely focussing on stoichiometric glass compositions. Real glasses do not always possess fully polymerized networks. In several instances networks demix into two or more backbones, i.e., nanoscale phase separate (NSPS) [3.43]. Such structure information is encoded in the glass composition variation of T_g. For example, the maximum of T_g in binary Ge_xSe_{1-x} glasses near $x \sim 1/3$ is due to NSPS [3.44, 3.45], and results because Ge-Ge bonds formed at $x > 0.32$ apparently demix from the network backbone. Such a maximum in T_g is *not observed* in corresponding Si_xSe_{1-x} glasses suggesting that in this case Si-Si bonds formed at x

> 1/3 apparently *do form part of* the network backbone. Finally, in comparing T_gs of binary Ge_xS_{1-x} glasses with Ge_xSe_{1-x} ones, one finds the ratio to be 1.13 and nearly independent of x once $x > 1/5$. The T_g ratio of 1.13 reflects a chemical bond-strength scaling of glassy networks possessing the same connectivity. We note that the Pauling [3.46] single Ge-S bond strength of 55.52 kcal/mole exceeds that of the Ge-Se bond strength of 49.08 kcal/mole by 13%. These ideas permit one to separate the role of chemical bond strengths [3.47] from network connectivity in determining T_g of glassy systems. Structure information on network *connectivity* accessible from thermal probes complements in a significant way results on glass structure accessible from traditional methods -diffraction and local probes.

3.5 Calorimetric Measurement of the Glass Transition Temperature and Related Thermal Properties

Differential Scanning Calorimetry (DSC) has been traditionally used to measure glass transitions [3.48] in a wide array of disciplines including food science, pharmaceuticals, and materials science. In a typical experiment, a few tens of milligrams of a sample encapsulated in an Al pan with a lid, and a second identical pan and lid with no sample used as a reference pan, are heated at a fixed scan rate typically of 10°C/min. The difference in heat flow between a sample and a reference displays an endotherm near a glass transition event as a sample softens and atomic mobility increases. We provide in Fig. 3. 3 an example of a DSC scan of a stoichiometric bulk $GeSe_2$ glass.

If the scan rate is lowered from 10°C/min to 3°C/min, the endothermic signal strength decreases by a factor of 3/10 and shifts to a lower temperature. The inflexion point of the endotherm is generally used to define T_g. The lowering of scan rate dT/dt (where t is the time) lowers the signal strength, but it also lowers T_g because of the kinetic nature of the glass transition. The rate of heat flow can be written as

$$dH/dt = m\, C_p^{\,k}\, (dT/dt) \tag{3. 6}$$

where H designates the quantity of heat flow to a sample of mass m with a kinetic specific heat $C_p^{\,k}$. To increase dH/dt, it is customary to increase the scan rate (dT/dt). This has the effect of up-shifting glass transition temperatures as well. Such shifts have been used to extract activation energies for enthalpy relaxation near T_g using model descriptions. On the other hand, changes in T_g brought about by those in sample chemistry alone are difficult to isolate using DSC not only due to finite scan rates used but also because the observed heat flow signal has both ergodic and non-ergodic contributions.

Modulated DSC is a more recent variant of DSC [3.49] and it makes use of a programmed *T* profile that includes a sinusoidal temperature variation superposed on the linear *T*-ramp. From the envelope of the modulated heat flow, one obtains the part of the total heat flow that tracks the sinusoidal modulation, and it is called the *reversing* heat flow. The average of the modulated heat flow gives the *total heat flow*, and by subtracting the *reversing heat flow* from the *total heat flow* one obtains the *non-reversing heat flow*. Fig. 3. 3 shows an m-DSC scan of a similar $GeSe_2$ glass sample taken at 3 °C/min scan rate and a 0.01 °C/s modulation rate.

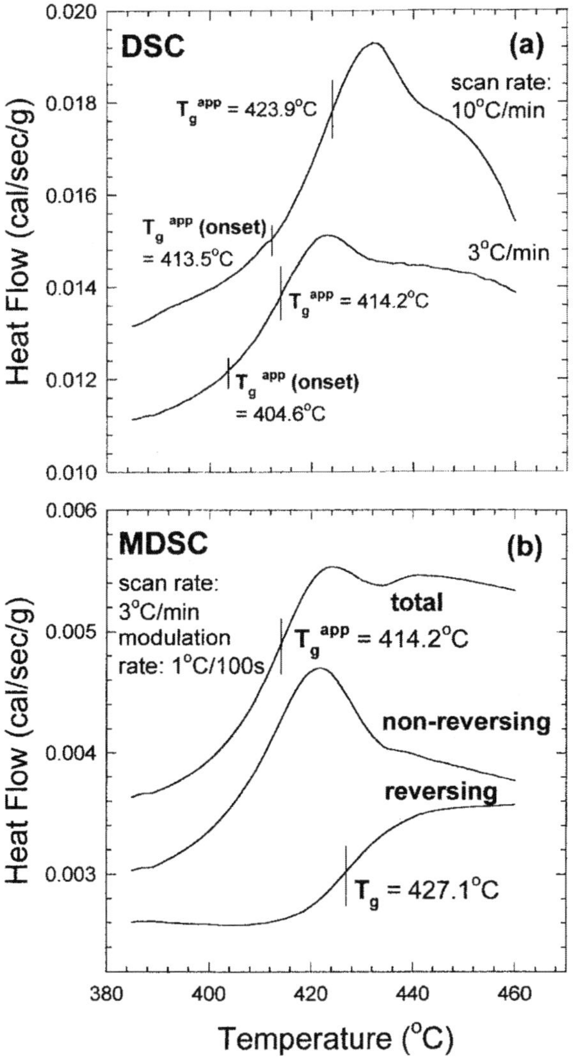

Fig. 3.3 Illustrates in (a) DSC scans of $GeSe_2$ glass taken at 10°C/min and at 3°C / min. and in (b) m-DSC scan of the same sample taken at 3°C / min scan rate and 1°C / 100 s modulation rate. Reprinted with permission from [3.44].

One finds that the total heat flow signal in the m-DSC scan is the same as in the DSC scan (compare Fig. 3. 3a and b). Furthermore, the *reversing* heat flow signal shows a rounded step, and one extracts the glass transition (T_g) from the inflexion point of the step. From the height of the step one obtains the thermodynamic specific heat difference ΔC_p between the liquid (C_p^l) and the glassy state (C_p^g), $\Delta C_p = C_p^l - C_p^g$ unpolluted from kinetic effects. The *non-reversing* heat flow signal usually displays a Gaussian profile and the integrated area under the profile yields the *non-reversing enthalpy* (ΔH_{nr}) associated with T_g. Since the $H = H_r + H_{nr}$ where H_r is the reversing and H_{nr} is the non-reversing enthalpy, one can recast Eq. (3.6) above as

$$dH/dt == mC_p dT/dt + dH_{nr}/dt \tag{3.7}$$

where C_p designates the quasi-thermodynamic specific heat, dH_{nr}/dt the *non-reversing* heat flow, and dH/dt the *total* heat flow term. It is usual to program m-DSC scans for a scan down following a scan up in T across T_g. The difference signal, ΔH_{nr}(up)- ΔH_{nr}(down) = ΔH_{nr} yields the frequency corrected *non-reversing enthalpy* which is independent of the modulation frequency [3.49]. The AC nature of m-DSC experiments permits phase sensitive detection of signals, resulting in higher sensitivity than possible in a DSC experiment. This has the consequence that m-DSC runs can be undertaken at rather low scan rates (1 °C/min or lower) and even under quasi-static conditions to study glass transitions [3.49]. For scan rates of 3 °C/min used in our experiments, the typical difference between T_g(up) and T_g(down) is about 2 °C. By taking the mean value of these transitions, one then obtains a scan rate independent T_g. Results on $GeSe_2$ glass reveal T_g = 427(1)°C. m-DSC derived glass transition temperatures thus have made it possible to track changes in T_g brought about by sample composition or on account of sample heterogeneity brought about by incomplete sample processing [3.50]. Experiments on covalent glasses have shown that aging effects, sample purity (including water traces) and heterogeneity effects are reflected in the non-reversing heat flow term exclusively. The reversing heat flow signal remains largely independent of thermal history including aging and sample purity. Compositional trends in T_g have served a useful role in elucidating the nature of nanoscale phase separation in glasses.

Consider Fig. 3.4 that shows trends in T_g in three covalent systems studied as a function of their mean coordination number $\langle r \rangle$. Note that in the two binary As-Se and Ge-Se glasses a global maximum in T_g occurs near $\langle r \rangle \sim 2.4$ and $\langle r \rangle \sim 2.67$ respectively, the chemical thresholds. On the other hand, in the ternary glass system no T_g maximum is observed even though a chemical threshold exists near $\langle r \rangle \sim 2.55$. Homopolar bonds once nucleated in the binary glasses near $\langle r \rangle \sim 2.40$ in As-Se and near $\langle r \rangle \sim 2.67$ in Ge-Se, apparently segregate from the backbone to form separate nanophases. On the other hand in the ternary glass system homopolar bonds form part of the main backbone [3.51], and steadily increase network cross-linking as reflected in the almost linear increase of T_g.

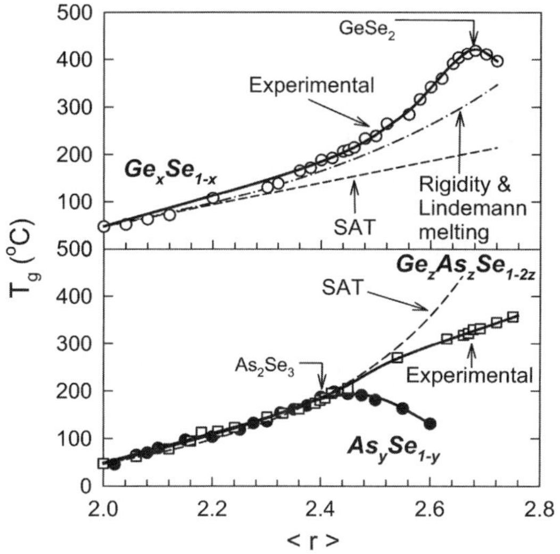

Fig. 3.4. T_g variation in titled chalcogenides with composition [3.22]. The mean coordination number, $\langle r \rangle = 2(1 + x)$ for Ge_xSe_{1-x} glasses (○), $2+y$ for As_ySe_{1-y} glasses (●) and $2 + 3z$ for $Ge_zAs_zSe_{1-2z}$ glasses (□). Top panel gives predictions of equ. (3.3), and (3.5) for Ge_xSe_{1-x}, and bottom panel equ. (3.3) for the Ge-As-Se ternary. SAT-Stochastic Agglomeration Theory. Reprinted with permission from [3.12].

Ternary $Ge_zAs_zSe_{1-2z}$ glasses are ideal test systems to probe glass forming tendency, nature of glass transitions and the elastic behavior of underlying backbones because their connectivity steadily increases as their chemical composition z is increased. DSC measurements on DC (direct current) magnetron sputtered amorphous $Ge_xSb_yTe_{1-x-y}$ thin-films at several compositions were investigated by Kalb et al. [3.52] in both the as deposited (virgin) and thermally annealed state. Although no T_g was observed in the virgin state, these authors could observe a T_g in the thermally annealed state just before crystallization. These experiments on phase change materials highlight the presence of structural relaxation effects in amorphous thin-films.

3.6 Three Generic Classifications of Glasses and Glass Transitions

The physics of the *glassy state* and specially the nature of *glass transitions* can be elegantly probed using ternary chalcogenide glasses. In such glasses one has the luxury of systematically and reproducibly changing network connectivity (mean coordination number $\langle r \rangle$) by compositional tuning. These materials are available in rather pure chemical form, and homogeneous bulk glass samples can be synthesized by appropriate handling and processing. The rich database on covalent systems reveals *glass transitions* can be classified into *three distinct* types based on the *elastic behavior* of their underlying backbones.

In the $Ge_xAs_xSe_{1-2x}$ ternary, variation of the non-reversing enthalpy, $\Delta H_{nr}(x)$ reveals [3.51] three distinct regimes. One observes the enthalpy to show a global minimum (~ 0 cal/g) in the $9\% < x < 16\%$ range, and to increase at $x > 16\%$, and to increase at $x < 9\%$. These three regimes represent respectively at low x (< 9%) the *flexible phase*, at intermediate x ($9\% < x < 16\%$) the *rigid but unstressed phase (IP)*, and at high x (> 16%) the *rigid but stressed phase*. These identifications are further supported by Raman scattering measurements, which reveal vibrational mode frequency ($\nu_{CS}(x)$) of corner-sharing tetrahedral $Ge(Se_{1/2})_4$ units to increase with x displaying thresholds at respective phase boundaries (Fig. 3.5a).

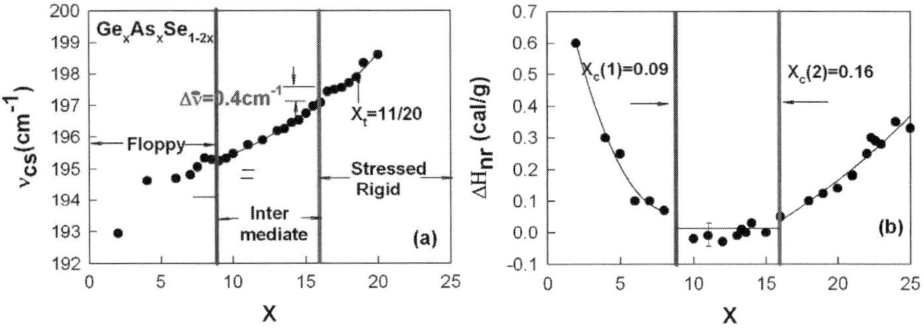

Fig. 3.5. Variations in (a) the frequency of Raman active corner-sharing $GeSe(_{1/2})_4$ units and (b) non-reversing enthalpy near T_g in $Ge_xAs_xSe_{1-2x}$ glasses examined as a function of x. These data reveal three distinct regimes of behavior separated by two elastic phase boundaries; rigidity transition near $x = 9\%$ and stress transition near 16%. Reprinted with permission of the MRS, Warrendale, PA, from [3.51].

Furthermore, optical elasticity (ν_{CS}^2) variation as a function of x displays [3.53] characteristic power-laws in the intermediate and stressed-rigid phases. The observed elasticity power-law in the stressed-rigid regime is in harmony with numerical predictions [3.54] based on random network models. The striking correlation between Raman scattering and calorimetric data shown here is observed in several other glasses as well.

In Fig. 3.6, we show m-DSC results on a companion ternary glass system; $P_xGe_xSe_{1-2x}$. Perhaps the most striking feature [3.55] of these data is the minuscule ΔH_{nr} term in the $9\% < x < 14.5\%$ range, the IP, and the absence of aging in the IP and the presence of aging for compositions outside the IP, both in the flexible and stressed-rigid phases. The structure of $P_xGe_xSe_{1-2x}$ ternary differs from that of the $As_xGe_xSe_{1-2x}$ ternary in one important respect; some of the P atoms segregate in the form of P_4Se_3 molecules as x increases to ~ 20%. In Raman scattering sharp intra-molecular stretch modes of the molecule are observed [3.53]. Segregation of P_4Se_3 molecules leads to some loss of network backbone, and that is reflected in not only the compositional dependence of T_g (x) but also in that of the non-reversing enthalpy (Fig. 3.6). The satellite feature labeled NSPS (Fig. 3.6) is due to nanoscale phase separation of P_4Se_3 molecules.

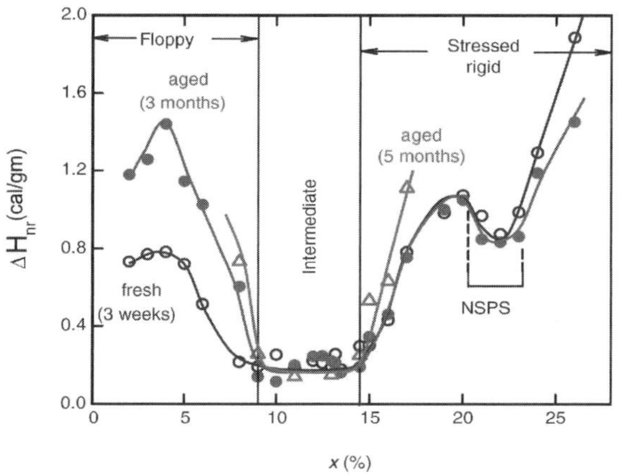

Fig. 3.6. Variation in the non- reversing enthalpy of $Ge_xP_xSe_{1-2x}$ glasses studied as a function of x in their fresh (○), 3-month aged (●) and 5 month aged (Δ) state. These data reveal the lack of aging of the enthalpy term in the IP and its presence outside the IP. Reprinted with permission from the Institute of Physics, from [3.55].

A close examination of the *glass transition endotherm* in the three elastic regimes displays the following characteristic features. In the *elastically flexible* phase, one finds the non-reversing enthalpy at T_g , i.e., ΔH_{nr} term to be generally narrow (~ 20°C wide typically), symmetric and to steadily increase with waiting time in a stretched exponential fashion. In the IP, glass endotherms possess a minuscule ΔH_{nr} term, and the term furthermore does not to age much. In this range glass transitions become *thermally reversing* in character. In the stressed-rigid phase, one finds glass transition endotherms to be wider (~40 °C), asymmetric with a high-*T* tail and the heat flow term to age. An example of such a behavior is seen in $GeSe_2$ (Fig. 3.3). Glass transition studies supported by spectroscopic experiments have permitted constructing global phase diagrams, delineating the three phases as for example for the case of the Ge-P-Se ternary shown in Fig. 3.7.

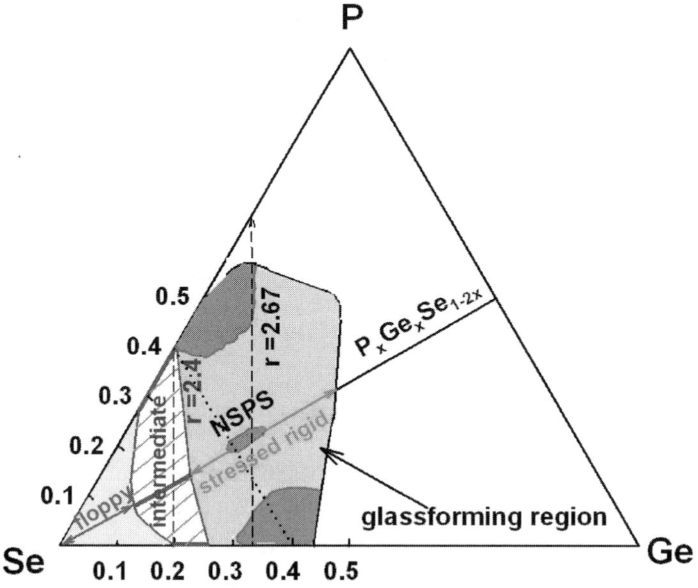

Fig. 3.7. Global phase diagram of Ge-P-Se ternary illustrating the three elastic phases. De-mixing of network backbone near compositions of $GeSe_2$, P_4Se_6 and $P_2Ge_2Se_6$ is reflected by darker grey regions. Reprinted with permission from Wiley from [3.56].

Se-rich glasses are flexible because Se-chains dominate the structure of such glasses. Progressive cross-linking of chains of selenium by P- and Ge-cations leads to the formation of the intermediate phase (hashed region). Increased P- and Ge-alloying leads next to the formation of stressed-rigid glasses (grey region) in which selected regions show a propensity to demix (darker grey regions).

3.7 Elastic Phases in Ionic and Super-ionic Glasses

Recently, the physical properties of ionically bonded network glasses, such as alkali silicates [3.57] and germanates [3.58] have been examined in thermal, optical, and mechanical measurements. Experiments also reveal that traces of water influence their physical property in significant ways, particularly the non-reversing enthalpy near T_g. In $(Na_2O)_x(GeO_2)_{1-x}$ glasses one finds a wide, sharp and deep reversibility window in the $14\% < x < 19\%$ soda range; the composition range defines the *IP*. Glass compositions at $x < 14\%$ are in the *stressed-rigid phase* while those at $x > 19\%$ are in the *flexible phase*. The stress-free character of glasses in the IP has been strikingly demonstrated in birefringence measurements. Mass densities (molar volumes) reveal a broad maximum (minimum) near $x = 18\%$ [3.59], a result that is identified with the germanate anomaly. The overlapping of the molar volume minimum with the IP suggests that multiscale structural self-organization most likely is responsible for the space filling nature of the IP and that the germanate anomaly is of an elastic origin.

Recently, the molecular structure and thermal properties of the fast-ion conducting glass system, $(AgPO_3)_{1-x}(AgI)_x$ were examined [3.60] in Raman scattering, infrared reflectance and m-DSC experiments. The thermal measurements show a rather strikingly wide, deep and sharp reversibility window in the $9\% < x < 37.8\%$ AgI range. This observation in conjunction with Raman optical elasticity fixes the three elastic phases in this solid electrolyte system, a conclusion similar to the one described above for sodium germanate glasses. In both systems particular care to exclude water [3.61] in sample synthesis is necessary to observe these intrinsic effects.

3.8 Ideal Glasses and Self-organization of Networks

The minuscule non-reversing enthalpy (ΔH_{nr}) of glass compositions in the thermally reversing window suggests that such glass compositions find themselves configurationally quite close to their liquid counterparts. They can form glasses even when liquids are cooled very slowly since small changes in structure underlie structural arrest. These compositions represent the sweet spots of glass formation. On an entropy plot, $S(T)$ (Fig. 3.1) such glasses exist in states characterized by the highest entropy. This is due to the fact that there are multitudes of energetically equivalent global configurations in which such stress-free networks can exist. The stress-free character [3.13] of such networks was recently demonstrated [3.62] in Raman pressure experiments on binary Ge-Se glasses. Measurements of mass densities independently show that such glass compositions usually form *space filling networks*. Such glass compositions are "*ideal glasses*" and their structures co-operatively adapt to self-organize.

Fig. 3.8 provides a summary of reversibility windows observed in several families of inorganic glasses.

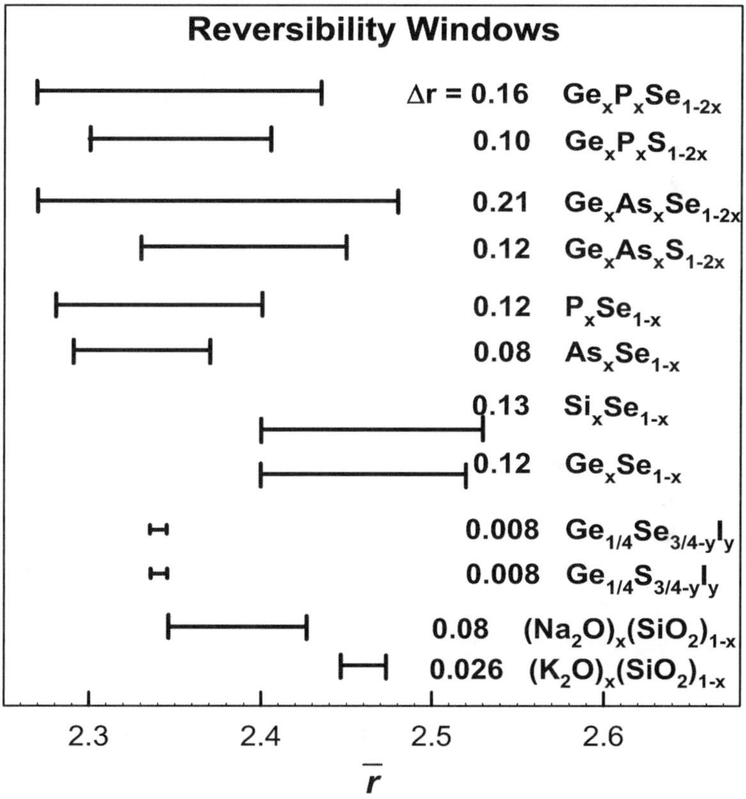

Fig. 3.8. Observed reversibility windows in titled glasses shown as a bar chart in $\langle r \rangle$-space. Window centroids in group IV selenides are at $\langle r \rangle > 2.40$, and in group V-selenides at $\langle r \rangle < 2.40$, while ternary alloys encompass both regions of $\langle r \rangle$ space.

Here the horizontal bars represent the region in $\langle r \rangle$ -space across which reversibility windows are manifested in m-DSC experiments. There are aspects of local and intermediate range structure that control centroids and widths of these windows

[3.63]. Glasses of the group IV-selenides have window centroids moved to <*r*> > 2.40, while glasses of the group V- selenides have window centroids moved to <*r*> < 2.40. Ternary alloys comprising of both group IV and V–selenides display much wider windows encompassing both regions. Network structural variance in several forms may be controlling location of IPs, and it is an issue of ongoing discussions in the field.

3.9 Does the View Below the Glass Transition Temperature Correlate with the View above the Glass Transition Temperature?

The dynamics of glass forming liquids can be conveniently represented on a plot of log η against $(1/T)$ (Fig. 3.2) with the slope near $T = T_g$ providing an activation energy (ΔE_a) for viscosity relaxation (Fig. 3.9).

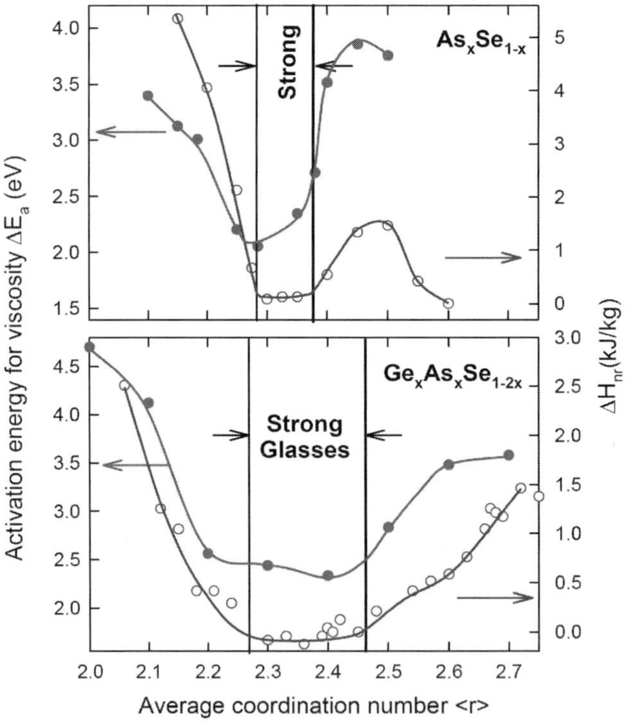

Fig. 3.9. Variations in the activation energy for viscosity (•) and non-reversing enthalpy (○) in (a) binary As_xSe_{1-x} glasses, and (b) ternary $Ge_xAs_xSe_{1-2x}$ glasses as a function of their mean coordination number <*r*>. Global minima in $\Delta H_{nr}(r)$ coincide with a window in ΔE_a for these covalently bonded networks. Reprinted with permission from [3.7, 3.64].

The dimensionless activation energy $\Delta E_a/KT_g$ becomes the fragility (m). Liquids characterized by a small ($m \sim 12$) fragility show an Arrhenius variation of η and are labeled as *strong* liquids while those characterized by a large fragility $m > 16$

are labeled as *fragile*. Subsequently, calorimetric, light scattering, dielectric relaxation and shear measurements were brought to bear [3.65] as measures of liquid fragilities through a measurement of relaxation times τ. Fairly complete viscosity measurements on binary As_xSe_{1-x} and ternary $Ge_xAs_xSe_{1-2x}$ liquids are available in the literature [3.7]. Compositional trends in $\Delta E_a(x)$, and the non-reversing enthalpies, $\Delta H_{nr}(x)$ of these glasses (Fig. 3.9), reveal that the global minimum in $\Delta E_a(x)$ coincides with that in $\Delta H_{nr}(x)$. These data serve to correlate the *strong-fragile* classification of glass forming *liquids* with the *flexible-intermediate stressed-rigid* classification of corresponding *glasses*. The correlation unequivocally shows that in covalent systems IP glasses give rise to *strong liquids*, while both *flexible and stressed-rigid* glasses give rise to *fragile liquids*. These results demonstrate that in the chemically strongly bonded network glasses there are aspects of liquid structure that carry over to the glassy state. Do such correlations exist in other types of networks such as the oxides, superionic solid electrolytes, hydrogen bonded glasses, organic polymers and proteins? We comment on some of these issues next.

3. 10 Glass Formation in Hydrogen Bonded Networks

Several carbohydrates including polyalcohols and saccharides are known to be good glass formers, and their liquid properties such as fragility (m) and stretched exponents (β) and glassy behavior including T_g and Boson peak characteristics have been the subject of earlier work in the field [3.66]. Their glass transitions have recently been examined in m-DSC experiments [3.67]. These measurements permit enthalpic relaxation studies to be extended to much lower frequencies than accessible to dielectric studies, and thus approaching T much closer to T_g than hitherto possible. Modulated-DSC results of m and β correlate well with A.C. calorimetric, shear and light scattering measurements. On the other hand dielectric measurements yield m and β parameters that differ noticeably from thermal results [3.68].

Recently Phillips [3.10] has extended constraint counting algorithms to H-bonded systems, and has suggested that the remarkably high T_g of trehalose and its bioprotective properties may be the result of a tandem repeat structure of this disaccharide. In this structure weakly and strongly bound glucose rings alternate. His enumeration of constraints suggests that trehalose could represent the case of an optimally constrained network glass if the two gearing constraints associated with torsional motion of the two flaps are intact. We have examined the glass transition in trehalose samples prepared by dehydrating crystalline trehalose dihydrate. Fig. 3.10 gives an m-DSC scan of such a sample and reveals a T_g of 117 °C, and a vanishing (ΔH_{nr} ~0 cal/g) non-reversing enthalpy. The magnitude of ΔH_{nr} places trehalose glass as belonging to an optimally coordinated network, in good agreement with the prediction of Phillips [3.10]. The fragility of liquid trehalose is rather high, m = 107 [3.67] and it clearly suggests that as $T > T_g$, weak H-bonds

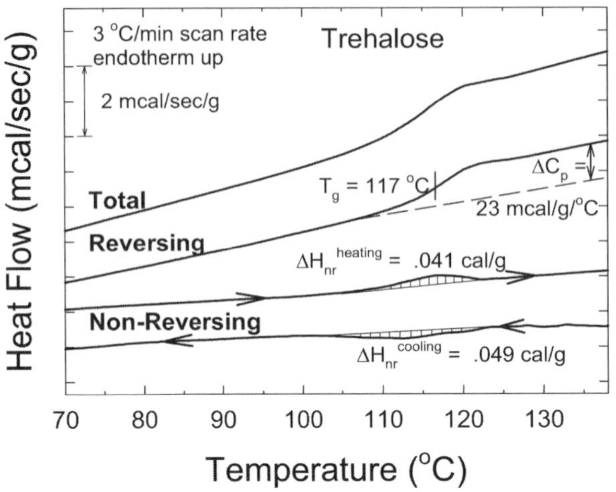

Fig. 3.10. Modulated-DSC scan of trehalose showing a T_g of 117°C, and the heating and cooling non-reversing enthalpies to cancel.

(O-H, ~ 3kcal/mole) constraining the tandem repeats must be progressively broken. Thus, in this H-bonded network we have an example of an IP glass that gives rise to a fragile liquid. One can expect a similar circumstance to prevail in all carbohydrates providing a basis to understand why the fragilities of sugars and alcohols are quite high. In the base oxides such as SiO_2, GeO_2, B_2O_3 composed of network forming cations, one expects their liquids to display a strong behavior (m ~ 20); their networks are built of strong chemical bonds and there are no weak links. The situation alters dramatically when the base oxide networks are modified by alkali-oxide additives. Weak alkali atom-non-bridging oxygen ionic bonds form, and as $T > T_g$, these weak links cease to constrain networks, and alkali-atom mobilities rapidly increase contributing to liquid fragilities. Note that sodium-disilicate and -trisilicate are more fragile than SiO_2 (Fig. 3.2), even though the trisilicate composition is near the IP. In general, in the modified oxides as in the H-bonded systems, one does not expect the glass-liquid correlation to uphold as in the chalcogenides (Fig. 3.9). In the case of the covalently bonded networks such as $Ge_{14}As_{14}Se_{72}$ there are no weak bonds, and one expects this IP glass to give rise to a strong liquid as is found to be the case (Fig. 3.9) and as discussed earlier in section 3.9. Finally, presence of weak H-bonds in trehalose gives rise to a rather large and measurable temperature change in the specific heat (dC_p/dT) in the glassy ($T < T_g$) as well as in the liquid state ($T > T_g$) of about 1 mcal/gm/°C^2 . The dC_p/dT term is fixed by the slope of the reversing heat flow signal (Fig .3.10). In covalent glasses the dC_p/dT term in the glassy and liquid states are lower by about an order of magnitude than the value obtained here for trehalose, which is as it should be. The temperature variation of vibrational entropy in a H-bonded network can be expected to be larger than in a covalently bonded one.

3.11 Epilogue

Experiments supported by theory have shown that network glasses broadly exist in *three* generic elastic phases: *flexible*, *intermediate* and *stressed-rigid,* which are determined by the connectivity of their backbones. *IP glasses* are composed of rigid but unstressed networks that are in a state of *quasi-equilibrium* and do not age much. They form space filling networks that are neither random, nor crystalline, but their structures adapt to expel stress and to become *self-organized.* Understanding the structure of IPs in real glasses is an exciting but challenging problem. Multi-scale structural adaption leading to self-organization is an exponentially complex problem. And resolution of the problem will undoubtedly impact basic understanding of the disordered state of matter with consequences on protein folding, and engineering applications of these materials in a variety of existing and evolving technologies.

Acknowledgements. It is a pleasure to acknowledge the participation of Daniel Georgilv, Fei Wang, Swapnajit Chakravarty, Deassy Novita, Min Zhang and Tao Qu in the experimental program at University of Cincinnati. We are also pleased to acknowledge discussions with Professor Bernard Goodman, Professor Jim Phillips and Professor Darl McDaniel during the course of this work. This work is supported by US National Science Foundation grant DMR- 04-56472.

References

[3.1] Zachariasen, W.H.: The atomic arrangement in a glass. J. Am. Chem. Soc. **54**, 3841-3851 (1932)

[3.2] Lucovsky, G., Baker, D.A., Paesler, M.A., Phillips, J.C.: Spectroscopic and electrical detection of intermediate phases and chemical bonding self-organizations in (i) dielectric films for semiconductor devices, and (ii) chalcogenide alloys for optical memory devices. J. Non-Cryst. Solids **353**, 1713-1722 (2007)

[3.3] Matthews, J.N.A.: Semiconductor Industry Switches to Hafnium-Based Transistors. Physics Today **61**, 25-26 (2008)

[3.4] Macfarlane, A., Martin, G.: Glass : a world history. University of Chicago Press, Chicago (2002)

[3.5] Kerner, R., Phillips, J.C.: Quantitative principles of silicate glass chemistry. Solid State Commun. **117**, 47-51 (2000)

[3.6] Selvanathan, D., Bresser, W.J., Boolchand, P.: Stiffness transitions in Si_xSe_{1-x} glasses from Raman scattering and temperature-modulated differential scanning calorimetry. Phys. Rev. B **61**, 15061-15076 (2000)

[3.7] Boolchand, P., Lucovsky, G., Phillips, J.C., Thorpe, M.F.: Self-organization and the physics of glassy networks. Phil. Mag. **85**, 3823-3838 (2005)

[3.8] Barre, J., Bishop, A.R., Lookman, T., Saxena, A.: Adaptability and "intermediate phase" in randomly connected networks. Phys. Rev. Lett. **94**, 208701-4 (2005)

[3.9] Phillips, J.C.: Universal intermediate phases of dilute electronic and molecular glasses. Phys. Rev. Lett. **88**, 216401-4 (2002)

[3.10] Phillips, J.C.: Ideally glassy hydrogen-bonded networks. Phys. Rev. B **73**, 024210-10 (2006)

[3.11] Rader, A.J., Hespenheide, B.M., Kuhn, L.A., Thorpe, M.F.: Protein unfolding: Rigidity lost. Proceedings of the National Academy of Sciences of the United States of America **99**, 3540-3545 (2002)

[3.12] Boolchand, P., Georgiev, D.G., Goodman, B.: Discovery of the intermediate phase in chalcogenide glasses. J. Optoelectron. Adv. Mater. **3**, 703-720 (2001); Micoulaut, M., Phillips, J.C.: Onset of rigidity in glasses: From random to self-organized networks. J. Non-Cryst. Solids **353**, 1732-1740 (2007); Brière, M.A., Chubynsky, M.V., Mousseau, N.: Self-organized criticality in the intermediate phase of rigidity percolation. Phys. Rev. E **75**, 56108 (2007)

[3.13] Thorpe, M.F., Jacobs, D.J., Chubynsky, M.V., Phillips, J.C.: Self-organization in network glasses. J. Non-Cryst. Solids **266**, 859-866 (2000)

[3.14] Ovshinsky, S.R.: Reversible Electrical Switching Phenomena in Disordered Structures. Phys. Rev. Lett. **21**, 1450–1453 (1968)

[3.15] Yamada, N., Ohno, E., Nishiuchi, K., Akahira, N., Takao, M.: Rapid-phase transitions of GeTe-Sb_2Te_3 pseudobinary amorphous thin films for an optical disk memory. J. Appl. Phys. **69**, 2849-2856 (1991)

[3.16] Kolobov, A.V., Fons, P., Frenkel, A.I., Ankudinov, A.L., Tominaga, J., Uruga, T.: Understanding the phase-change mechanism of rewritable optical media. Nature Materials **3**, 703–708 (2004); Baker, D.A., Paesler, M.A., Lucovsky, G., Agarwal, S.C., Taylor, P.C.: Application of bond constraint theory to the switchable optical memory material $Ge_2Sb_2Te_5$. Phys. Rev. Lett. **96**, 255501-3 (2006); Wuttig, M.: Phase-change materials: Towards a universal memory? Nat Mater **4**, 265-266 (2005); Lankhorst, M.H.R., Ketelaars, B.W.S.M.M., Wolters, R.A.M.: Low-cost and nanoscale non-volatile memory concept for future silicon chips. Nat Mater **4**, 347-352 (2005)

[3.17] Kauzmann, W.: The Nature of the Glassy State and the Behavior of Liquids at Low Temperatures. Chem. Rev. **43**, 219-256 (1948)

[3.18] Debenedetti, P.G., Stillinger, F.H.: Supercooled liquids and the glass transition. Nature **410**, 259-267 (2001)

[3.19] Angell, C.A., Ngai, K.L., McKenna, G.B., McMillan, P.F., Martin, S.W.: Relaxation in glassforming liquids and amorphous solids. J. Appl. Phys. **88**, 3113-3157 (2000)

[3.20] Kerner, R., Micoulaut, M.: On the glass transition temperature in covalent glasses. J. Non-Cryst. Solids **210**, 298-305 (1997)

[3.21] Micoulaut, M.: The slope equations: A universal relationship between local structure and glass transition temperature. European Physical Journal B **1**, 277-294 (1998)

[3.22] Boolchand, P., Georgiev, D.G., Micoulaut, M.: Nature of glass transition in chalcogenides. J. Optoelectron. Adv. Mater. **4**, 823-836 (2002)

[3.23] Anderson, P.W.: Through the glass lightly. Science **267**, 1615-e-1616 (1995)

[3.24] Binder, K., Kob, W.: Glassy Materials And Disordered Solids, An Introduction to Their Statistical Mechanics. World Scientific, Singapore (2005)

[3.25] Angell, C.A.: Structural instability and relaxation in liquid and glassy phases near the fragile liquid limit. J. Non-Cryst. Solids **102**, 205-221 (1988)

[3.26] Tammann, G., Hesse, W.: Die Abhängigkeit der Viscosität von der Temperatur bie unterkühlten Flüssigkeiten. Z. Anorg. Allg. Chem. **156**, 245-257 (1926); Fulcher, G.S.: Analysis of recent measurements of the viscosity of glasses. J. Am. Ceram. Soc. **8**, 339-355 (1925); Vogel, H.: Physik. Zeitschrift **22**, 645-646 (1921)

[3.27] Cugliandolo, L.F.: Dynamics of glassy systems. arXiv:cond-mat/0210312v2 (2002)

[3.28] O'Hern, C.S., Langer, S.A., Liu, A.J., Nagel, S.R.: Force Distributions near Jamming and Glass Transitions. Phys. Rev. Lett. **86**, 111-114 (2001)

[3.29] Langer, S.A., Liu, A.J.: Sheared foam as a supercooled liquid? EPL (Europhysics Letters) **49**, 68-74 (2000)

[3.30] Giovambattista, N., Stanley, H.E., Sciortino, F.: Potential-Energy Landscape Study of the Amorphous-Amorphous Transformation in H_2O. Phys. Rev. Lett. **91**, 115504-4 (2003); Angell, C.A.: Glass formation and the nature of the glass transitions. In: Boolchand, P. (ed.) Insulating and Semiconducting Glasses, pp. 1-51. World Scientific, Sin-

gapore; River Edge, NJ (2000); Stillinger, F.H.: A Topographic View of Supercooled Liquids and Glass Formation. Science **267**, 1935-1939 (1995)
[3.31] Phillips, J.C.: Topology of covalent non-crystalline solids I: Short-range order in chalcogenide alloys. J. Non-Cryst. Solids **34**, 153-181 (1979)
[3.32] Azoulay, R., Thibierge, H., Brenac, A.: Devitrification characteristics of Ge_xSe_{1-x} glasses. J. Non-Cryst. Solids **18**, 33-53 (1975); Fang, C.-Y., Yinnon, H., Uhlmann, D.R.: A kinetic treatment of glass formation. VIII: Critical cooling rates for Na_2O-SiO_2 and K_2O-SiO_2 glasses. J. Non-Cryst. Solids **57**, 465-471 (1983)
[3.33] Boolchand, P., Thorpe, M.F.: Glass-forming tendency, percolation of rigidity, and onefold-coordinated atoms in covalent networks. Phys. Rev. B **50**, 10366-10368 (1994)
[3.34] Mitkova, M., Boolchand, P.: Microscopic origin of the glass forming tendency in chalcohalides and constraint theory. J. Non-Cryst. Solids **240**, 1-21 (1998)
[3.35] Zhang, M., Boolchand, P.: The Central Role of Broken Bond-Bending Constraints in Promoting Glass-Formation in the Oxides. Science **266**, 1355-1357 (1994)
[3.36] Mysen, B., Richet, P.: Silicate glasses and melts: properties and structure. Elsevier, Amsterdam; Boston (2005); Richet, P.: Viscosity and configurational entropy of silicate melts. Geochim. Cosmochim. Acta **48**, 471-483 (1984)
[3.37] Naumis, G.G.: Variation of the glass transition temperature with rigidity and chemical composition. Phys. Rev. B **73**, 172202-4 (2006)
[3.38] Tabor, D.: Gases, liquids, and solids : and other states of matter. Cambridge University Press, Cambridge; New York (1991)
[3.39] Phillips, W.A., Buchenau, U., Nücker, N., Dianoux, A.J., Petry, W.: Dynamics of glassy and liquid selenium. Phys. Rev. Lett. **63**, 2381 (1989)
[3.40] Gibbs, J.H., DiMarzio, E.A.: Nature of the glass transition and the glassy state. J. Chem. Phys. **28**, 373-383 (1958)
[3.41] Micoulaut, M., Naumis, G.G.: Glass transition temperature variation, cross-linking and structure in network glasses: A stochastic approach. Europhys. Lett. **47**, 568-574 (1999)
[3.42] Boolchand, P., Bresser, W., Zhang, M., Wu, Y., Wells, J., Enzweiler, R.N.: Lamb-Mössbauer factors as a local probe of floppy modes in network glasses. J. Non-Cryst. Solids **182**, 143-154 (1995)
[3.43] Boolchand, P., Georgiev, D.G., Qu, T., Wang, F., Cai, L.C., Chakravarty, S.: Nanoscale phase separation effects near $\langle r \rangle = 2.4$ and 2.67, and rigidity transitions in chalcogenide glasses. Comptes Rendus Chimie **5**, 713-724 (2002)
[3.44] Boolchand, P., Bresser, W.J.: The structural origin of broken chemical order in $GeSe_2$. Phil. Mag. B **80**, 1757-1772 (2000)
[3.45] Boolchand, P.: The maximum in glass transition temperature (T_g) near x = 1/3 in Ge_xSe_{1-x} glasses. Asian J. of Phys. **9**, 709 (2000)
[3.46] Pauling, L.: The Nature of the Chemical Bond. Cornell University, Ithaca, NY (1960)
[3.47] Tichý, L., Tichá, H.: Covalent bond approach to the glass-transition temperature of chalcogenide glasses. J. Non-Cryst. Solids **189**, 141-146 (1995)
[3.48] Wunderlich, B.: The tribulations and successes on the road from DSC to TMDSC in the 20th century the prospects for the 21st century. J. Therm. Anal. Calorim. **78**, 7-31 (2004)
[3.49] Thomas, L.C.: Modulated DSC Technology (MSDC-2006). T.A. Instruments, Inc (www.tainstruments.com), New Castle, DE (2006)
[3.50] Cai, L.C., Boolchand, P.: Nanoscale phase separation of GeS_2 glass. Phil. Mag. B **82**, 1649-1657 (2002)
[3.51] Qu, T., Georgiev, D.G., Boolchand, P., Micoulaut, M.: The intermediate phase in ternary $Ge_xAs_xSe_{1-2x}$ glasses. In: Egami, T., Greer, A.L., Inoue, A., Ranganathan, S. (eds.) Supercooled Liquids, Glass Transition and Bulk Metallic Glasses, p. 157. Materials Research Society **754** (2003)
[3.52] Kalb, J.A., Wuttig, M., Spaepen, F.: Calorimetric measurements of structural relaxation and glass transition temperatures in sputtered films of amorphous Te alloys used for phase change recording. J. Mater. Res. **22**, 748-754 (2007)

[3.53] Boolchand, P., Jin, M., Novita, D.I., Chakravarty, S.: Raman scattering as a probe of intermediate phases in glassy networks. Journal of Raman Spectroscopy **38**, 660-672 (2007)
[3.54] He, H., Thorpe, M.F.: Elastic Properties of Glasses. Phys. Rev. Lett. **54**, 2107-2110 (1985)
[3.55] Chakravarty, S., Georgiev, D.G., Boolchand, P., Micoulaut, M.: Ageing, fragility and the reversibility window in bulk alloy glasses. J. Phys. Condens. Matter **17**, L1-L7 (2005)
[3.56] Chakravarty, S.: Self-Organization and Aging in Network Glasses. In: Electrical and Computer Engineering, University of Cincinnati, MS Thesis (2003)
[3.57] Vaills, Y., Qu, T., Micoulaut, M., Chaimbault, F., Boolchand, P.: Direct evidence of rigidity loss and self-organization in silicate glasses. J. Phys. Condens. Matter **17**, 4889-4896 (2005)
[3.58] Rompicharla, V., Novita, D.I., Chen, P., Boolchand, P., Micoulaut, M., Huff, W.: Abrupt boundaries of intermediate phases and space filling in oxide glasses. J. Physics Condensed Matter **20**, 202101-4 (2008)
[3.59] Henderson, G.S.: The Germanate Anomaly: What do we know? J. Non-Cryst. Solids **353**, 1695-1704 (2007)
[3.60] Novita, D.I., Boolchand, P., Malki, M., Micoulaut, M.: Fast-ion conduction and flexibility of glassy networks. Phys. Rev. Lett. **98**, 195501-4 (2007)
[3.61] Novita, D.I., Boolchand, P.: Synthesis and structural characterization of dry $AgPO_3$ glass by Raman scattering, infrared reflectance, and modulated differential scanning calorimetry. Phys. Rev. B **76**, 184205-12 (2007). Also see ArXiv 08081154
[3.62] Wang, F., Mamedov, S., Boolchand, P., Goodman, B., Chandrasekhar, M.: Pressure Raman effects and internal stress in network glasses. Phys. Rev. B **71**, 174201-8 (2005)
[3.63] Micoulaut, M., Phillips, J.C.: Rings and rigidity transitions in network glasses. Phys. Rev. B **67**, 104204-9 (2003)
[3.64] Wang, Y., Boolchand, P., Micoulaut, M.: Glass structure, rigidity transitions and the intermediate phase in the Ge-As-Se ternary. Europhys. Lett. **52**, 633-639 (2000)
[3.65] Carpentier, L., Desprez, S., Descamps, M.: From strong to fragile glass- forming systems: a temperature modulated differential scanning calorimetry investigation. Phase Transitions **76**, 787 - 799 (2003)
[3.66] Sokolov, A.P., Rössler, E., Kisliuk, A., Quitmann, D.: Dynamics of strong and fragile glass formers: Differences and correlation with low-temperature properties. Phys. Rev. Lett. **71**, 2062-2065 (1993)
[3.67] DeGusseme, A., Carpentier, L., Willart, J.F., Descamps, M.: Molecular Mobility in Supercooled Trehalose. J. Phys. Chem. B **107**, 10879-10886 (2003)
[3.68] Macdonald, J.R., Phillips, J.C.: Topological derivation of shape exponents for stretched exponential relaxation. J. Chem. Phys. **122**, 074510-074510 (2005)

4. Structure of Amorphous Ge-Sb-Te Solids

Stephen R. Elliott

Abstract This chapter reviews what is currently known about the atomic structure of Ge-Sb-Te phase-change materials in the amorphous phase. An introduction is first given to aspects of structural order in amorphous solids at different length scales, and brief reviews of experimental techniques for structure determination and of computer-simulation methods are outlined. A discussion is then given of the available experimental data on the structure of Ge-Sb-Te amorphous solids, primarily $Ge_2Sb_2Te_5$, together with the results of various computer-simulation studies of these materials.

4.1 Introduction

The word 'amorphous' literally means without shape or form, that is lacking the *external* shape of a crystal in terms of crystalline facets. Thus, an amorphous material lacks the long-range translational and orientational order characteristics of the ideal crystalline state [4.1]. However, the term amorphous should not be construed as implying that there is *no* structural order present whatsoever: the atomic positions in amorphous materials are never truly random in a statistical sense. Instead, various types of atomic structural order can occur in amorphous materials at different length scales.

Amorphous materials are invariably thermodynamically metastable: they generally have a higher configurational entropy than the lowest free-energy state, the corresponding crystal, and their production is kinetically controlled. For example, kinetic arrest of atomic diffusion occurs at the glass-transition temperature, T_g, in the supercooled liquid, associated with the exponential increase of viscosity with decreasing temperature, where atomic diffusion effectively ceases on the experimental time-scale. This leads to the formation of the glassy (or vitreous) state on rapidly quenching a liquid through T_g (see Chapter 3). Alternatively, even more rapid quenching is associated with the deposition of amorphous thin films onto (cooled) substrates from the vapor phase (e.g. via RF (radio frequency) or DC (direct current) sputtering, thermal, electron-beam or (pulsed) laser-induced evaporation, chemical-vapor deposition etc.).

S. Raoux, M. Wuttig (eds.), *Phase Change Materials*,
DOI 10.1007/978-0-387-84874-7_4,

These (dominant) different methods of producing amorphous materials, namely melt-quenched vitrification and vapor-phase amorphous-film deposition, are *not* expected to lead to the *same* amorphous state of a given material with a particular composition. In general, vapor deposition leads to a more structurally-disordered material than does melt-quenching because of the much higher rates of cooling involved: vapor-deposited adatoms on a substrate have only a very limited surface mobility. In addition, rather well-ordered structural motifs (e.g. rings or clusters of atoms) may be present in the liquid state, and can be frozen into the glassy state on quenching, whereas atom-by-atom vapor-phase deposition methods (particularly sputtering) will produce amorphous films essentially devoid of an excess of such structural moieties. (However, those vapor-phase deposition methods that proceed via an intermediate *liquid* state, *viz.* evaporation techniques, in which liquid and vapor are in near-thermal equilibrium, can also lead to the trapping of ordered structural units, e.g. atomic clusters, in the condensed amorphous solid film, if these stably occur in the vapor phase.)

The above considerations of the effects of the fabrication route on the atomic structure of the resulting amorphous and glassy materials are particularly relevant to phase-change (PC) Ge-Sb-Te (GST) and related materials, where the material, for example for optical data-storage media, is initially deposited in amorphous thin-film form by vapor deposition (sputtering) prior to subsequent crystallization initialization. Laser-written dots in such optical media are believed to be glassy as a result of the local melting of the starting crystalline material and its subsequent ultra-rapid quenching. These two amorphous states of GST, *viz.* as-deposited and laser-amorphized, need not (and probably do not) have the same atomic structure, and consequently can exhibit different behavior (e.g. in subsequent crystallization, or in property contrast between amorphous and crystalline states).

4.2 Structural Order in Amorphous Materials

Amorphous solids have no long-range translational or orientational order, by definition. Yet, such materials can exhibit several distinct types of marked, non-random structural correlations at a variety of length-scales. It is convenient to sub-divide such types of structural order according to the characteristic length scale at which they can occur in amorphous materials.

4.2.1 Short-range Order

Short-range order (SRO) occurs in the 2 - 4 Å length scale and is associated with nearest-neighbor (nn) and next-nearest-neighbor (nnn) atomic correlations. These characterize nearest-neighbor bond lengths and bond angles, and hence are determined by 2-body and 3-body correlation functions, respectively (Fig. 4.1). Such correlations are associated, respectively, with the first and second peaks in the radial distribution function (RDF). Thus, structural quantities characterizing SRO include average (nearest-neighbor) bond lengths, coordination numbers, and bond angles. Disorder in amorphous materials is manifested in fluctuations in these quantities. In stoichiometric covalent *crystal* structures, for example, heteropolar (cation-anion) chemically-ordered bonding is the norm, whereas in amorphous materials another manifestation of structural disorder is the presence of chemical disorder, e.g. homopolar bonds. In the case of GST materials, for example, Ge-Te and Sb-Te atomic pairs comprise the heteropolar bonds, whereas Ge-Ge, Sb-Sb, Te-Te homopolar bonds, and also Ge-Sb bonds, are instances of chemical disorder.

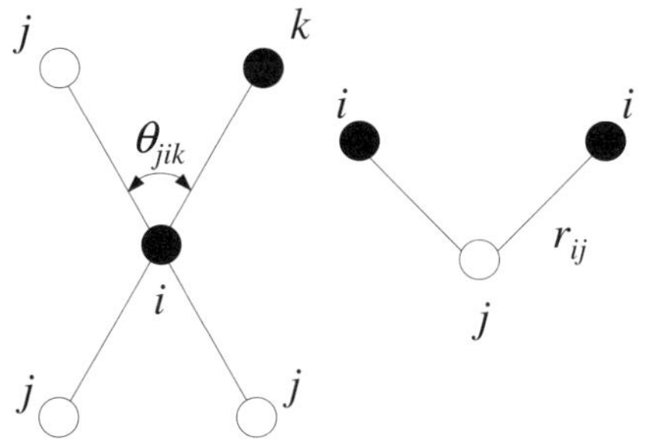

Fig. 4.1. Short-range order: 2-body and 3-body correlations associated with distributions of nearest-neighbor bond lengths, r_{ij}, and bond angles, θ_{jik}, between atoms *i, j, k*. Reprinted with permission from [4.1].

4.2.2 Medium-range Order

Medium-range order (MRO) – or intermediate-range order (IRO) – is more difficult to define precisely than is the case for SRO: it characterizes non-random structural correlations at distances beyond those characterizing SRO, i.e. say 4-10 Å [4.1]. As such, it constitutes atom-atom correlations beyond the 2- and 3-body correlations characterizing SRO. Thus, non-random distributions of dihedral angles, $P(\phi)$, with preferred values of ϕ (the angle of twist around a common bond to bring the attached bonds into projected coincidence), is an example of MRO at the shortest length scales since it corresponds to 4-body correlations (see Fig. 4.2). At somewhat larger distances, MRO can be characterized by (conditional) probabilities relating adjacent dihedral angles (*viz.* governed by 5-body correlations): given a (preferred) value of dihedral angle with a particular sign (say $+\phi_0$), corresponding to a particular sense of rotation, into projected coincidence, of bonds around the common bond, what is the probability of an adjacent dihedral angle also being $+\phi_0$, or the anti-value $-\phi_0$? Structural configurations containing a series of *alternating* signs of dihedral angle, i.e. $+\phi_0$, $-\phi_0$, $+\phi_0$, $-\phi_0$, …., can lead to *chain-like* conformations, whereas those having a series of the *same* sign of dihedral angle can lead to *closed rings,* the preferred value ϕ_0 dictating the ring size (together with the value of the average bond angle). Thus, preferred values of ring sizes, in an otherwise broad distribution of sizes, is another form of MRO in amorphous solids: in a typical crystal, in contrast, only one or two sizes of irreducible rings are generally present.

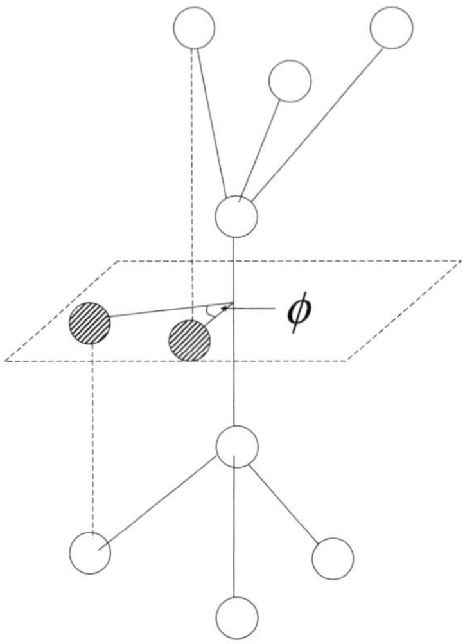

Fig. 4.2. Medium-range order: 4-body correlations associated with distributions of the dihedral angle, ϕ. Reprinted with permission from [4.1].

4.2.3 Long-range Structure

Although amorphous materials exhibit no long-range translational periodicity, by definition, that is not to say that they cannot possess some form of structural order at distances longer than that (≈ 10 Å) characterizing MRO. Instances of such long-range structure can include the existence of layering (albeit distorted), e.g. vestiges of crystalline order. It can be difficult to observe such long-range structure. In a structural model, this can be accomplished by performing a 3D Fourier transform of the atomic positions [4.2]. In a crystal, where all the atoms lie on lattice planes, only a few Fourier components (corresponding to Bragg peaks in a diffraction pattern) exist. Disordered materials, such as glasses, are characterized by a broad distribution of Fourier components, but if preferential layering is also present, this will correspond to the existence of a dominant Fourier component.

4.3 Experimental Structural Probes

A number of experimental techniques can be used to probe the atomic structure of disordered materials [4.1]. Dominant amongst these is diffraction, of neutrons, x-rays or electrons. These three techniques are complementary in some respects. Neutrons are very penetrating, being only weakly scattered by the atomic nuclei, and the atomic scattering amplitude, varying randomly between elements and isotopes (but generally having comparable values) is independent of the scattering angle, 2θ, or equivalently of the modulus of the scattering wave vector transfer:

$$Q = (4\pi \sin\theta) / \lambda \tag{4.1}$$

where Q is the scattering vector and λ is the wavelength of the neutrons. X-rays and electrons are more strongly scattered (by the electron distribution in atoms), and in these cases the atomic scattering factor is a strongly decreasing function of Q, and for x-rays having a $Q = 0$ value equal to the number of electrons in an atom, i.e. the atomic number. In weakly-scattering diffraction (e.g. neutrons), single-scattering events are predominant, and so only 2-body correlations are probed. A related technique is fluctuation electron microscopy in which the intensity fluctuations ("speckle") in a dark-field electron-diffraction image are analyzed. Ultimately, these fluctuations result from 4-body-correlation scattering events, and hence can give information on MRO. In more strongly scattering diffraction (e.g. electrons), multiple-scattering events can be important, potentially also giving information on higher-order atomic correlations. The suitably normalized scattering intensity, the structure factor, $S(Q)$, can be Fourier transformed to give the real-space pair-distribution function, e.g. the RDF.

X-ray absorption spectroscopy (XAS) is a related technique, in which x-rays are absorbed by electrons in atoms, thereby emitting photoelectrons which are "internally diffracted" by atoms surrounding the absorbing atoms. The resulting interference between outgoing and backscattered electron waves leads to a photoelectron energy-, or wave vector (k)-, dependent oscillation in the x-ray absorption coefficient. At larger values of k, the process is dominated by single-scattering events (giving the so-called extended x-ray absorption fine structure, EXAFS, in the x-ray absorption), whereas at energies close to the absorption edge, multiple electron scattering processes dominate (x-ray absorption near-edge structure, XANES). Fourier transformation of the EXAFS is good at providing accurate nearest-neighbor bond-length information, but accompanying estimates of nearest-neighbor coordination numbers are less accurate. XANES can provide conformational (i.e. 3- and 4-body correlation) information through multiple-scattering calculations of the XANES spectrum for model structures.

Another element-specific structural probe is nuclear magnetic resonance (NMR), in which transitions between nuclear-spin energy levels are produced by (RF) electromagnetic radiation. Different isotopes have different nuclear-spin states, whose degeneracy is lifted by the application of an external magnetic field via the Zeeman interaction, and whose energies are further altered by additional interactions, such as local-electron shielding (chemical shift), dipolar and quadrupolar interactions. Information on local conformations around the resonant atomic nucleus, e.g. on the nature of coordination polyhedra, can be obtained from chemical-shift values by comparison with spectra for standard structures.

4.4 Structural Modeling

Unlike the structure determination of (single) crystals, which entails only the unambiguous solving of the atomic positions in a small unit cell (by diffraction methods), in contrast the atomic structure of amorphous materials can only be obtained experimentally using the methods outlined in the previous section in a *statistical* sense, i.e. as a distribution of bond lengths or angles, or an average coordination number, etc.

In view of this inherent uncertainty in experimental determination, it is often very helpful to construct atomic models of the structure of amorphous materials as an aid to understanding their structure more thoroughly. Two types of modeling can be distinguished in this regard, one type being derived from existing experimental evidence, and the other type being simulated *ab initio*.

The former type of modeling is exemplified by the reverse Monte Carlo (RMC) procedure [4.3], in which atoms are placed in a simulation box and moved by Monte Carlo steps (subject to a minimum interatomic-separation criterion), a step being accepted if the structural characteristics of the model (e.g. RDF) agrees better with experiment, and rejected otherwise. The best RMC results are obtained if more than one experimental data-set is used, e.g. x-ray and neutron diffraction (ND) and also EXAFS data, as this greatly improves the constraints on the fit to experimental data. The method suffers from there being, in general, no energetic basis for the fits to the data, but an advantage is that the RMC algorithm is quick, so that large models (several thousands of atoms) can be simulated readily.

The second type of simulation procedure is molecular dynamics (MD), in which atoms, placed in a simulation box (subject to periodic boundary conditions to eliminate surface effects), evolve dynamically with time under the influence of an interatomic potential. This potential may be empirical (having a simple algebraic form), in which case simulations are fast and models containing many thousands of atoms can be simulated for long times (many ns). Alternatively, the interatomic forces can be calculated quantum mechanically (albeit employing certain approximations, such as the local-density approximation (LDA) in density-functional theory (DFT), see also chapter 2). This *ab initio* molecular-dynamics (AIMD) simulation method has the merit that the interatomic forces are calculated as accurately as is feasible, but the disadvantage that, because the method is so computationally intensive, only small models (typically a few hundred atoms at most) can be simulated for short times (typically a few tens of ps). All types of MD simulations have the distinct advantage over RMC modeling, as the name implies, that the full dynamical history of the model is simulated, as well as knowing the atomic positions: starting in the liquid state at elevated temperatures, for example, the dynamical behavior (diffusion, viscosity, vibrations) can be followed as a function of time during a thermal quench to the glassy state.

4.5 The Structure of Amorphous Phase-change Materials

There have been relatively few experimental studies on the atomic structure of the amorphous state of PC materials, and fewer still computer simulations, and these only on GST materials, notably $Ge_2Sb_2Te_5$ (225). In the following, results from experimental investigations, and from computer simulations, will be discussed separately.

4.5.1 Experimental Studies

There have been a few neutron and x-ray diffraction studies of 225. For example, Kohara et al. [4.4, 4.5] performed synchrotron x-ray diffraction experiments on crystalline, liquid and as-sputtered amorphous 225 (and GeTe). The corresponding structure factors are shown in Fig. 4.3(a) and the total pair correlation function, *T(r)*, obtained by Fourier transformation, in Fig. 4.3(b).

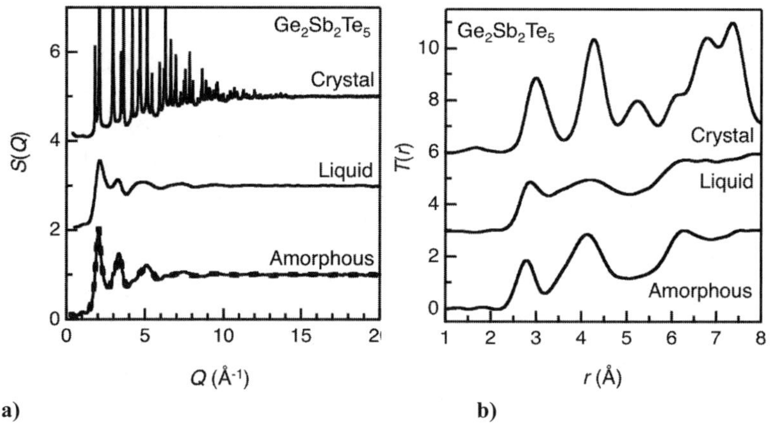

Fig. 4.3. a) X-ray diffraction structure factors *S(Q)* for crystalline, liquid and amorphous $Ge_2Sb_2Te_5$ (225); b) Corresponding Fourier-transformed total correlation functions *T(r)* [4.4, 4.5]. Reprinted with permission from [4.5].

It can be seen that *S(Q)* and *T(r)* are qualitatively similar for the amorphous and liquid states, although evidently the latter is considerably more structurally disordered than the former by inspection of the peak widths. Jovari et al. [4.6] have also performed synchrotron x-ray diffraction experiments, as well as neutron-diffraction experiments on as-sputtered a-225 (amorphous) and a-$GeSb_2Te_4$ (124). The structure factors for a-225 are shown in Fig. 4.4; it can be seen that the x-ray and neutron *S(Q)* curves are rather similar, with the exception of a more pronounced first sharp diffraction peak (FSDP) at $Q_1 \sim 1$ Å^{-1}, indicative of MRO, in the neutron *S(Q)*.

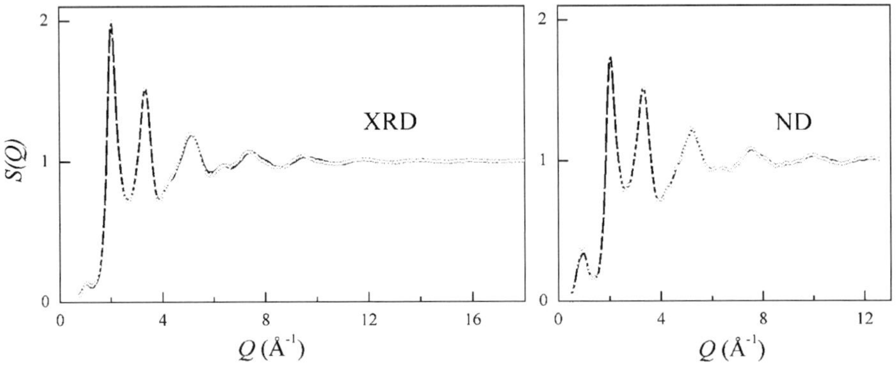

Fig. 4.4. XRD and ND structure factors (circles) for a-225, together with RMC fits (lines). Reprinted with permission from the Institute of Physics, from [4.6].

The values of the neutron-scattering lengths *b* (*b*(Ge) = 8.185; *b*(Sb)=5.57; *b*(Te)=5.80 fm), compared with the corresponding atomic numbers *Z* determining the x-ray scattering factors (*Z*(Ge)=32; *Z*(Sb)=51; *Z*(Te)=52) mean that Ge-centered correlations are primarily responsible for the appearance of the ND FSDP; the value of Q_1 indicates that pseudoperiodic atomic-density fluctuations, with a period $D \sim 2\pi/Q_1 \approx 6$ Å are present in the amorphous phase. It might be significant that this is the value of the cubic unit-cell parameter, *a*, corresponding to the sum of Ge-Te and Sb-Te bond lengths, each ~ 3 Å in the crystal, in the metastable rocksalt crystalline structure of 225 [4.7]. The presence of vacancies in a-225 may also accentuate such atomic-density fluctuations.

The diffraction-derived RDFs of the amorphous phase of 225 indicate that there is an appreciable shortening, compared with the crystal values, of the average Ge(Sb)-Te nearest-neighbor bond length to ~ 2.78 Å [4.4, 4.5]. In ternary compounds, such as 225, there are a total of 6 partial pair-correlation functions (Ge-Te, Sb-Te etc.), and so a single diffraction experiment cannot provide all such structural information. For this purpose, modeling (e.g. RMC) is required (see Sect. 4.5.2).

The element-specific probe of EXAFS permits some improvement on this situation; EXAFS spectra taken at, say, the Ge K-edge contain information on possible Ge-Te, Ge-Ge and Ge-Sb correlations, but not on Sb-Te, Sb-Sb and Te-

Te correlations, and likewise for the other 2 edges. There have been several EXAFS measurements on a-225, a few [4.6, 4.8-12] on as-sputtered material, and one [4.13] on both this and (laser) re-amorphized material (Fig. 4.5).

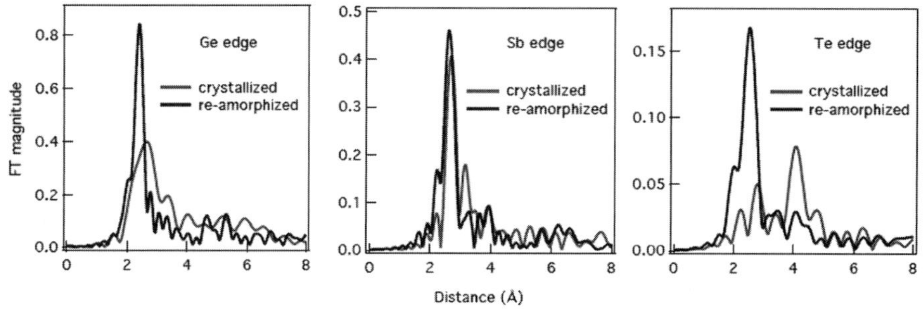

Fig. 4.5. Fourier-transformed EXAFS data taken at the Ge, Sb and Te K-edges for both metastable crystalline (rocksalt) and re-amorphized 225. Reprinted by permission from Macmillan Publishers Ltd: Nature Materials [4.13], copyright (2004).

Jovari et al. [4.11] also studied a-124 by EXAFS at the Ge, Sb, and Te K-edges. Measurements at the Ge K-edge for a-225 confirm the shortening of the Ge-Te bond length in the amorphous state compared with that of the metastable rocksalt crystal. Furthermore, Kolobov et al. [4.13] stressed the marked difference in peak amplitude in the EXAFS Ge-edge pair distribution function (PDF) between amorphous and crystalline states, and inferred the local Ge coordination to be *tetrahedral* (i.e. 4-fold coordinated with a coordination number $N_{Ge} \sim 4$ and an average Ge bond angle of $\theta_{Ge} \sim 109^o$) in contrast with the local structure in the metastable rocksalt crystalline structure, where $N_{Ge} \sim 6$ and $\theta_{Ge} \sim 90^o$. There is also a large change in the Te K-edge EXAFS spectra, but little change for the Sb data, between amorphous and crystalline phases of 225 [4.13]. Furthermore, the inclusion of homopolar Ge-Ge bonds was found to give the best fitting to the Ge K-edge EXAFS spectra [4.6, 4.8, 4.9, 4.11]. Conversely, no evidence was found by Jovari et al. [4.6] for the presence of Sb-Sb or Te-Te homopolar bonds from fittings of the Sb and Te K-edge spectra, but the presence of Sb-Ge wrong bonds was also found to be essential in fitting the Sb EXAFS data [4.6, 4.11], whereas Baker et al. [4.8] found evidence for Sb-Sb bonding, and later postulated the existence of short $Sb\text{-}Te^+$ bonds (2.51 Å) [4.10], from fits to their Sb K-edge data. Such discrepancies point up the difficulties encountered in interpreting EXAFS data in terms of coordination number (and type).

Finally, experimental studies of the *electronic* structure of (amorphous) PC materials can give indirect information on the atomic structure. Klein et al. [4.14] reported results from ultraviolet photoemission spectroscopy (UPS) of the valence-band density of states (DOS), and from x-ray photoemission spectroscopy (XPS) of Ge, Sb and Te core-level states from as-sputtered amorphous, and thermally-annealed crystalline, films of 124. From the observation of rather subtle asymmetric changes in Ge 2p and Sb 3d XPS core-level peaks on crystallizing amorphous films, it was inferred that two different Ge and Sb local environments coexist. One component of the core-level peaks is very similar in chemical shift between crystalline (c-) and amorphous (a-) phases, implying that some Ge and Sb atoms in a-124 have the same locally octahedral-like structural environment as in the rocksalt structure of c-124. The other component in a-124, at higher/lower (Ge/Sb) binding energies, corresponding to atomic states with higher/lower partial oxidation states, was tentatively ascribed to local configurations having shorter/longer Ge-Te bonds (assuming that Ge and Sb form bonds only to Te). Such configurations were assumed to correspond to tetrahedral environments, as inferred from EXAFS measurements [4.13]. However, the analysis of EXAFS data of a-124 by Jovari et al. [4.11], discussed above, indicates that, in addition, Ge-Sb wrong bonds can also exist, and so such configurations (not considered by Klein et al. [4.14]) could equally be the origin of the extra core-level peak components in a-124.

4.5.2 Simulational Studies

As discussed previously, there are two different approaches to modeling the structure of amorphous solids in general, and PC amorphous materials in particular, namely reverse Monte Carlo (RMC) simulations using experimental (e.g. diffraction and EXAFS) data, and *ab initio* molecular dynamics (AIMD). There have been three such RMC studies for as-sputtered a-225 using XRD data [4.4, 4.5], XRD and EXAFS data (at the Ge, Sb and Te K-edges) [4.12], and XRD, ND and EXAFS data (again for all three edges) [4.6, 4.11], respectively. In addition, Jovari et al. [4.11] have undertaken a similar study of a-124. In this way, the 6 partial pair-distribution functions, (Ge-Te, Sb-Te, Ge-Ge, Sb-Sb, Te-Te and Sb-Ge), were obtained by Jovari et al. [4.6] from a 36,000-atom RMC model – e.g. for a-225 (Fig. 4.6). (The sharp cut-offs in the Te-Te and Sb-Sb PDFs result from the minimum atomic separation imposed for such correlations.) The partial PDFs shown in Fig. 4.6 are similar to those found by Kohara et al. [4.4, 4.5] in their RMC-modeling analysis of XRD data alone, especially for the Ge-Te, Sb-Te, Sb-Sb and, to a lesser extent, Te-Te correlations. However, the Ge-Sb and Ge-Ge PDFs differ appreciably in the height of the first peak. The nearest-neighbor coordination information so obtained by Jovari et al. [4.6] is given in Table 4.1.

It should be noted that these average coordination numbers N obtained from this RMC simulation of a-225 are extremely close to the values expected from the 8-N rule [4.1] (Ge:4, Sb:3, Te:2), and are very different from the near-octahedral coordination for all types of atoms characteristic of the rocksalt (but vacancy-containing) metastable crystal structure of 225 and 124 [4.7].

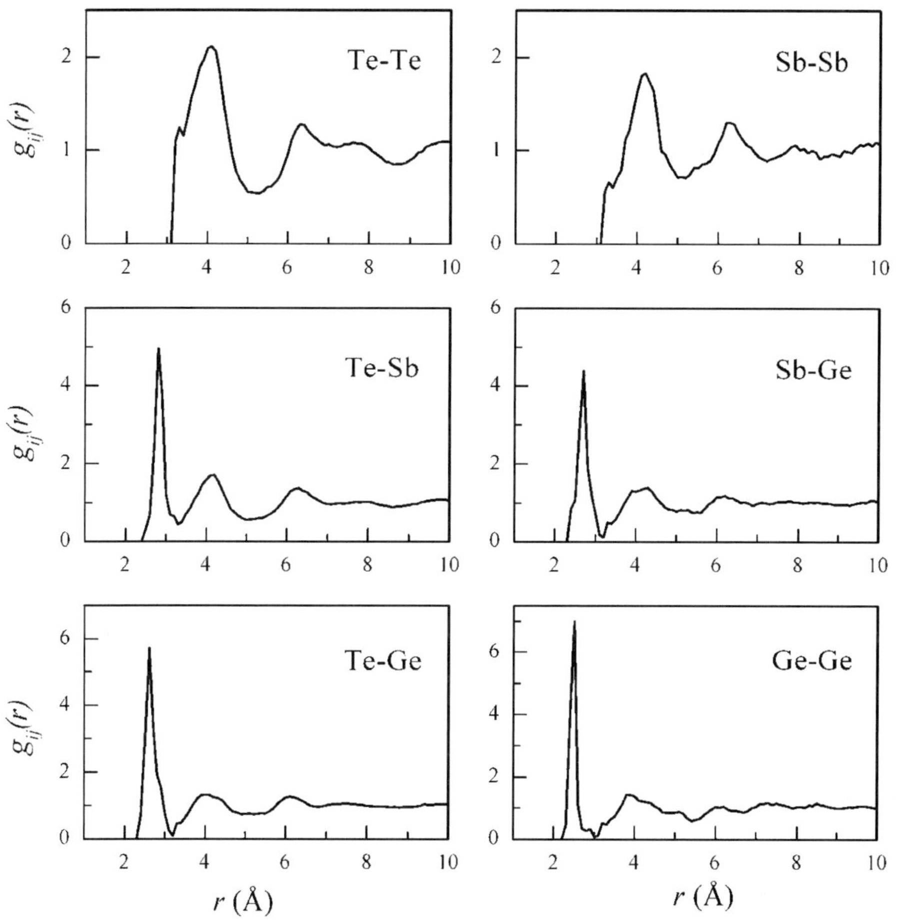

Fig. 4.6. Partial pair distribution functions for a-225 obtained by RMC modeling using XRD, ND and EXAFS data. Reprinted with permission from the Institute of Physics, from [4.6].

Table 4.1. Average bond lengths and coordination numbers in as-sputtered a-225 obtained from RMC simulation. Reprinted with permission from the Institute of Physics, from [4.6].

Atomic Pairs	Te-Sb	Te-Ge	Ge-Sb	Ge-Ge	N_{Ge}	N_{Sb}	N_{Te}
Bond length (Å)	2.83	2.64	2.69	2.48	-	-	
Coord. No.	0.96	1.08	0.83	0.69	4.24	3.22	2.04

In particular, it should be noted that 4-fold coordination of Ge atoms does *not* necessarily imply solely *tetrahedral* coordination, as assumed by Kolobov et al. [4.13].

Kohara et al. [4.4, 4.5] showed that the RMC bond-angle distributions, for all types of atom taken as origin, were very similar, all peaking at $\theta \sim 90^{\circ}$. Of course, this is the value of bond angle characterizing the ideal rocksalt crystal structure, which indicates that such a local defectively octahedral structural environment might also be predominant in a-225. In their latest RMC study, using XRD, ND and 3 EXAFS data-sets, Jovari et al. [4.11] found an average Te-Sb-Te bond angle of ~ 90° but, in contrast, the Te-Ge-Te bond-angle distribution was found to peak at $\theta \sim 109°$, indicative of predominantly tetrahedral coordination of Ge atoms, but with a variety of local tetrahedral units, $Ge(GeTe_3)$ being predominant, and with $Ge(SbTe_3)$, $Ge(GeSbTe_2)$ and the chemically-ordered $Ge(Te_4)$ units all being approximately equally probable in a-225. In the case of the more Te-rich a-124 material, $Ge(Te_4)$ units were the most probable in the RMC model [4.11]. Kohara et al. [4.4, 4.5] also made a ring-size analysis of their RMC models. In the ideal rocksalt structure, only 4-membered rings exist; the introduction of atomic vacancies in the Ge/Sb sub-lattice in the metastable rocksalt-like structure of c-225 causes the appearance of some 6-membered rings as well [4.5]. A ring-size analysis of RMC models of a-225 showed a broad distribution of ring sizes, extending up to 12-fold and centered on 6-fold rings, with even-membered rings obviously only being present for chemically-ordered models, and a small number of odd-membered rings also occurring when wrong/homopolar bonds were included in the models.

Three *ab initio* MD simulational studies have been carried out so far on a-225, exploring different aspects of the structure and dynamics.

Caravati et al. [4.15] used the Car-Parrinello (CP) MD package, modified by Kühne et al. [4.16] to include large integration time steps, to produce a 270-atom model of a-225, quenched from the liquid state at 990 K to 300 K in 18 ps. This model gave rather good agreement with the x-ray structure factor measured by Kohara et al. [4.4]. The bond-angle distributions of all types of atoms were found to peak at $\theta \sim 90°$, in agreement with the RMC findings [4.4, 4.5] and indicative of a defective locally octahedral environment of all atoms. However, there was also evidence in the model for about 33% of the Ge atoms being in tetrahedral environments. A local order parameter is q, defined as [4.17]:

$$q = 1 - \frac{3}{8} \sum_{i>k} \left(\frac{1}{3} + \cos\theta_{ijk} \right)^2 , \tag{4.2}$$

where the sum runs over couples of atoms bonded to a central atom, j: $q = 1$ for ideal tetrahedral geometry, $q = 5/8$ for a 4-coordinated defective octahedral site and $q = 0$ for an ideal octahedral 6-cordinated site. The distribution of q for Ge atoms is shown in Fig. 4.7, where this mixed coordination geometry is clearly seen.

Such locally octahedral-like geometry was also found in the parent liquid state (albeit of 124) by Bichara et al. [4.18]. Note also from Fig. 4.7 that the ring-size distribution of the model shows a greater proportion of 4-membered rings than were found in the RMC model [4.4, 4.5].

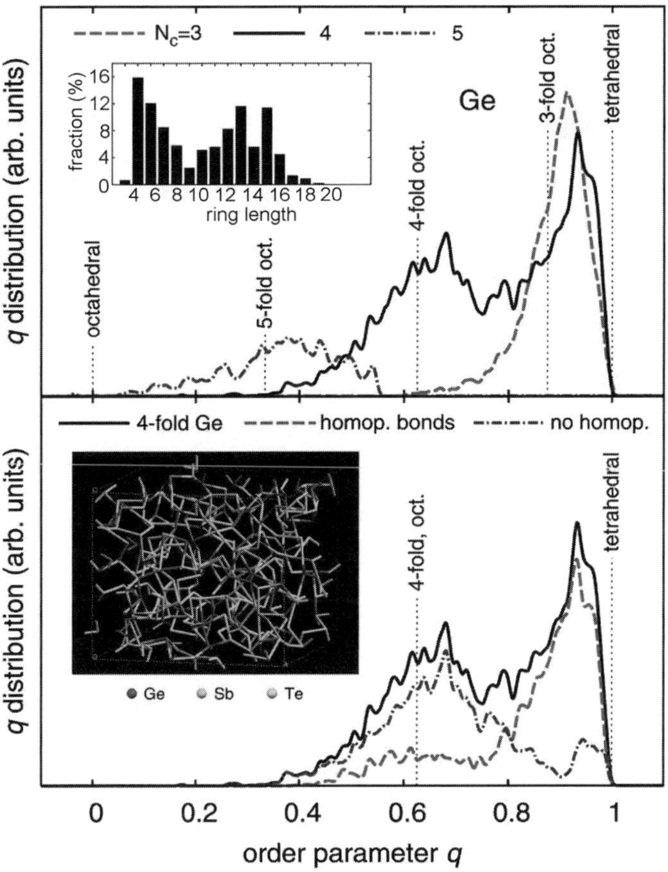

Fig. 4.7. Distribution of the local order parameter, q (Eq. 4.2) in a-225 for Ge with different coordination numbers, N_c (top), and for 4-coordinated Ge with, and without, wrong bonds (bottom). The top inset shows the model ring-size distribution and the bottom inset shows an image of the model. Reprinted with permission from [4.15]. Copyright [2007], American Institute of Physics.

Average coordination numbers in the model were found to be N_{Ge} = 3.82, N_{Sb} =4.03 and N_{Te} = 2.87; these Sb and Te coordinations are appreciably higher than the values found from RMC modeling of XRD, ND and EXAFS data [4.6] – see Table 4.1.

Much larger-scale (460-atom) MD simulations of a-225 (and a-GeTe) have been performed by Akola and Jones [4.19], also using the CPMD code; their model was quenched to 300 K from the liquid state at 900 K in 139 ps, and thence to 100 K in 74 ps. Atomic coordination numbers found in this model of a-225 are N_{Ge} = 4.2, N_{Sb} = 3.7 and N_{Te} = 2.9, in qualitative agreement with those found in the MD simulation of Caravati et al. [4.15]. The ring-size distribution of this model is also very similar to that found by Caravati et al. [4.15], showing a preponderance of 4-membered (square-like) rings, most of which are chemically ordered (i.e. with ABAB alternation). Akola and Jones [4.19], however, also made a study of the vacancy structure of their model of a-225. It was found that the vacancies are surrounded mainly by Te atoms (Fig. 4.8a), as in c-GST phases, where the atomic vacancies are confined to the Ge/Sb sub-lattices.

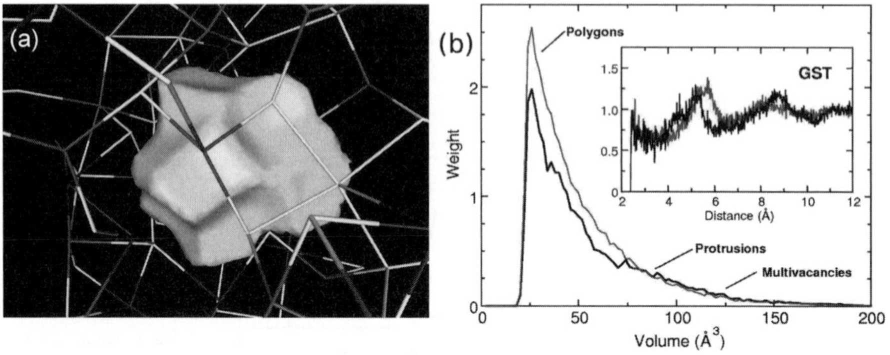

Fig. 4.8. (a) Image of vacancy in a model of a-225; (b) distribution of atomic vacancies in a-225 with inset, the vacancy-vacancy correlation function (black lines, a-225 at 300L; grey lines, liquid 225 at 900K). Reprinted with permission from [4.19]. Copyright (2006) by the American Physical Society.

The volume distribution of vacancies is shown in Fig. 4.8b, together with the vacancy-vacancy correlation function, which shows appreciable ordering out to at least 12 Å (half the simulation-box length).

Finally, Hegedüs and Elliott [4.2] have performed AIMD simulations on GST materials, including 225 and 124, using the VASP code [4.20]. Although these simulations were for somewhat smaller models (up to 90 atoms), they were for considerably longer simulation times (up to 1.3 ns) than the other AIMD simulations. It was found that, whereas rapidly-quenched (dT/dt = -15 K/ps, where T is the temperature and t is the time) liquid models invariably, and unsurprisingly [4.15, 4.19], produced amorphous structures (Fig. 4.9c), much slower cooling (say dT/dt = -0.45 K/ps) resulted in *crystallization* of the models to the metastable rocksalt structures (Fig. 4.9a, b). Moreover, thermal annealing of the rapidly-quenched amorphous model also produced crystallization to the rocksalt structure in a similar time-scale (~ 1 ns) – see Fig. 4.9d.

It was found that there were an appreciable number of (distorted) 4-membered square-like rings in models of liquid 225 that become quenched into the amorphous state (Fig. 4.10a). This amorphous 225 model, especially when energy relaxed using the CASTEP code [4.21], gives an extremely good fit to the total ND PDF [4.6], i.e. the Ge-Te bond length is correctly reproduced – see Fig. 4.10(b). Furthermore, it was found that the distribution of nearest-neighbor Ge distances is narrower than in the simulated distorted rocksalt crystal, in agreement with EXAFS results [4.13]. The bond-angle distribution for this model, shown in Fig. 4.10(c), indicates that the dominant *average* bond angle is $\theta \approx 90^\circ$ for all types of atoms, indicative of defective octahedral local geometry, in agreement with the findings for other MD models [4.15, 4.19], but a clear subsidiary peak appears in

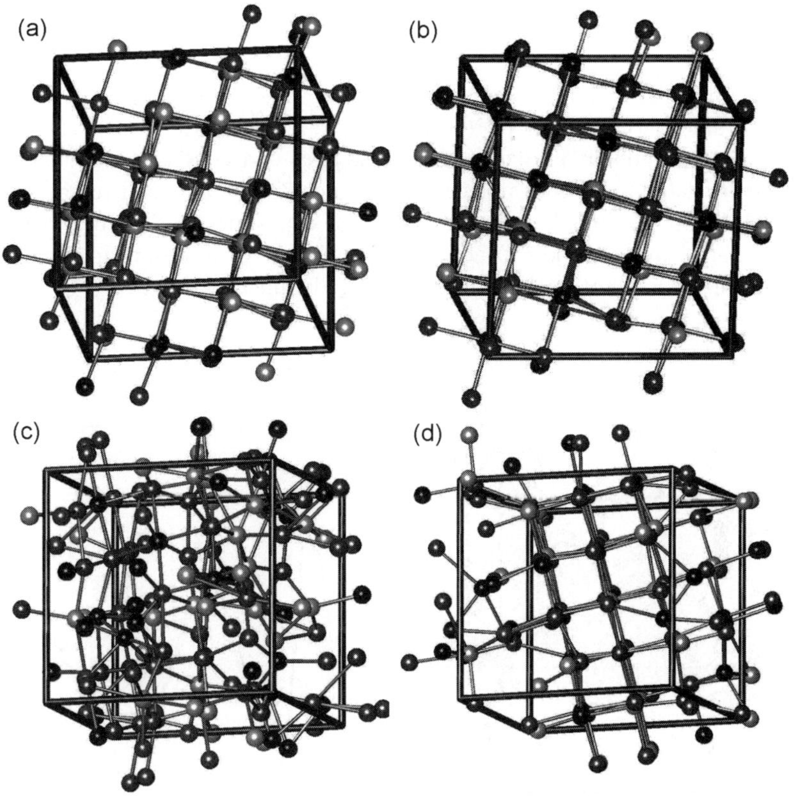

Fig. 4.9. Model configurations showing the result of simulated crystallization of GST materials obtained either by slow cooling of the liquid of: (a) 225 or (b) 124, or by thermally annealing at 900 K of the rapidly-quenched amorphous state (c) of 225 to form the crystal (d). Color coding of atoms: Sb - black, Ge - white, Te –gray. Reprinted by permission from Macmillan Publishers Ltd: Nature Materials [4.2], copyright (2008).

the Ge-centered distribution at $\theta \approx 109^{\circ}$, indicative of simultaneous tetrahedral coordination. This tetrahedral peak is much more pronounced than found in previous AIMD studies [4.15, 4.19], (but the recent RMC study of Jovari et al. [4.11] found that the Ge bond-angle distribution peaked at $\theta \approx 109^{\circ}$ in their model). An analysis of the average local coordination of Ge, Sb and Te atoms in the model of a-225, made in terms of clustering of differences in bond lengths, indicates that the basic coordination obeys the 8-N rule, i.e. $N_{Ge} = 4$, $N_{Sb} = 3$ and $N_{Te} = 2$, in accord with RMC results ([4.6] – see Table 4.1).

4.6 Summary

A number of experimental structural studies have been performed on samples of (sputtered) amorphous $Ge_2Sb_2Te_5$ (a-225) and, to a lesser extent, on a-124. In addition, various computer simulations have also been carried out on a-225, and also on a-124, including *ab initio* molecular-dynamics simulations and reverse-Monte Carlo (RMC) simulations based on a variety of experimental structural data-sets, e.g. from x-ray and neutron diffraction and extended x-ray absorption fine structure at all three x-ray absorption edges (i.e. Ge, Sb and Te K-edges).

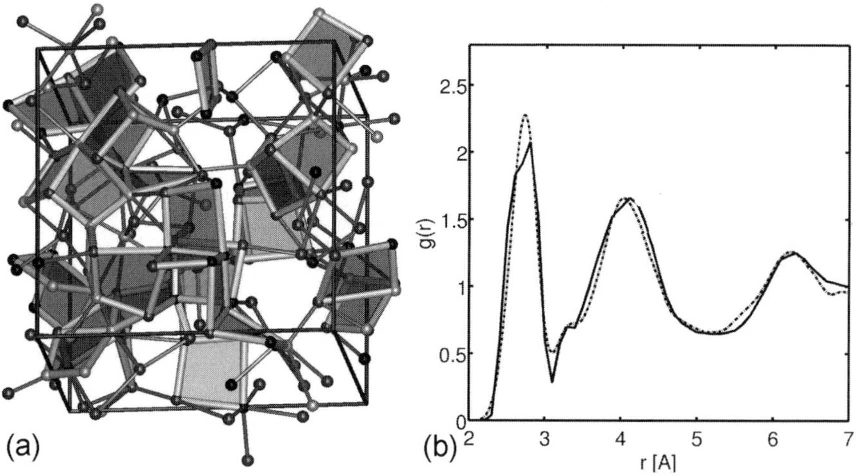

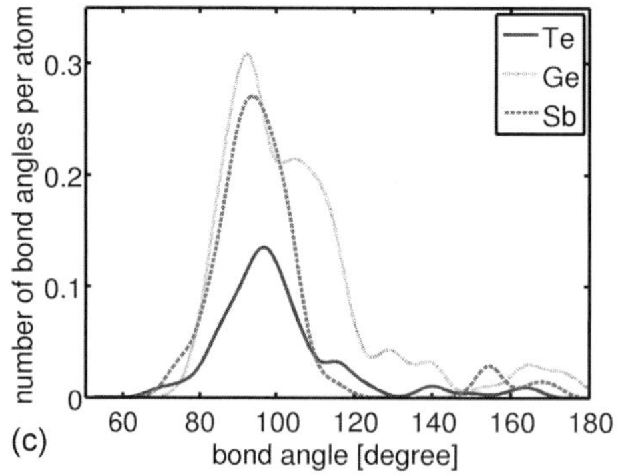

Fig. 4.10. Energy-relaxed structure of a model of a-225 : a) atomic configuration showing the presence of (distorted) 4-rings (the color coding of atoms is the same as in Fig. 4.9); b) the PDF of the model (dashed curve) compared with experimental ND data (solid curve – [4.6]); c) Ge-, Sb- and Te-centered bond-angle distributions. Reprinted by permission from Macmillan Publishers Ltd: Nature Materials [4.2], copyright (2008).

A number of structural features seem to be generally agreed, following these experimental and computational studies. The Ge-Te bond length is shorter, and its distribution is narrower, in a-225 than in the disordered rocksalt metastable structure. The Ge environment is 4-fold coordinated in the amorphous phase, but there is still disagreement between the various modeling studies as to whether tetrahedral or distorted octahedral (4-fold coordinated) configurations are dominant. The Sb and Te coordinations are still disputed somewhat in relation to how closely they adhere to the 8-*N* rule.

References

[4.1] Elliott, S.R., Physics of Amorphous Materials, 2nd ed. (Longman: 1990)
[4.2] Hegedüs, J. and Elliott, S.R.: Microscopic origin of the fast crystallization ability of Ge–Sb–Te phase-change memory materials. Nature Materials **7**, 399-405 (2008)
[4.3] McGreevy, R.L. and Pusztai, L.: Reverse Monte Carlo Simulation: A new technique for the determination of disordered structures. Mol. Simul. **1**, 359-367 (1988)
[4.4] Kohara, S., Kato, K., Kimura, S., Tanaka, H., Usuki, T., Suzuya, K., Tanaka, H., Moritomo, Y., Matsunaga, T., Yamada, N., Tanaka, Y., Suematsu, H. and Takata, M.: Structural basis for the fast phase change of $Ge_2Sb_2Te_5$: Ring statistics analogy between the crystal and amorphous states. Appl. Phys. Lett., **89**, 201910(1-3) (2006)
[4.5] Kohara, S., Kimura, S., Tanada, H., Yasuda, N., Fukuyama, Y., Murayama, H., Kim, J., Takata, M., Kato, K., Tanaka, Y., Usuki, T., Suzuyu, K., Tanaka, H., Moritomo, Y., Matsunaga, T., Kojima, R. and Yamada, N.: Structural basis for fast phase change of DVD-RAM -Topological order in the amorphous phase. Proc. European Phase Change and Ovonic Science Symposium 2007, http://www.epcos.org/library/Library2007.htm
[4.6] Jovari, P., Kabau, I., Steiner, J., Beuneu, B., Schöps, A. and Webb, A.: 'Wrong bonds' in sputtered amorphous $Ge_2Sb_2Te_5$. J. Phys. Cond. Matter **19**, 335212 (9pp) (2007)
[4.7] Wuttig, M., Lüsebrink, D., Wamwangi, D., Welnic, W., Gillessen, M. and Dronskowski, R.: The role of vacancies and local distortions in the design of new phase-change materials. Nature Mater. **6**, 122-U7 (2007)
[4.8] Baker, D.A., Paesler, M.A., Lucovsky, G. and Taylor, P.C.: EXAFS study of amorphous $Ge_2Sb_2Te_5$. J. Non-Cryst. Sol. **352**, 1621-1623 (2006a)
[4.9] Baker, D.A., Paesler, M.A., Lucovsky, G., Agarwal, S.C. and Taylor, P.C.: Application of bond constraint theory to the switchable optical memory material $Ge_2Sb_2Te_5$. Phys. Rev. Lett. **96**, 255501(1-3) (2006b)
[4.10] Paesler, M.A., Baker, D.A., Lucovsky, G., Edwards, A.E. and Taylor, P.C.: EXAFS study of local order in the amorphous chalcogenide semiconductor $Ge_2Sb_2Te_5$. J. Optoel. Adv. Mater. **8**, 2039-2043 (2006)
[4.11] Jovari, P., Kaban, I., Steiner, J., Beuneu, B., Schöps, A. and Webb, M.A.: Local order in amorphous $Ge_2Sb_2Te_5$ and $GeSb_2Te_4$. Phys. Rev. B**77**, 035202(1-6) (2008)
[4.12] Arai, T., Sato, M and Umesaki, N.: Structural change of crystalline and amorphous-$Ge_2Sb_2Te_5$ by reverse Monte Carlo analysis of x-ray diffraction data and extended x-ray absorption fine structure data. J. Phys. Cond. Matter. **19**, 335213 (11pp) (2007)
[4.13] Kolobov, A.V., Fons, P., Frenkel, A.I., Ankudinov, A.L., Tominaga, J. and Uruga, T.: Understanding the phase-change mechanism of rewritable optical media. Nature Mater. **3**, 703-708 (2004)
[4.14] Klein, A., Dieker, H., Späth, B., Fons, P., Kolobov, A., Steiner, C. and Wuttig, M.: Changes in electronic structure and chemical bonding upon crystallization of the phase change material $GeSb_2Te_4$. Phys. Rev. Lett. **100**, 016402(1-4) (2008)
[4.15] Caravati, S., Bernasconi, M., Kuhne, T.D., Krack, M. and Parrinello M.: Coexistence of tetrahedral- and octahedral-like sites in amorphous phase change materials. Appl. Phys. Lett., **91**, 171906(1-3) (2007)

[4.16] Kühne, T.D., Krack, M., Mohamed, F.R. and Parrinello, M.: Efficient and accurate Car-Parrinello-like approach to Born-Oppenheimer molecular dynamics. Phys. Rev. Lett. **98**, 066401(1-4) (2007); http://cp2k.berlios.de

[4.17] Errington, J.R. and Debenedetti, P.G., Relationship between structural order and the anomalies of liquid water. Nature **409**, 318-321 (2001)

[4.18] Bichara, C., Johnson, M. and Gaspard, J.P.: Octahedral structure of liquid $GeSb_2Te_4$ alloy: First-principles molecular dynamics study. Phys. Rev. B**75**, 060201(R)(1-4) (2007)

[4.19] Akola, J. and Jones, R.O.: Structural phase transitions on the nanoscale: The crucial pattern in the phase-change materials $Ge_2Sb_2Te_5$ and GeTe. Phys. Rev. B**76**, 235201 (10 pages) (2007); http://www.cpmd.org

[4.20] Kresse, G. and Hafner, J.: Ab initio molecular-dynamics for liquid-metals. Phys. Rev. B**47**, 558-561 (1993)

[4.21] Segall, M.D., Lindan, P.J.D., Probert, M.J., Pickard, C.J., Hasnip, P.J., Clarke, S.J. and Payne, M.C.: First-principles simulation: ideas, illustrations and the CASTEP code. J. Phys. Cond. Matt. **14**, 2717-2744 (2002)

5. Experimental Methods for Material Selection in Phase-change Recording

Liesbeth van Pieterson

Abstract Phase-change materials and their crystallization behavior are qualitatively described. This crystallization behavior is important for the applicability of materials in optical and electrical recording. We describe an experimental method to determine the crystallization rate at ambient temperatures and as a result the archival life stability of recorded data. Furthermore, we explain two experimental methods to determine the crystallization rate at elevated temperatures, which is related to the data rate of rewritable recording. We illustrate that systematic research for materials with improved crystallization characteristics has led to very stable and fast-crystallizing phase-change compositions.

5.1 Introduction

Since the discovery of reversible electric-field induced [5.1] and laser-induced [5.2] crystallization of amorphous semiconductor films in the late 1960s and early 1970s by Ovshinsky and co-workers, a lot of research has been devoted to these so-called phase-change materials. The research on phase-change materials even increased when they found commercial application as the data-containing layer in rewritable optical recording, such as CD-RW (Compact Disc - ReWritable), DVD±RW (Digital Versatile Disc - ReWritable) and Blu-Ray disc (BD) [5.3] (see Chapter 12). Now, investigations are continuing, since new possibilities may exist for use in electrical, solid-state memory devices (Phase Change Random Access Memory - PCRAM) [5.4]. This chapter has been written with these applications, and mostly optical recording, in mind.

For phase-change materials to be useful as an information layer in CD, DVD, BD, or other technologies, it is important that data can be written with sufficient contrast, either optically or electrically. In the next section, we will shortly discuss the switching between the amorphous and crystalline phase and the concomitant change in optical and electrical contrast. This will be elucidated in much more detail in Chapter 9. Then, we will introduce the chalcogenide type of materials that are typically applied. It appears that when data are written in these materials, crystallization can occur in various ways, i.e. the crystallization process is dominated

S. Raoux, M. Wuttig (eds.), *Phase Change Materials*,
DOI 10.1007/978-0-387-84874-7_5, © Springer Science+Business Media, LLC 2009

either by nucleation or by growth from the crystalline-amorphous edges. Each of these mechanisms can have its advantages and disadvantages as will be explained in Section 5.3.

Naturally, data stored in a memory device should be sufficiently stable, i.e. they should not be accidentally lost over time. Experimental methods that give a quick estimate of a material's archival life stability are described in Section 5.4. Furthermore, for practical applications in rewritable recording it is important that data can be overwritten at high speeds, which implies that amorphous marks should re-crystallize sufficiently fast. This is described in Section 5.5. Finally, section 5.6 describes ways to optimize materials in order to fulfill all requirements by means of materials optimization.

5.2 Reversible Switching

In 1971, Feinleib et al. reported on "high-speed crystallization of amorphous semiconductor films and the reversal of this crystallization back to the amorphous state using short pulses of laser light and evidenced by a sharp change in optical transmission and reflection" [5.2]. They mentioned that this "optical switching behavior is analogous to the memory type electrical switching effect that was reported before" [5.1]. These sentences contain a lot of important information. First of all a transition between the amorphous and crystalline state of an amorphous semiconductor film is described. This transition is reversible and can be triggered both optically and electrically. In order to go back to the amorphous state, short laser pulses should be applied. And finally, the transition is accompanied by a change in optical and electrical properties. In this section we will shortly discuss the meaning and implications of these observations.

Figure 5.1 shows a Transmission Electron Microscopy (TEM) picture of amorphous dots in a crystalline background. The amorphous areas are grey and lack structure, whereas small crystallites are visible in the crystalline area. When semiconductor films are applied, for example by sputtering, they are often amorphous. They can be crystallized by supplying heat, e.g. in an oven or with a laser. The resulting structure of the crystallites can be dependent on the applied temperature profile, as is discussed in Ref. [5.5], but in general heating above the crystallization temperature (but below the melting temperature) will result in the formation of the crystalline phase.

Formation of the amorphous phase is slightly more complicated. In this case, the material should be heated to temperatures above the melting temperature and then quickly cooled to below the glass transition temperature. This leads to a drastic reduction of atomic mobility, which is negligible at room temperature. Hence the atoms do not return to the stable, crystalline state but are trapped, or 'frozen', in a metastable, amorphous state. The process of crystallization and amorphization is schematically depicted in Fig. 5.2.

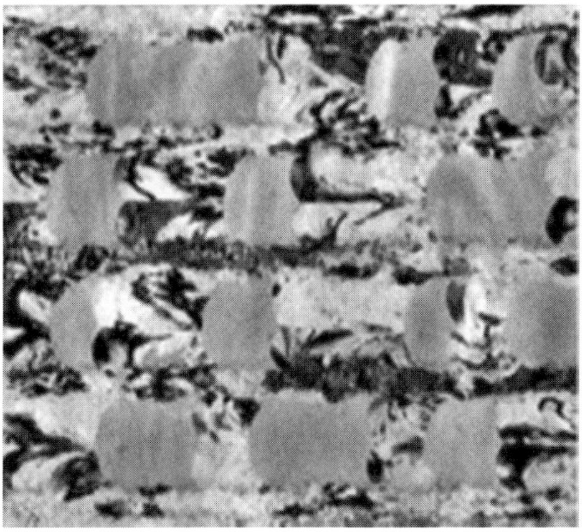

Fig. 5.1. TEM picture of amorphous data in a crystalline phase-change layer. Reprinted from [5.3] with kind permission of Springer Science and Business Media.

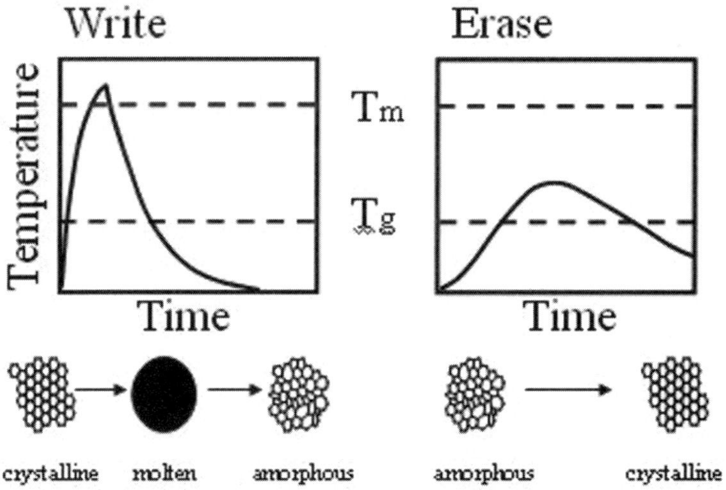

Fig. 5.2. Schematic of the temperature-time profiles associated with formation of amorphous dots (left panel) and crystallization (right panel) of amorphous marks in a crystalline layer. Reprinted from [5.3] with kind permission of Springer Science and Business Media.

In optical or electrical memories, fast cooling is achieved by the small volume that is amorphized and a special stacking of layers in the device. As an example, Fig. 5.3 shows schematically the structure of an optical disc. The phase-change layer is switched between two dielectric layers, usually $ZnS\text{-}SiO_2$, and as such only separated by tens of nanometers from a thicker silver layer.

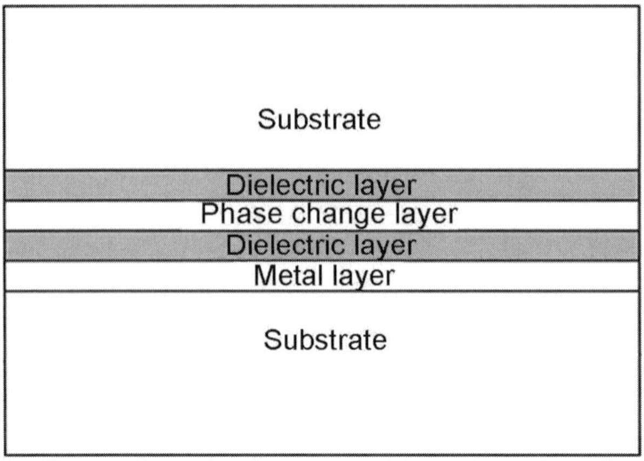

Fig. 5.3. Schematic of a phase-change stack for DVD recording.

This silver layer serves as a heat sink to facilitate fast cooling after applications of short laser pulses that heat the phase-change layer. The layer stack in an optical disc has other functions as well. Very important is the amplification of the optical contrast between the crystalline and amorphous state. While a majority of materials can be amorphized if cooled rapidly enough, just a few show a pronounced difference in the optical properties between the amorphous and crystalline states [5.6]. The presently used phase-change materials are a result of a 30-year and still continuing period of empirical optimization of materials. We will come back to this in the next section. Now, by optimizing the thicknesses of the dielectric layers and using the silver layer as mirror, optical contrasts approaching 100% can be achieved. This makes read-out of data easy, as can be intuitively felt from Fig. 5.4.

5.3 Phase-change Materials

Over the past thirty years, many materials have been investigated for their suitability as phase-change materials. Table 5.1 gives a short historical overview of the most important discoveries. $Te_{48}As_{30}Si_{12}Ge_{10}$ and $Te_{81}Ge_{15}Sb_2S_2$ are the materials that Ovshinsky used for his first electrical and optical switching experiments, respectively. Variations of $Ge_2Sb_2Te_5$, (Ag, In) doped $Sb_{69}Te_{31}$ and Ge-In-Sn-Sb are currently used in the various optical recording formats.

A phase-change material that is used in commercial applications like optical recording should be able to form glasses, and is characterized by a sufficiently low melting point that is attainable by current laser diodes. Furthermore, it should have a stable amorphous phase, characterized by high activation energy towards crystallization.

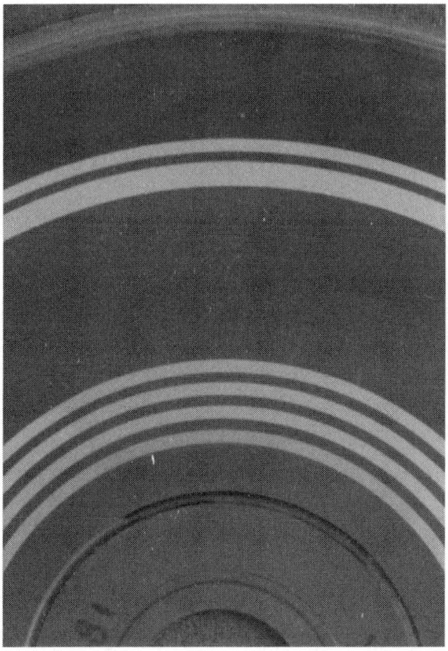

Fig. 5.4. Picture of crystalline stripes in amorphous DVD+RW disc. The stack has a contrast close to 100% and is optimized for red laser light.

Table 5.1. Historic overview of compositions used / studied as phase change materials.

Year	Composition
1968	$Te_{48}As_{30}Si_{12}Ge_{10}$ [5.1]
1971	$Te_{81}Ge_{15}Sb_2S_2$ [5.2]
1986	GeTe [5.7]
1987	GeTe-Sb_2Te_3,($Ge_2Sb_2Te_5$, $GeSb_2Te_4$) [5.8]
1991	(Ag,In) doped $Sb_{69}Te_{31}$ [5.9]
2003	Ge-In-Sn-Sb [5.10]

It should have a high contrast between the amorphous and crystalline state (optical and/or electrical) and should be able to show fast (laser-induced) crystallization. Materials that fulfill all these requirements often are composed of antimony, tellurium and germanium in various ratios, and with possible additions of elements like indium. Compositions are usually given by means of a 'composition triangle', as depicted in Fig. 5.5.

Often, the phase-change materials are described as being divided into two classes with slightly different compositions, based on their crystallization mechanism [5.11]. Below, we will discuss the crystallization properties of both phase-change classes.

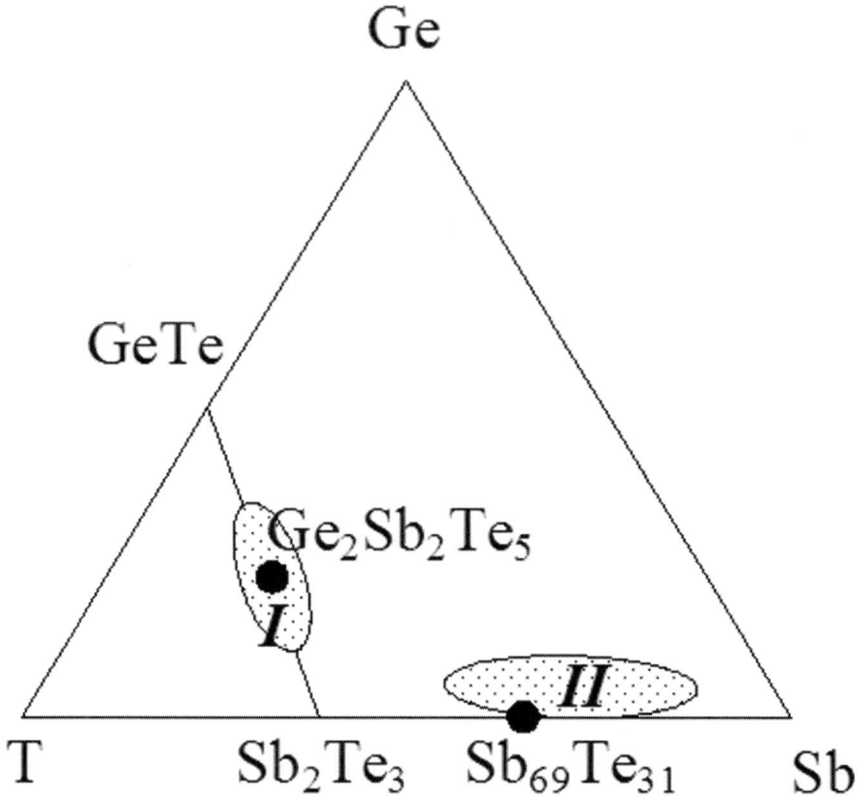

Fig. 5.5. Composition triangle of Sb, Te and Ge. Compositions with a nucleation-dominated crystallization mechanism (class I) and with a growth-dominated crystallization mechanism (class II) are indicated.

5.3.1 Crystallization by Nucleation and Growth

Phase-change materials with compositions on the tie-line GeTe-Sb_2Te_3 were discovered in 1987 and are still frequently used, for example in DVD-RAM [5.8]. The best known composition is $Ge_2Sb_2Te_5$, which combines all requirements of fast crystallization, amorphous phase stability and excellent contrast.

When amorphous areas in a crystalline layer of GeTe-Sb_2Te_3 material are subjected to sufficient heat, crystallization will occur. A schematic representation of this process is given in Fig. 5.6(a). Crystallization is characterized by nucleation followed by growth of the nuclei over a small distance, until they impinge on other crystallites. Nucleation can be fast and can occur in tens of nanoseconds [5.11].

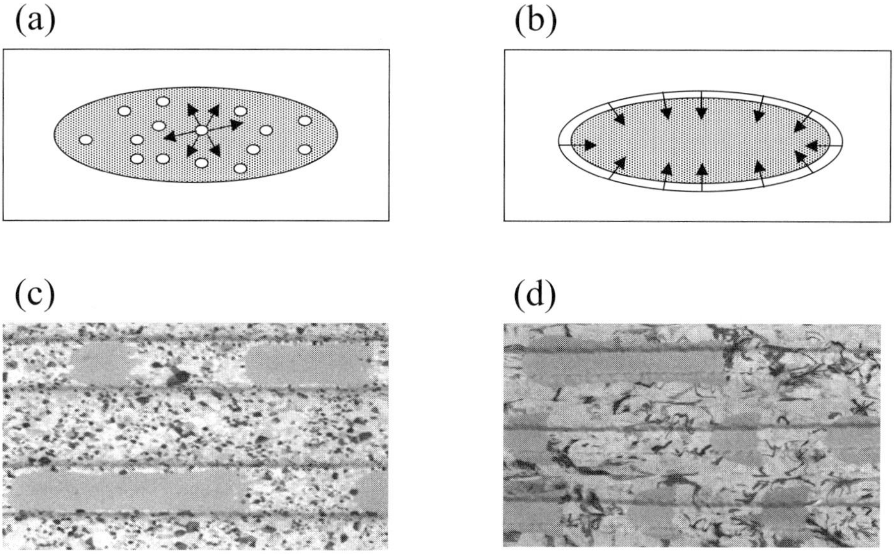

Fig. 5.6. (a), (b) Schematic representation of the crystallization of an amorphous mark in a crystalline phase-change layer of a material with a nucleation-dominated crystallization mechanism (a) and with a growth-dominated crystallization mechanism (b). (c), (d) Transmission electron microscope images of amorphous marks in a crystalline phase-change layer. In (c) crystallization occurs via nucleation and growth, in (d) the crystallization process is dominated by crystal growth. The nature of the crystallization process can be clearly observed in the resulting crystalline texture. Reprinted from [5.3] with kind permission of Springer Science and Business Media.

Materials that show this type of crystallization are sometimes called nucleation-dominated materials (NDMs).

The large nucleation probability has implications for the morphology of the crystalline layer, as can be observed in Fig. 5.6(c). A large number of small crystallites are observed. This is something to take into account when writing amorphous marks; this process involves melt quenching of the phase-change material and some re-crystallization of the molten state may occur. This can affect the shape of recorded marks. To prevent the formation of strangely shaped areas as a result of incomplete re-crystallization, influencing the nucleation probability of the phase-change material may be necessary.

The nucleation probability (and therewith the crystallization rate) can be influenced in several ways [5.12]. First, the crystallization rate increases with increasing phase-change layer thickness. This can be understood as a result of competition between the contributions of interface and bulk effects; the crystallization rate is dominated by the interface if the phase-change layer is thin and by the bulk if the layer becomes thicker than a critical value. Apparently, for $Ge_2Sb_2Te_5$ the nucleation probability in the bulk is higher than at the phase-change – dielectric interface. Nucleation in the bulk can be promoted by addition of nitrogen or oxygen. Furthermore, adding nucleation promoting interface layers, such as SiC, can enhance the nucleation probability.

5.3.2 Crystallization Dominated by Crystal Growth

Whereas crystallization in fully amorphous layers always starts with the formation of a nucleus, this is not needed for amorphous dots in a further crystalline layer. For materials with compositions around the eutectic point $Sb_{69}Te_{31}$, the probability for nucleation is much lower than for crystal growth as is depicted in Fig. 5.7.

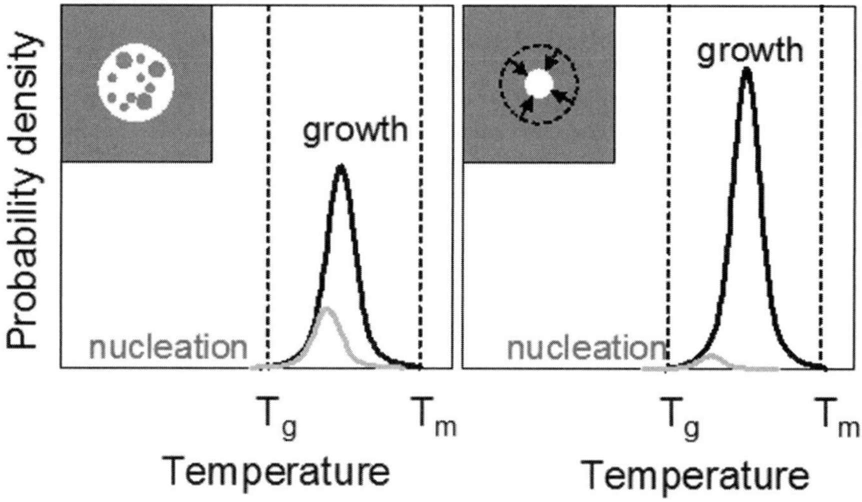

Fig. 5.7. Probability for nucleation and growth as a function of temperature and amorphous marks in a crystalline phase-change layer. In (a) crystallization occurs by a process of nucleation and growth, in (b) crystallization proceeds by crystal growth from the amorphous-to-crystalline interface to the center of the amorphous area. Reprinted from [5.3] with kind permission of Springer Science and Business Media.

As this is the case, crystallization will start from the amorphous-crystalline interface and proceed inwards. Materials with this type of crystallization behavior are sometimes called growth-dominated materials (GDMs).

Figure 5.6(d) shows a TEM picture of a crystalline Ge-doped Sb-Te layer. Large and irregular shaped crystallites are observed, characteristic for the growth-dominated crystallization mechanism. It has been shown that for GDMs with a given composition and hence a given crystallization rate the time needed for recrystallization of an amorphous mark decreases with the mark size [5.13]. This can easily be understood from Fig. 5.6(b); it will take less time to crystallize a small area. For this reason, GDMs are interesting options to achieve high data rates (short crystallization time) in high data density formats (small amorphous marks), such as DVDs and Blu-ray discs.

The properties of the class II materials can be adjusted by slight variations in the composition [5.14]. This is largely an empirical process and will be described in Section 5.6.

5.4 Archival Life Stability

The stability of the amorphous phase towards crystallization is an important materials parameter, especially from the application point of view. If we store data on a medium we do not want to loose them over time. When studying the amorphous phase stability of a material, we need to distinguish between the fully amorphous state and amorphous areas (or data) in a crystalline phase-change layer. For the fully amorphous state, crystallization takes place by nucleation and growth. For materials with a growth dominated crystallization mechanism, the situation is different for amorphous areas (*i.e.* data) in an already crystalline layer. Here, crystallization occurs by growth from the crystalline-amorphous interface, and nucleation effects can be neglected.

The theory of crystallization has been described by Johnsen, Mehl and Avrami [5.15]. Here, we will describe an experimental method to study crystallization behavior of amorphous data marks in phase-change materials. The method is based on isothermal measurements in which the reflection is monitored as a function of time. In this way, it is possible to characterize the volume fraction that has been crystallized, denoted as ξ, at time t. We can calculate $\xi(t)$ from the reflection (R) as follows:

$$\xi(t) = \frac{R(t) - R_{begin}}{R_{end} - R_{begin}} \tag{5.1}$$

In this expression R_{begin} is the reflection of the crystalline layer stack with amorphous data marks and R_{end} is the reflection when the amorphous data have recrystallized. The reflection R can be expressed as

$$R = \frac{R_c A_c + R_a A_a}{A} \tag{5.2}$$

where R_c and R_a represent the reflection of the stack in the crystalline and amorphous state, respectively. A is the total surface and A_c and A_a represent the surface occupied by the crystalline and amorphous state, respectively.

Figure 5.8 shows isothermal reflection measurements on crystalline phase-change materials with regularly repeating amorphous dots.

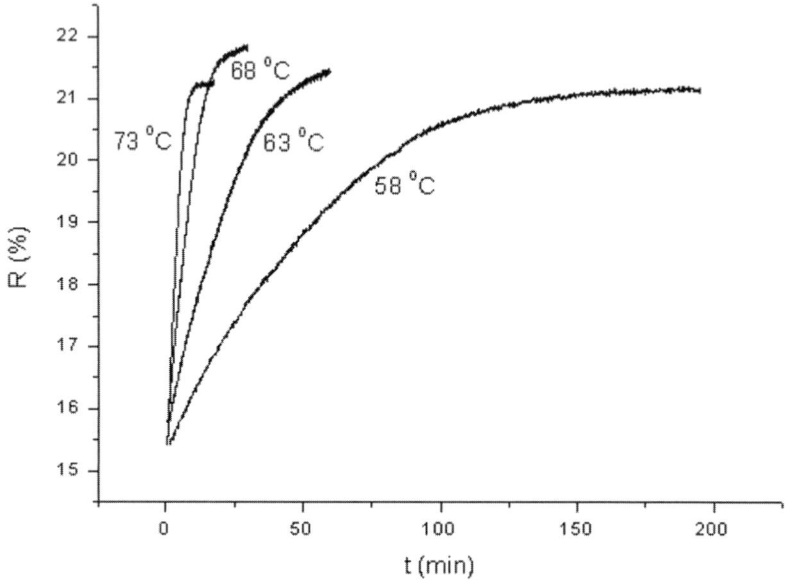

Fig. 5.8. Reflection as function of time of a crystalline Ga-doped SbTe layer with written amorphous marks.

From these curves we can determine the time v needed for complete erasure of amorphous dots. Using

$$\frac{X_0}{Y} = v \tag{5.3}$$

with X_0 being the initial mark radius, we can determine the growth rate Y.

However, from an application point of view it is more interesting to determine the archival life stability of the amorphous data marks. Using

$$v = v_0 e^{-\frac{Q}{RT}} \tag{5.4}$$

with R being the molar constant and T the temperature, we can determine the activation energy Q for crystallization of amorphous areas in a crystalline phase-change layer from isothermal reflection curves at various temperatures. If we extrapolate the amorphous phase stability at elevated temperatures to room temperature (or 50 ^{0}C), we get an indication of the archival life stability. Generally, it is assumed that amorphous data cannot be read anymore when their diameters have decreased by about 4%.

Figure 5.9 shows the archival life stability of various doped SbTe phase-change materials. It can be observed that the archival life greatly depends on the composition.

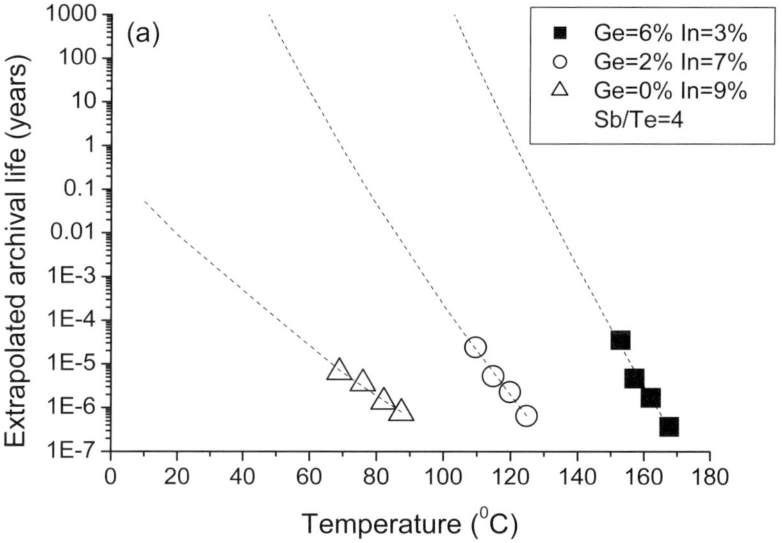

Fig. 5.9. Archival life stability of amorphous dots in a crystalline doped Sb-Te phase-change layer (markers represent experimental data), for various dopant concentrations. The dotted line shows the extrapolation to room temperature. Reprinted with permission from [5.14].

This will be explained in more detail in section 5.6. Accelerated crystallization experiments as described above are important to get a quick estimate of materials amorphous phase stability. For commercial applications also extended tests at slightly elevated temperature and humidity are performed (e.g. 50 - 85 ^{0}C).

5.5 Crystallization Rate

While a phase-change material should have a negligible crystallization rate at ambient temperatures, it is preferable that at elevated temperatures, such as achieved by pulsing with laser diodes or electrical stimulation, crystallization is fast. In this section, two methods to determine the crystallization rate in this high-temperature regime are described, namely static tester [5.11] and erasability [5.14] measurements. Both methods originate from optical recording research, but give valuable information for other research fields as well.

Static testing is optical recording on a stationary sample: a static tester therefore reads, writes and erases a single bit of information on the active layer of the optical recording medium. The optical response (*i.e.* changes in the reflection) of one area of the phase-change layer to pulses from a focused laser beam is measured. By varying the power and length of the pulse, it can be studied when the phase-change material crystallizes or amorphizes. Static testing therefore provides a rapid and informative tool in the study of phase-change materials; the information gained not only being directly applicable to material choice, but also providing much insight into the processes involved in the recording.

A typical static tester experiment involves a sequence of laser pulses. First, the reflectivity of the crystalline phase-change layer is measured by a low power laser pulse. Then, a laser pulse of higher power is applied that could possibly melt the phase-change material. Again, the reflectivity is measured. In this way, amorphous dots of specific size, but typically with radius of about 125 nm, are 'written' in the phase-change layer [5.14]. Finally, laser pulses are applied to crystallize the amorphous dots, and again the reflectivity is measured. An example of a static tester experiment is given in Fig. 5.10. The minimum time to crystallize an amorphous area, independent of laser power, is called the complete erasure time and is characteristic for the crystallization speed of a material.

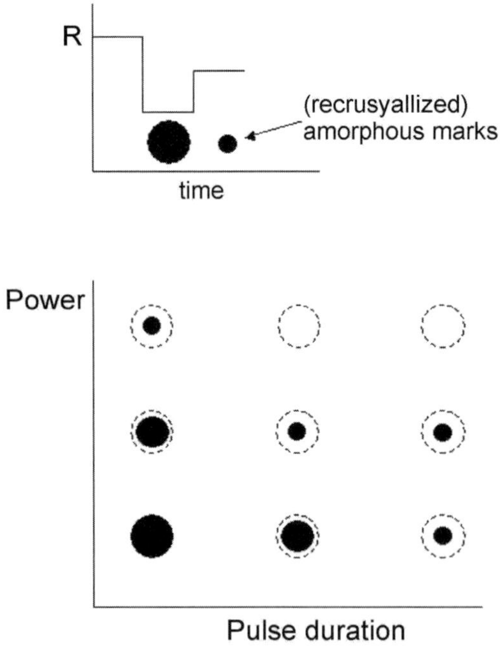

Fig. 5.10. Top: Reflection as a function of time showing steps in a static tester experiment. For a fully crystalline layer, a high reflection is measured. Then, a laser pulse is applied that melts the material, resulting in a lower reflection. A subsequent laser pulse partially recrystallizes the material. Bottom: The static tester writes and erases an array of effects, where the duration and power of the pulses is varied. This picture shows the effect on amorphous dots in a crystalline layer.

Although static tester experiments give valuable insight in the crystallization and amorphization of phase-change materials under influence of laser pulses, direct relation of the crystallization rate to the conditions in a disc drive is difficult, as here rotation of the disc influences the heat distribution in the material stack. For that reason, as a next step so-called erasability experiments can be performed. In such an experiment, long amorphous marks, or stripes, are written with a sufficiently high contrast. The maximum linear disc velocity at which the amorphous areas can be re-crystallized is determined. For a typical size of the data marks and typical DC (direct current) erase power [5.10], this is the disc velocity at which these marks are reduced by 25 dB. Fig. 5.11 shows the DC erasability for two phase-change materials, denoted as 4x and 16x [5.10]. It can be observed that for the 4x material, -25 dB DC erasability is attained up to linear velocities of about 14 m/s, whereas the 16x material shows good erasability up to velocities of over 58 m/s (!).

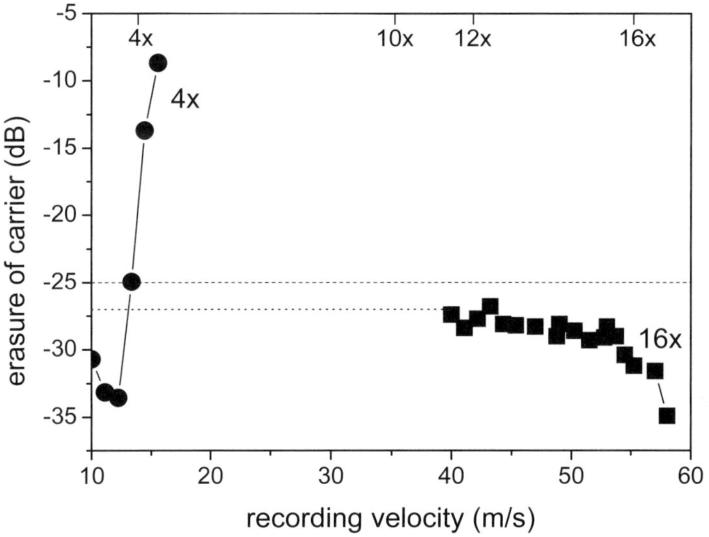

Fig. 5.11. DC-erasability as a function of the linear velocity for Ge-Sn-Sb based phase-change compositions. Reprinted with permission from [5.11].

5.6 Material Optimization

In the previous sections, we have introduced several types of phase-change materials and explained some experimental methods to determine important characteristics such as amorphous phase stability and crystallization rate. It was mentioned that for a material to be useful in applications, such as optical recording, it should be able to form glasses, have a stable amorphous phase, have a high contrast between the amorphous and crystalline state (optical and/or electrical) and should be able to show fast (laser-induced) crystallization. In the past twenty years, a lot of research has focused on optimization of materials to fulfill all these requirements. To illustrate this, we describe in this section systematic efforts to optimize phase-change materials for high-speed optical recording. We will focus on materials with a growth dominated crystallization mechanism like compositions close to $Sb_{69}Te_{31}$.

Doped Sb-Te compositions like AgInSbTe are widely applied in CD-RW and DVD+RW [5.9]. Optimization of these materials is largely an empirical process. By adding dopants to the phase-change material, variations in optical contrast, crystallization rate and amorphous stability can be obtained [5.14]. Note that in phase-change recording the term dopant is generally used when atoms are added in concentrations of typically 2-10 %, as long as the addition of these foreign atoms does not influence the crystal structure and crystallization mechanism.

Pure antimony films show explosive crystallization [5.16]. This effect was explained by an exothermic amorphous-crystalline phase transition in which the heat of crystallization that is released from the crystallized region induces crystallization of the surrounding amorphous regions [5.17]. However, crystallization takes place already at room temperature, making pure Sb films unsuitable for recording applications. Alloying with other elements can increase the amorphous stability. Good examples are the (doped) Sb-Te compositions. For Sb_2Te, a crystallization temperature of about 103 ^{0}C was measured, and adding dopants like germanium to the material can increase the crystallization temperature even further. This is shown in Table 5.2, which summarizes the crystallization temperatures of the as-deposited (sputtered) amorphous phase of many GD phase-change materials (column 1) [5.18]. An estimate of the amorphous stability can be made *a priori* by calculating the glass transition temperature [5.19], which can be considered as the lower limit of the crystallization temperature. The glass transition temperature, and hence the crystallization temperature, increases by adding elements that increase the average coordination number and/or the average bond enthalpy. Ge has a high bonding strength and a relatively large coordination number of 4.

Table 5.2 also summarizes data on the crystallization temperature of amorphous dots in a further crystalline layer. It is observed that the crystallization temperature of the as-deposited amorphous layer is generally higher than that of layers with recorded amorphous marks. This can be understood by realizing that crystallization in the as-deposited layer occurs by nucleation followed by growth, whereas in the recorded layer crystal growth occurs directly from the crystalline-amorphous boundaries of the written marks. Nucleation in these GD materials is a slower process. It appears that the difference in crystallization temperature of the written phase and the sputtered amorphous phase is very large for Sb-Te compositions doped with Ga, Sn or In. Apparently, these dopants stabilize the as-deposited amorphous phase, but not the melt-quenched recorded phase. Possibly, these elements influence the nucleation process rather than the growth process.

Table 5.2. Overview of crystallization temperatures T_c (for as deposited amorphous layers and amorphous dots in a crystalline layer), activation energies E_{act} for crystallization of amorphous dots and extrapolated archival life stabilities at 50 ^{0}C, of various GD phase-change compositions. RT is room temperature. Reprinted with permission from [5.18]. Copyright [2005], American Institute of Physics.

Phase-change composition	*T_c (as deposited) (^{0}C)*	*T_c (dots) (^{0}C)*	*E_{act} (kJ/mol)*	*Archival life @ 50 ^{0}C*
Sb_2Te	103			
$Ge_8Sb_{72}Te_{20}$	222	192		
$Ag_8Sb_{72}Te_{20}$	175	120	196	100 days
$In_8Sb_{72}Te_{20}$	200	93	121	1 hour
$Ga_8Sb_{72}Te_{20}$	223	88	184	2 hours
$Ga_8Sb_{77}Te_{15}$	230	84	155	0.5 hours
$Sn_8Sb_{72}Te_{20}$	153			<few days at RT
$Ge_6In_2Sb_{72}Te_{20}$	208	176	465	$4x10^{13}$ years
$Ge_2In_7Sb_{72}Te_{20}$	196	133	290	$4x10^2$ years

A quick and good estimate of the archival life stability can be obtained by determining the activation energy for crystallization and then using this value to estimate the lifetime of recorded marks at ambient temperature. This was explained in more detail in Section 5.4. The activation energy for crystallization can be derived from an Arrhenius plot, in which the isothermal crystallization time of the amorphous state is plotted as a function of the inverse temperature. To determine the isothermal crystallization time, we measured the reflection of many amorphous dots (approximately 200 nm radius) in a crystalline layer as a function of time at a specific temperature for a standard disc and determined the time needed for complete re-crystallization of the marks. This experiment was repeated at four typical temperatures ranging from 10 to 25 ^{0}C below the crystallization temperature of the phase-change material. Assuming an activation energy independent of temperature and using the relation $erasure\ time = Ae^{E_{act}/RT}$, with constant A and activation energy E_{act} as determined from the Arrhenius plot, the time for complete erasure of amorphous marks at ambient temperature can be calculated. The archival life is defined now as the time in which the diameter of the amorphous area has been reduced by 4% due to re-crystallization at a temperature of 50 ^{0}C, and is also included in Table 5.2. It can be observed that for a given Sb-Te alloy, the archival life stability decreases with increasing Sb-content. The archival life stability is also dependent on the dopant. For example, for doped Sb-Te compositions the archival life stability decreases in the order Ge>Ag>Ga, In>Sn.

As was described in Section 5.5, also the crystallization rate at elevated temperatures is important, e.g. to re-crystallize data by laser pulses. In fact, this is the rate-limiting step in rewritable optical recording and therefore a lot of research has focused on optimizing materials to make them crystallize faster. Figure 5.12 shows static tester experiments for doped Sb-Te compositions with varying Sb/Te ratios.

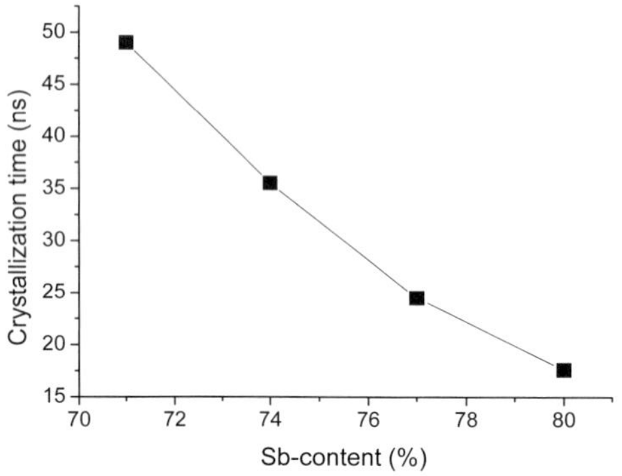

Fig. 5.12. Minimum laser-induced crystallization time of an amorphous mark of 125 nm radius, measured on a static tester set-up, as a function of Sb-content for In-doped Sb-Te compositions. Reprinted with permission from [5.18]. Copyright [2005], American Institute of Physics.

It can be observed that the crystallization time decreases, and hence the crystallization rate increases, with increasing Sb-content. Given a certain Sb-content, the nature of the dopant has an influence on the crystallization rate. For example, for doped Sb-Te phase-change materials, crystallization times decrease in the order Ag>Ge>In>Ga>Sn [5.18]. The influence of the phase-change composition on the crystal growth velocity has been explained by the effect of dopants on the mobility of the atoms [5.20]. The mobility can be related to the experimentally available viscosity via the Stokes-Einstein relation. For example, the increase in viscosity with Ge-content may explain the decreased growth velocity.

Besides influencing crystallization rate and amorphous phase stability, the material composition effects the nucleation time and optical constants [5.18]. For growth dominated phase-change materials, the nucleation rate is small compared to the growth rate. However, nucleation probability becomes important when reading amorphous dots numerous times (repeated reading of data), as the formation of nuclei inside the amorphous area, followed by crystal growth, will result in misinterpretation of the data. An indication of nucleation times can also be obtained using a static tester, by monitoring the onset of a reflection increase when exposing the amorphous material to laser pulses of variable power and duration. For doped Sb-Te compositions, a dependence of nucleation time on dopant and dopant concentration can be observed. Longest nucleation times are usually observed when Ge is used as a dopant. This may be due to a large activation energy for nucleation because of the large binding energy of Ge [5.19].

Very important in phase-change optical recording is the optical contrast between the amorphous and crystalline state. Usually, optical stacks are designed in such a way that the reflectivity of the amorphous marks is as low as possible, whereas the crystalline surroundings have a high reflectivity. We determined optical parameters from reflection and transmission measurements on phase-change layers of varying thickness on a custom-made spectrometer. The optical constants *n* and *k* were extracted by fitting the data to Lorentz equations that describe the dispersion dependency of the phase-change media [5.21, 5.22]. Table 5.3 shows optical constants of doped Sb-Te phase-change compositions at a wavelength of 658 nm.

Table 5.3. Optical constants of some doped Sb-Te phase-change layers. Reprinted with permission from [5.18]. Copyright [2005], American Institute of Physics.

Phase-change composition	*Amorphous phase*		*Crystalline phase*		*Crystalline reflection at* C_{max}*~99 (%)*
	n	k	n	k	
Ge-doped Sb-Te	4.6	2.6	3.9	5.2	18
In-doped Sb-Te	4.5	2.4	3.0	4.6	20

Also given is the crystalline reflection for an optimized phase-change stack (consisting of a phase-change layer sandwiched between two dielectric layers and with a metal mirror) with a contrast $C = (R_c - R_a)/R_c$ of 99 %. Good optical contrast in the red can be obtained for all doped Sb-Te phase-change compositions by minimizing the reflectivity of the amorphous phase. Furthermore, it is observed that In and Sn-additions to these growth dominated phase-change materials can enhance the crystalline reflection. A sufficiently high crystalline reflectivity is important for backwards compatibility of recorded DVD+RW media in standard DVD-players.

5.7 Outlook

In this section we have described experimental methods to determine crystallization-related properties of phase-change materials. We have shown that these measurements have helped systematic research towards improved materials for phase-change optical recording. Driven by industry, materials research in optical recording has focused in the past 30 years on higher data rates and higher density formats. With the realization of the high-speed DVD+RW and the Blu-Ray disc standard, research on optical phase-change materials seems to slow down. However, at the same time two interesting developments can be observed. On the one hand, experimental research into electrical properties of phase-change materials is increasing. Here, a lot can be learned from the optical research from the past, but also many new opportunities exist. On the other hand, there is an increasing effort to improve understanding of the fundamental properties of phase-change materials. This may lead to new insights that will help to explore these exciting materials even further.

References

[5.1] Ovshinsky, S. R.: Reversible electrical switching phenomena in disordered structures. Phys.Rev. Lett. **21**, 1450-1453 (1968)

[5.2] Feinleib, J., deNeufville, J., Moss, S. C., Ovshinsky, S. R.: Rapid reversible light-induced crystallization of amorphous semiconductors. Appl. Phys. Lett. **18**, 254-257 (1971)

[5.3] Meinders, E. R., Mijiritskii, A. V., van Pieterson, L., Wuttig, M., Optical Data Storage – Phase-change Media and Recording. Springer, The Netherlands (2006)

[5.4] Lankhorst, M. H. R., Ketelaars, B. W. S. M. M., Wolters, R. A. M.: Low-cost and nano-scale non-volatile memory concept for future silicon chips. Nat. Mat. **4**, 347-352 (2005)

[5.5] Verheijen, M. A., Mijiristkii, A. V., Kooij, B. J.: TEM study of InSbTe crystal morphology as a function of crystallization conditions. MRS 2003 Proceedings, **803**, 161-166 (2003)

[5.6] Yamada, N.: Erasable phase-change optical materials. MRS Bull. **21**, 48 (1996)

[5.7] Chen, M., Rubin, K. A., Barton, R. W.: Compound materials for reversible, phase-change optical data storage. Appl. Phys. Lett. **49**, 502-504 (1986)

[5.8] Miyamoto, M., Ushiyama, J., Umezawa, K., Kashiwakura, A., Tamura, R.: High-transfer-rate 4.7-GB DVD-RAM. 2002 International symposium on optical memory and optical data storage topical meeting ISOM/ODS Technical Digest, 416-418 (2002)

[5.9] Kageyama, Y., Iwasaki, H., Hariagaya, M., Ide, Y.: Compact disc erasable (CD-E) with Ag-In-Sb-Te phase-change recording material. Jpn. J. Appl. Phys. **35**, 500-501 (1996)

[5.10] van Pieterson, L., Rijpers, J. C. N., Hellmig, J.: Phase-change media for ultrahigh-speed digital versatile disc recording. Jpn. J. Appl. Phys. **43**, 4974-4977 (2004)

[5.11] Zhou, G. F., Borg, H. J., Rijpers, J. C. N., Lankhorst, M. H. R., Horikx, J. J. L: Crystallization behavior of phase-change materials: comparison between nucleaction- and growth-dominated crystallization. Proc. of SPIE **4090,** 108-115 (2000)

[5.12] Zhou, G. F.: Materials aspects in phase change optical recording. Mat. Sci. Eng. A, **A304-306,** 73-80 (2001)

[5.13] Borg, H. J., Blom, P. W. M., Jacobs, B. A. J., Tieke, B., Wilson, A. E., Ubbens, I. P. D., Zhou, G. F.: AgInSbTe materials for high-speed phase-change recording. Proc. SPIE **3864**, 191 (1999)

[5.14] Lankhorst, M. H. R., van Pieterson, L., van Schijndel, M., Jacobs, B. A. J., Rijpers, J. C. N.: Prospects of doped Sb-Te phase-change materials for high-speed recording. Jpn. J. Appl. Phys. **42,** 863-868 (2003)

[5.15] Christian, J. W.: The theory of transformations in metals and alloys. Pergamon Press (1965)

[5.16] Wickersham, C. E., Bajor, G., Greene, J. E.: Impulse stimulated 'explosive' crystallization of sputter deposited amorphous (In,Ga)Sb films. Solid State Comm. **27**, 17-20 (1978)

[5.17] Kaiser, N.: Crystallization of amorphous antimony films. Thin Solid Films **116**, 259-265 (1984)

[5.18] van Pieterson, L., Lankhorst, M. H. R., van Schijndel, M., Kuiper, A. E. T., Roosen, J. H. J.: Phase-change recording materials with a growth-dominated crystallization mechanism: A materials overview. J. Appl. Phys. **97**, 083520-1-083520-7 (2005)

[5.19] Lankhorst, M. H. R.: Modeling glass transition temperatures of chalcogenide glasses. Applied to phase-change optical recording materials. J. Non-Cryst. Solids, **297**, 210-219 (2002)

[5.20] Gravesteijn, D. J.: Materials developments for write-once and erasable phase-change optical recording. Appl. Opt. **27**, 736-738 (1988)

[5.21] Wooten, F.: Optical Properties of Solids. Academic Press, New York (1972)

[5.22] Sokolov, A. V.: Optical Properties of Metals. Blackie & Son limited, Glasgow (1961)

6. Scaling Properties of Phase Change Materials

Simone Raoux

Abstract. Optical storage based on phase change materials has been so successful because the data density was increased from generation to generation. Phase Change Random Access Memory will only be a viable technology when this trend of increased storage density can continue for several future lithography generations. This chapter reviews the scaling properties of the phase change materials themselves and explores the limit when size effects start to play a role influencing the crystallization temperature, melting temperature, crystallization speed and other material parameters that are vital for this technology.

6.1 Introduction

Moore's law, as stated in 1965 [6.1], predicted that the number of devices in a given chip area would double every year. However, Kurzweil [6.2] pointed out that Moore's law had been followed by calculating machines long before it was ever stated, ranging from mechanical calculating devices in the 1890s through the vacuum-tube-based computers that pre-date transistors. Thus the semiconductor industry has been following Moore's law of exponential growth (the actual time for the number of transistors to double is closer to 18 month), or Kurzweil's law of accelerating returns, during its entire existence. However, considerations about the ultimate limits of computing tell us that, if nothing else, the laws of physics at least will put an end to Moore's law. The ultimate computer is estimated to perform 10^{51} operations per second on 10^{31} bits [6.3], compared to 10^{10} operations per second on 10^{10} bits of today's laptops. This would indeed be an extreme machine, since the energy required for such a computer to work would correspond to operation at a billion degrees Kelvin, or such a computer might look like a black hole [6.3].

Thus, a critical aspect of any new semiconductor technology, such as phase change random access memory (PCRAM) technology in this case, is its capability for continuing to extend Moore's law, both explicitly (smaller memory devices) and implicitly (higher performance through faster devices). In this chapter the limits of scaling for phase change memory technology for the next generations of

S. Raoux, M. Wuttig (eds.), *Phase Change Materials*,
DOI 10.1007/978-0-387-84874-7_6, © Springer Science+Business Media, LLC 2009

lithographic technologies are explored. Scalability has many aspects to it. For example, for a PCRAM cell, the scalability of the access device (a diode or more typically a transistor) used to address the device can be as important as the scalability of the actual memory cell itself. In fact, in most PCRAM memory devices the cell size is determined by the access device, not the memory cell (see Chapters 15-17). For optical memory based on phase change materials (DVDs and rewritable CDs, see Chapter 12) the storage density is limited (among other factors such as the crystallization speed) by the wavelength of the laser light and the rotation speed of the discs, and these will not shrink (for the wavelength) or grow (for the speed) exponentially. Other optical methods that circumvent the limits of the wavelength are being investigated (Chapter 13). In this chapter it is assumed that tools and processes for the next lithographic technology nodes will be developed to drive feature sizes to smaller and smaller dimensions and circuits to faster and faster speed along the roadmap developed by the semiconductor industry and described in the International Technology Roadmap for Semiconductors (ITRS) [6.4]. This chapter concentrates only on the scaling properties of the phase change materials themselves. The other aspects of scaling are investigated in later chapters of this book and in [6.5-8]. It will be investigated how material properties are influenced when one dimension is reduced in size (thin films), two dimensions are reduced in size (nanowires), and all three dimensions are reduced in size (nanoparticles). Finally, it will be studied how far switching *speed* can be scaled.

6.2 Thin Films of Phase Change Materials

One advantage of using very thin films of phase change material for PCRAM applications is the fact that in the dimension of the film thickness the cell design is no longer limited by lithographic patterning. Thin films can be deposited with atomic layer precision through deposition techniques such as atomic layer deposition, but even simpler techniques such as sputter deposition can lead to almost atomically flat films with accurate thickness control. The later has been used to fabricate PCRAM devices with cross-sections of the phase change material which are limited only in one dimension by lithography, enabling very small cross-sectional areas of 3 nm x 20 nm [6.9]. The thickness dependence of the switching behavior of thin films of phase change materials was studied by a number of authors. It was found that several material properties such as the crystallization temperature T_x [6.10-12], crystallization rate [6.13-15], optical constants [6.16], electrical parameters and incubation times [6.11] change as film thickness is reduced. We will look at these changes in more detail in the next sections.

6.2.1 Crystallization Temperature as a Function of Film Thickness

The crystallization temperature is one of the most important properties of a phase change material as it has an impact on the data retention and thermal stability as well as the power required to crystallize the material. It depends on several factors such as the heating ramp rate, the surrounding materials (substrate and capping layer), the method used to measure it (sudden drop in resistivity, appearance of x-ray diffraction peaks, sudden change in reflectivity, appearance of an exothermic peak in a differential scanning calorimetry measurement etc.). For scalability studies, the dependency of T_x on the film thickness is important.

The crystallization temperature T_x of the amorphous semiconductors *a*-Ge and *a*-Si was studied as a function of film thickness using multilayer structures [6.17-24]. It was found that the crystallization temperatures are very strong functions of the material(s) that the amorphous semiconductor was sandwiched between. In many cases (when the semiconductor was sandwiched between oxides or nitrides), a strong increase in T_x was observed when the film thickness of the semiconductor was reduced below a certain thickness, typically < 20nm. This increased T_x can be fitted by an exponential function [6.17]

$$T_x = T_{ax} + \left(T_m - T_{ax}\right) e^{-d/C} \tag{6.1}$$

where T_{ax} is the crystallization temperature of the bulk material or a thick film, T_m is the melting point of the material, d is the film thickness, and C is a fitting constant.

Such a behavior was reported for *a*-Si/SiO_x [6.17, 6.18, 6.22], *a*-Si:H/Si_3N_4:H [6.23], *a*-Ge:H/GeN_x:H and *a*-Ge/GeN_x [6.20], and *a*-Ge/SiO_x [6.18, 6.19]. For very thin films, e.g. 1.5 nm of Ge sandwiched between SiO_x [6.19], no crystallization was observed up to 760 °C annealing. Similarly, for 5 nm thick Ge sandwiched between SiO_x no crystallization was observed up to 800 °C annealing (T_x of bulk Ge and thick films is about 500 °C) [6.18]. Zacharias and Streitenberger [6.18] assumed in their model for the crystallization of thin films that the crystallization starts in the middle of the thin semiconductor film. They proposed that the oxide or nitride interfaces inhibit the crystallization of a very thin adjacent interfacial layer in the semiconductor. If the total thickness of the semiconductor film comes close to twice the thickness of the interfacial layer, no crystallization will occur. Based on calculations of the Gibbs free energy they calculated the radius of the critical nucleus and the crystallization temperature as a function of film thickness resulting in (see [6.18] for the complete derivation of Eq. (6.2))

$$T_x = T_{ax}\left(1 + \frac{\gamma_{oc} - \gamma_{ac} - \gamma_{oa}}{\gamma_{ac}}\, e^{-d/4l_0}\right) \tag{6.2}$$

where T_x is the crystallization temperature of thin film, T_{ax} is the crystallization temperature of the bulk amorphous material or a thick film, γ_{oc} is the specific free energy of an oxide/crystalline interface, γ_{ac} is the specific free energy of an amorphous/crystalline interface, γ_{oa} is the specific free energy of an oxide/amorphous interface, d is the film thickness and l_0 is an average screening or bonding length related to the range of inter-atomic forces typical for the oxide and crystalline material. This Eq. (6.2) has the same form as the empirical Eq. (6.1) with

$$C = 4l_0 \tag{6.3}$$

A very different result was obtained by sandwiching the semiconductor between metals such as Al, Ag, Au, Bi, Cd, Cu, Pb, Sb, or Zn [6.21, 6.24]. In this case a drastic decrease in T_x was observed. For example T_x of Ge capped by the above mentioned metals decreases to 75 – 300 °C depending on the metal [6.24], and T_x was further decreased when the Ge film thickness was reduced below 10 nm [6.21]. The adjacent metal layer acts as a crystallization promoter providing nucleation sites. Thus, for thin amorphous films of Si and Ge, T_x is a strong function of film thickness and a very strong function of the adjacent material. Different adjacent materials can lead to a substantial increase or decrease of T_x as film thickness is reduced.

Thin films of Sb are typically crystalline as deposited. It was observed however that in the beginning of the deposition the films were amorphous and underwent an amorphous-crystalline phase transition at a critical film thickness that depends on the substrate, the substrate temperature, the capping layer material, the deposition rate, and the deposition pressure [6.25-27]. This critical thickness is in the range between very few to tens of nanometers. Similarly to Ge and Si, T_x is a function of the film thickness for Sb. It was found that capping layers of Cu, Ag, Au or Ni stimulate the crystal growth by forming compounds with Sb, capping layers of Bi, Pb, or Sn contribute to the crystallization of Sb without forming compounds, and Ti, Cr, Fe, Co, Ni-Fe alloy, Al, or Ge capping layers have little influence on the crystallization of thin Sb films [6.26].

The crystallization temperature as a function of film thickness was studied also for materials relevant for PCRAM and optical storage applications. Houle and co-workers [6.12] measured the resistivity as a function of temperature using thin films of $Ge_2Sb_2Te_5$ (GST) of various thicknesses on Si substrates coated with a 1 μm thick SiO_2 film and capped with 10 nm SiO_2 to prevent oxidation. It was found that the amorphous-fcc phase transition temperature increases from about 150 °C for 50 nm thick films to about 250 °C when the GST film thickness was reduced, and that the phase transition was more gradual for thinner films. The fcc-hexagonal phase transition was observed at about 340 °C for thick films, and this transition temperature did not change very much with film thickness (about 320 °C for thinner films). The thinnest film studied was 2.5 nm and it still showed a clear phase transition.

The crystallization behavior of thin layers of GST, nitrogen-doped GST, Sb_2Te, $Ge_{15}Sb_{85}$, and Ag- and In-doped Sb_2Te (AIST) sandwiched between SiO_2 and Al_2O_3 was studied using time-resolved x-ray diffraction (XRD) during sample annealing [6.10, 6.28]. Similarly to Ge and Si sandwiched between oxides and nitrides, it was found that T_x was increased exponentially as the film thickness was reduced below 20 nm. Figure 6.1 shows T_x as a function of film thickness for these materials.

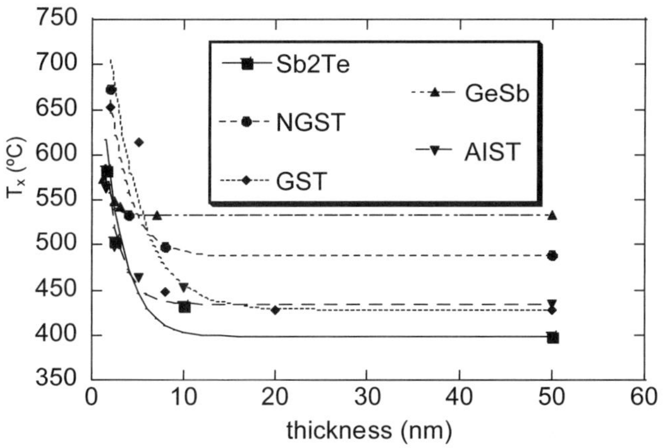

Fig. 6.1. Crystallization temperature T_x as a function of film thickness for GST, N-GST, Sb_2Te, $Ge_{15}Sb_{85}$, and AIST. Films were deposited on 1 μm thick SiO_2 film on Si, and capped with 10 nm Al_2O_3. T_x was determined by the appearance of x-ray diffraction peaks during heating at 1K/s in a He atmosphere.

The lines in Fig. 6.1 are fits of the data to Eq. (6.1). The thinnest films that still crystallized were between 1.3 – 2 nm depending on the material. While for most materials the crystal structure of the ultra-thin films was similar to thick films (typically with broadened XRD peaks due to the small film thickness and grain size, and sometimes with a change in texture), ultra-thin films of GST and nitrogen-doped GST showed a more complex behavior.

As the GST and nitrogen-doped GST films became thinner, the XRD peak intensity of the fcc phase became weaker and weaker, and for the thinnest films (below 5 nm) the amorphous films transformed directly into the hexagonal phase without going through the metastable fcc phase.

A similar exponential increase in T_x with reduced film thickness was observed for GST sandwiched between ZnS-SiO_2 layers by measuring the resistivity as a function of temperature [6.11]. It was also found that the phase transition is more gradual for thinner films and that incubation times are considerably longer for thinner films. Figure 6.2 shows the resistivity as a function of time for GST films isothermally heated at 143.5 °C. It is evident that the incubation time for crystallization (time between start of heating and start of reduction of resistivity) was increased by orders of magnitude as film thickness was reduced.

From measurements of T_x as a function of heating rate at a fixed film thickness it was found that the activation energy is also increased as film thickness is reduced, corresponding to an increase in the crystallization temperature.

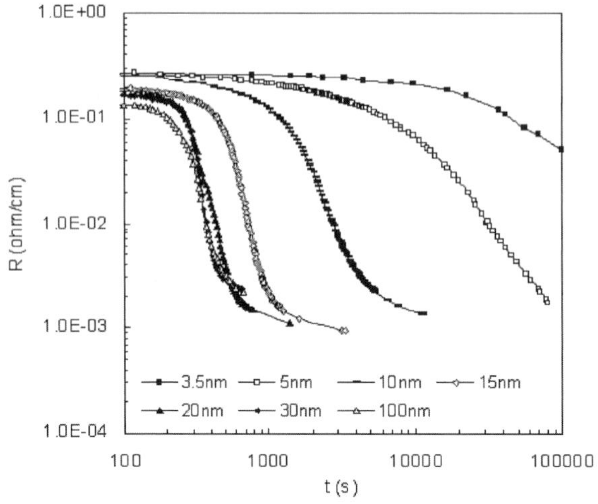

Fig. 6.2. Resistivity as a function of time at different thickness when GST thin films were heated at 143.5 °C. The incubation time increases from 100 s for 100 nm thick films to about 15 min for 10 nm to about 3 h for 3.5 nm thick films. The resistivity of the as-deposited films also increases as film thickness is reduced. Reprinted with permission from [6.11].

With reduced film thicknesses, interfaces become more and more important. It was mentioned already that interfaces have a large influence on the crystallization temperature for amorphous Ge and Si thin films. For Ge-Sb-Te films the effect of substrate and capping layer was studied by Ohshima [6.29]. It was found that different capping layers can either accelerate nucleation (Ta_2O_5, Si_3N_4), inhibit nucleation (SiO_2), or promote nucleation even in the grain growth phase (ZnS, ZnS-SiO_2). For $Ag_5In_6Sb_{59}Te_{30}$ films it was found that capping layers had a substantial effect on the activation energies for crystallization, either increasing it from 3.03 eV of the uncapped film to 3.24 eV for a Si_3N_4 capping layer, or decreasing it to 2.39 eV for a ZnS-SiO_2 cap layer [6.30]. In both cases the crystallization temperature was not strongly influenced by the capping layers.

Interdiffusion with adjacent materials was observed also for Ti/GST interfaces which will become more detrimental as film thicknesses will shrink [6.31, 6.32]. On the other hand it is possible to tailor the interfaces by adding a very thin TiO_x layer on the surface of the TiN electrodes [6.33], or introducing a very thin Ta_2O_5 layer at the phase change material– electrode interface [6.34] to improve device performance. The role of interfaces will certainly increase as film thicknesses are reduced.

Besides the changes in the crystallization behavior, changes in the structural relaxation (drift of the resistivity of the amorphous phase to higher values with time [6.35]) may occur with decreasing film thickness. Experiments on this aspect of scaling still need to be performed.

All these data must be considered with caution if the findings are to be implemented in actual optical storage or PCRAM devices. It has been shown that the properties of phase change materials such as the values for T_x or the crystallization speed can be considerably different for crystallization of the amorphous, as-deposited material compared to the re-crystallization of amorphous, melt-quenched material [6.36]. Bridge-type devices were fabricated on 1 μm thick SiO_2 on Si using 30 nm thick GST films capped by Al_2O_3. The electron-beam lithographic fabrication process was performed in such a way that the process temperatures did not reach T_x of GST, so that the devices were in the amorphous state after fabrication. The devices were 5 μm long and 500 nm wide. A second set of devices was heated above T_x so that the devices were in the fcc crystalline state. A laser pulse was used to produce a melt-quenched, amorphous mark of about 1 μm length (diameter of the laser beam) in the middle of the device. Measurements of resistivity as a function of time showed that T_x of the as-deposited device was about 150 ºC as expected from measurements on thin films, but T_x of the melt-quenched device was reduced by about 30 ºC. The incubation time of the melt-quenched devices was also remarkably reduced from 4500 s for as-deposited devices to 100 s for melt-quenched devices (at 110 ºC). In applications (both optical and PCRAM) the relevant material is the melt-quenched, amorphous state, and thus results obtained with as-deposited materials can give misleading answers if applied to optical storage or solid-state memory devices. One reason for the different behavior of as-deposited and melt-quenched films can be found in the increased structural order of the melt-quenched phase similarly to the so-called "primed" phase [6.37, 6.38]. Fluctuation electron microscopy [6.39] studies indicate the existence of nanocrystallites of a few nanometers in diameter that act as supercritical nuclei, and a higher medium range order in melt-quenched, amorphous films compared to as deposited, amorphous films [6.37]. High-resolution transmission electron microscopy (TEM) studies comparing as deposited and melt-quenched phase change material ($Ge_5Sb_{70}Te_{25}$) also showed very small nanocrystallites in both phases, but more frequently in the melt-quenched phase [6.40]. The existence of these nuclei can strongly influence the crystallization behavior.

6.2.2 Crystallization Rate as a Function of Film Thickness

Thickness dependent crystallization rates were observed for growth-dominated SbTe-based alloys [6.13, 6.15] and also for nucleation-dominated GeSbTe-based alloys [6.14, 6.15], and were studied by measuring complete erasure times (CET) of amorphous marks formed on previously initialized (crystallized) optical discs. It should be noted that CET is a function of mark size for growth-dominated materials, since the crystal growth occurs from the border between the amorphous mark and the surrounding crystalline matrix. In contrast, CET is not a strong function of mark size for nucleation-dominated materials because re-crystallization occurs by the formation of new crystals everywhere within the mark.

Martens and co-workers [6.13] found an optimum film thickness of 9 nm for a SbTe-based phase change material having the shortest CET of about 10 ns for a mark of about 60 nm diameter. This correlates to a maximum crystal growth velocity of about 5 m/s for this film thickness. The phase change material was sandwiched between the dielectric material ZnS-SiO_2 typically used in optical storage discs. For small marks (< 60 nm), the authors observed a linear dependence of CET as a function of mark radius, i.e. a constant crystal growth velocity, while for larger marks CET depended more strongly on the mark radius. This was explained by variations in temperature over the area of larger marks, leading to a temperature at the mark edges lower than the temperature of fastest crystal growth.

Zhou and Jacobs [6.14] found that CET values increased as film thicknesses were reduced below 30 nm for various nucleation-dominated materials ($Ge_2Sb_2Te_5$, $Ge_1Sb_2Te_4$, $Ge_1Sb_4Te_7$). No minimum in CET was observed here as a function of film thickness. By introducing a thin crystallization acceleration layer (SiC) between the phase change material and the ZnS-SiO_2 layer, it was possible to further reduce CET, from about 40 ns down to about 25 ns for 30 nm thick films. The fact that the adjacent material influences the crystallization speed is not a surprise considering the very strong influence these materials can have on the crystallization temperature and possibly also the actual crystallization mechanism (homogeneous nucleation versus heterogeneous nucleation, see Chapter 7).

In another paper, Zhou [6.15] compared the effect of film thickness on CET for both nucleation-dominated Ge-Sb-Te based materials as well as growth-dominated Ag-In-Sb-Te materials (both sandwiched between ZnS-SiO_2). These materials showed opposite thickness-dependence trends: while CET was increased when film thickness was reduced for Ge-Sb-Te based materials, CET was reduced for thinner Ag-In-Sb-Te materials. Unlike the 9 nm thick Sb-Te based films studied by Martens [6.13], a minimum CET was not observed, but this might imply that the thinnest films studied by Zhou were not thin enough to reach the minimum. This minimum might also be at a different thickness depending on the material. It is known from x-ray diffraction studies [6.10] that a minimum thickness of about 1.5 – 2 nm is required for phase change materials to crystallize at all, and it is reasonable to assume that close to this thickness the crystallization speed might be reduced, correlated also to an increased T_x. Again, it was found that the introduction of a thin SiC and also GeN layer at the interface between Ge-Sb-Te and ZnS-SiO_2 reduced CET considerably, by about a factor of 2.

Here again a word of caution. Besides large influences of the crystallization speed caused by layer thickness and interfaces, there is again a very big difference between the "first crystallization" of an as-deposited material, and subsequent re-crystallizations of melt-quenched material in terms of switching time [6.36]. This difference is more pronounced for growth-dominated materials, because for a "first crystallization" critical nuclei must first be produced during a long incubation time. However, in any practical re-crystallization experiment, the melt-quenched amorphous material will most likely border crystalline material (in a laser experiment and also in a PCRAM device), so that crystal growth can happen very fast from this interface. Figure 6.3 shows the difference in crystallization time for as-deposited and melt-quenched $Ge_{15}Sb_{85}$, a growth dominated material [6.41].

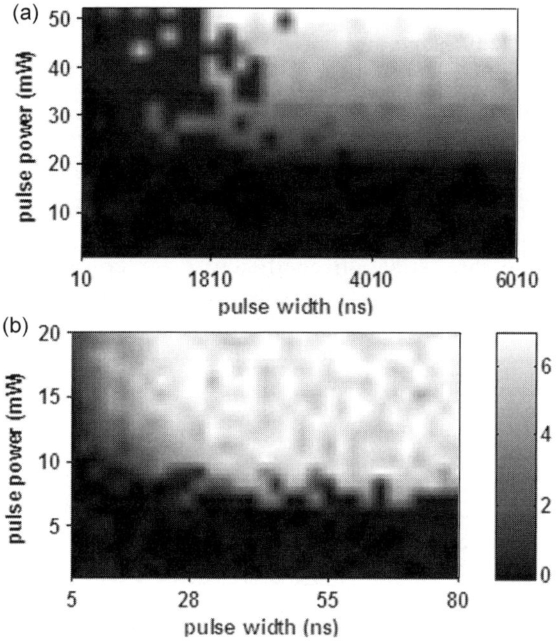

Fig. 6.3. Change in reflectivity as a function of laser power and duration (a) for as deposited $Ge_{15}Sb_{85}$, and (b) $Ge_{15}Sb_{85}$ film first annealed to 300 °C for 10 min above its T_x = 250 °C, exposed to a first laser pulse of fixed power of 40 mW and duration of 50 ns to produce a melt-quenched, amorphous mark, and exposed to a second laser pulse of variable pulse length and power attempting to re-crystallize the amorphous mark. In (a) there is a clear increase in reflectivity for times longer than about 2 μs and powers higher than 25 mW connected with first crystallization. The onset of crystallization is somewhat stochastic since this material has typically only one nucleation site per laser spot. In (b) we find very fast re-crystallization for times longer than 10 ns and power between 10 – 20 mW.

Shown in Figure 6.3a is the change in reflectivity after the as-deposited, amorphous sample was exposed to laser pulses of increasing power and duration. The increase in reflectivity for times longer than about 2 μs and powers higher than 25 mW was caused by the first crystallization. In Figure 6.3b data are shown for a sample that was first crystallized in a furnace, then exposed to identical laser pulses that produced melt-quenched marks, and then to second pulses of variable durations and powers to re-crystallize the melt-quenched spots. Very fast re-crystallization for times longer than 10 ns and powers between 10 – 20 mW was observed. There is a difference in crystallization speed by two orders of magnitude between the as-deposited and melt-quenched amorphous material. Crystallization speed data obtained on as-deposited films can be very misleading when judging a material for its applicability for data storage. The relevant experiment needs to be performed with the material of the same thickness and in contact with the same interfaces as it will be used in the actual discs or devices.

6.2.3 Change in Optical Constants and Electrical and Thermal Parameters as a Function of Film Thickness

Also the optical, electrical and thermal properties of phase change materials change when their thickness is reduced. Miao and co-workers [6.16] studied the optical properties of $Ge_1Sb_2Te_4$ as a function of film thickness. The optical parameters (refractive index n and extinction coefficient k) were obtained from ellipsometry measurements using a classical dispersion model for the data fitting. It was found that below about 50 nm film thickness the optical parameters are not constants but depend on the film thickness. As-deposited, amorphous and crystalline $Ge_1Sb_2Te_4$ films were investigated.

Figure 6.4 shows as an example the extinction coefficient k for crystalline $Ge_1Sb_2Te_4$ as a function of wavelength for various film thicknesses.

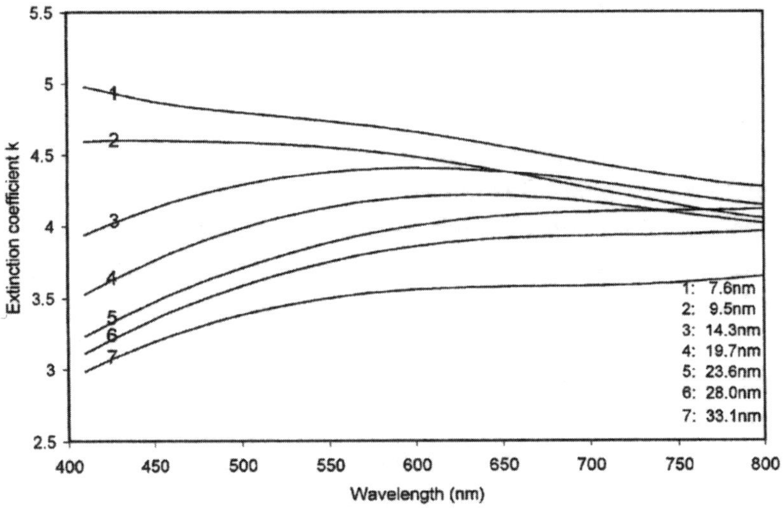

Fig. 6.4. Extinction coefficient of crystalline $Ge_1Sb_2Te_4$ as a function wavelength for different film thicknesses. Reprinted with permission from [6.16].

The extinction coefficient was higher for thinner films, and in particular at lower wavelengths, those of the operation of Blu-ray technology (see Chapter 12). The refractive index was also higher for thinner films for all wavelengths studied (400 - 800 nm).

Very little is known about the electrical properties of phase change films as the film thickness is reduced. From the work of Wei and co-workers [6.11] one can see that the resistivity of the thinnest amorphous, as-deposited films of GST is increased by a factor of about 2 compared to thick films (see Fig. 6.2). From the measurements of resistivity as a function of temperature shown in this paper one can also see that the thinner films (5 nm) have actually a lower resistivity in the crystalline phase compared to thicker films (30 nm) when heated to 190 ºC.

Houle and co-workers [6.12] observed a similar increase in the resistivity of the as-deposited, amorphous phase, but they found an increase of the resistivity of the crystalline phases also measured on GST. More work is needed to study the electrical properties of thin films. It is important that care is taken to investigate the influence of interfaces on the electrical properties since they become more and more important as film thickness is reduced.

Another concern is the behavior of the threshold switching as film thickness is reduced. The threshold switching is assumed to be related to the density of traps (see Chapter 14 for details), and if the phase change film becomes very thin the number of traps may become too small to allow for threshold switching. On the other hand, doping might help to increase the trap density to the required levels for threshold switching to occur.

Even less is known about the thermal conductivity of phase change materials as their thickness is reduced. Reifenberg and co-workers [6.42] studied the thermal conductivity of Ge-Sb-Te with thicknesses between 60 and 350 nm using nanosecond laser heating and thermal reflectance measurements. They found a substantial decrease in the thermal conductivity as film thickness is reduced (from 0.29, 0.42, 1.76 W/mK in the amorphous, fcc, and hexagonal phase for 350 nm thick films to 0.17, 0.28, 0.83 W/mK for 60 nm thick films). The influence of the stoichiometry on the thermal conductivity was found to be rather weak.

6.2.4 Limits of Storage Density in Thin Films

Very thin films of 6 nm Sb_2Te_3 and GeTe have been used for optical discs in a superlattice–like structure combining the properties of Sb_2Te_3 (fast switching) and GeTe (good thermal stability) to form a medium with excellent recording properties [6.43]. For PCRAM devices the thinnest films achieved were 3 nm of doped $Ge_{15}Sb_{85}$ in a bridge type cell structure with 20 nm width; these ultra-thin film bridge devices could repeatedly be switched and showed very good thermal stability properties [6.9]. A theoretical study of the capabilities of scanning probe based storage found that 10 – 30 nm small marks should be possible to write, read, and erase with relatively low energies of 100 pJ (crystallization) and 300 pJ (melt-quenching) for an optimized layer stack leading to storage densities of about 1.5 Tb/in^2 [6.44]. Indeed, Hamann and co-workers [6.45] demonstrated experimentally an even higher storage density of 3.3 Tb/in^2 using a heated atomic force microscope (AFM) tip on an 18 nm thick GST film.

Very small crystalline marks of 10 - 100 nm diameter (Fig. 6.5) were produced in an amorphous GST film as thin as 1 nm using an AFM [6.46]. The smallest marks had a very short lifetime of a few minutes while larger marks showed a much better thermal stability. Amorphous marks in a crystalline GST film were

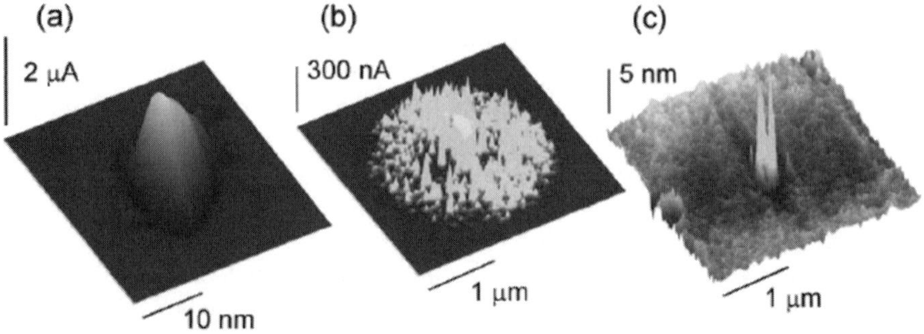

Fig. 6.5. Three-dimensional current images of marks produced in (a) a 1 nm thick film using a pulse of 0.6 V and 200 ns and (b) a 25 nm thick film using a pulse of 4 V and 10 ms. (c) is a topographic image of the mark in (b). Images obtained within 3 min after producing mark. Reprinted with permission from [6.46].

also produced using an AFM and a scanning tunneling microscope with mark sizes between 10 – 100 nm with very short pulses of 5 – 100 ns [6.47]. All these data show the very promising scaling capabilities of phase change materials.

From the thin film research we can already answer a number of important scaling questions. With shrinking dimensions we will find that:

- the crystallization temperature and activation energy for crystallization will be increased for as-deposited, amorphous phase change materials sandwiched between oxides or nitrides
- the incubation time will increase substantially for amorphous, as-deposited phase change materials sandwiched between oxides or nitrides
- at a certain films thickness in the range of 1 – 2 nm phase change materials will not crystallize anymore
- the crystallization speed for nucleation dominated materials (sandwiched between dielectrics) will be reduced
- the crystallization speed for growth-dominated materials (sandwiched between dielectrics) will first be increased until it reaches a maximum at an optimum thickness in the 10 nm thickness range
- the refractive index and extinction coefficient will increase
- at a certain size in the 10 nm range crystalline marks in an amorphous matrix or amorphous marks in a crystalline matrix will not be stable anymore over long periods of time.

All of these statements might not be true for all materials under all circumstances, but these are the results of the limited set of data available at the moment. In particular data are missing about the properties of very thin phase change films sandwiched between metals that might for example be used as the electrodes in PCRAM devices.

6.3 Phase Change Nanowires

Nanowires are very useful for studying the scaling properties of phase change materials because they can be grown in a controlled fashion with variable diameter, well-defined composition, and in most cases they are single-crystalline. The size of the nanowires can be small enough that size effects, for example, on the melting temperature, are observed. Thus, they offer an ideal tool to investigate scalability.

Nanowires of different phase change materials were grown by a number of groups (GeTe – [6.48-50], GeTe and Sb_2Te_3 – [6.51], In_2Se_3 – [6.52], GST – [6.53-55]). GeTe nanowires (Fig. 6.6) with thicknesses between 40 – 80 nm were grown by the thermal evaporation method using the vapor-liquid-solid mechanism with Au nanoparticles as the catalyst [6.48].

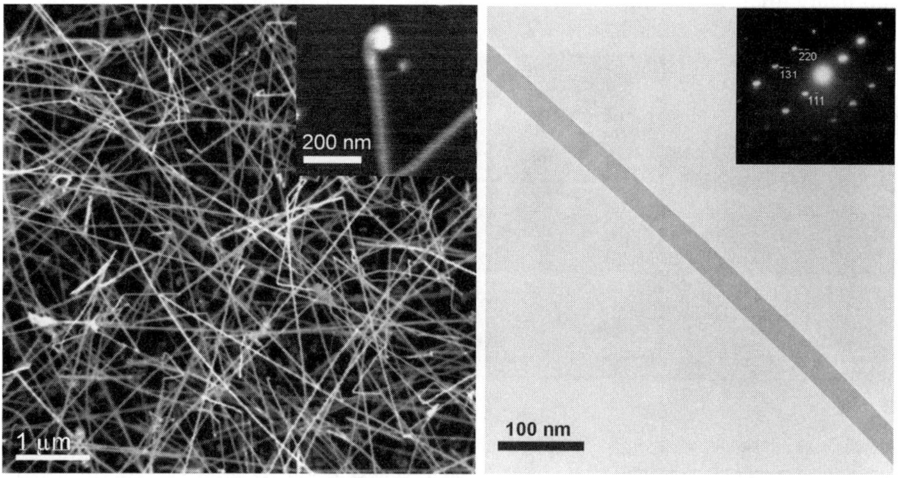

Fig. 6.6. Left: Scanning electron microscope (SEM) image of GeTe nanowires grown using 20 nm Au nanoparticles as catalyst. The scale bar presents 1 µm. Inset: Close-up view showing Au nanoparticle at tip of nanowire. The scale bar presents 200 nm. Right: High-magnification TEM of GeTe nanowires with 40 nm diameter. Inset: Selected area electron diffraction pattern for fcc lattice structure. Reprinted in part with permission from [6.48]. Copyright (2007) American Chemical Society.

The nanowires were single crystalline in the face-centered cubic (fcc) form with a lattice constant of 6.01 Å (the rhombohedral phase was not formed because the growth temperature was above the rhombohedral-fcc phase transition temperature) and contained only Ge and Te in a ratio close to 1:1. The melting temperatures of GeTe nanowires were measured in situ in a TEM and it was found that nanowires start to melt at 390 °C which is a substantially lower value than the melting temperature of bulk GeTe of 725 °C. This size effect on the melting point is advantageous for the application in PCRAM devices because the highest current to switch a device is required for the melt-quenching (RESET operation), and a reduced melting temperature will reduce the required RESET current. Interestingly, it was observed that the GeTe nanowires were covered by a 1 – 3 nm thin amorphous GeO_2 layer leading to an empty GeO_2 nanotube after the GeTe core was molten and evaporated during heating.

In_2Se_3 nanowires grown by the same technique [6.52] were also single-crystalline in an hexagonal lattice structure with lattice constants of a = 4.0 Å and c = 19.2 Å consistent with thin film data. The 40 – 80 nm diameter nanowires again showed a reduction in the melting point from 890 °C for bulk In_2Se_3 to 680 °C for nanowires.

Meister and co-workers [6.51] also grew GeTe nanowires by the vapor-liquid-solid method using Au nanoparticles as catalyst and obtained two types of GeTe nanowires: a straight type and a helical type (Fig. 6.7), both identified to be single-crystalline in the rhombohedral phase with diameters between 50 – 200 nm and many microns in length.

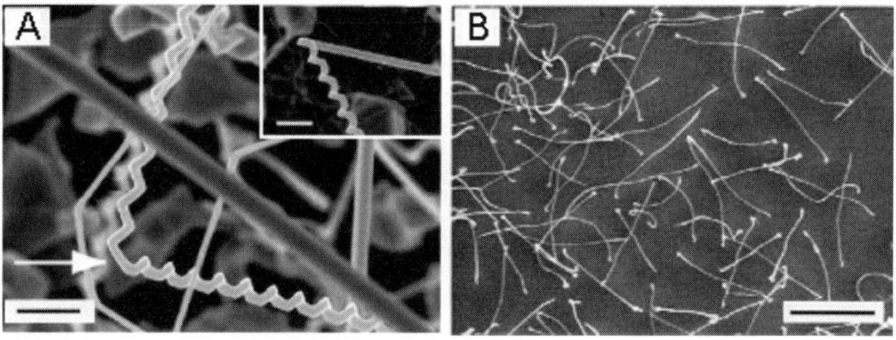

Fig. 6.7. A – High-resolution SEM of helical nanowires with two different pitch sections, the arrow indicates the pitch change point. Inset: Straight-helical nanowire. Scale bars are 500 nm. B – SEM image of thin and curly GeO_2 nanowires, scale bar represents 2 μm. Reprinted with permission from [6.51]. Copyright (2006) American Chemical Society.

The helical wires showed left and right chirality with equal probability and sometimes changed pitch. The presence of the rhombohedral phase was probably due to the lower growth temperature (370 °C for the best nanowires) that was below the rhombohedral-fcc phase transition temperature (432 °C for large single crystals of GeTe [6.56]) compared to 450 °C used by Sun and co-workers [6.48].

In addition to GeTe nanowires, curly nanowires of amorphous GeO_2 were also observed. Single-crystalline Sb_2Te_3 nanowires in the rhombohedral phase were also grown with diameters of about 70 nm.

Besides measurements of the melting point as a function of size, nanowires allow studying of their switching behavior. GST nanowires grown by the same vapor-liquid-solid technique were found to be either in the fcc phase [6.48, 6.53] or the hexagonal phase [6.54, 6.55] depending on the growth temperature. GST nanowires in the fcc phase were used to encode information on the wires by locally melt-quenching the wire using the scanning focused electron beam of a TEM [6.53]. It was possible to produce 25 nm long amorphous marks on a crystalline 50 nm diameter wire.

The ultimate test of the applicability of nanowires to data storage is the fabrication and testing of nanowire memory devices. Devices comprised of nanowires (rhombohedral GeTe [6.49, 6.50], hexagonal GST [6.54, 6.55]) were fabricated (and tested) by transferring the as-deposited nanowires to an oxidized Si substrate and using a focused ion beam to produce Pt contacts [6.49, 6.54, 6.55], or electron-beam lithography to produce Ni contacts [6.50]. It was possible to repeatedly switch the nanowire devices between a high and low resistance state for more than 10^5 times [6.55], and the devices showed the typical behavior of PCRAM devices such as threshold switching (Fig. 6.8).

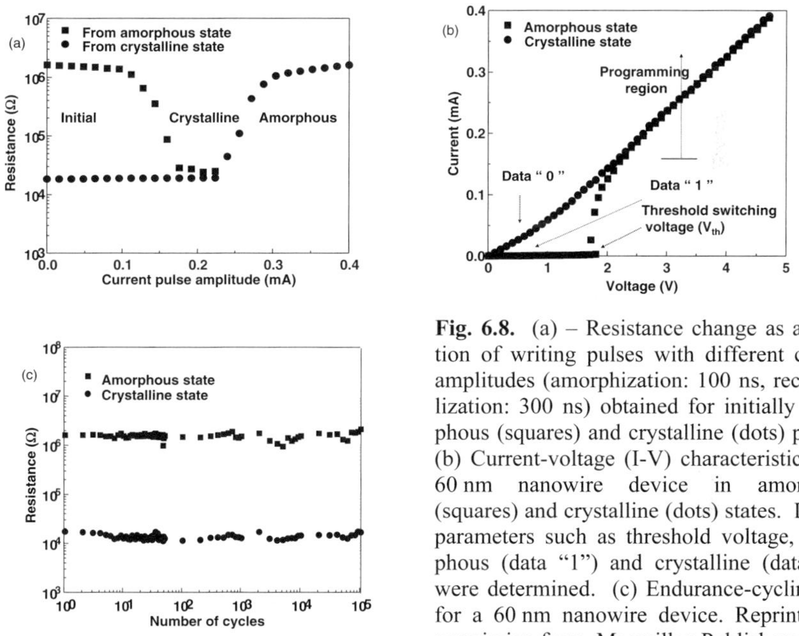

Fig. 6.8. (a) – Resistance change as a function of writing pulses with different current amplitudes (amorphization: 100 ns, recrystallization: 300 ns) obtained for initially amorphous (squares) and crystalline (dots) phases. (b) Current-voltage (I-V) characteristics of a 60 nm nanowire device in amorphous (squares) and crystalline (dots) states. Device parameters such as threshold voltage, amorphous (data "1") and crystalline (data "0") were determined. (c) Endurance-cycling test for a 60 nm nanowire device. Reprinted by permission from Macmillan Publishers Ltd: Nature Nanotechnology [6.55], copyright (2007).

The RESET current as function of nanowire diameter was determined and the data showed the expected reduction in RESET current as wire dimensions were reduced. The lowest RESET current was 0.16 mA for a 30 nm nanowire device. Data retention was also studied by measuring the activation energy for crystallization as a function of nanowire diameter. A substantial size effect was observed for the incubation time for recrystallization and for the related activation energies. While larger nanowire devices (200 nm) show an activation energy of about 2.34 eV the activation energy is reduced for smaller wires down to 1.98 eV for 30 nm nanowires devices. This corresponds to around 1800 years of data retention for 200 nm nanowire devices at 80 °C and around 3 years for 30 nm nanowires devices at 80 °C [6.55]. This is in contrast to thin film data [6.11] where an increase of the activation energy for crystallization was found with reduced film thickness. This difference may stem from the fact that the films were amorphous, as deposited while the nanowires were melt-quenched amorphous. The same is true for the incubation times that were found to be shorter for thinner nanowires [6.55] but longer for thinner films [6.11].

From the nanowire research we can answer again a number of important scaling questions. With shrinking dimensions we will find that:

- the melting temperature will be reduced
- the incubation time for re-crystallization will be shortened for melt-quenched amorphous nanowires
- the activation energy for re-crystallization will be reduced for melt-quenched amorphous nanowires connected with a deteriorated data retention
- the switching currents and powers in PCRAM devices will be reduced.

6.4 Phase Change Nanoparticles

Nanoparticles of phase change materials offer another tool to study scaling. One method to produce nanoparticles is by pulsed laser ablation [6.57-60]. Laser ablation is a relatively easy method to form phase change nanoparticles of a desired composition but it leads to nanoparticles with a relatively wide size distribution (5 – 25 nm [6.57], 4 – 30 nm [6.58], 5 - 50 nm [6.60], all for GST nanoparticles). Depending on their thermal treatment the GST particles can be either amorphous, in the fcc or the hexagonal phase. While the literature describes rather consistent data for the phase transitions in thick GST films (amorphous-fcc transition around 150 °C and fcc-hexagonal transition around 350 °C [6.10, 6.61, 6.62]), it is less consistent for GST nanoparticles. Yoon and co-workers [6.60] found that particles treated at 100 °C, 200 °C, 300 °C and 400 °C all showed the fcc crystalline phase. Choi and co-workers [6.58] on the other hand found that particles heated to 200 °C were still amorphous, particles heated to 300 °C showed both the fcc and hexagonal phase, while particles heated to 400 °C and 500 °C were only in the fcc phase. Similarly, it was observed by Park and co-workers [6.57] that as deposited nanoparticles were mostly amorphous, particles annealed at 100 °C were mostly hexagonal, and particles annealed at 200 °C were predominantly in the fcc phase. These findings are surprising because, as mentioned above, for thick GST films the fcc (metastable) phase is formed first during a heating cycle and the hexagonal (stable) phase later at higher temperatures. Very thin films however showed only the hexagonal phase forming at around 370 ºC [6.10, 6.62]. Clearly more work is needed to fully understand the crystallization behavior of nanoscaled GST. However it is promising that nanoparticles can be made that show phase transitions, and can exists in the amorphous and crystalline phase.

Devices have also been fabricated using laser-ablated GST nanoparticles as the storage medium [6.59]. A nanoparticle layer of about 100 nm consisting of multilayers of nanoparticles with an average diameter of 10 nm was deposited between two electrodes (Al and highly p-doped Si), and threshold switching was demonstrated.

Phase change nanoparticles were also fabricated from sputter deposited thin films using electron beam lithography [6.62-64]. A lithographic process was chosen in such a way that the process temperatures did not exceed the crystallization temperatures of the respective phase change materials, and the nanoparticles were in the amorphous state after fabrication. Phase change nanoparticles from GST, nitrogen-doped GST, $Ge_{15}Sb_{85}$, Sb_2Te, and AIST [6.62-64] were fabricated over large area (2 x 5 mm) and their crystallization behavior was investigated using time-resolved x-ray diffraction (XRD). The particle sizes were between 20 – 80 nm and encapsulated by about 8 nm Al_2O_3. Figure 6.9 shows a high-resolution TEM image of a $Ge_{15}Sb_{85}$ nanoparticle after heating in a purified He atmosphere to 450 ºC and one can see that the particle is polycrystalline. The Al_2O_3 encapsulation layer is also visible in Fig. 6.9. As fabricated particles were amorphous as confirmed by TEM, and all heated particles studied here were polycrystalline with crystal sizes of about 10 nm.

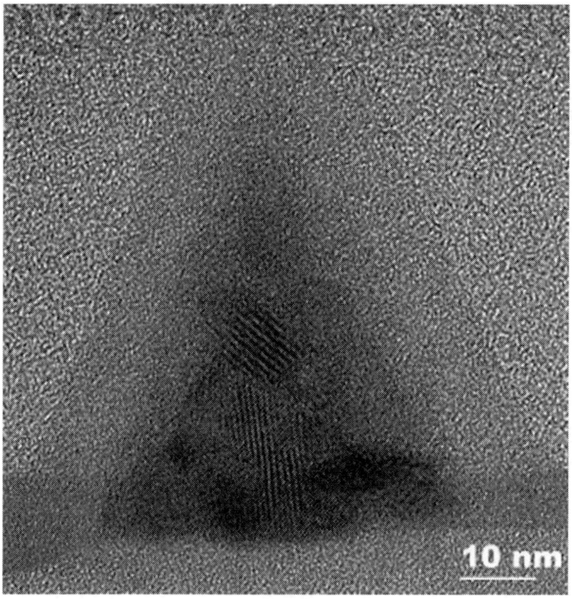

Fig. 6.9. High-resolution bright-field TEM image of annealed $Ge_{15}Sb_{85}$ nanoparticle. Reprinted with permission from [6.63]. Copyright (2007). American Institute of Physics.

Time-resolved XRD revealed that all nanoparticle arrays studied crystallized at temperatures which were similar to thick blanket films of the same material except Sb_2Te nanoparticles which crystallized at a temperature of about 40 °C higher (170 °C) compared to blanket film (130 °C). The crystalline structure of the nanoparticles was very similar to blanket films for $Ge_{15}Sb_{85}$, Sb_2Te, and N-GST (with some change in texture with the nanoparticles showing a texture more resembling a powder spectrum texture). For the smallest GST nanoparticles the fcc phase was formed but no hexagonal phase was observed for temperatures up to 450 °C while larger particles behaved similar to thick blanket film. This again is in contrast to ultra-thin blanket films showing the formation of the hexagonal phase only when films are thinner than 5 nm [6.10], but in agreement with some of the results on particles formed by laser ablation [6.60].

In an attempt to fabricate phase change nanoparticles that are smaller than the limits of electron-beam lithography self-assembly based techniques were used [6.65-68]. In one case the nanoparticles were produced by deposition of a self-assembling polymer (polystyrene-b-polyvinylpyridine) layer on a thin layer of the phase change material $Ge_{15}Sb_{85}$, and the pattern was transferred into the phase change layer forming isolated phase change nanodots [6.65, 6.66]. In a second approach [6.67] a self-assembling polymer (polystyrene-b-poly(methylmethacrylate)) was deposited on a Si substrate, followed by removal of the poly(methylmethacrylate) domains, sputter deposition of AIST into the holes followed by lift-off. Figure 6.10 shows nanodots made from $Ge_{15}Sb_{85}$ using the first method and nanodots made from AIST using the second method.

Fig. 6.10. Left - SEM image of isolated $Ge_{15}Sb_{85}$ nanoparticles fabricated by transferring the self-assembled diblock-copolymer pattern into the underlying $Ge_{15}Sb_{85}$ phase change material. Right – Nanoparticles of AIST fabricated by deposition into a diblock-copolymer-based template, and lift-off. Reprinted with permission from EPCOS from [6.67].

The nanoparticles were in the 15 nm size range (smaller than the smallest particles made by electron-beam lithography), and cross-sectional TEM confirmed that they were well isolated. Using the same self-assembling polymer and fabrication principles newly developed spin-on phase change material GeSbSe (in detail described in Chapter 11.3) was deposited and nanoparticles were fabricated again by lift-off [6.68].

The crystallization behavior of all these nanoparticle arrays was studied using time-resolved XRD. The nanoparticles were found to be amorphous after fabrication and it was demonstrated that these nanoparticles all showed clear amorphous-crystalline phase transitions. The crystallization temperature was about 15 ºC lower than thick films for $Ge_{15}Sb_{85}$ nanodots, and about 15 ºC higher for nanodots made from AIST. The nanoparticles made from spin-on GeSbSe crystallized at 220 ºC compared to thick films at 250 ºC.

The smallest phase change nanoparticles (Fig. 6.11) were described in a presentation by Caldwell and co-workers [6.69].

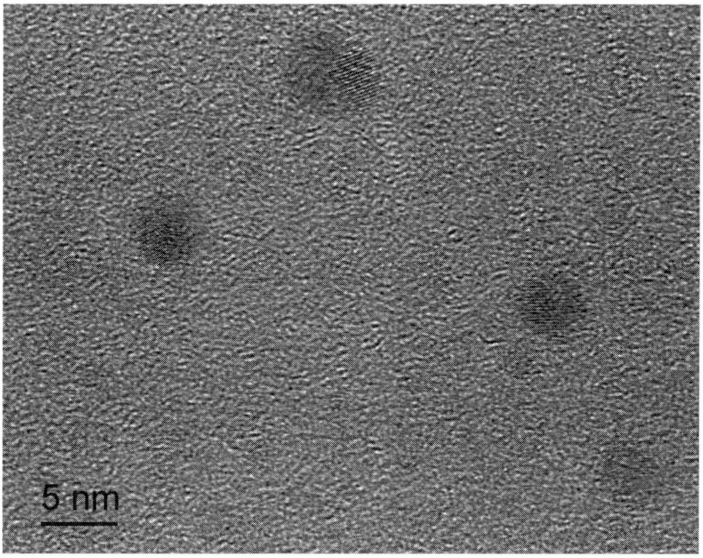

Fig. 6.11. TEM image of germanium selenide nanoparticles synthesized by solution-based chemistry [6.69].

These GeSe nanoparticles were synthesized using a solution-chemistry-based approach, and they were in the 1 – 8 nm diameter range. XRD studies showed that these particles were crystalline after synthesis.

GeTe nanoparticles of similar sizes (in the few nm range) were also synthesized using solution-based chemistry [6.70]. It was shown that the melting temperature of these particles depends on their size and is reduced for smaller particles compared to bulk GeTe. This is similar to the reduction of the melting temperature observed in GeTe nanowires [6.48] and GeTe thin films [6.71].

This is the size range where size effects should play an important role, and future studies of these very small nanoparticles will give further insight into the scaling behavior of these materials.

From the nanoparticle research we can answer again a number of scaling questions. With shrinking dimensions we will find that:

- nanoparticles show still good crystallization behavior for dimensions down to 15 nm with crystallization temperatures that are not too dramatically different from thick bulk films
- spin-on phase change materials offer an alternative to sputter-deposited phase change materials
- nanoparticles as small as a few nanometers can be crystalline, offering potential for devices scaled to very small dimensions.

6.5 Scaling in Time – Switching Speed of Phase Change Materials

When phase change random access memory was first discovered in the 1960s [6.72] the phase change memory alloys such as $Te_{48}Ar_{30}Si_{12}Ge_{10}$ described in the first papers and inventions had a relatively long crystallization time on the order of tens of microseconds. Melt-quenching is an inherently fast process, the limits are determined by the time in which the material can be melted (determined by the electrical pulse or by the power and duration of the laser pulse, and by the thermal environment), and the quenching time is determined also by the fall time of the electrical pulse and the thermal environment. It has been demonstrated that melt-quenching can be performed in sub-nanosecond time scales in PCRAM devices [6.73]. In addition to conventional thermal melting it has been demonstrated that intense femtosecond laser pulses can lead to ultra-fast, non-thermal melting within times of less than a picosecond caused by production of a dense electron-hole plasma leading to ultra-fast atomic disordering [6.74].

The relatively long crystallization (SET) times led to a period without much activity in the development of phase change technology until fast crystallizing alloys based on materials on the GeTe-Sb_2Te_3 pseudo-binary line were discovered that were found to crystallize in about 50 ns [6.75]. Fastest SET times in PCRAM devices have been recently reported in the 1 – 10 ns range [6.73]. Picosecond laser pulses (30 ps) were used to measure crystallization speed and melt-quenching times of GST thin films in an optical disc multilayer stack, and it was also found that melt-quenching can be performed at times shorter than 1 ns while re-crystallization can be achieved in about 13 ns [6.76]. Essential to these short switching times was the optimization of the multilayer structure. Again, there was a large difference observed between the minimum crystallization time for as-deposited (100 ns) and melt-quenched (10 ns) GST thin films [6.77].

Afonso and co-workers studied Sb and non-stoichiometric Sb-rich $Ge_{1-x}Sb_x$ alloys with $x > 0.85$ (above the eutectic point corresponding to $x = 0.85$) using fast laser pulses in the 170 fs to 8 ns range [6.78-81]. These alloys were chosen because they crystallize at very short time scales below 15 ns so that melt-quenching in many cases was not possible which confined these materials to write-once applications [7.82]. It was shown that very fast pulses of 400 fs can lead to an increase in reflectivity or decrease caused by ablation for higher pulse powers [6.78]. A more detailed study [6.79] indicated that the initial increase in reflectivity within less than a picosecond is caused by melting of the surface (the molten material has the same reflectivity as the crystalline state). Between 1 ps and 29 ns the irradiated spot underwent a transition regime with a reflectivity between the value of the amorphous and crystalline or molten phase, and that a final stage was reached only after 29 ns corresponding to a crystalline mark in the as-deposited, amorphous film. However, a later study [6.83] suggested that this ultrafast transition produced a nonequilibrium disordered state rather than the thermodynamically stable crystalline phase. By optimizing film thickness and substrate Solis and Afonso [6.80] demonstrated the amorphization and re-crystallization of Sb-rich materials within about 400 ps using a 30 ps laser for the switching. Wiggins et al. [6.81] showed that the crystallization and amorphization times depend not only on the films thickness, substrate, and laser pulse duration, but also on the wavelength, and for optimized conditions crystallization times of 650 ps could be achieved.

Wang and co-workers [6.84] demonstrated a very large increase in reflectivity (about 250%) in a very short time (2 ps) after irradiation of a 80 nm thick GST film by a 130 fs laser pulse. This transient non-equilibrium state lasts for hundreds of ps and a crystalline state is finally reached with an increased reflectivity of about 20% typical for many phase change alloys.

The fastest electrical switching that has been achieved so far for PCRAM devices was also reported by Wang and co-workers [6.85]. They fabricated PCRAM devices using near-field optical lithography and $Ge_2Sb_2Te_5$ as the phase change material. The contact diameter of the phase change material to the lower electrode was between hundreds of nanometers and 19 nm which is well below the limits of conventional lithography. It was found that both, the fastest SET and RESET pulses able to switch the cells, were function of the cell size and decreased as cell size was reduced. The fastest SET / RESET pulses varied from 90 ns / 1.5 ns for 500 nm cells to only 2.5 ns / 0.4 ns for 19 nm cells.

From the time-resolved switching studies we can conclude that:

- phase change materials can be molten in a non-thermal melting process within times of about 1 ps
- phase change materials can be found that have crystallization times of about 500 ps under optimum conditions in terms of thickness, substrate, laser power, duration, and wavelength
- much faster data rates could potentially be achieved for the writing process (amorphization) in optical discs as well as for the RESET operation in PCRAM devices, while the SET operation seems to be limited to times in

the hundreds of picoseconds. So far, no material has been found that crystallizes in much shorter times than that (switched optically or electrically). Much shorter laser pulses can however be used to induce crystallization (even though the actual crystallization occurs after the laser pulse) so that the data transfer rate for erasure could potentially also be increased (with a "dead" time before reading can be performed).

All experiments directed towards scaling indicate that PCRAM and optical storage based on phase change materials will not be limited by the performance of the phase change materials themselves for several future technology nodes.

References

[6.1] Moore, G.: Cramming more and more components onto integrated circuits. Electronics **38**, No. 8, April 19 (1965)

[6.2] Kurzweil, R.: The Age of Spiritual Machines. Penguin Books, New York (1999)

[6.3] Lloyd, S.: Ultimate physical limits to computation. Nature **406**, 1047-1054 (2000)

[6.4] http://www.itrs.net/. Accessed 30 November 2007

[6.5] Raoux, S., Burr, G. W., Breitwisch, M. J., Rettner, C. T., Chen, Y.-C., Shelby, R. M., Salinga, M., Krebs, D., Chen, S. H., Lung, H.-L., Lam, C. H.: Phase change random access memory — a scalable technology. IBM J. Res. Develop. (2008), in print

[6.6] Pirovano, A., Lacaita, A. L., Benvenuti, A., Pellizzer, F., Hudgens, S., Bez, R: Scaling analysis of phase-change memory technology. Int. Electron Devices Meeting, Washington, DC (2003)

[6.7] Shi, L. P, Chong, T. C.: Nanophase change for data storage applications. J. Nanosci. Nanotechnol. **7**, 65-93 (2007)

[6.8] Raoux, S., Rettner, C. T., Jordan-Sweet, J. L., Chen, Y.-C., Zhang, Y., Caldwell, M., Wong, H.-S. P., Milliron,. D., Cha, J.: Scaling properties of phase change materials. Non-Volatile Memory Symposium, Albuquerque, pp. 30- 34 (2007)

[6.9] Chen, Y.-C., Rettner, C. T., Raoux, S., Burr, G. W., Chen, S. H., Shelby, R. M., Salinga, M., Risk, W. P., Happ, T. D., McClelland, G. M., Breitwisch, M., Schrott, A., Philipp, J. P., Lee, M. H., Cheek, R., Nirschl, T., Lamorey, M., Chen, C. F., Joseph, E., Zaidi, S., Yee, B., Lung, H. L., Bergmann, R., Lam, C.: Ultra-thin phase-change bridge memory device using GeSb. Int. Electron Devices Meeting, Technical Digest, San Francisco, CA, pp. 777-780 (2006)

[6.10] Raoux, S., Jordan-Sweet, J. L., Kellock, A. J.: Crystallization properties of ultra-thin phase change films. J. Appl. Phys. **103**, 114310 (2008)

[6.11] Wei, X., Shi, L., Chong, T. C., Zhao, R., Lee, H. K.: Thickness-dependent nanocrystallization in $Ge_2Sb_2Te_5$ films and its effect on devices. Jpn. J. Appl. Phys. **46**, 2211-2214 (2007)

[6.12] Houle, F. A., Raoux, S., Shelby, R., Kellock, A., Deline, V. A., Chen, Y.-C., Rettner, C. T.: Chemical structure and switching behavior of ultrathin GeSbTe phase change films. Mater. Res. Soc. Spring Meeting, San Francisco (2006)

[6.13] Martens, H. C. F., Vlutters, R., Prangsma, J. C.: Thickness dependent crystallization speed in thin phase change layers used for optical recording. J. Appl. Phys. **95**, 3977-3983 (2004)

[6.14] Zhou, G.-F., Jacobs, B. A. J.: High performance media for phase change optical recording. Jpn. J. Appl. Phys. **38**, 1625-1628 (1999)

[6.15] Zhou, G.-F.: Materials aspects in phase change optical recording. Mater. Sci. Eng. A **304-306**, 73-80 (2001)

[6.16] Miao, X. S., Chong, T. C., Huang, Y. M., Lim, K. G., Tan, P. K., Shi, L. P.: Dependence of optical constants on film thickness of phase-change media. Jpn. J. Appl. Phys. **38**, 1638-1641 (1999)

[6.17] Zacharias, M., Bläsing, J., Veit, P., Tsybeskov, L., Hirschman, K., Fauchet, P. M.: Thermal Crystallization of amorphous Si/SiO_2 superlattices. Appl. Phys. Lett. **74**, 2614-2616 (1999)

[6.18] Zacharias, M. and Streitenberger, P.: Crystallization of amorphous superlattices in the limit of ultrathin films with oxide interfaces. Phys. Rev. B **62**, 8391-8396 (2000)

[6.19] Williams, G. V. M., Bittar, A., Trodahl, H. J.: Crystallization and diffusion in progressively annealed *a*-Ge/SiO_x superlattices. J. Appl. Phys. **67**, 1874-1878 (1990)

[6.20] Honma, I., Hotta, H., Kawai, K., Komiyama, H., Tanaka, K.: The structural stability of reactively-sputtered amorphous multilayer films. J. Non-Cryst. Solids **97/98**, 947-950 (1987)

[6.21] Homma, H., Schuller, I. K., Sevenhans, W., Bruynseraede, Y.: Interfacially initiated crystallization in amorphous germanium films. Appl. Phys. Lett. **50**, 594-596 (1987)

[6.22] Persans, P. D., Ruppert, A., Abeles, B.: Crystallization kinetics of amorphous Si/SiO2 superlattice structures Source: J. Non-Cryst. Solids **102**, 130-135 (1988)

[6.23] Miyazaki, S., Ihara, Y., Hirose, M.: Structural stability of amorphous semiconductor superlattices. J. Non-Cryst. Solids **97/98**, 887-890 (1987)

[6.24] Oki, F., Ogawa, Y., Fujiki, Y.: Effect of deposited metals on the crystallization temperature of amorphous germanium film. Jpn. J. Appl. Phys. **8**, 1056 (1969)

[6.25] Stiddard, M. H. B.: This films of antimony on metal substrates: crystallite orientation and critical thickness for the occurrence of the amorphous-crystalline phase transition. J. Mater. Sci. Lett. **4**, 1157-1159 (1985)

[6.26] Hashimoto, M., Niizeki, T., Kambe, K.: Effect of substrate temperature on crystallization of amorphous antimony film. Jpn. J. Appl. Phys. **19**, 21-23 (1980)

[6.27] Hashimoto, M. and Hamano, T.: The stability of the amorphous phase in an Sb layer vacuum deposited on the air- and vacuum-cleaved NaCl and the effects of Sb thickness and overdeposits of Ag, Au, Sn, and Pb. Vacuum **40**, 445-448 (1990)

[6.28] Raoux, S., Jordan-Sweet, J. L. and Kellock, A.: Thickness-dependent crystallization behavior of phase change materials. Mater. Res. Soc. Spring Meeting, San Francisco, CA, March 2008

[6.29] Ohshima, N.: Crystallization of germanium-antimony-tellurium amorphous thin film sandwiched between various dielectric protective films. J. Appl. Phys. **79**, 8357-8363 (1996)

[6.30] Njoroge, W. K., Dieker, H., Wuttig, M.: Influence of dielectric capping layers on the crystallization kinetics of $Ag_5In_6Sb_{59}Te_{30}$ films. J. Appl. Phys. **96**, 2624-2627 (2004)

[6.31] Alberici, S. G., Zonca, R., Pashmakov, B.: Ti diffusion in chalcogenides: a TooF-SIMS depth profile characterization approach. Appl. Surf. Sci. **231-232**, 821-825 (2004)

[6.32] Cabral, Jr, C., Chen, K. N., Krusin-Elbaum, L.: Irreversible modification of Ge2Sb2Te5 phase change material by nanometer-thin Ti adhesion layers in a device-compatible stack. Appl. Phys. Lett. **90**, 051908 (2007)

[6.33] Kang, D.-H., Kim, I. H., Jeong, J.-H., Cheong, B.-K., Ahn, D.-H., Lee, D., Kim, H.-M. and Kim, K.-B.: An experimental investigation on the switching reliability of a phase change memory device with oxidized TiN electrode. J. Appl. Phys. **100**, 054506 (2006)

[6.34] Matsui, Y., Kurotsuchi, K., Tonomura, O., Morikawa, T., Kinoshita, M., Fujisaki, Y., Matsuzaki, N., Hanzawa, S., Terao., M., Takaura, N., Moriya, H., Iwasaki, T., Moniwa, M. and Koga, T.: Ta_2O_5 interfacial layer between GST and W plug enabling low power operation of phase change memories. IEDM Tech. Dig., 769-772 (2006)

[6.35] Ielmini, D., Lavizzari, S., Sharma, D. And Lacaita, A.: Physical interpretation, modeling and impact on phase change memory (PCM) reliability of resistance drift due to chalcogenide structural relaxation. IEDM Tech. Dig. 939-942 (2007)
[6.36] Chen, Y.-C., Rettner, C. T., Raoux, S., Burr, G. W., Shelby, R., Salinga, M.: Crystallization kinetics of as-deposited and melt-quenched phase-change materials. Mat Res. Soc. Spring Meeting, San Francisco (2007)
[6.37] Kwon, M.-H., Lee, B.-S., Bogle, S. N., Nittala, L. N., Bishop, S. G., Abelson, J. R., Raoux, S., Cheong, B.-K., Kim, K.-B.: Nanometer-scale order in amorphous $Ge_2Sb_2Te_5$ analyzed by fluctuation electron microscopy. Appl. Phys. Lett. **90**, 021923 (2007)
[6.38] Lee, B.-S., Raoux, S., Shelby, R. M., Rettner, C. T., Burr, G. W., Bogle, S., Bishop, S. G., Abelson, J. R.: Detecting nuclei in phase change materials by Fluctuation Electron Microscopy (FEM): An experimental proof of nucleation theory. Europ. Phase Change and Ovonic Sci. Symp., Zermatt, Switzerland, September 2007
[6.39] Voyles, P. M. and Abelson, J. R.: Medium-range order in amorphous silicon measured by fluctuation electron microscopy. Sol. Energy Mater. Sol. Cells **78**, 85-113 (2003)
[6.40] Naito, M., Ishimaru, M., Hirotsu, Y., Takashima, M.: Local structure analysis of Ge-Sb-Te phase change materials using high-resolution electron microscopy and nanobeam diffraction. J. Appl. Phys. **95**, 8130-8135 (2004)
[6.41] Shelby, R. M., Houlse, F. A., Raoux, S.: Phase-change dynamics of eutectic GeSb alloy. Mat. Res. Soc. Spring Meeting, San Francisco, April 2006
[6.42] Reifenberg, J. P., Panzer, M. A., Kim, S.-B., Gibby, A. M., Zhang, Y., Wong, S., Wong, H.-S. P., Pop, E. And Goodson, K. E.: Thickness and stoichiometry dependence of the thermal conductivity of GeSbTe films. Appl. Phys. Lett. **91**,111904 (2007)
[6.43] Chong, T. C., Shi, L. P., Qiang, W., Tan, P. K., Miao, X. S., Hu, X.: Superlattice-like structure for phase change optical recording. J. Appl. Phys. **91**, 3981-3987 (2002)
[6.44] Wright, D., Armand, M., Aziz, M. M.: Terabit-per-square-inch data storage using phase-change media and scanning electrical nanoprobes. IEEE Trans. Nanotechnol. **5**, 50-61 (2006)
[6.45] Hamann, H. F., O'Boyle, M., Martin, Y. C., Rooks, M., Wickramasinghe, H. K.: Ultra-high-density phase-change storage and memory. Nature Mater. **5**, 383-387 (2006)
[6.46] Gotoh, T., Sugawara, K., Tanaka, K.: Minimal phase-change marks produced in amorphous $Ge_2Sb_2Te_5$. Jpn. J. Appl. Phys. **43**, L818-L821 (2004)
[6.47] Satoh, H., Sugawara, K., Tanaka, K.: Nanoscale phase changes in crystalline Ge2Sb2Te5 films using scanning probe microscopy. J. Appl. Phys. **99**, 024306 (2006)
[6.48] Sun, X., Yu, B., Ng, G., Meyyappan, M.: One-dimensional phase-change nanostructure: Germanium telluride nanowires. J. Phys. Chem C **111**, 2421-2425 (2007)
[6.49] Lee, S.-H., Ko, D.-K., Jung, Y., Agarwal, R.: Size-dependent phase transition memory switching behavior and low writing currents in GeTe nanowires. Appl. Phys. Lett. **89**, 223116 (2006)
[6.50] Yu, D., Wu, J., Gu, Q., Park, H.: Germanium telluride nanowires and nanohelices with memory-switching behavior. J. Am. Chem. Soc. **128**, 8148-8149 (2006)
[6.51] Meister, S., Peng, H., McIlwrath, K., Jarausch, K., Zhang, X. F., Cui, Y.: Synthesis and characterization of phase-change nanowires. Nano Lett. **6**, 1514-1517 (2006)
[6.52] Sun, X., Yu, B., Ng, G., Nguyen, T. D., Mayyappan, M.: III-VI compound semiconductor indium selenide (In_2Se_3) nanowires: Synthesis and characterization. Appl. Phys. Lett. **89**, 233121 (2006)
[6.53] Sun, X., Yu, B., Meyyappan, M.: Synthesis and nanoscale thermal encoding of phase-change nanowires. Appl. Phys. Lett. **90**, 183116 (2007)
[6.54] Jung, Y., Lee, S.-H., Ko, D.-K., Agarwal, R.: Synthesis and characterization of $Ge_2Sb_2Te_5$ nanowires with memory switching effect. J. Am. Chem. Soc. **128**, 14026-14027 (2006)

[6.55] Lee, S.-H., Jung, Y., Agarwal, R.: Highly scalable non-volatile and ultra-low power phase-change nanowires memory. Nature Nanotechnol. **2**, 626-630 (2007)

[6.56] Chattopadhyay, T., Boucherle, J. X., von Schnerig, H. G.: Neutron diffraction study on the structural phase transition in GeTe. J. Phys. C: Solid State Phys. **20**, 1431-1440 (1987)

[6.57] Park, G.-S., Kwon, J.-H., Jo, W., Kim, T. K., Zuo, J.-M., Khang, Y.: Crystalline and amorphous structures of Ge-Sb-Te nanoparticles. J. Appl. Phys. **102**, 013524 (2007)

[6.58] Choi, H. S., Seol, K. S., Takeuchi, K., Fujita, J. and Ohki, Y.: Sythesis and size-controlled $Ge_2Sb_2Te_5$ nanoparticles. Jpn. J. Appl. Phys. **44**, 7720-7722 (2005)

[6.59] Suh, D.-S., Lee, E., Kim, K. H. P., Noh, J.-S., Shin, W.-C., Kang, Y.-S., Kim, C., Khang, Y.: Nonvolatile switching characteristics of laser-ablated $Ge_2Sb_2Te_5$ nanoparticles for phase-change memory applications. Appl. Phys. Let.. **90**, 023101 (2007)

[6.60] Yoon, H. R., Jo, W., Lee, E. H., Lee, J. H., Kim, M., Lee, K. Y. And Khang, Y.: Generation of phase-change Ge-Sb-Te nanoparticles by pulsed laser ablation. J. Non-Crystalline Solids **351**, 3430-3434 (2005)

[6.61] Friedrich, I., Weidenhof, V., Njoroge, W., Franz, P., Wuttig, M.: Structural transformations of Ge2Sb2Te5 films studied by electrical resistance measurements. J. Appl. Phys. **87**, 4130-4134 (2000)

[6.62] Raoux, S., Rettner, C. T., Jordan-Sweet, J. L., Deline, V. R., Philipp, J. B., Lung, H.-L.: Scaling properties of phase change nanostructures and thin films. Europ. Phase Change and Ovonic Science Symp., Grenoble, France (2006)

[6.63] Raoux, S., Rettner, C. T., Jordan-Sweet, J. L., Kellock, A. J., Topuria, T., Ride, P. M., Miller, D.: Direct observation of amorphous to crystalline phase transitions in nanoparticle arrays of phase change materials. J. Appl. Phys. **102**, 094305 (2007)

[6.64] Raoux, S., Rettner, C. T., Jordan-Sweet, J. L., Salinga, M., Toney, M.: Crystallization behavior of phase change nanostructures. Europ. Phase Change and Ovonic Science Symp., Cambridge, UK (2005)

[6.65] Zhang, Y., Wong, H.-S. P., Raoux, S., Cha, J. N., Rettner, C. T., Krupp, L. E., Topuria, T., Milliron, D., Rice, P. M., Jordan-Sweet, J. L.: Phase change nanodots arrays fabricated using self-assembly diblock copolymer approach. Appl. Phys. Lett. **91**, 013104 (2007)

[6.66] Cha, J., Zhang, Y., Wong, H.-S. P., Raoux, S., Rettner, C., Krupp, L. and Deline, V.: Biomimetic approaches for fabricating high-density nanopatterned arrays. Chem. Mater. **2007**, 839-843 (2007)

[6.67] Raoux, S., Zhang, Y., Milliron, D., Cha, J. Caldwell, M. Rettner, C. T., Jordan-Sweet, J. L., Wong, H.-S. P.: X-ray diffraction studies of the crystallization of phase change nanoparticles produced by self-assembly-based techniques. Europ. Phase Change and Ovonic Science Symp., Zermatt, Switzerland (2007)

[6.68] Milliron, D. J., Raoux, S., Shelby, R. M., Jordan-Sweet, J.: Solution-phase deposition and nanopatterning of GeSbSe phase-change materials. Nature Mater. **6**, 352-356 (2007)

[6.69] Caldwell, M., Raoux, S., Milliron, D. J., Wong, H.-S. P.: Synthesis and characterization of germanium chalcogenide nanoparticles via single-source precursors and coprecipitation. 234th Am. Chem Soc. Meeting, Boston (2007)

[6.70] Milliron, D.: Solution-phase deposition of phase change material. Mater. Res. Soc. Spring Meeting, San Francisco (2007)

[6.71] Raoux, S., Shelby, R. M., Jordan-Sweet, J., Munoz, B., Salinga, M., Chen, Y.-C., Shih, Y.-H., Lai, E.-K. and Lee, M.-H.: Phase change materials and their application to Random Access Memory Technology. Europ. Mater. Res. Soc. Spring Meeting, Strasbourg, France (2008)

[6.72] Ovshinsky, S. R.: Reversible electrical switching phenomena in disordered structures. Phys. Rev. B **21**, 1450-1453 (1968)

[6.73] Shi, L. P., Chong, T. C., Zhao, R., Wei, X. Q., Wang, W. J., Li, J. M., Lim, K. G., Yang, H. X., Lee, H. K.: Investigation on high density and high speed phase change random access memory. Non-Volatile Memory Symposium, Albuquerque, pp. 129- 130 (2007)

[6.74] Rousse, A., Rischel, C., Fourmaux, S., Uschmann, I., Sebban, S., Grillon, G., Balcou, Ph., Förtser, E., Geindre, J. P., Audebert, P., Gauthiers, J. C., Hulin, D.: Non-thermal melting in semiconductots measured at femtosecond resolution. Nature **410**, 65-68 (2001)

[6.75] Yamada, N., Ohno, E., Nishiuchi, K., Akahira, N., Takao, M.: Rapid-phase transitions of GeTe-Sb_2Te_3 pseudobinary amorphous thin films for an optical disk memory. J. Appl. Phys. **69**, 2849-2856 (1991)

[6.76] Siegel, J., Schropp, A., Solis, J., Alfonso, C. N.: Rewritable phase change optical recording in $Ge_2Sb_2Te_5$ films induced by picosecond laser pulses. Appl. Phys. Lett. **84**, 2250-2252 (2004)

[6.77] Weidenhof, V., Friedrich, I., Ziegler, S., Wuttig, M.: Laser induced crystallization of amorphous $Ge_2Sb_2Te_5$ films. J. Appl. Phys. **89**, 3168-3176 (2001)

[6.78] Solis, J., Afonso, C. N., Hyde, S. C. W., Barry, N. P., French, P. M. W.: Existence of electronic excitation enhanced crystallization in GeSb amorphous thin films upon ultrashort laser pulse irradiation. Phys. Rev. Lett. **76**, 2519-2522 (1996)

[6.79] Sokolowski-Tinten, K., Solis, J., Bialkowski, J., Siegel, J., Afonso, C. N., von der Linde, D.: Dynamics of ultrafast phase changes in amorphous GeSb films. Phys. Rev. Lett. **81**, 3679-3682 (1998)

[6.80] Solis, J. and Afonso, C. N.: Ultrashort-laser-pulse-driven rewritable phase-change optical recording in Sb-based films. Appl. Phys. A **76**, 331-338 (2003)

[6.81] Wiggins, S. M., Bonse, J., Solis, J., Afonso, C. N., Sokolowsi-Tinten, K., Temnov, V. V., Zhou, P., van der Linde, D.: The influence of wavelength on phase transformations induced by picosecond and femtosecond laser pulses in GeSe thin films. J. Appl. Phys. **98**, 113518 (2005)

[6.82] Gravesteijn, D. J.: Materials developments for write-once and erasable phase-change optical recording. Appl. Otics **27**, 736-738 (1988)

[6.83] Callan, J. P., Kim, A. M.-T., Roeser, C. A. D., Mazur, E., Solis, J., Siegel, J., Afonso, C. N. and de Sande, J. C. G.:. Ultrafast laser-induced phase transitions in amorphous GeSb films. Phys. Rev. Lett. **86**, 3550-3653 (2001)

[6.84] Wang, Q. F., Shi, L., Huang, S. M., Mioa, X. S., Wong, K. P. And Chong, T. C.: Dynamics of ultrafast crystallization in as-deposited $Ge_2Sb_2Te_5$ films. Jpn. J. Appl. Phys. **43**, 5006-5008 (2004)

7. Crystallization Kinetics

Johannes A. Kalb

Abstract The classical theory of steady state crystal nucleation is discussed, as originally developed by Gibbs, Volmer, Weber, Becker, Döring, Turnbull and Fisher. A particular focus is drawn on the implications of heterogeneous nucleation sites, which can increase the homogeneous nucleation rate by many orders of magnitude. Classical theory of crystal growth is covered as well.

In Sect. 7.2, these theories are applied to measurements of nucleation and growth parameters in amorphous and liquid phase change materials by calorimetry and microscopy. The results contribute to a better understanding of the kinetics of the phase transformation in these materials, which helps to develop next-generation phase change media and to scale them to smaller dimensions.

7.1 Theory

Two processes are involved in the crystallization of a liquid or an amorphous solid (hereafter: parent phase):

1. First, crystallization is initiated by crystal nucleation. In the simplest case, this occurs in the interior of the parent phase, i. e., without the involvement of a "foreign" substance. This is called *homogeneous nucleation* (Sect. 7.1.1). If the parent phase is in contact with a foreign substance that acts as a preferred nucleation site (like an impurity or a container wall), nucleation is called *heterogeneous* (Sect. 7.1.2).
2. Subsequently, a stable crystal cluster grows to macroscopic size (crystal growth, Sect. 7.1.3).

7.1.1 Homogeneous Crystal Nucleation

The basic concept for nucleation theory was provided by Gibbs in 1878 (Sect. 7.1.1.1, [7.1-7.3]). This early treatment is still purely thermodynamic and describes cluster formation of a new phase inside the parent phase. Using this approach, the first *kinetic* model for nucleation was proposed by Volmer and Weber in 1926 (Sect. 7.1.1.2, [7.2-7.4]) and has served as a basis for a further improve-

S. Raoux, M. Wuttig (eds.), *Phase Change Materials*,
DOI 10.1007/978-0-387-84874-7_7,

ment by Becker and Döring in 1935 (Sect. 7.1.1.3, [7.2, 7.3, 7.5]). Finally, in 1949, Turnbull and Fisher obtained an expression for the pre-exponential factor of the nucleation rate in a condensed phase (Sect. 7.1.1.4, [7.2, 7.3, 7.6]). Today, the theory is collectively known as the *classical nucleation theory* by Volmer, Weber, Becker, Döring, Turnbull and Fisher.

7.1.1.1 Thermodynamics of Cluster Formation (Gibbs, 1878)

In the parent phase, atoms approach each other statistically, forming crystalline clusters by thermodynamic fluctuations. For simplicity, these clusters are assumed spherical with radius r. The equilibrium cluster distribution is then given by Boltzmann statistics [7.7],

$$N^{equ}(r) = N_0 \cdot \exp\left(-\frac{\Delta G_{cluster}(r)}{k_B T} \right), \tag{7.1}$$

where $\Delta G_{cluster}(r)$ is the reversible work for crystal cluster formation, k_B the Boltzmann constant, T the absolute temperature, N_0 the total number of atoms in the liquid and $N^{equ}(r)$ the number of clusters of radius r at equilibrium (N_0 and $N^{equ}(r)$ are normalized per unit volume). $\Delta G_{cluster}(r)$ can be expressed as a sum of two contributions [7.1-7.3]:

$$\Delta G_{cluster}(r) = -\Delta G_{lc,V} \cdot \frac{4}{3}\pi r^3 + \sigma \cdot 4\pi r^2 . \tag{7.2}$$

$\Delta G_{lc,V}(T)$ is the (Gibbs) free energy difference between the parent and the crystalline phase per unit volume. It is zero at the melting temperature T_m and defined to be positive below T_m. $\Delta G_{lc,V}(T)$ increases with increasing undercooling below T_m, i. e., it increases with decreasing temperature T.[1] The second term in Eq. (7.2) results from the creation of an interface between the cluster and the liquid. This term is positive and therefore energetically not favorable. The quantity $\sigma > 0$ is the interfacial free energy (hereafter: interfacial free energy). An additional elastic strain energy term due to the density change upon crystallization can be neglected in Eq. (7.2) unless the parent phase is very viscous [7.2].

Figure 7.1 qualitatively displays the evolution of $\Delta G_{cluster}(r)$. The curve passes through a maximum, which can be obtained by solving $\frac{\partial \Delta G_{cluster}(r)}{\partial r} = 0$. The

[1] Note that a complete description for alloys would employ the liquidus temperature T_l instead of the melting temperature T_m, which would go beyond the scope of this chapter [7.8].

maximum occurs because the surface-to-volume ratio is large for small clusters. The position of the maximum is (Fig. 7.1)

$$r_c = \frac{2\sigma}{\Delta G_{lc,V}}, \tag{7.3}$$

and the height is

$$\Delta G_c := \Delta G_{cluster}(r_c) = \frac{16\pi}{3}\frac{\sigma^3}{\left(\Delta G_{lc,V}\right)^2}. \tag{7. 4}$$

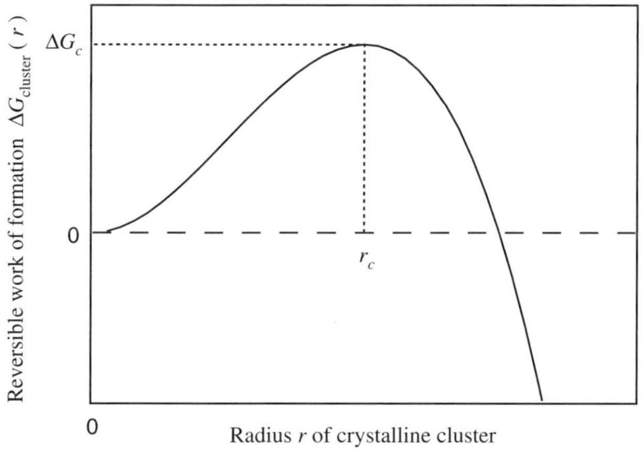

Fig. 7.1. Reversible work $\Delta G_{cluster}$ (r) for the formation of crystalline clusters of radius r in the parent phase ($T < T_m$). Clusters decay for $r < r_c$ and grow for $r > r_c$.

The quantity r_c is called the *critical radius*, which is on the order of nanometers [7.2, 7.3]. A cluster of radius r_c is called a *critical cluster* and ΔG_c (T) the *critical work for cluster formation.* $\Delta G_{cluster}$ (r) increases for $r < r_c$. Hence, clusters of size $r < r_c$ are energetically not favorable and spontaneously decay. However, for $r > r_c$, clusters grow due to a gain in free energy. Therefore, ΔG_c can be considered an activation barrier against crystallization. The existence of this barrier enables undercooling of a liquid below T_m without immediate crystallization. Simply speaking, nucleation in this early Gibbs treatment is the formation of post-critical clusters of size $r > r_c$.

7.1.1.2 Model Based on Equilibrium Distribution of Clusters (Volmer and Weber, 1926)

Volmer and Weber developed the first *kinetic* model for nucleation. Their model utilizes the equilibrium cluster distribution [Eq. (7.1)]. However, since $N^{equ}(r)$

becomes unphysical for $r > r_c$ (cluster number increases with increasing radius r, dashed in Fig. 7.2), it was ignored and set to zero: $N^{equ}(r > r_c) = 0$ [7.2, 7.3].

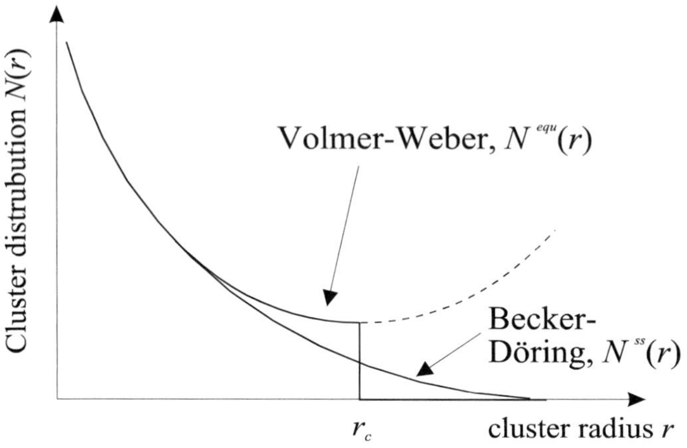

Fig. 7.2. Statistical cluster distribution function. $N^{equ}(r)$: Volmer-Weber model [equilibrium, Eqs. (7.1) and (7.2)]. $N^{ss}(r)$: Becker-Döring model (steady state). r_c is the radius of the critical cluster. $N^{ss}(r_c)$=1/2·$N^{equ}(r_c)$. The dashed part of the curve was ignored by Volmer and Weber: $N^{equ}(r > r_c) = 0$.

For $r = r_c$,

$$N^{equ}(r_c) = N_0 \cdot \exp\left(-\frac{\Delta G_c}{k_B T}\right). \tag{7.5}$$

Volmer and Weber assumed that nucleation occurs when a critical cluster acquires one more atom. The nucleation rate I^{equ} (dimension: 1/m^{3}s) is then given by [7.2, 7.3]

$$I^{equ} = s_c \cdot k \cdot N^{equ}(r_c) = s_c \cdot k \cdot N_0 \cdot \exp\left(-\frac{\Delta G_c}{k_B T}\right), \tag{7.6}$$

where k is the arrival rate of parent phase atoms to the critical crystalline cluster (number of arrivals per atom on the cluster surface per unit time, dimension: 1/s). s_c is the number of surface atoms in the critical cluster.

7.1.1.3 Steady State Model (Becker and Döring, 1935)

The Volmer-Weber model assumes that a critical cluster grows to macroscopic size as soon as it becomes post-critical by the addition of one more parent phase atom. Its main deficiency is that it neglects that post-critical clusters ($r > r_c$) can still decay with a certain probability (though they are more likely to grow), and that critical clusters ($r = r_c$) grow or shrink with equal probability (since $\partial \Delta G_{cluster}(r) / \partial r|_{r=r_c} = 0$). Becker and Döring have argued that the true cluster distribution $N^{ss}(r)$ in steady state (Fig. 7.2) does not abruptly fall to zero at $r = r_c$, but takes a value of $N^{ss}(r_c) = 1/2\, N^{equ}(r_c)$ and decreases gradually to zero for large cluster sizes [7.2, 7.3, 7.5]. For small cluster sizes, N^{ss} approaches N^{equ}.

After a few additional assumptions, Becker and Döring obtain the following expression for the steady state nucleation rate I^{ss} [7.2, 7.3]:

$$I^{ss} = s_c \cdot k \cdot N_0 \cdot \underbrace{\frac{1}{i_c} \cdot \left(\frac{\Delta G_c}{3\pi k_B T} \right)^{\frac{1}{2}}}_{\Gamma_z} \cdot \exp\left(-\frac{\Delta G_c}{k_B T} \right), \tag{7.7}$$

where i_c is the number of atoms in the critical cluster. Equation (7. 7) differs from the Volmer-Weber equation [Eq. (7.6)] only by the Zeldovich factor Γ_z, which only has a weak temperature dependence. Based on analytical and numerical studies, Γ_z is between 1/100 and 1/10 in most cases [7.2, 7.3, 7.9].

As the nucleation rate is far more sensitive to slight changes in ΔG_c than to the exact value of the pre-exponential factor [7.2], the Becker-Döring expression [Eq. (7.7)] is essentially identical to the Volmer-Weber expression [Eq. (7.6)] for all practical purposes. However, the importance of the Becker-Döring theory is that the kinetic problem has been treated correctly [7.2].

7.1.1.4 The Kinetic Pre-factor of the Nucleation Rate (Turnbull and Fisher, 1949)

Volmer, Weber, Becker and Döring originally developed their theories for the case of a *gaseous* parent phase (i. e., vapor condensation). For this case, the arrival rate k [Eq. (7.7)] is readily obtained by the theory of gases (not discussed here, [7.2]). However, Turnbull and Fisher were the first to evaluate the pre-exponential factor in Eq. (7.7) for crystal nucleation in an undercooled liquid or an amorphous phase [7.2, 7.3, 7.6]. They differentiated between two limiting cases: diffusion-limited and collision-limited crystallization kinetics. In both cases, the composition of the liquid and the crystalline cluster are the same. "Diffusion-limited" therefore refers to the nature of a *local* rearrangement (diffusive jump of an atom across the liquid-crystalline interface), *not* to the presence of a long-range diffusion field.

1. For diffusion-limited crystallization, changes of neighbors and/or coordination number are necessary for crystallization. This usually applies to metallic alloys, ionic materials, covalent materials and is also observed for phase change materials as discussed in more detail in Sect. 7.2. The frequency of diffusive

jumps k [Eq. (7.7)] across the interface per interface atom is according to Turnbull and Fisher

$$k = \frac{6D}{\lambda^2} \qquad \text{(diffusion-limited)} \qquad (7.8)$$

where D is associated with the diffusivity in the liquid or amorphous phase (not in the crystal). The parameter $\lambda = \Omega^{1/3}$ is the average interatomic distance (Ω= atomic volume).

2. For collision-limited crystallization, atomic neighbors generally do not have to change by diffusive rearrangements upon crystallization. Instead, atomic movement from the liquid to the crystalline cluster can be accomplished by thermal vibration. Hence, crystallization is governed by the collision of the atoms. This usually occurs only in pure metals and in van der Waals bonded materials but not in phase change materials as discussed in more detail in Sect. 7.2. The arrival rate constant k [Eq. (7.7)] is then equal to the collision rate at which the atoms attempt to join the crystalline cluster:

$$k = \frac{u_{\text{sound}}}{\lambda} \qquad \text{(collision-limited).} \qquad (7.9)$$

u_{sound} is the sound velocity in the liquid or amorphous phase, which is characteristic for the vibrational motion of the atoms.

Substituting Eq. (7.8) into Eq. (7.7) gives the nucleation rate for diffusion-limited crystallization:

$$I^{ss} = s_c \cdot \frac{6D}{\lambda^2} \cdot N_0 \cdot \Gamma_z \cdot \exp\left(-\frac{\Delta G_c}{k_B T}\right) \qquad \text{(diffusion-limited).} \qquad (7.10)$$

Approximating the diffusivity D locally with an Arrhenius equation, $D \propto \exp\left[-E_D / (k_B T)\right]$, where $E_D = -\partial \ln D / \partial\left[1/(k_B T)\right]$ is the (local) activation energy of the diffusivity and k_B the Boltzmann constant, gives the (local) activation energy $E_{I^{ss}}$ of the steady-state nucleation rate I^{ss}:

$$E_{I^{ss}} = E_D + \Delta G_c \qquad \text{(diffusion-limited).} \qquad (7.11)$$

It is often helpful to express I^{ss} in terms of the liquid shear viscosity η, which is easier to obtain experimentally than the liquid diffusivity D. Using the Stokes-Einstein equation,

$$\eta D = \frac{k_B T}{3\pi\lambda}, \tag{7.12}$$

which relates D and η, gives

$$I^{SS} = s_C \cdot \frac{2k_B T}{\eta\pi\lambda^3} \cdot N_0 \cdot \Gamma_Z \cdot \exp\left(-\frac{\Delta G_c}{k_B T}\right) \quad \text{(diffusion-limited).} \tag{7.13}$$

Note that Eq. (7.12) implies that the (local) activation energies of diffusivity E_D and viscosity E_η are equal if the viscosity is approximated (locally) by an Arrhenius equation, $\eta \propto \exp\,[E_\eta / (k_B T)]$:

$$E_D = E_\eta. \tag{7.14}$$

The linear term in T on the right side of Eq. (7.12) varies slowly with T compared to the exponential terms on the left side and therefore does not contribute to Eq. (7.14). The Stokes-Einstein equation has been found to hold for a large variety of undercooled liquids. Violations of this relation have been reported only for some fragile liquids [7.10-7.13]. Substituting $N_0 = 1/\lambda^3 = 1/\Omega$ and estimating the pre-exponential factor by $N_0 \sim 10^{28}$ m^{-3}, $s_c \sim 10$, $T \sim 1000$ K, and $\Gamma_z \sim 1/100$ [7.2], this gives

$$I^{SS} = \frac{10^{36}}{\eta} \exp\left(-\frac{\Delta G_c}{k_B T}\right) \frac{1}{\text{m}^3\text{s}} \quad \text{(diffusion-limited),} \tag{7.15}$$

where η is the numerical value of the liquid shear viscosity in units of poise.[2]

Substituting Eq. (7.9) into Eq. (7.7) gives the nucleation rate for collision-limited crystallization,

$$I^{SS} = 10^{39} \exp\left(-\frac{\Delta G_c}{k_B T}\right) \frac{1}{\text{m}^3\text{s}} \quad \text{(collision-limited),} \tag{7.16}$$

where $\lambda \sim 3$ Å and $u_{\text{sound}} \sim 1000$ ms^{-1} have been used as a typical example [7.2].

The uncertainty of the pre-exponential factors in Eqs. (7.15) and (7.16) is about two to four orders of magnitude [7.2]. However, as the exponential term varies so rapidly with ΔG_c upon undercooling, the value of ΔG_c that is required to give a fixed nucleation rate at a specific temperature is insensitive to the exact

---

[2] 1 Pa s =10 poise.

value of the pre-exponential factor [7.2, 7.3]. This applies even more to the interfacial energy σ, since it is raised to the third power in the exponential.

Equations (7.15) and (7.16) coincide for practical purposes at low undercooling (i. e., at a temperature slightly below T_m), where the viscosity is roughly independent of temperature (Fig. 7.3) and usually on the order of 10^{-1} - 10^{-3} poise. Both equations have been used widely and successfully to model experimental data on crystal nucleation in undercooled liquids [7.2, 7.8, 7.14-7.18].

The nucleation rate I^{ss} for diffusion-limited crystallization [Eq. (7.15)] becomes negligibly small close to the melting point T_m, where $\Delta G_c\ (T_m) = \infty$ [since $\Delta G_{lc,V}\ (T_m) = 0$, Eq. (7.4)], and close to the glass transition temperature T_g, where the viscosity strongly increases upon cooling from the liquid (Fig. 7.3).

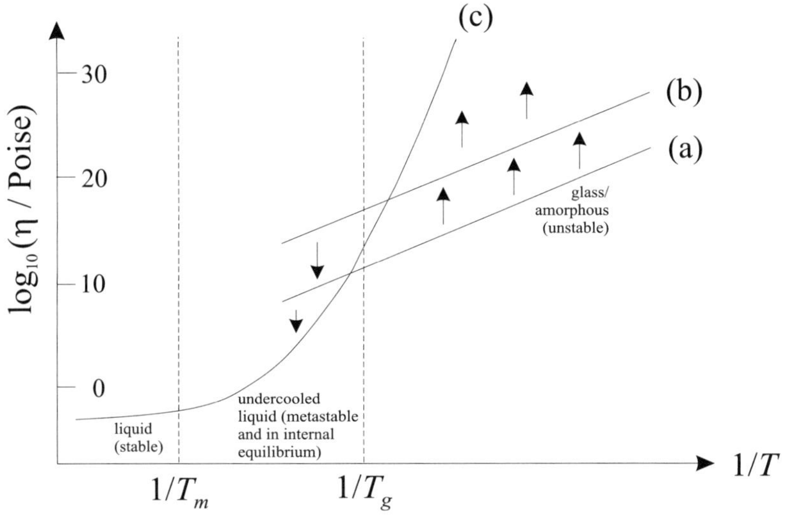

Fig. 7.3. Viscosity η in various stability regimes (T_g: glass transition temperature, which marks the boundary between a glass (amorphous phase) and an undercooled liquid; T_m: melting temperature). Stable equilibrium for $T > T_m$, metastable equilibrium (undercooled liquid) for $T_g < T < T_m$ and amorphous (frozen isoconfigurational states, non-equilibrium) for $T < T_g$. The amorphous states are unstable with respect to structural relaxation, i. e., they alter their structure towards the equilibrium structure of the undercooled liquid, which is indicated by the arrows. The slower the cooling rate, the longer the metastable equilibrium can be maintained upon cooling: (a) Fast cooling. (b) Slow cooling. (c) Hypothetically infinitively slow cooling, which maintains metastable equilibrium at all temperatures [7.19-7.21]. Therefore, T_g depends on the timescale of the experiment, but it usually occurs at a temperature at which the viscosity adopts a value on the order of 10^{12} Pa s = 10^{13} poise [7.22].

Therefore, I^{ss} exhibits a pronounced maximum at a temperature between T_g and T_m. Such a maximum is not present for collision-limited kinetics, where I^{ss} increases continuously upon cooling [Eq. (7.16)], which is not observed for phase change materials (Sect. 7.2).

7.1.2 Heterogeneous Crystal Nucleation

Homogeneous nucleation is an *intrinsic* process. In practice, however, homogeneous nucleation is difficult to identify specifically. Usually, foreign phases like container walls and impurities aid in the nucleation process and thereby increase the nucleation rate. In this case, nucleation is called *heterogeneous*. Heterogeneous nucleation is therefore an *extrinsic* process and can be influenced by the experimental conditions [7.9].

The simplest model for heterogeneous nucleation is due to Volmer [7.23, 7.24]. The model is based on the Gibbs model (Sect. 7.1.1.1) but with a flat substrate, which acts as a heterogeneous nucleation site (Fig. 7.4).

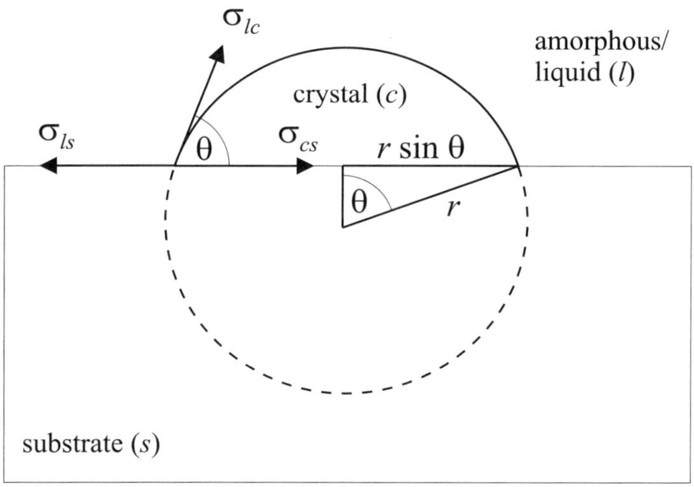

Fig. 7.4. Volmer's spherical cap model for heterogeneous cluster formation. The exposed crystal cluster has the shape of a spherical cap, whose volume is $f(\theta) \cdot 4/3\pi r^3$. The quantities σ_{cs}, σ_{ls} and σ_{lc} are the interfacial energies between the three substances involved. The dashed circle segment is a guide for the eye.

Under the assumption that the phases are isotropic, the interface between the crystalline cluster and the liquid (or amorphous) parent phase must have the same curvature everywhere. Hence, the crystalline cluster grows on the flat substrate like a spherical cap of radius r (Fig. 7.4). As a function of the wetting angle θ (Fig. 7.4), the exposed volume fraction $0 \leq f(\theta) \leq 1$ relative to a sphere of the same radius r is [7.24]

$$f(\theta) = \frac{(2 + \cos\theta)\,(1 - \cos\theta)^2}{4}. \tag{7.17}$$

Based on Gibbs' approach [Eq. (7.2)], Volmer [7.23, 7.24] could show that heterogeneous nucleation is preferred over homogeneous nucleation if

$$\sigma_{cs} - \sigma_{ls} < \sigma_{lc}, \tag{7.18}$$

where σ_{cs}, σ_{ls} and σ_{lc} denote, respectively, the crystal-substrate, liquid-substrate and liquid-crystal interfacial energies (Fig. 7.4). In this case, the reversible work for cluster formation *per atom*, $\Delta G_{\text{cluster}}$ (Fig. 7.1), is lower than if the circular cluster surface $\pi\,(r \sin\theta)^2$ were exposed to the liquid. The critical work for heterogeneous cluster formation is then reduced to

$$\Delta G_c^{\text{het}} = \underbrace{\frac{16\pi}{3}\frac{\sigma^3}{\left(\Delta G_{lc,V}\right)^2}}_{\Delta G_c \equiv \Delta G_c^{\text{hom}}} \cdot f(\theta), \tag{7.19}$$

whereas the critical radius remains unchanged[3]: $r_c^{\text{het}} = r_c^{\text{hom}}$. If there is no wetting (homogeneous nucleation, $\theta = 180°$, $f = 1$), Eq. (7.19) reduces to Eq. (7.4), i. e., the substrate does not aid in the nucleation process.

The classical theory for homogeneous nucleation (Sect. 7.1.1) can equally be applied to the case of heterogeneous nucleation [7.2, 7.9]. The only difference is the lower work for critical cluster formation [Eq. (7.19)] and the reduced number of parent phase atoms that can act as a nucleation site: While any parent phase atom can act as a nucleation site for homogeneous nucleation, only those atoms in contact with the impurity can act as a nucleation site for heterogeneous nucleation [7.2, 7.9, 7.25]. If ε is the fraction of parent phase atoms in contact with the heterogeneity (usually $\varepsilon \ll 1$), then the steady state nucleation rate for heterogeneous nucleation and diffusion-limited kinetics, $I^{ss,\text{het}}$, is

$$I^{ss,\text{het}} = \varepsilon \cdot s_c \cdot \frac{6D}{\lambda^2} \cdot N_0 \cdot \Gamma_z \cdot \exp\left(-\frac{\Delta G_c^{\text{het}}}{k_B T}\right) \quad \frac{1}{\text{m}^3\text{s}} \tag{7.20}$$

[cf. Eq. (7.10)] with a (local) activation energy of

$$E_{Iss,\text{het}} = E_D + \Delta G_c^{\text{het}}. \tag{7.21}$$

Since $\Delta G_c^{\text{het}} < \Delta G_c$ for $f(\theta) < 1$ [Eq. (7.19)] it follows that $E_{Iss,\text{het}} < E_{Iss}$ [Eq. (7.11)]. The ratio of heterogeneous to homogenous nucleation rate (dimensionless) at a given temperature is then

---

[3] However, note that the *number of atoms* in the critical cluster is reduced by the factor $f(\theta)$ for heterogeneous nucleation compared to homogeneous nucleation for the same critical radius r_c (Fig. 7.4).

$$\frac{I^{ss,\mathrm{het}}}{I^{ss,\mathrm{hom}}} = \varepsilon \cdot \exp\left(\frac{\Delta G_c}{k_B T} \cdot \left[1 - f(\theta)\right] \right), \tag{7.22}$$

where $I^{ss,\mathrm{hom}} \equiv I^{ss}$. Equation (7.22) applies to both diffusion-limited and collision-limited crystallization. Figure 7.5 shows a plot of Eq. (7.22) for a reasonable parameter of $\varepsilon = 10^{-6}$ [7.9] for three values of the critical work for *homo*geneous cluster formation, ΔG_c.

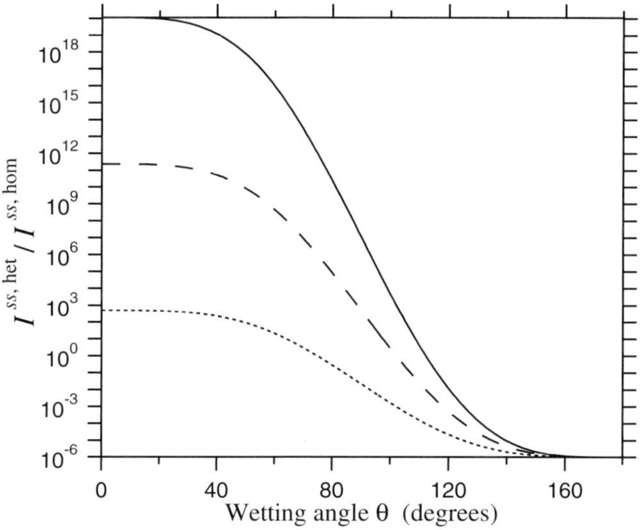

Fig. **7.5.** Ratio of heterogeneous to homogenous nucleation rate $I^{ss,\mathrm{het}}$ / $I^{ss,\mathrm{hom}}$ (dimensionless) at constant temperature, calculated from Eqs. (7.22) and (7.17) for $\varepsilon = 10^{-6}$.

Solid line: $\Delta G_c = 60\ k_B T$.

Dashed line: $\Delta G_c = 40\ k_B T$.

Dotted line: $\Delta G_c = 20\ k_B T$.

This illustrates the drastic influence of heterogeneities on the nucleation rate by many orders of magnitude. Usually, heterogeneous nucleation rates are observed to be far higher than homogeneous nucleation rates [7.2, 7.3], which implies that θ must be small (Fig. 7.5). In any case, since heterogeneous nucleation occurs *in addition* to homogeneous nucleation, $I^{ss,\mathrm{total}} = I^{ss,\mathrm{het}} + I^{ss,\mathrm{hom}}$, the nucleation rate is always lowest for purely homogeneous nucleation ($I^{ss,\mathrm{het}} = 0$).

7.1.3 Crystal Growth

The nucleation theory describes the formation of stable crystal clusters. Those can grow to macroscopic size. There are two possibilities: interface-controlled growth (Sect. 7.1.3.1) and growth controlled by a long-range diffusion field (Sect. 7.1.3.2).

7.1.3.1 Interface-controlled Growth

If there is no phase separation upon crystallization, the composition of parent phase and crystal cluster are the same at all times. Therefore, growth is controlled by rearrangement processes *only* at the liquid-crystalline interface and is therefore called *interface-controlled* [7.2]. Such a rearrangement process may either involve a diffusive jump in the case of diffusion-limited kinetics or an atomic collision in the case of collision-limited kinetics (Sec. 7.1.1.4). The rearrangement frequency is in both cases independent of the interface position so that the growth velocity u is time-independent. The crystal growth velocity is then [7.2]

$$u = \gamma_S \cdot \lambda \cdot k \cdot \left[1 - \exp\left(-\frac{\Delta G_{lc,\text{atom}}(T)}{k_B T} \right) \right] \qquad (T \leq T_m) \qquad (7.23)$$

where $0 \leq \gamma_s \leq 1$ is the fraction of sites where a new atom can be incorporated. λ is the average interatomic distance (i. e., the distance that the interface moves by each rearrangement), and $\Delta G_{lc,\text{atom}} > 0$ is the (Gibbs) free energy difference between parent phase and crystal *per atom*. The square bracket term in Eq. (7.23) results from thermally activated atomic transfer across the liquid-crystalline interface in *both* directions. The value of the rate constant k is the same as for nucleation [Eqs. (7.8) and (7.9)]. For diffusion-limited kinetics (as observed for phase change materials), this gives [7.2]

$$\begin{aligned} u &= \gamma_S \cdot \frac{6D}{\lambda} \left[1 - \exp\left(-\frac{\Delta G_{lc,\text{atom}}(T)}{k_B T} \right) \right] \\ &= \gamma_S \cdot \frac{2 k_B T}{\eta \pi \lambda^2} \left[1 - \exp\left(-\frac{\Delta G_{lc,\text{atom}}(T)}{k_B T} \right) \right] \qquad (T \leq T_m), \qquad (7.24) \end{aligned}$$

where Eq. (7.12) has been used. At the melting temperature T_m, the growth velocity u is zero since $\Delta G_{lc,\text{atom}}\ (T_m) = 0$. Close to T_g, the square bracket term in Eq. (7.24) varies slowly with temperature compared to the diffusivity D or viscosity η (Fig. 7.3) and can therefore be set as a constant. Therefore:

$$u \propto D \propto \frac{1}{\eta} \qquad (T \approx T_g, \text{ diffusion-limited}), \qquad (7.25)$$

which implies that the (local) activation energies of the crystal growth velocity E_u and the diffusivity E_D are equal:

$$E_u = E_D \qquad (T \approx T_g, \text{ diffusion-limited}). \qquad (7.26)$$

Comparing Eq. (7.26) with Eq. (7.11) or (7.21) gives

$$E_{Jss} - E_u = \Delta G_c \qquad (7.27a)$$

$$E_{Jss,\text{het}} - E_u = \Delta G_c^{\text{het}}, \qquad (7.27b)$$

i. e., the activation energies of nucleation rate and growth velocity differ by the critical work for cluster formation.

Since $u\ (T_m) = 0$ and since u becomes negligibly small close to T_g [Eq. (7.25), Fig. 7.3], it exhibits a maximum between T_g and T_m, which is usually located at higher temperature than the maximum for the nucleation rate [Eqs. (7.10) and (7.20)].

For collision-limited kinetics, u increases continuously upon cooling (not observed for phase change materials) and is limited by the velocity of sound [Eqs. (7.9) and (7.23)]:

$$u = \gamma_S \cdot u_{\text{sound}} \left[1 - \exp\left(-\frac{\Delta G_{lc,\text{atom}}}{k_B T} \right) \right] \qquad (T \leq T_m). \qquad (7.28)$$

7.1.3.2 Growth Controlled by Long-range Diffusion

If there is a composition change upon crystallization, *long range* diffusive atomic transport controls the growth velocity because the liquid depletes in certain components close to the liquid-crystalline interface. This depletion becomes more pronounced with increasing time, so that the growth velocity u must decrease with time t. Dimensional analysis of the diffusion equation [7.19, 7.26] gives:

$$u \propto \left(\frac{D}{t} \right)^{\frac{1}{2}}. \qquad (7.29)$$

A time-dependent crystal growth velocity has not been observed in phase change materials that are commonly used for phase change recording (cf. Sect. 7.2). Therefore, this growth mode is not discussed further here.

7.2 Measurements

As discussed in Sect. 7.1, crystallization of an undercooled liquid proceeds on a relatively long timescale just below the melting temperature T_m and close to or below the glass transition temperature T_g. The fastest crystallization is therefore observed at an intermediate temperature T_{int} between T_g and T_m. In optical (or electronic) phase change media, the laser power (or the current) is usually optimized to give high data transfer rates, which implies that crystallization probably occurs somewhere around T_{int}. It is well-established that crystallization in phase change materials near $T = T_{int}$ occurs on timescales of less than 100 ns [7.27-7.31]. Such a short timescale makes it impossible to perform systematic measurements of crystal nucleation rate and crystal growth velocity as a function of temperature around T_{int}. Therefore, crystallization is usually studied either around T_g (Sect. 7.2.1, [7.30-7.42]) or slightly below T_m (Sect. 7.2.2, [7.43]), where crystallization is slow enough to be observed on a laboratory timescale. In some cases, the measured parameters can then carefully be extrapolated to the temperature regime around T_{int}.

7.2.1 Crystallization Parameters Around the Glass Transition Temperature

Frequently, crystallization of amorphous phase change films is studied experimentally around T_g either by Kissinger analysis [7.44], where the crystallization temperature is measured as a function of heating rate [7.32, 7.35, 7.37, 7.38, 7.42], or by Johnson-Mehl-Avrami analysis [7.45, 7.46], where crystallization is monitored isothermally [7.30, 7.35, 7.41]. Both methods determine an *effective* activation energy for crystallization, which includes contributions of both crystal nucleation and crystal growth. Unfortunately, the relative contribution of these two processes remains unknown with these methods. It is therefore more meaningful to measure crystal nucleation rate and crystal growth velocity *independently*, which can only be performed by *direct* observation of crystal size and number as a function of time in an isothermal experiment [7.34, 7.40, 7.41, 7.47, 7.48]. In-situ transmission electron microscopy (TEM) studies have been performed to accomplish this task [7.40], but the drawbacks of this method are imprecise temperature control [7.41] and that the electron beam can influence the crystallization due to additional localized sample heating [7.49]. The experiment described below avoids these difficulties by using ex-situ atomic force microscopy (AFM) in combination with a precise furnace of a power-compensated differential scanning calorimeter (DSC). 30 nm-thin films of composition $Ge_2Sb_2Te_5$, $Ge_4Sb_1Te_5$, $Ge_1Sb_2Te_4$, and $Ag_{0.055}In_{0.065}Sb_{0.59}Te_{0.29}$ (hereafter: AgIn-Sb_2Te) were sputter-deposited on a Si wafer by direct-current magnetron sputtering and annealed isothermally in the DSC around the glass transition temperature T_g. [T_g has been determined in a separate study by DSC to be about 150-200° depending on the alloy (e. g., Fig. 7.6)].

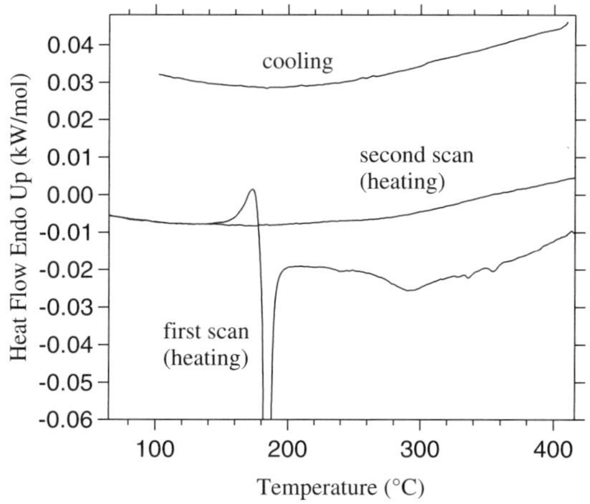

Fig. 7.6. Heat flow as a function of temperature for $Ge_2Sb_2Te_5$ measured by differential scanning calorimetry (DSC). Lowest curve: first scan for the initially fully amorphous sample (heating). The second scan (re-scan of the crystallized sample, heating) is shown for comparison and serves as a baseline for the first scan. Top curve: cooling signal (reproduced in additional scans). The scan rate was ± 40 K/min. The onset of the glass transition temperature T_g, which is usually observed as an endothermic step [7.22], occurs at around 170°. This step cannot be entirely resolved since crystallization interferes (large exothermic peak at around 180-185°, not entirely shown to make the onset of T_g more visible). Details: [7.19, 7.51]. Reprinted from [7.51] with permission of the MRS.

Due to the mass density increase upon crystallization, which induces a reduction in film thickness on the order of 5 % [7.39, 7.50], crystals could be directly observed as depressions in the not-yet-crystallized amorphous film [7.47]. Several AFM scanning and annealing cycles were alternately performed, and the annealing temperature remained the same for the same sample in subsequent anneals (details: [7.19, 7.47, 7.48]). Comparing number density and sizes of crystals on subsequent AFM scans at the same sample location revealed the (heterogeneous) steady-state crystal nucleation rate, $I^{ss,\mathrm{het}}$, and the crystal growth velocity, u. The experiment was repeated at different temperatures to determine their temperature dependencies [7.19, 7.47, 7.48]. Results are displayed in Figs. 7.7 and 7.8. For simplicity, the crystal nucleation rate in Fig. 7. 8 was normalized per unit area (not per unit volume) since cross-sectional TEM has shown that crystals nucleate only heterogeneously at the film surface [7.19, 7.33, 7.48]. Both $I^{ss,\mathrm{het}}$ and u are observed to *increase* with increasing temperature. Hence, as usually expected in alloys, crystallization proceeds diffusion-limited (not collision-limited, Sect. 7.1.1.4 and 7.1.3.1), i. e., Eqs. (7.10), (7.11), (7.20) and (7.21) should apply. For the case of collision-limited kinetics, $I^{ss,\mathrm{het}}$ and u would *decrease* with increasing temperature [Eqs. (7.16) and (7.28)], which is not observed. As a further indication, experience indicates that the crystal growth velocities in Fig. 7.7 are too low to be the result of collision-limited kinetics [7.52].

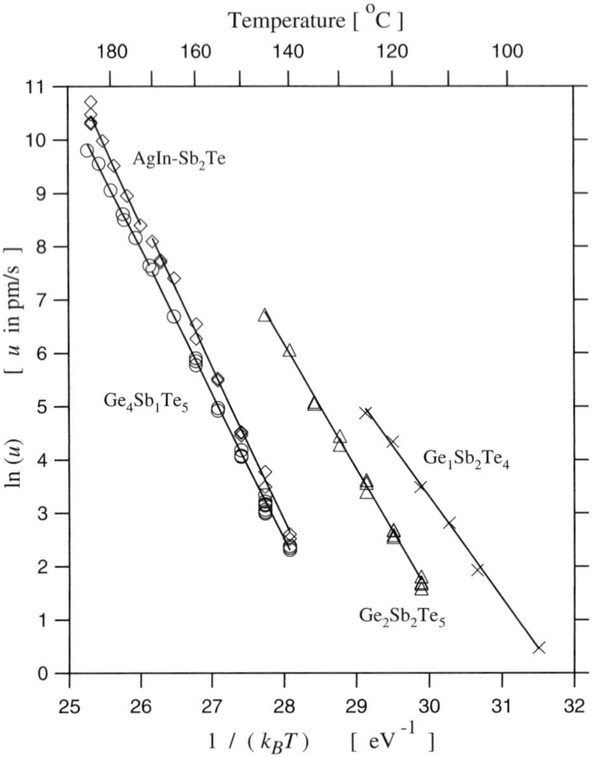

Fig. 7.7. Crystal growth velocity u as a function of temperature T: AgIn-Sb_2Te (squares), $Ge_4Sb_1Te_5$ (circles), $Ge_2Sb_2Te_5$ (triangles) and $Ge_1Sb_2Te_4$ (crosses). The error bars on the velocity are approximately equal to the size of the symbols. The data were fitted with an Arrhenius equation (fit parameters: Table 7.1). For $T > 170°$, a different annealing technique was used, and therefore, the data in this regime were fitted separately (details: [7.19, 7.47]). Reprinted with permission from [7.47]. Copyright [2004], American Institute of Physics.

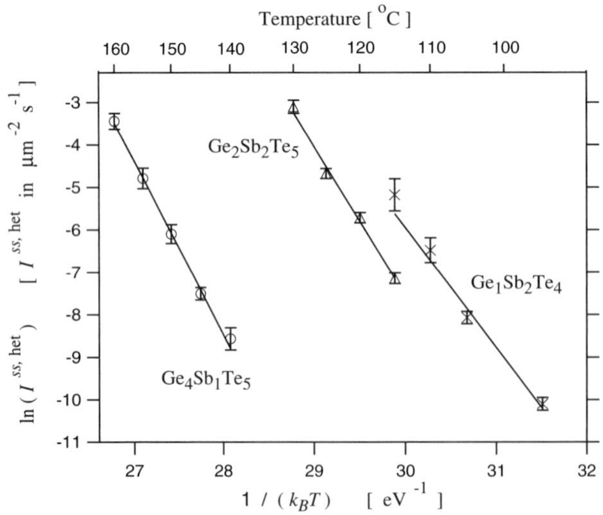

Fig. 7.8. Heterogeneous steady-state crystal nucleation rate $I^{ss,\mathrm{het}}$ as a function of temperature T: $Ge_4Sb_1Te_5$ (circles) $Ge_2Sb_2Te_5$ (triangles) $Ge_1Sb_2Te_4$ (crosses). The data were fitted with an Arrhenius equation (fit parameters: Table 7. 1). Reprinted with permission from [7.48]. Copyright (2005). American Institute of Physics.

Table 7.1. Activation energies E_u and $E_{I^{ss},\text{het}}$ for the crystal growth velocity u and the heterogeneous steady-state nucleation rate $I^{ss,\text{het}}$, respectively. E_u and $I^{ss,\text{het}}$ were obtained from an Arrhenius fit to the data in Figs. 7.7 and 7.8, respectively: $u \propto [- E_u / (k_B T)]$ and $I^{ss} \propto [- E_{I^{ss},\text{het}} / (k_B T)]$. The critical work for heterogeneous cluster formation, ΔG_c^{het}, was obtained from Eq. (7.27). The activation energy for the isoconfigurational viscosity in the amorphous phase, E_η, was taken from [7.53, 7.54].

Alloy	E_u (eV)	$E_{I^{ss},\text{het}}$ (eV)	ΔG_c^{het} (eV)	E_η (eV)
$Ge_4Sb_1Te_5$	2.74±0.03	4.09±0.20	1.35±0.23	1.94±0.09
$Ge_2Sb_2Te_5$	2.35±0.05	3.50±0.17	1.15±0.22	1.76±0.05
$Ge_1Sb_2Te_4$	1.89±0.05	2.82±0.18	0.93±0.23	– [a]
AgIn-Sb_2Te	2.90±0.05	n/a [b]	n/a [b]	1.33±0.09

[a] No data available.

[b] Steady-state nucleation not observed for this alloy.

A time dependence of the crystal growth velocity u could not be observed within error between subsequent anneals of the same sample at the same temperature (Fig. 7.7). This implies that crystal growth occurs interface-controlled (as opposed to growth controlled by long-range diffusion, Sect. 7.1.3), i. e., Eqs. (7.24) – (7.27) should apply [as opposed to Eq. (7.29)].

In the initial stage of crystallization (at short times, i. e., during the first annealing cycle), the crystal nucleation rate for all GeSbTe alloys increased with time due to transient effects (not shown in Fig. 7.8, details: [7.19, 7.48]). This effect occurs since the steady-state cluster distribution N^{ss} (Fig. 7.2) has not yet been developed in the initial stage of crystallization [7.2, 7.24]. After the so-called *incubation time*, the cluster distribution adopts a steady-state value, and therefore, the crystal nucleation rate is time-independent as shown in Fig. 7.8 [cf. Eq. (7.20)]. For AgIn-Sb_2Te, no data are shown in Fig. 7.8 since a steady state could not be observed. Apart from that, the nucleation rate for AgIn-Sb_2Te was far lower than for the GeSbTe alloys (details: [7.19, 7.47, 7.48]). AgIn-Sb_2Te is known to exhibit growth-dominated crystallization upon laser heating, i. e., an amorphous mark in a crystalline matrix re-crystallizes by the growth from the rim of the amorphous mark [7.28, 7.55]. This is in contrast to the GeSbTe alloys, which re-crystallize nucleation-dominated, i. e., an amorphous mark re-crystallizes upon laser heating predominantly by nucleation inside the amorphous mark [7.27]. This suggests that the different recrystallization mechanisms observed upon laser heating can be ascribed to the significant qualitative difference in crystal nucleation behavior rather than to the smaller difference in crystal growth velocity (qualitatively similar behavior for all alloys).

The activation energy for the heterogeneous steady state nucleation rate, $E_{I^{ss},\text{het}}$, and for the growth velocity, E_u, is determined from the slope of each straight line in Figs. 7.7 and 7.8 and given in Table 7. 1. The critical work for heterogeneous cluster formation, $\Delta G_c^{\mathrm{het}}$ [Eq. (7.19)], is obtained from Eq. (7.27b). Since nucleation is heterogeneous, $\Delta G_c^{\mathrm{het}}$ is a lower limit for the critical work for homogeneous cluster formation ΔG_c [Eq. (7.4)].

Table 7.1 also shows the activation energy E_η of the shear viscosity η [Eq. (7.14)], which was obtained from stress relaxation experiments in thin films by wafer curvature measurements in a temperature range between 60 °C and 100 °C [7.53, 7.54]. According to Eq. (7.14), E_u and E_η should be equal under the assumption that the Stokes-Einstein equation [Eq. (7.12)] is valid for the phase change materials. However, since E_u is larger than E_η (Table 7. 1) and since the glass transition temperature T_g is usually accompanied by a discontinuity in activation energy (Fig. 7.3, activation energies above T_g are higher than below T_g), the data presented in Fig. 7.7 appear to be taken above T_g in the undercooled liquid. On the other hand, the viscosity was measured below T_g in the amorphous phase. T_g depends on the timescale of the experiment [7.19, 7.51] and should be significantly lower in the isothermal experiments (Figs. 7.7 and 7.8) than in a scanning experiment at constant heating rate of 40 K/min (Fig. 7.6). That the crystal growth velocity in Fig. 7.7 is time-independent points in the same direction: Only in the amorphous phase, but not in the undercooled liquid, would a time dependence of the atomic transport rates be expected due to structural relaxation (Fig. 7.3).

7.2.2 Crystallization Parameters Close to the Melting Temperature

Crystallization kinetics slightly below the melting temperature have rarely been investigated due to the high volatility of phase change materials at elevated temperatures (high vapor pressure of Sb and Te). However, in a recent study [7.19, 7.43], a lower limit for the crystal-melt interfacial free energy σ [Eq. (7.2)] and an upper limit for the *homo*geneous steady-state crystal nucleation rate I^{ss} [Eqs. (7.10) or (7.15)] was estimated by undercooling liquid droplets of phase change materials below the liquidus temperature T_l in a differential thermal analyzer (DTA) and measuring the undercooling, $\Delta T_n := (T_l - T_n) > 0$, at a constant cooling rate.[4] T_n is the temperature at which nucleation initiates upon cooling and was detected in the DTA by recalescence (re-heat of the droplet by the release of the heat of crystallization). Note that purely homogeneous nucleation gives the lowest possible nucleation rate, and that heterogeneities increase the nucleation rate substantially (Sect. 7.1.2). As a consequence, the undercooling ΔT_n is maximized for purely homogeneous nucleation, whereas the presence of any heterogeneity re-

---

[4] Since melting in alloys occurs over a range of temperatures, the liquidus temperature T_l is used instead of the melting temperature T_m in this section. T_l is the high-temperature end of the equilibrium melting range upon heating [7.19, 7.43, 7.56].

duces the value of ΔT_n. Therefore, in order to maximize ΔT_n and to approach the limit of homogeneous nucleation as closely as possible, the phase change material was embedded in a liquid flux of B_2O_3: This helped to isolate the droplet from the DTA crucible walls, which could act as heterogeneous nucleation sites. Additionally, B_2O_3 eliminates nucleants from the surface of the droplet by dissolution and inclusion [7.19, 7.43]. Moreover, the B_2O_3 prevented evaporation of the liquid volatile phase change material. Many heating and cooling cycles were performed in order to maximize ΔT_n further by additional fluxing. The undercooling data were then analyzed by *assuming* that the highest measured value for ΔT_n corresponds to the limit of *homo*geneous crystal nucleation. This substantially simplified the data analysis since $f(\theta)$ could be set to unity [Eq. (7.19)]. However, since it is unknown if this assumption is true, the obtained value for σ is only a *lower* limit for *homo*geneous nucleation, whereas the obtained curve for I^{ss} is only an *upper* limit for *homo*geneous nucleation (details: [7.19, 7.43]). Figure 7.9 shows the result.

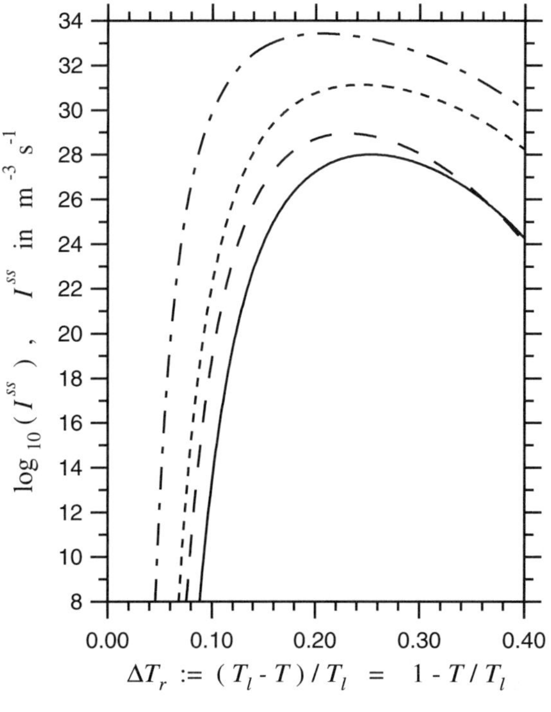

Fig. 7.9. Upper limit for the *homo*geneous steady state crystal nucleationrate I^{ss} [Eqs. (7.10) or (7.15)] as a function of *relative* undercooling $\Delta T_r := (T_l - T)/T_l$ for $Ge_2Sb_2Te_5$ (dot-dashed), $Ge_4Sb_1Te_5$ (dotted), AgIn-Sb_2Te (dashed), and $Ge_{12}Sb_{88}$ (solid). T_l is the liquidus temperature. Reprinted with permission from [7.43]. Copyright [2005]. American Institute of Physics.

As explained in Sect. 7.1.1.4, the nucleation rate is negligibly small close to the melting (or liquidus) temperature, then increases rapidly upon cooling to a maximum, and then decreases again rapidly at the glass transition temperature T_g, which occurs (*relative* to the liquidus temperature) at $\Delta T_{rg} := 1 - T_g/T_l = 0.45$ - 0.55 depending on the alloy [7.19, 7.43, 7.51, 7.57]. The nucleation rates are higher for the GeSbTe alloys, which exhibit nucleation-dominated crystallization [7.27], than for the Sb-rich alloys AgIn-Sb_2Te and $Ge_{12}Sb_{88}$, which exhibit growth-dominated crystallization [7.28, 7.29, 7.55]. This is a consequence of the fact that the Sb-rich alloys are characterized by a higher *relative* glass transition temperature, $T_{rg} = T_g/T_l$, and a higher entropy of fusion than the GeSbTe alloys (details: [7.19, 7.43, 7.51]).

The steady state nucleation rates in Fig. 7.9 are too high to allow amorphization in both optical and electronic phase change media under operating conditions. Nucleation interferes at the highest attainable cooling rates, which are on the order of 10^{10} Ks^{-1} [7.43, 7.58]. For optical data storage, the bit volume is on the order of $V_b = 1\,\mu m \times 1\,\mu m \times 10$ nm $=10^{-20}$ m^3 [7.59] (limited by the wavelength of the laser light). For a cooling rate of 10^{10} Ks^{-1}, the temperature decreases by 100 K during a time of $t_c := 10$ ns $= 10^{-8}$ s. Therefore, nucleation would interfere upon amorphization if the nucleation rate were $I^{ss} > V_b^{-1}\, t_c^{-1} = 10^{28}$ m^{-3} s^{-1} over a range of 100 K. This is certainly the case for the GeSbTe alloys, but most likely also for the Sb-rich alloys: Since the B_2O_3 flux has eliminated heterogeneous nucleation sites, the fluxing technique should have approached the limit of homogeneous nucleation (which gives the lowest possible nucleation rate) far closer than the nucleation rates in phase change media, for which heterogeneous nucleation is expected to dominate strongly due to the thin-film nature of the device. Moreover, nucleation in phase change media is frequently enhanced (i. e., heterogeneities are added) by nucleation-promoting dielectric capping layers [7.37, 7.60]. For electronic phase change media, programmable volume sizes as small as $V_b = 10^{-23}$ m^3 are reported [7.61, 7.62]. Hence, nucleation would interfere if $I^{ss} > 10^{31}$ m^{-3} s^{-1} over a range of 100 K, which is at least the case for $Ge_2Sb_2Te_5$, which is often reported as a prototypic material for phase change random access test devices [7.63]. Therefore, it can be concluded that:

1. Melt quenching under operating conditions occurs during the incubation time for crystal nucleation, when the steady-state cluster distribution N^{ss} (Fig. 7.2) is not yet formed, so that the nucleation rate remains far smaller than its steady state value I^{ss} [Eqs (7. 10), (7. 15) or (7. 20) [7.2, 7.24]]. As the incubation time is independent of the bit volume V_b, this statement applies equally to optical and electronic media for all alloys investigated. Consistently, it was reported by Kelton and Greer [7.64] that transient effects become increasingly important with increasing quench rate: While the steady state nucleation rate is readily maintained for cooling rates on the order of 1 K s^{-1} as used in the present experiments or in conventional metallurgical solidification, deviations from the steady state are already large at cooling rates in rapid solidification techniques, such as melt-spinning ($\sim 10^6$ K s^{-1}), and must be even larger for cooling rates on the order of 10^{10} K s^{-1} which occur under operating conditions of phase change media.
2. Amorphization would not be possible if the incubation time for crystal nucleation were absent. Hence, the existence of an incubation time makes phase change recording possible. This statement should apply for optical data stor-

age to all alloys investigated and for electronic data storage at least to $Ge_2Sb_2Te_5$. Indeed, incubation times were reported upon crystallization on the timescale of minutes around T_g [7.36, 7.40, 7.41, 7.48], as well as upon laser crystallization on the nanosecond timescale [7.27, 7.30, 7.65-7.67].

Acknowledgement. Frans Spaepen is gratefully acknowledged for critical proof-reading and valuable input to this chapter.

References

[7.1] Gibbs, J.: The scientific papers of J. Willard Gibbs. Dover Publications, New York (1961)

[7.2] Christian, J.: Transformation in metals and alloys, 2nd edn. Pergamon Press, Oxford (1975)

[7.3] Kelton, K.: Crystal nucleation in liquids and glasses. Solid State Physics **45**, 75-177 (1991)

[7.4] Volmer, M. and Weber, A.: Keimbildung in übersättigten Gebilden. Zeitschrift für Physikalische Chemie **119**, 277 (1926)

[7.5] Becker, R. and Döring, W.: Kinetische Behandlung der Keimbildung in übersättigten Dämpfen. Annalen der Physik **24**, 719 (1935)

[7.6] Turnbull, D. and Fisher, J.: Rate of nucleation in condensed systems. J. Chem. Phys. **17**, 71-73 (1949)

[7.7] Landau, L. and Lifshitz, E.: Statistical Physics. Pergamon Press, Oxford (1969)

[7.8] Thompson, C. and Spaepen, F.: Homogeneous crystal nucleation in binary metallic melts. Acta Metallurgica **31**, 2021-2027 (1983)

[7.9] Herlach, D.: Non-equilibrium solidification of undercooled metallic melts. Materials Science and Engineering R **12**, 177-272 (1994)

[7.10] Angell, C., Ngai, K., McKenna, G., McMillan, P. and Martin, S.: Relaxation in glass-forming liquids and amorphous solids. J. Appl. Phys. **88**, 3113-3157 (2000)

[7.11] Debenedetti, P. and Stillinger, F.: Supercooled liquids and the glass transition. Nature **410**, 259-267 (2001)

[7.12] Götze, W.: Liquids, freezing and the glass transition. Les Houches LI. North-Holland, Amsterdam (1991), p. 287

[7.13] Hodgdon, J. and Stillinger, F.: Stokes-Einstein violation in glass-forming liquids. Phys. Rev. E **48**, 207-213 (1993)

[7.14] Shao, Y. and Spaepen, F.: Undercooling of bulk liquid silicon in an oxide flux. J. App. Phys. **79**, 2981-2985 (1996)

[7.15] Spaepen, F.: The identification of the metallic glass state. In: Mat. Res. Soc. Symp. Proc., vol. **57**, p. 161-184 (1986)

[7.16] Thompson, C., Greer, A. L. and Spaepen, F.: Crystal nucleation in amorphous $(Au_{100-y}Cu_y)_{77}Si_9Ge_{14}$ alloys. Acta Metallurgica **31**, 1883-1894 (1983)

[7.17] Turnbull, D.: Kinetics of solidification of supercooled liquid mercury droplets. J. Chem. Phys. **20**, 411-424 (1952)

[7.18] Turnbull, D.: Under what conditions can a glass be formed? Contemp. Phys. **10**, 473-488 (1969)
[7.19] Kalb, J.: Crystallization kinetics in antimony and tellurium alloys used for phase change recording. Ph.D. thesis, RWTH Aachen, Germany (2006). URL www.kalb.eu/publications
[7.20] Spaepen, F.: Physics of Defects. Les Houches XXXV. North-Holland, Amsterdam (1981), pp. 133-174
[7.21] Spaepen, F. and Turnbull, D.: Metallic glasses. Ann. Rev. Phys. Chem. **35**, 241-263 (1984)
[7.22] Elliott, S.: Physics of amorphous materials, 2nd edn. Longman, London (1990)
[7.23] Volmer, M.: Über Keimbildung und Keimwirkung als Spezialfälle der heterogenen Katalyse. Zeitschrift für Elektochemie **35**, 555 (1929)
[7.24] Wu, D.: Nucleation theory. Solid State Physics **50**, 37-187 (1997)
[7.25] Holland-Moritz, D.: Short-range order and solid-liquid interfaces in undercooled melts. Int. Journ. Non-Equilibrium Processing **11**, 169-199 (1998)
[7.26] Crank, J.: The mathematics of diffusion, 2nd edn. Clarendon Press, Oxford (1975)
[7.27] Coombs, J., Jongenelis, A., van Es-Spiekman, W. and Jacobs, B.: Laser-induced crystallization phenomena in GeTe-based alloys. II. Composition dependence of nucleation and growth. J. Appl. Phys. **78**, 4918-4928 (1995)
[7.28] van Pieterson, L., Lankhorst, M., van Schijndel, M., Kuiper, A. and Roosen, J.: Phase-change recording materials with a growth-dominated crystallization mechanism: A materials overview. J. Appl. Phys. **97**, 083520 (2005)
[7.29] van Pieterson, L., van Schijndel, M., Rijpers, J. and Kaiser, M.: Te-free, Sb-based phase-change materials for high-speed rewritable optical recording. Appl. Phys. Lett. **83**, 1373-1375 (2003)
[7.30] Weidenhof, V., Friedrich, I., Ziegler, S. and Wuttig, M.: Laser induced crystallization of amorphous $Ge_2Sb_2Te_5$ films. J. Appl. Phys. **89**, 3168-3176 (2001)
[7.31] Yamada, N., Ohno, E., Nishiuchi, K., Akahira, N. and Takao, M.: Rapid phase transitions of GeTe-Sb_2Te_3 pseudobinary amorphous thin films for an optical disk memory. J. Appl. Phys. Physics **69**, 2849-2856 (1991)
[7.32] Friedrich, I., Weidenhof, V., Njoroge, W., Franz, P. and Wuttig, M.: Structural transformations of $Ge_2Sb_2Te_5$ films studied by electrical resistance measurements. J. Appl. Phys. **87**, 4130-4134 (2000)
[7.33] Jeong, T., Kim, M., Seo, H., Kim, S. and Kim, S.: Crystallization behavior of sputter-deposited amorphous $Ge_2Sb_2Te_5$ thin films. J. Appl. Phys. **86**, 774-778 (1999)
[7.34] Kooi, B. and De Hosson, J.: On the crystallization of thin films composed of $Sb_{3.6}Te$ with Ge for rewritable data storage. J. Appl. Phys. **95**, 4714-4721 (2004)
[7.35] Libera, M. and Chen, M.: Time-resolved reflection and transmission studies of amorphous Ge-Te thin-film crystallization. J. Appl. Phys. **73**, 2272-2282 (1993)
[7.36] Lu, Q. and Libera, M.: Microstructural measurements of amorphous GeTe crystallization by hot-stage optical microscopy. J. Appl. Phys. **77**, 517-521 (1995)
[7.37] Njoroge, W., Dieker, H. and Wuttig, M.: Influence of dielectric capping layers on the crystallization kinetics of $Ag_5In_6Sb_{59}Te_{30}$ films. J. Appl. Phys. **96**, 2624-2627 (2004)
[7.38] Njoroge W. and Wuttig, M.: Crystallization kinetics of sputter-deposited amorphous AgInSbTe films. J. Appl. Phys. **90**, 3816 (2001)
[7.39] Pedersen, T.L., Kalb, J., Njoroge, W., Wamwangi, D., Wuttig, M. and Spaepen, F.: Mechanical stresses upon crystallization in phase change materials. Appl. Phys. Lett. **79**, 3597-3599 (2001)
[7.40] Privitera, S., Bongiorno, C., Rimini, E. and Zonca, R.: Crystal nucleation and growth processes in $Ge_2Sb_2Te_5$. Appl. Phys. Lett. **84**, 4448-4450 (2004)
[7.41] Ruitenberg, G., Petford-Long, A. and Doole, R.: Determination of the isothermal nucleation and growth parameters for the crystallization of thin $Ge_2Sb_2Te_5$ films. J. Appl. Phys. **92**, 3116-3123 (2002)

[7.42] Wamwangi, D., Njoroge, W. and Wuttig, M.: Crystallization kinetics of $Ge_4Sb_1Te_5$ films. Thin Solid Films **408**, 310-315 (2002)
[7.43] Kalb, J., Spaepen, F. and Wuttig, M.: Kinetics of crystal nucleation in undercooled droplets of Sb-and Te-based alloys used for phase change recording. J. Appl. Phys. **98**, 054910 (2005)
[7.44] Kissinger, H.: Reaction kinetics in differential thermal analysis. Analyt. Chem. **29**, 1702 (1957)
[7.45] Avrami, M.: Kinetics of phase change. I. General theory. J. Chem. Phys. **7**, 1103-1112 (1939)
[7.46] Johnson, W. and Mehl, R.: Reaction kinetics in process of nucleation and growth. Trans. Amer. Inst. of Mining, Metallurgical and Petroleum Engineers **135**, 416 (1939)
[7.47] Kalb, J., Spaepen, F. and Wuttig, M.: Atomic force microscopy measurements of crystal nucleation and growth rates in thin films of amorphous Te alloys. Appl. Phys. Lett. **84**, 5240-5242 (2004)
[7.48] Kalb, J., Wen, C., Spaepen, F., Dieker, H. and Wuttig, M.: Crystal morphology and nucleation in thin films of amorphous Te alloys used for phase change recording. J. Appl. Phys. **98**, 054902 (2005)
[7.49] Kooi, B., Groot, W. and De Hosson, J.: In situ transmission electron microscopy study of the crystallization of $Ge_2Sb_2Te_5$. J. Appl. Phys. **95**, 924-932 (2004)
[7.50] Weidenhof, V., Friedrich, I., Ziegler, S. and Wuttig, M.: Atomic force microscopy study of laser induced phase transitions in $Ge_2Sb_2Te_5$. J. Appl. Phys. **86**, 5879-5887 (1999)
[7.51] Kalb, J., Wuttig, M. and Spaepen, F.: Calorimetric measurements of structural relaxation and glass transition temperatures in sputtered films of amorphous Te alloys used for phase change recording. J. Mater. Res. **22**, 748-754 (2007)
[7.52] Spaepen, F.: Private communication
[7.53] Kalb, J.: Stresses, viscous flow and crystallization kinetics in thin films of amorphous chalcogenides used for optical data storage. Diploma thesis, RWTH Aachen, Germany (2002). URL www.kalb.eu/publications
[7.54] Kalb, J., Spaepen, F., Pedersen, T.L. and Wuttig, M.: Viscosity and elastic constants of thin films of amorphous Te alloys used for optical data storage. J. Appl. Phys. **94**, 4908-4912 (2003)
[7.55] Borg, H., van Schijndel, M., Rijpers, J., Lankhorst, H., Zhou, G., Dekker, M., Ubbens, I. and Kuijper, M.: Phase-change media for high-numerical-aperture and blue-wavelength recording. Jpn. J. Appl. Phys. **40**, Part 1, 1592-1597 (2001)
[7.56] Shackelford, J.: Introduction to Materials Science for Engineers, 2nd edn. Macmillan, New York (1988)
[7.57] Kalb, J., Spaepen, F. and Wuttig, M.: Calorimetric measurements of phase transformations in thin films of amorphous Te alloys used for optical data storage. J. Appl. Phys. **93**, 2389-2393 (2003)
[7.58] Peng, C., Cheng, L. and Mansuripur, M.: Experimental and theoretical investigations of laser-induced crystallization and amorphization in phase-change optical recording media. J. Appl. Phys. **82**, 4183-4191 (1997)
[7.59] Kaiser, M., van Pieterson, L. and Verheijen, M.: In situ transmission electron microscopy analysis of electron beam induced crystallization of amorphous marks in phase-change materials. J. Appl. Phys. **96**, 3193-3198 (2004)
[7.60] Ohshima, N.: Crystallization of germanium-antimony-tellurium amorphous thin film sandwiched between various dielectric protective films. J. Appl. Phys. **79**, 8357-8363 (1996)
[7.61] Haring-Bolivar, P., Merget, F., Kim, D.H., Hadam, B. and Kurz, H.: European Symposium on Phase Change and Ovonic Science (EPCOS), Balzers, Liechtenstein, unpublished (2004)
[7.62] Lankhorst, M., Ketelaars, B., Wolters, R.: Low-cost and nanoscale non-volatile memory concept for future silicon chips. Nature Materials **4**, 347-352 (2005)

[7.63] Hudgens, S. and Johnson, B.: Overview of phase-change chalcogenide nonvolatile memory technology. Materials Research Society Bulletin **29**, 829-832 (2004)

[7.64] Kelton, K., Greer, A.: Transient nucleation effects in glass formation. Journal of Non-Crystalline Solids **79**, 295-309 (1986)

[7.65] Friedrich, I., Weidenhof, V., Lenk, S. and Wuttig, M.: Morphology and structure of laser-modified $Ge_2Sb_2Te_5$ films studied by transmission electron microscopy. Thin Solid Films **389**, 239-244 (2001)

[7.66] Wöltgens, H.W., Detemple, R., Friedrich, I., Njoroge, W., Thomas, I., Weidenhof, V., Ziegler, S. and Wuttig, M.: Exploring the limits of fast phase change materials. In: Materials Research Society Symposia Proceedings, vol. **674**, p. V1.3 (2001)

[7.67] Ziegler, S.: Rekristallisationskinetik von Phasenwechselmedien. Ph.D. thesis, RWTH Aachen, Germany (2005)

8. Short and Long-Range Order in Phase Change Materials

Paul Fons

Abstract This chapter offers a brief historical review of phase change materials as well as an overview of more recent findings. The chapter mainly focuses on the prototypical phase change materials in the Ge-Sb-Te and doped Sb-Te systems as these are currently the two composition spaces that are being applied both in the current generation of optical storage as well as future generations of electrical memory. In many ways, the structures of both the crystalline and amorphous phase change materials are atypical; the crystalline phase is often metastable, while the amorphous phase defies description as a typical random covalent network. We explore both long and short-range order of some prototypical compositions in the hope of providing a baseline from which deeper interpretations can be made.

8.1 Historical Background

The historical origins of what are contemporarily referred to as phase-change materials (PCM) have their origins in the study of amorphous materials, specifically chalcogenides alloys. Chalcogenide based glass has a long history dating back to the pioneering work in the 1950s by B. T. Kolomiets [8.1] in which the importance of short-range order on determining semiconducting properties was first recognized. Ovshinsky [8.2] first proposed use of the crystalline-amorphous phase transition in chalcogenide materials as a memory element in an electrical switch. Research on chalcogenide glasses remained mainly of academic interest until materials with high switching speeds became commercially viable. As the number of viable compositions of phase-change alloys usable for either optical or electrical switching is a small subset of the entirety of amorphous materials, we choose to focus here on this subset in the hope of enlightening the reader of the characteristics which have proved to be necessary for high-speed switching. Among the important requirements of phase-change memory alloys is that they switch rapidly and reversibly in both directions between crystalline and amorphous phases, and exhibit reasonably large changes in electronic properties (as manifested by electrical resistivity or optical reflectivity differences between phases). The first require-

S. Raoux, M. Wuttig (eds.), *Phase Change Materials*,
DOI 10.1007/978-0-387-84874-7_8, © Springer Science+Business Media, LLC 2009

ment in practice means switching on the order of hundreds of nanoseconds (or less). While forming the amorphous state does not lead to serious restrictions, the need for rapid crystallization is quite severe and eliminates good glass formers as candidates for use in phase-change memory applications where speed is paramount.

8.1.1 Glass Formation Process

A schematic of the glass formation process is shown in Fig. 8.1.

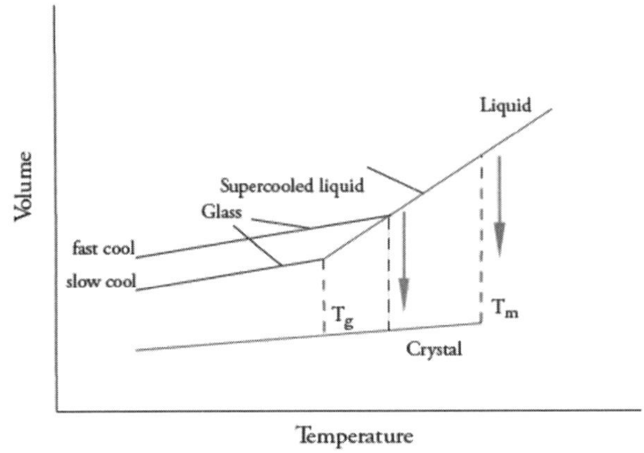

Fig. 8.1. A schematic view of the glass formation diagram.

Starting from the right of Fig. 8.1, we see that the PCM is in a liquid state. As the PCM cools from the liquid state (upper right), it may transform into the lower free energy crystalline state and undergo a sharp change in volume. If, however, the characteristic time constants (e.g. the kinetics) of the crystallization process are slow compared to the cooling rate, the PCM may become supercooled and enter a glassy state in a narrow temperature range near the glass transition temperature T_g. For temperatures below T_g, the ability of the system to rearrange itself becomes long compared to experimentally accessible times and the material becomes a solid (glass). Unlike the crystalline state, the glass undergoes no sharp changes in volume and is in a *metastable* state whose volume and free energy depend on the cooling rate; the value of T_g thus also changes with cooling rate. In contrast, the speed of the crystallization process is determined by a combination of the mobility of the atoms (e.g. temperature) and the free energy difference between the starting and ending phases. At the melting point T_m, the liquid and crystalline states are in equilibrium hence the free energy difference is zero. This implies that the maximum crystallization rate occurs at an intermediate value between the T_g and T_m. It has been suggested that the reduced glass transition temperature

$T_{rg} = T_g / T_m$ can serve as an indicator of nucleation rate. T_{rg} assumes values between 0.4 and 0.84 for known glass formers with PCM displaying T_{rg} values between 0.45 and 0.55 [8.3]. Good PCM candidates are thus poor glass formers and unlike typical glass materials show a tendency not to assume random covalent network like topologies, but display distinct short range order as will be seen in Section 8.3.

The above distinction narrows somewhat the materials that can be considered to be PCM, but the distinction is necessarily fuzzy and to date much of the progress in finding good PCM for a given application has been carried out experimentally. It is only recently that a theoretical understanding of some aspects of the switching process has allowed theoretical approaches [8.4] to be applied to finding the most effective PCM for a given application.

The materials system of choice for both optical and electrical storage has been the Ge-Sb-Te system with the composition being selected to match the end use requirements, although other compounds have been used, also. In this sense, it is difficult to define exactly what constitutes a phase change alloy. The compositions that lie along the GeTe-Sb_2Te_3 pseudo-binary tie line have received much attention as the switching speeds of alloys along the tie line have been found to be fast and the materials show single-phase behavior. As might be expected, the understanding of the structure of the crystalline phases is much more complete than that of the amorphous phases and along these lines we present structural information as a function of the coherence length of the observation technique.

8.2 Long-Range Order

Most of the early reports on the structure of PCM were based upon x-ray diffraction. X-ray diffraction probes the two-body correlation function over a characteristic length scale typically on the order of several tens of nanometers or longer. Initial structural reports on the GeTe-Sb_2Te_3 pseudo-binary phase diagram were carried out by several Soviet groups in the early 1960s before the phenomena of threshold switching was reported and the technological significance of Ge-Sb-Te alloys was known [8.5–7]. In 1964, Abrikosov et al. published the equilibrium phase diagram along the GeTe-Sb_2Te_3 pseudo-binary tie-line. Samples were prepared by annealing at 500 °C for ~1000 hours [8.5]. This work culminated in the pseudo-binary system GeTe-Sb_2Te_3 with many of the early efforts dating back to the late 1960s. It was not until the end of the 1960s that the possibilities of utilizing phase-change materials as solid state switches was realized with the publishing of a paper by Ovshinsky on the observation of the phenomenon of threshold switching in amorphous chalcogenide materials [8.2, 8.14]. The bulk of structural data on PCM has been obtained using x-ray diffraction. The fact that many of the crystalline phases of PCM are only available in metastable films a few tens of nanometers thick presents significant challenges to unambiguous identification of structural features. This fact in conjunction with presence of significant site disorder in some of the most important PCMs such as Ge-Sb-Te and Ag-In-Sb-Te, makes the need for complementary techniques such as Extended X-ray Absorption Fine Structure (EXAFS) spectroscopy or transmission electron microscopy paramount. The amorphous phase, remains almost completely inaccessible to diffraction based techniques with the possible exception of total scattering techniques such as powder based pair-distribution function analysis.

8.2.1 GeTe

The binary compound GeTe has a long history of involvement with PCM applications. Early research in GeTe was stimulated in part by theoretical calculations of the origins of the A7 like structure of GeTe [8.8]. At room temperature, GeTe forms in a rhombohedral structure with space group $R3m$ and lattice constants a = 4.21 Å and c = 10.60 Å (hexagonal setting). The GeTe lattice can be visualized (see Fig. 8.2) as a distorted rocksalt structure with Ge and Te uniquely located on fcc sublattices with origins at (¼, ¼, ¼) and (-¼, -¼, -¼) that undergo a distortion (u, u, u) and (-u, -u, -u) along the [111] direction with u = 0.237 leading to the angle α = 88.35° deviating from 90° [8.9].

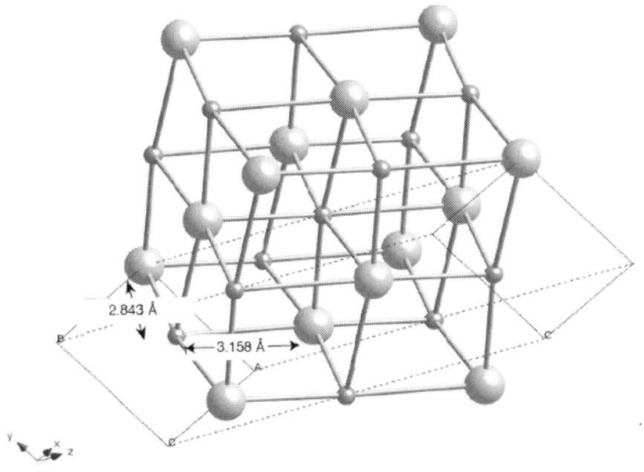

Fig. 8.2. The rhombohedral structure of room temperature GeTe can be thought of as a rocksalt structure distorted along the [111] direction.

Upon heating GeTe remains in the distorted rocksalt structure until a temperature T_c ~ 420 °C whereupon it transforms into a higher symmetry cubic phase with a volume reduction of approximately 1 % [8.10]. The quantum mechanical origin of the distortion has been suggested to be a form of Peierls distortion (a spontaneous lower of symmetry distortion that lowers the free energy of the system similar to the Jahn-Teller effect in molecular systems) and to be general to Te-based chalcogenides [8.9].

Locally, each atom in GeTe is nominally octahedrally coordinated with three short (2.843 Å) bonds and three longer (3.158 Å) bonds. GeTe is also an example of a material that undergoes a displacive ferroelectric transition as its net dipole moment vanishes at T_c with the change in symmetry which can be understood as temperature activated lifting of the Peierls distortion [8.9]. Experimental data supporting a displacive transition has been obtained using Raman scattering with the vanishing of the $\Gamma(A_1)$ TO mode which *dynamically* corresponds to the distortion in the rocksalt structure [8.11].

Figure 8.3 shows the equilibrium phase diagram of the GeTe system [8.12].

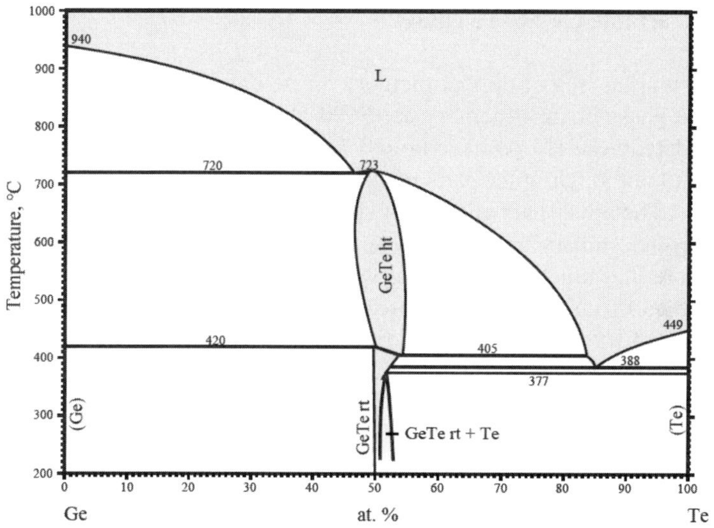

Fig. 8.3. Equilibrium phase diagram for the binary alloy Ge-Te. Reprinted from [8.12], copyright (1988), with permission from Elsevier.

Here 'GeTe rt' represents the distorted rocksalt room temperature phase while 'GeTe ht' indicates the high-temperature cubic phase. In the lower temperature region, it can be seen that there is a slight solubility of Te in stoichiometric GeTe. In optical device applications, this leads to severe implications. While the minimum time required for laser-induced crystallization of stoichiometric GeTe is on the order of ~ 30 ns, composition deviations as small as 5 % can lead to minimum crystallization times as much as ten times larger [8.13]. This sensitivity led to the idea of alloying with Sb [8.14] to reduce the effects of the compositional fluctuations inevitable in manufacturing processes. Before closing this section, it is worth reporting on recent experiments on the structure of liquid GeTe which suggest the presence of a *reentrant Peierls distortion* in the liquid state [8.15]. While the presence of a Peierls distortion in the room temperature phase of GeTe disappears above T_c, neutron diffraction measurements indicate the presence of a Peierls distortion that appears above the melting point and only gradually disappears with increasing temperature. The application of external work in the form of pressure to room temperature GeTe can also induce a transition to the cubic phase at approximately 3 GPa [8.16].

8.2.2 Ge-Sb-Te Alloys

8.2.2.1 Metastable Ge-Sb-Te Alloys

One of the workhorses of optical memory as well as in the rapidly evolving field of electrical phase change memory are PCM alloys from the Ge-Sb-Te phase diagram. To date, these efforts have largely focused along the pseudo-binary tie line GeTe-Sb_2Te_3, although other parts of the phase diagram are being investigated as well [8.17]. The upper part of Fig. 8.4 shows the equilibrium alloy phase diagram along the pseudo-binary tie-line between GeTe and Sb_2Te_3 [8.5, 8.18]. It is interesting to note that much of the initial work on the properties of Ge-Sb-Te alloys was performed by curiosity driven research in the former Soviet Union, it was not until the end of the 1960s when the first paper on chalcogenide based switching appeared in the literature [8.2].

It was the pioneering effort of Yamada et al. at the beginning of the 1990s that recognized the utility of $GeSb_2Te_4$ and $Ge_2Sb_2Te_5$ alloys for optical recording both due to their long-term stability at ambient temperatures as well as their fast (50 ns) minimum crystallization times under the influence of fast laser irradiation. Surprisingly, it was found that under the influence of fast laser irradiation, these alloys transformed not into the equilibrium crystalline phases shown in the upper part of Fig. 8.4, but rather into simpler, higher symmetry fcc like phases [8.19]. It was suggested that substantially longer crystallization times would be required to obtain the nine layer hexagonal equilibrium structure. Both $GeSb_2Te_4$ and $Ge_2Sb_2Te_5$ alloys display surprisingly large changes in optical constants in transforming from the amorphous to the metastable crystalline phases from $(4.7 + i\,1.3)$ to $(6.9 + i\,2.6)$ and from $(5.0 + i\,1.3)$ to $(6.5 + i\,3.5)$, respectively [8.19].

It was suggested that the three stoichiometric phases $GeSb_4Te_7$, $GeSb_2Te_4$ and $Ge_2Sb_2Te_5$ would all be possible candidates for optical PCM alloys, however, the $Ge_2Sb_2Te_5$ composition was found to exhibit the greatest stability in the amorphous to crystalline transition with an activation energy of more than 2.2 eV as determined by a Kissinger analysis of differential scanning calorimetry data (DSC) [8.20]. A powder diffraction-based investigation of the average structure of $GeSb_2Te_4$ and $Ge_2Sb_2Te_5$ laser-crystallized alloys determined that they formed in the rocksalt structure. The rocksalt structure can be visualized as consisting of two interpenetrating fcc sublattices, one located at the origin and the other at (¼, ¼, ¼). Rietveld analysis under the assumption of isotropic temperature factors for each of the two unique sites in the lattice suggested that Te fully occupies one fcc sublattice while Ge and Sb substituting randomly on the other fcc sublattice with

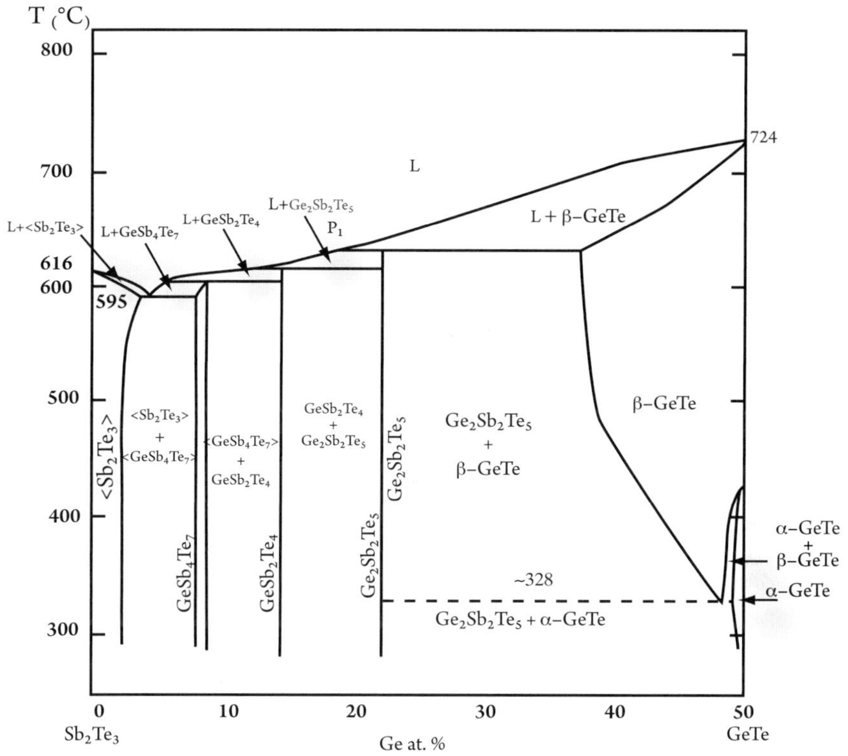

Fig. 8.4. Equilibrium phase diagram for alloys along the pseudo-binary tie-line GeTe-Sb_2Te_3 (adapted from references [8.5, 8.18]. In the lower section of the figure, the existence of metastable phases is also shown. Reprinted with permission from [8.19]. Copyright 1991, American Institute of Physics.

either 25 % ($GeSb_2Te_4$) or 20 % ($Ge_2Sb_2Te_5$) vacancies. The fact that x-ray diffraction of a site disordered system can only see the average structure, however, obscured the presence of Peierls like distortion in the metastable alloys as was subsequently shown both experimentally (see Section 8.3) and theoretically [8.4, 8.21]. Recently, it has been confirmed that the distorted rocksalt structure exists through the GeTe-Sb_2Te_3 pseudo-binary phase diagram with a Ge/Sb site vacancy concentration that is proportional to $(1\text{-}x)/(3\text{-}2x)$ for $(GeTe)_x(Sb_2Te_3)_{1\text{-}x}$ [8.22]. It was also reported that Bi substitutes on Sb sites in the metastable distorted rocksalt structure similar to that found for the Ge-Sb-Te system [8.23, 8.24]. Similar reports were made for Sn [8.25].

8.2.2.2 High-Pressure Effects on Metastable Ge-Sb-Te Alloys

The high concentration of vacancies and their role in the stability of the metastable structures of $GeSb_4Te_7$, $GeSb_2Te_4$, and $Ge_2Sb_2Te_5$ have been shown to be a necessity based upon EXAFS determination of local structure [8.26] as well as theoretically [8.4]. That vacancies are an intrinsic part of the structure of the metastable phase also follows from simple electron counting arguments. In the ideal rocksalt structure, there are six saturated covalent bonds, each populated by two electrons. For this to occur each atom must have six valence electrons. Sb requires one addition electron, while Ge needs two. These missing electrons are donated to Ge/Sb from the Te orbitals pointing into vacancies allowing all octahedral bonds to be saturated. It was also suggested that pressure may be a significant factor in the formation of the laser-induced amorphous phase as localized laser heating of a PCM embedded in a transparent dielectric matrix can lead to the transient generation of peak pressures of about 5 GPa [8.27]. Experimentally it was observed in diamond anvil diffraction experiments that hydrostatic compression of $Ge_2Sb_2Te_5$ leads to pressure-induced amorphization [8.27]. Fig. 8.5 shows a series of x-ray diffraction patterns of the metastable crystalline phase of $Ge_2Sb_2Te_5$ as a function of applied pressure.

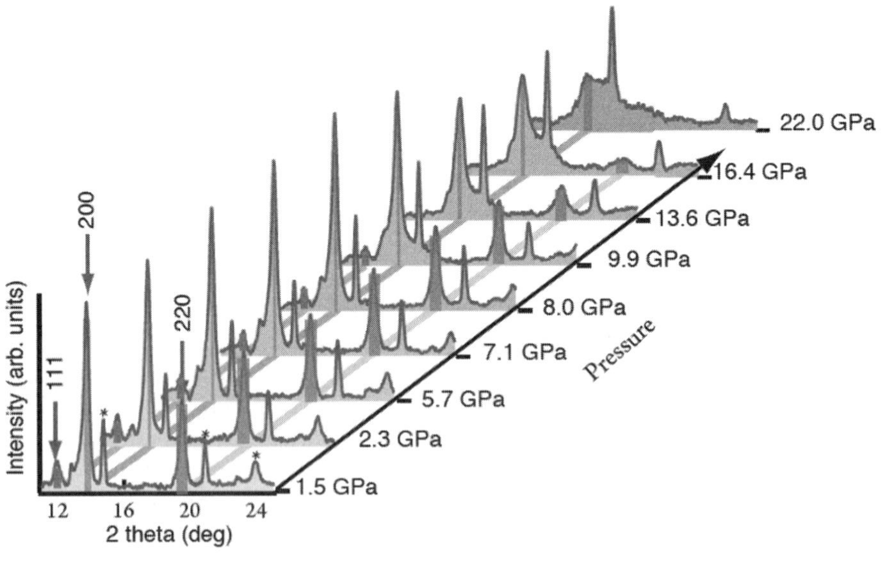

Fig. 8.5. A series of diffraction patterns from the low-temperature crystalline phase of $Ge_2Sb_2Te_5$ as a function of pressure. Data were taken in a diamond anvil cell. The '*' marks diffraction peaks originating from the NaCl pressure marker. Reprinted with permission from [8.27]. Copyright (2007) by the American Physical Society.

It can be seen that for pressures greater than approximately 22 GPa, the crystalline peaks vanish (the remaining peaks are of the NaCl pressure marker). EXAFS results showed that the amorphization initiated at about 10 GPa when the Sb atoms of the distorted rocksalt structure reach the center of the cell leaving vacancies next to be compressed [8.28]. For pressures above 10 GPa the anisotropy of atomic displacements from the selective annihilation of the vacancies leads to the preferential preservation of the (200) planes to higher pressures than the (220) planes.

It was found that the system remained in the amorphous phase when decompressed to room pressure. It was also reported that the $GeSb_2Te_4$ phase amorphized at lower pressures than $Ge_2Sb_2Te_5$ consistent with its higher concentration of vacancies. Near-edge x-ray absorption measurements also suggested that the local order in the laser-quenched and pressure amorphized were similar [8.29]. In addition, x-ray diffraction observations of in-situ annealing showed that the pressure-amorphized $Ge_2Sb_2Te_5$ samples crystallized into the distorted rocksalt metastable structure at approximately the same temperature as as-deposited amorphous films.

8.2.2.3 Ge-Sb-Te Equilibrium Structures

The equilibrium structure of GeTe-Sb_2Te_3 pseudo-binary alloys has been extensively explored in the literature starting with the discovery by several groups dating back to the late 1980s of a series of homologous compounds of the form $(GeTe)_n(Sb_2Te_3)_m$ where the number of layers is given in the unit cell by the expression $Z(2n+5m)$ where $Z = 1$ if $(2n + 5m)$ is a multiple of three (space group $P3m1$) otherwise $Z = 3$ (space group $R3m$) [8.30-33]. The existence of the series has been confirmed for a wide range of compositions $n \leq 4$ and $m \leq 4$, however, the extremely long unit cell ($GeSb_8Te_{13}$ ($n = 1$, $m = 4$) for example has 66 stacking layers in a unit cell) that can result will likely require long thermal treatments at high temperature to allow sufficient diffusion time [8.33]. In the current section, all references sited utilized either electron diffraction or single-crystal diffraction for analysis avoiding the ambiguities that arise from refinement of powder patterns. The structures can be described as close packed with each atom type, Ge, Sb, and Te possessing a different stacking sequence, for example for $GeSb_2Te_4$ the stacking sequence can be described as *AγBaCBa / CβAcBAc / BαCbACb/* where the capital letters represent Te layers, the small letters Sb layers, and the Greek letters Ge layers [8.30]. In addition to the long stacking sequences present in the bulk phases, all sites are fully occupied, c.f. there are no vacancies. Figure 8.6 shows the equilibrium structure of $Ge_2Sb_2Te_5$. It should also be mentioned here that the $(BiTe)_n(Sb_2Te_3)_m$ pseudo-binary has been found to occur in similar structures [8.30, 8.32, 8.34-36].

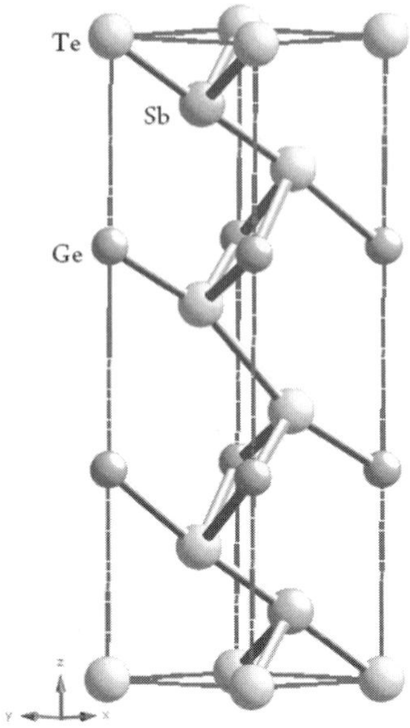

Fig. 8.6. The nine layer bulk equilibrium structure of the PCM alloy $Ge_2Sb_2Te_5$. All sites have unity occupancy.

8.2.2.4 Sb-Te Alloys

The Sb-Te alloys are also widely utilized as optical PCM specifically in the form of DVD-RW/DVD+RWs (Re-Writable Digital Versatile Discs), see also Chapter 12. The typical composition of these PCM alloys is approximately a Sb_2Te composition doped with Ag and In [8.37], although more recent mass production of DVD-RW discs has substituted a small amount of Ge in lieu of Ag and In. The addition of other dopants besides Ag and In allow a very large number of rewrite cycles [8.38]. As for the case of Ge-Sb-Te alloys, there is a low temperature crystalline phase of doped Sb_2Te that assumes the A7 structure with all atom types having equal probability of occupying a crystal site [8.39, 8.40]. The (Peierls distorted) A7 structure has rhombohedral symmetry and can be thought of as a distorted rocksalt structure in which the unit cell has been stretched along the [111] direction giving rise to a split octahedral coordination with three shorter bonds and three longer bonds. The equilibrium phase diagram [8.41] indicates the Sb_2Te phase field extends from approximately 18 to 36 % Te concentration. The crystal structure of the stoichiometric compound Sb_2Te has been solved using a single crystal [8.6] to be a nine layer structure.

The existence of a series of intermetallic compounds in the Sb-Te binary system has been recently demonstrated [8.42] despite the published phase diagrams indicating that the Sb_2Te (δ) phase field is a eutectic. As the scattering power of Sb and Te are nearly indistinguishable in x-ray diffraction, synchrotron radiation and anomalous dispersion was used. Much like the $(GeTe)_n(Sb_2Te_3)_m$ series, the unit cell for the formula unit $Sb_{2n}Te_3$ has a stacking number of $(2n+3)$ with an inverse relationship of stacking period and Te concentration. All atoms have a well-defined positions within the structure with unity occupancy. The number of layers with a single unit cell is given by $M = 2n + 3 = 3 \times 100/C_{Te}$, where C_{Te} is the Te concentration. For instance, this would imply that a $Sb_{72}Te_{28}$ composition would have a 11 layer sequence *Te-Sb-Te-Sb-Sb-Sb-Sb-Sb-Sb-Te-Sb-* which is then repeated three times based upon the ABC stacking of a cubic close-packed structure. As a consequence, the unit cell can become very long with for example the 11R structure having a c-lattice constant of ~63.89 Å. Some examples of the long repeat structure are shown in Fig. 8.7 along with the short and long bond distances of the Peierls distorted structure.

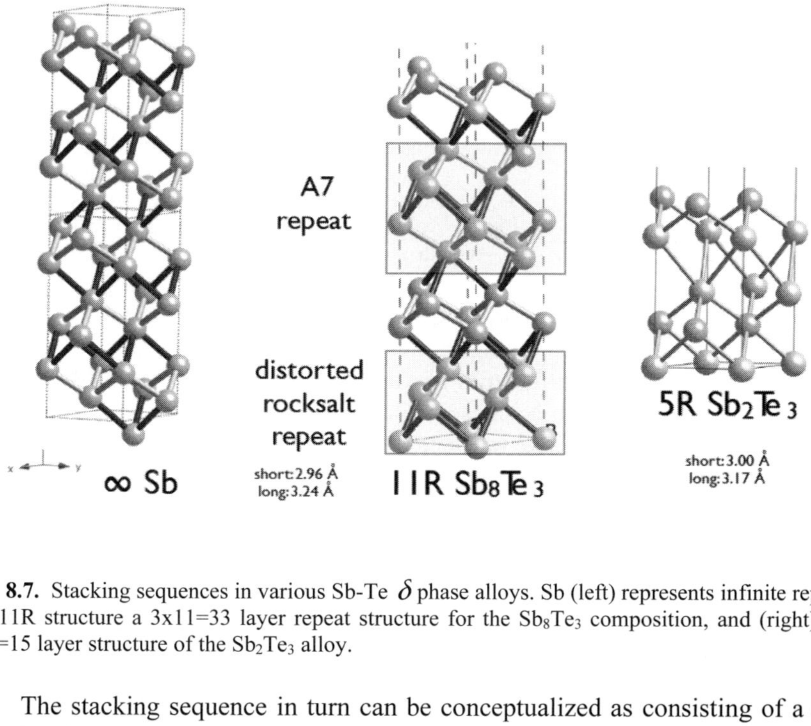

Fig. 8.7. Stacking sequences in various Sb-Te δ phase alloys. Sb (left) represents infinite repeat, the 11R structure a 3x11=33 layer repeat structure for the Sb_8Te_3 composition, and (right) the 3x5=15 layer structure of the Sb_2Te_3 alloy.

The stacking sequence in turn can be conceptualized as consisting of a distorted rocksalt subunit *Te-Sb-Te-Sb* followed by an A7 subunit of *Sb Sb-Sb-Sb-Sb-Sb-Sb* which is in turn followed by another distorted rocksalt subunit. As the Te concentration decreases, the number of Sb A7 subunits between distorted rocksalt-like subunits increases until the ultimate limit of pure Sb where the spacing between distorted rocksalt units goes to infinity. Considering the complicated stack sequence of the pure Sb_2Te phase, it is fascinating to note that doping with a small amount of Ge (or Ag and In) cause the structure to revert to a site-randomized distorted rocksalt structure. The small simple structural unit of doped Sb_2Te can be deduced to be an essential condition for its high speed of crystallization which is a necessity for high-speed PCM applications.

8.3 Short-Range Order

As mentioned in the Section 8.1.1, the local order present in some PCM materials, specifically the Ge-Sb-Te system, is quite different from that of a traditional glass. In a supercooled liquid, the inter-atomic spacing tends to be larger than in its crystalline counterpart. This is a consequence of the nature of the inter-atomic potentials where the repulsive force rapidly increases for inter-atomic spacings less than the equilibrium (crystalline) distance r_0, while gradually approaching zero for distances larger than r_0. A schematic view of this is shown in Fig. 8.8.

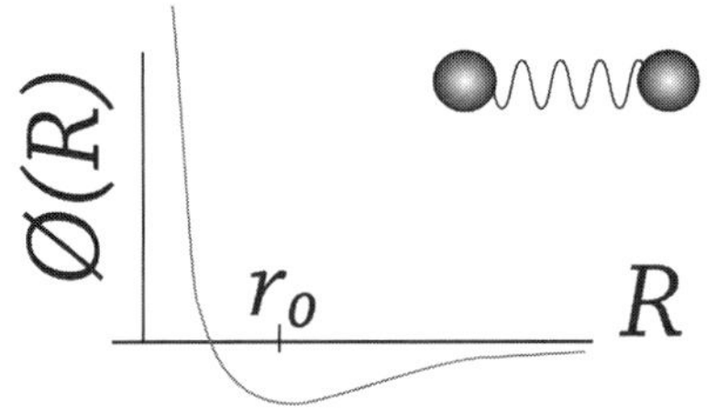

Fig. 8.8. A schematic view of the inter-atomic potential in a solid. For distances less than the equilibrium spacing r_0, the potential rises rapidly, while it falls off more slowly for distances greater than r_0.

Recent measurements of the short-range order in PCM, however, have shown the rather surprising result of exhibiting shorter distances in the amorphous than in the crystalline state. Subsequent neutron based pair distribution function (PDF) studies of the metastable crystalline phase of $Ge_2Sb_2Te_5$ confirmed a significant difference between the average structure determined by Rietveld refinement and that determined by a PDF analysis [8.43]. The combination of atomic selectivity and sensitivity to local order of extended x-ray absorption analysis (EXAFS) allowed a more definitive description of local order to be made [8.26].

8.3.1 X-ray Absorption

X-ray absorption fine structure (XAFS) utilizes the details of how x-rays are absorbed by an atom in the vicinity of a core-level absorption edge. The variation in the absorption probability is a consequence of the local environment of the atom. XAFS is especially sensitive to coordination chemistry, oxidation state, the types and distances of neighbor atoms, and the coordination number about the selected element. XAFS is a consequence of the quantum mechanically interference of the outgoing photoelectron originating from the absorbing site.

This can be described mathematically as a transition from a localized core state in the presence of a photon to a final state with a core hole and a photoelectron or:

$$\mu(E) \propto \left|\langle i|H|f\rangle\right|^2 \tag{8.1}$$

where $|i>$ represents the initial state consisting of an x-ray photon and the core electron, the final state $|f>$ consists of a photoelectron and core-hole, and $H = \vec{p} \cdot \vec{A}$ represents the electron-photon coupling where $\vec{A}$ is the vector potential and $\vec{p}$ is the photoelectron momentum. While the initial state $|i>$ consisting of a deep core electron feels little effect from its environment, the final state $|f>$ is affected due to the presence of neighboring atoms and can be rewritten as

$$|f\rangle = |f_0\rangle + |\Delta f\rangle \tag{8.2}$$

where $|f_0>$ represents the isolated atom and $|\Delta f>$ represents the effects of the neighboring atom. Using first-order perturbation theory yields

$$\mu(E) \propto \left|\langle i|H|f_0\rangle\right|^2 \left(1 + \langle i|H|\Delta f\rangle \frac{\langle f_0|H|i\rangle^*}{\left|\langle i|H|f_0\rangle\right|^2} + C.C.\right) \tag{8.3}$$

where $C.C.$ is the complex conjugate. This yields a relationship between the absorption $\mu(E)$ and $\chi(E)$ of the form

$$\mu(E) = \mu_0\left[1 + \chi(E)\right] \tag{8.4}$$

and $\chi(E)$ represents the *fine structure* (oscillatory part) of the x-ray absorption spectrum

$$\chi(E) = \frac{\mu(E) - \mu_0(E)}{\Delta\mu(E)}. \tag{8.5}$$

The fine structure $\chi(E)$ can thus be seen to be

$$(\chi(E) \propto \langle i|H|\Delta f\rangle. \tag{8.6}$$

In the dipole approximation, we can thus write

$$\chi(E) = \int dr\, \delta(r)\, e^{ikr}\, \Psi(r) = \Psi_{scatt}(0). \tag{8.7}$$

Here $\delta(r)$ is the Dirac delta function, k is magnitude of the momentum of the photoelectron, r is a radial vector centered on the absorbing atom, and Ψ_{scatt} represents the contributions to the final wavefunction at the origin of the absorbing atom. The core state has been approximated as a delta function to take into account its localization. This equation allows the simplified interpretation of the fine structure as the complex sum of the amplitudes of the backscattered photoelectron contributions from the neighboring atoms at the absorbing atom. A schematic outline of the x-ray absorption process is indicated in Fig. 8.9.

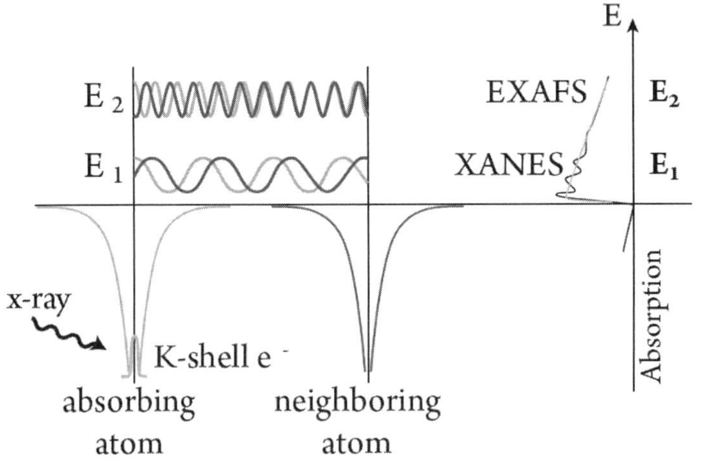

Fig. 8.9. Schematic view of the x-ray absorption process.

The interference is more easily interpreted in terms of the wavenumber k of the photoelectron

$$k = \sqrt{\frac{2m\,(E - E_0)}{\hbar^2}} \tag{8.8}$$

where E is the energy of the incident x-ray, m is the electron mass, $\hbar$ is the reduced Planck contant, and E_0 is the binding energy of the core electron.

In this example, an incoming x-ray is absorbed by a K-shell electron giving rise to a photoelectron. Due to the spherical symmetry of the K-state, the outgoing photoelectron expands as a spherical shell and in turn scatters off neighboring atoms. The phase-retarded backscattered components of the photoelectron contribute the to the final state giving rise to an expression of the form

$$\chi(k) = \sum_i \Im\left(\frac{N_i\, S_0^2\, F_i^{eff}(k)}{kR_i^2} \sin\left(2kR_i + \varphi_i(k)\right) e^{-2\sigma_i^2 k^2} e^{-2R_i/\lambda(k)} \right). \tag{8.9}$$

The sum is taken over the real part $\Im$ of contributions from shells of identical atoms at approximately the same distance from the absorbing atom and coordination number N_i. Here $F_i^{eff}(k)$ and $\varphi_i(k)$ are the scattering factors and phase shifts due to the atom i and σ_i^2 is mean-square relative displacement (MSRD) of the bond distance R_i (assuming a Gaussian atomic distribution). The remaining term the *amplitude reduction factor* S_0^2 representing the collective relaxation of the other electrons in the absorbing atom in response to the creation of a core-hole is usually taken to be a constant.

The effects of inelastic scattering and the finite core hole lifetime are included by approximating the outgoing photoelectron wavefunction as a damped spherical wave with damping factor λ. A typical plot of λ vs. k is shown in Fig. 8.10.

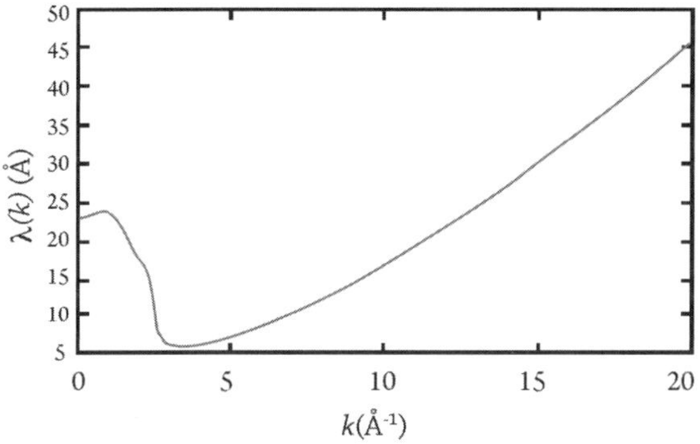

Fig. 8.10. A plot of the mean free path of the photoelectron as a function of momentum wavenumber k. Reprinted with permission from [8.44]. Copyright (2000) by the American Physical Society.

These two effects in conjunction with the inverse square distance term combine to limit the information available from XAFS to *short-range* order.

The discussion to this point has assumed that the scattering problem could be subdivided into contributions from two-atom scattering paths. For k values less than about 3 Å^{-1}, this binary approximation breaks down and the system must be solved using alternative full multiple scattering techniques; this region is commonly referred to as *x-ray absorption near-edge structure* (XANES). Due to the collective nature of the excitation process, XANES spectra contain indirectly three-dimensional information about the structure being examined. At the same time, for small k values the mean free path λ becomes significantly longer and spectra contain information out to distances as far as 15 Å. In contrast, the *extended x-ray absorption fine spectra* (EXAFS) region from > 3 Å is well served by the binary approximation and hence Eq. 8.9. The limited lifetime of the core-hole, inelastic scattering, and the k^2 dependence of the MSRD term conspire to limit information from EXAFS to typically values less than 6 Å from the absorbing atom.

Experimental fitting of EXAFS data thus consists of fitting the parameters R_i, N_i, σ^2 to experimental data Fourier transformed into real-space while the values of $F_i^{eff}(k)$ and $\varphi_i(k)$ are computed using first-principles multiple scattering theory code. It is important to note that in the plots of experimental data both the experimental and fitted data include the effects of the phase shift term δ hence plotted bond lengths are always shorter than true bond lengths. Fitting of XANES data is more difficult in that while a theoretical spectrum can be computed using first-principles multiple scattering theory for a given atom configuration, there is no simple relationship between a given atomic configuration and the corresponding XANES spectrum.

8.3.1.1 Short-range Order in Crystalline GeTe

Transmission XAFS experiments were carried out on GeTe sputter deposited on (x-ray transparent) Kapton foil. Thermally annealed (250 ºC for 3 hours) samples were then measured at both the Ge and Te *K*-edges at beamline BL01B1 of SPring-8 in Hyogo, Japan. X-ray fluorescence measurements confirmed the composition of both the as-deposited and annealed films were within one percent of the targeted GeTe composition. Fig. 8.11 (left) shows (non-phase corrected) Fourier transforms of k^3 weighted $\chi(R)$ data for the Te *K*-edge.

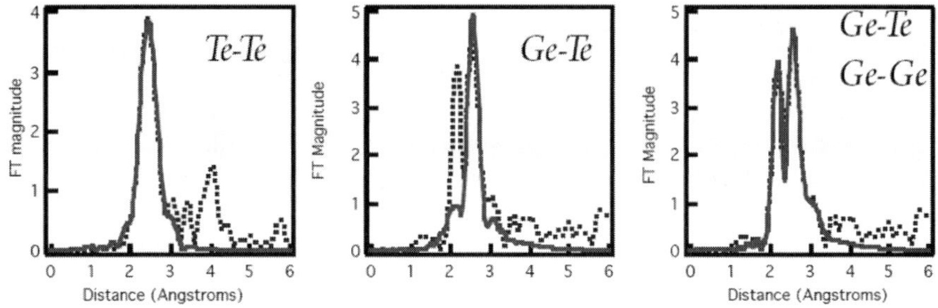

Fig. 8. 11. k^3 weighted spectra of Fourier transformed χ data and fitting results for (left) Te *K*-edge data with fitted Te scattering, (center) Ge *K*-edge data and fitted scattering assuming only Ge-Te bonds, and (right), Ge *K*-edge data with fitted scattering assuming both Ge-Te and Ge-Ge bonds (see text for details). Reprinted with permission from [8.45]. Copyright (2003) by the American Institute of Physics.

Also plotted is the fitted Te profile for the first coordination shell. Fig. 8.11 shows the first coordination shell Ge *K*-edge data and corresponding fit for the case (center) only Ge-Te bonds are assumed and (right) for the case when both Ge-Ge and Ge-Te bonds are assumed to exist. Both Ge and Te edges were fit simultaneously. The fit showed that both Ge and Te have six atoms of the opposite type located at two different distances (three at each distance). The shorter Ge-Te bond was 2.801 ± 0.003 Å while the longer bond was 3.146 ± 0.010 Å [8.45]. While the fit for the Te *K*-edge was excellent, the introduction of ~ 10 % Ge-Ge bonds was necessary to fit the *K*-edge data implying the presence of approximately 10 % vacancies on Ge sites.

8.3.1.2 Short-range Order in Amorphous GeTe

Short-range order in as-sputter deposited GeTe films were analyzed using XAFS at BL01B1 of SPring-8 at both the Ge and Te *K*-edges. The Ge-Te bond length was found to be 2.60 Å with coordination numbers of 3.3 ± 0.3 for Ge (with Ge-Te and Ge-Ge bond present) and 2.2 ± 0.2 for Te. As the lattice constant of crystalline GeTe is approximately 5.99 Å, a Ge atom located in a tetrahedral symmetry position would have a Ge-Te bond length of 2.60 Å in excellent agreement with the XAFS determined bond length. The previous discussion suggests that an overall 4(Ge):3(Te)-coordination is present in amorphous GeTe with possibly a small contribution from 3(Ge):3(Te)-coordinated structural units [8.46]. Raman measurements of the same films indicated mode hardening of the amorphous phase relative to the crystalline phase of GeTe consistent with the shorter Ge-Te bond length in the amorphous phase.

Fig. 8.12 shows experimental plots of Ge *K*-edge XANES data (left) taken on both amorphous and crystallized GeTe films and theoretical spectra (right) calculated using the full multiple scattering code Feff8 [8.44] assuming the distorted rocksalt structure for crystalline GeTe and Ge in a tetrahedral coordination for Ge in amorphous GeTe [8.46].

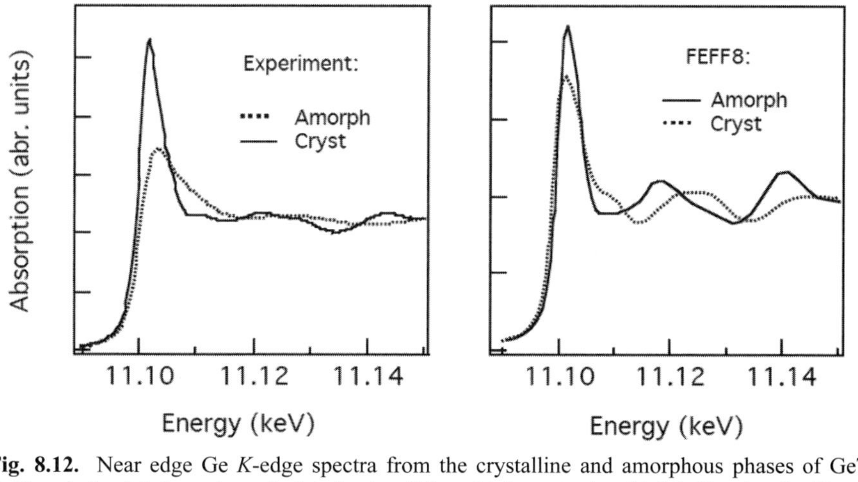

Fig. 8.12. Near edge Ge *K*-edge spectra from the crystalline and amorphous phases of GeTe (left) and simulated spectra calculated using full multiple scattering (right). Reprinted with permission from the Institute of Physics, from [8.46].

The simulated spectra qualitatively match the experimental data strongly suggesting 4(Ge)-coordination dominates in amorphous GeTe.

Raman was also used to examine the amorphous phase of GeTe [8.47]. Fig. 8.13 shows intensity reduced Raman spectra from amorphous and crystallized GeTe; the sample consisted of a 100 nm thick sputtered GeTe layer grown on Kapton with a $(ZnS)_{0.85}(SiO_2)_{0.15}$ encapsulation layer.

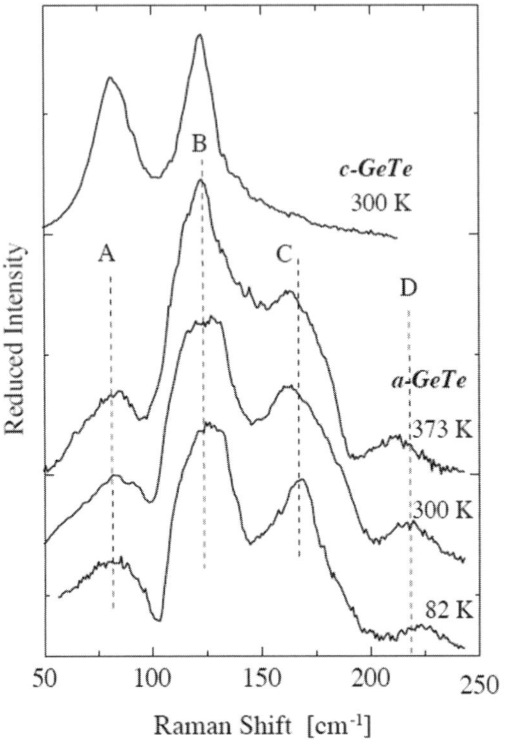

Fig. 8.13. Reduced intensity Raman data for crystalline and amorphous GeTe taken at three different temperatures. The marked peaks have been deconvoluted into vibrational modes of possible tetrahedral clusters present. Reprinted with permission from the Institute of Physics, from [8.47].

The experimental spectra for the amorphous phase were analyzed based upon identification of tetrahedral clusters. Group theory predicts for T_d symmetry that four Raman active modes should be present or $\Gamma(T_d) = A_1 + E + 2\ F_2$. The authors identified two components of peak A of Fig. 8.13 as originating from the $GeTe_4$ $A_1[\nu_4(F_2)]$ and $A_1[\nu_2(F_2)]$ bending modes at 63 and 86 cm^{-1}, respectively. Peak B components at 109 and 125 cm^{-1} were identified as $GeTe_4$ $B_1[\nu_1(A_1), n = 0]$ and $B_1[\nu_1(A_1), n = 1, 2]$ corner-sharing symmetric stretching modes. Peak C was identified as coming from $GeTe_4$ $C_1[\nu_1(A_1), n = 0]$ and $C_1[\nu_1(A_1), n = 1, 2]$ edge-sharing symmetric stretching modes at 160 and 180 cm^{-1}, while peak D originated from the $GeTe_4$ antisymmetric stretching mode $D_1[\nu_3(F_2)]$ at 211 cm^{-1}. The above scenario led to the $GeTe_{4-n}Ge_n$ tetrahedra structural model shown in Fig. 8.14. The figure shows an apex- and edge-bridged tetrahedra in the absence of Te-Te bonds. A key point of this model is that the configuration with small displacement changes can transform into the distorted rocksalt structure of crystalline GeTe. In particular, the Ge^c atom is surrounded by six Te atoms (thin circles) with three atoms lying closer and the other three slightly more distant. The Te^c atom on the other hand exists in an environment that offers local octahedral coordination of Ge atoms (thick circles).

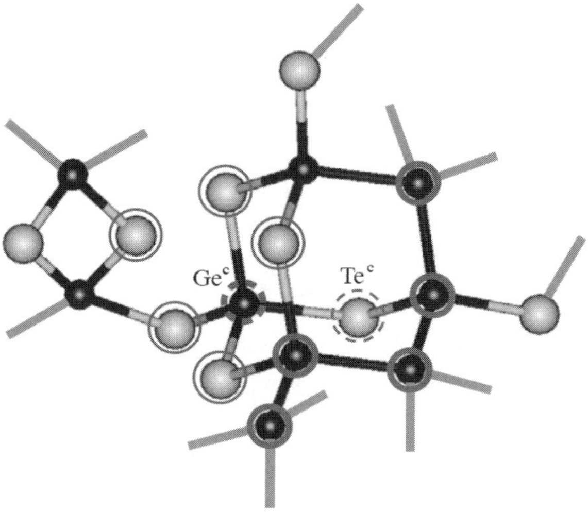

Fig. 8.14. Postulated short and medium range order structural model for amorphous GeTe. Small dark and larger gray spheres represent Ge and Te atoms, respectively. The Ge(Te) atoms circled in dark (light) represent the atoms that will form the first coordination spheres of the dashed encircled Ge and Ge atoms, respectively, after crystallization. Reprinted with permission from the Institute of Physics, from [8.47].

8.3.1.3 Short-range Order in Crystalline $Ge_2Sb_2Te_5$

Thin films of $Ge_2Sb_2Te_5$ were grown on optical disc substrates and regions were both laser-crystallized and amorphized using a commercial optical disc testing system. The dielectric encapsulation layers were then removed and XAFS was measured on device structure $Ge_2Sb_2Te_5$ films using conversion electron yield at BL01B1 of SPring-8. Samples were measured at the Ge, Sb, and Te *K*-edges and all data analysis was simultaneously carried out on all three edges [8.26]. It is significant that these data were taken on laser re-crystallized and re-amorphized $Ge_2Sb_2Te_5$ as the structure of the as-grown phase may not necessarily be the same as that of the laser re-amorphized phase. Fig. 8.15 shows XANES spectra for Ge, Sb, and Te *K*-edges of as-deposited (rf sputtered), laser re-crystallized, and laser re-amorphized $Ge_2Sb_2Te_5$ films. As multiple scattering processes dominate in XANES spectra and the electron mean free path is > 10 nm, XANES is often used as a type of *signature* for a given structure. Here it can be clearly seen that there are significant differences between the laser-reamorphized and as-deposited spectra and in particular the differences are dramatic for the Te edge implying large changes in the local bonding environment about the Te atoms. While it is often convenient to use as-deposited `amorphous' material to represent the re-amorphized phase in experiments, it is clear from this example that significant differences between re-amorphized materials can exist.

Fig. 8.15 shows k^3 weighted Fourier transformed $\chi(R)$ data for the Ge, Sb, and Te *K*-edges.

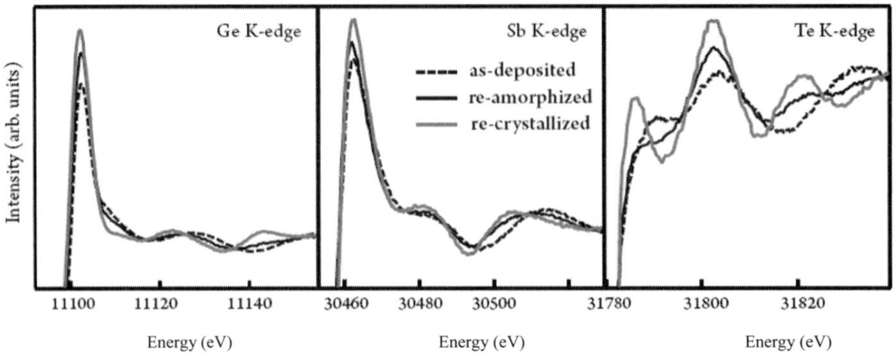

Fig. 8.15. XANES spectra for as-deposited (dashed), laser re-amorphized (black), and laser re-crystallized (gray) films for the Ge, Sb, and Te *K*-edges.

The data have not been corrected for the effect of the phase shift δ. The laser re-amorphized phase (solid line) is clearly stronger and peaks at lower values of *R* than the laser re-crystallized peak (gray line). It is also worth noting that the as-deposited structure (dashed) is different from that of the laser re-amorphized phase. A fit of the experimental data yields the following bond lengths: Te-Ge, 2.83 Å and Te-Sb, 2.91 Å. No Sb-Ge or Ge-Ge bonds were detected. A second nearest neighbor Te-Te peak was found to correspond to a distance of 4.26 Å. To accurately model the data, it was found necessary to introduce a bi-modal bond length distribution where shorter and longer Ge-Te and Sb-Te bonds were present; the bond-lengths of the longer bonds were found to be ~ 3.2 Å, a value longer than expected from the periodicity of the system suggesting the presence of a Peierls distortion like splitting in bond lengths. This led the authors to suggest a structural model for the crystalline phase in which due to the differences in the covalent radii of the constituent species, Ge, and to a lesser extent Sb, are shifted from their fcc sites, giving rise to a system of shorter and longer bonds in an overall buckled structure. The appearance of a second nearest neighbor peak in the Te spectra strongly suggests that the structural integrity of the distorted rocksalt structure may in large part be due to the Te sublattice.

In the amorphous phase bond lengths of Te-Ge and Te-Sb were found to become stronger and shorter, 2.61 Å and 2.85 Å respectively. The Te second nearest neighbor peak, although considerably weaker, was noted to remain. The MRSD was found to decrease from 0.02 $Å^2$ in the re-crystallized state to 0.008 $Å^2$ in the re-amorphized state. This was interpreted in terms of a molecular crystal analogy (for instance the case of Se) where the rupture of inter-molecular bonds results in the strengthening (shortening) of intra-molecular bonds. This suggested a model for the amorphous state in which the weaker, longer Te-Ge and Te-Sb bonds are broken upon laser excitation and consequentially intra-block bonds become shorter and stronger; the amorphous phase can thus be considered more ordered *locally* than the crystalline phase. This deduction is supported by Raman data of the re-crystallized and re-amorphized phases seen in Fig. 8.16 where it clear that the modes of the amorphous phase are more rigid (located at higher wavenumber) than the crystalline phase.

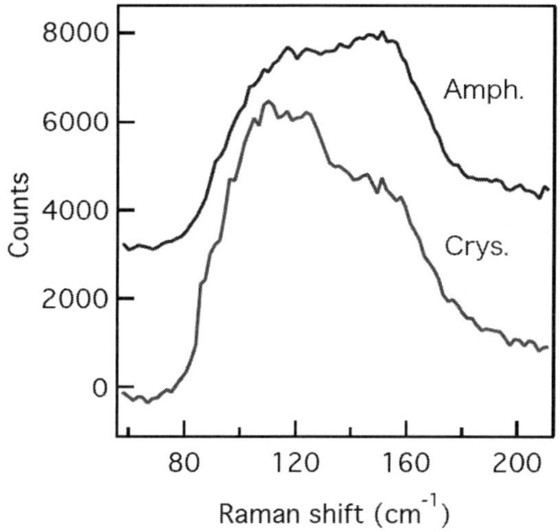

Fig. 8.16. Raman data taken from laser re-crystallized (bottom) and laser re-amorphized (top) $Ge_2Sb_2Te_5$ thin films. Note the mode hardening present in the re-amorphized films. Reprinted by permission from Macmillan Publishers Ltd. Nature Materials from [8.26], copyright (2004).

XANES data from the Ge *K*-edge for both the re-crystallized and re-amorphized films are shown in Fig. 8.17 (left) along with theoretical spectra calculated using full multiple scattering code (right).

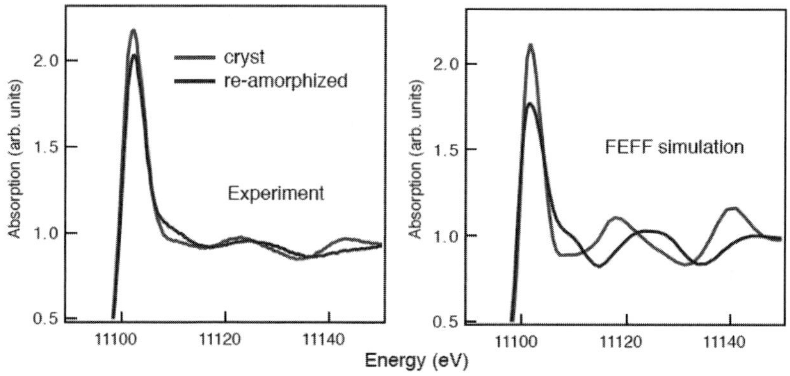

Fig. 8.17. Ge K-edge XANES data (left) for the laser re-crystallized and laser re-amorphized phases of $Ge_2Sb_2Te_5$ and theoretical (right) spectra of $Ge_2Sb_2Te_5$ assuming the distorted rocksalt model for the crystalline phase and Ge at a tetrahedral site in the amorphous phase. Reprinted by permission from Macmillan Publishers Ltd. Nature Materials from [8.26], copyright (2004).

It was found that the best agreement with experiment for the re-amorphized phase was found when Ge acquired its preferred tetrahedral environment. It was also noted that the Ge-Te bond length determined independently from EXAFS corresponds exactly to the Ge-Te distance calculated from crystallographic data.

These and other arguments led the authors to propose the so-called *umbrella flip model*. The transformation process is shown in Fig. 8.18 in which the local structure about Ge atoms is shown in re-crystallized (left) and re-amorphized (right) surroundings.

Fig. 8. 18. Schematic view of the switching process in $Ge_2Sb_2Te_5$. For simplicity only Te (light) and Ge (dark) atoms have been drawn. Ge (left) sits in an octahedral site in the crystalline phase while it switches to a tetrahedral site in the amorphous phase (right). Reprinted by permission from Macmillan Publishers Ltd. Nature Materials from [8.26], copyright (2004).

In the figure, a Ge atom is shown within an fcc structure formed by local Te atoms (Sb atoms are not shown for clarity). The Ge atom occupies octahedral and tetrahedral symmetry positions in the re-crystallized and re-amorphized states, respectively. The stronger (weaker) covalent bonds are indicated by thicker (thinner) lines. An intense laser pulse causes the longer bonds to break and the Ge atom moves into the tetrahedral symmetry position shown on the right. As three bonds remain intact during the transformation process, the authors suggested that the transition may not be molten in a conventional sense.

8.3.2 Short Range Order in Sb-Te Alloys

Few reports have been published on the short range order in Sb-Te alloys [8.48, 8.49]. Local structure in the crystallized phase of an Sb_2Te phase region alloy was reported to be consistent with that of the Sb_2Te_3 crystalline structure which in light of the more detailed subsequent study of long-period Sb-Te alloys by Kifune et al. is not surprising [8.42]. It was, however, observed that the local structure of the amorphous (as-deposited) phase was NaCl like with shorter bond lengths about the Sb site (2.86 Å) and Te site (2.83 Å) than the corresponding lengths in the crystallized structure of 2.89 Å and 2.87 Å, respectively possibly suggesting a similar distortion as seen for the case of Ge-Sb-Te alloys.

8.3.2.1 Conclusions

The above review of structure in PCM alloys is, of necessity, just a snapshot of current research and will undoubtedly continue to evolve. There are also certainly additional topics that could have been included were length and time constraints absent. A few active studies are now attempting to explore the limits of scaling of PCM materials [8.50, 8.51], see also Chapter 6. Another interesting trend is in the use of XAFS to probe local order on the sub-nanosecond time scale using x-ray pulses from a synchrotron and synchronized pump sources [8.52, 8.53]. One of the most challenging aspects of exploring the materials science of PCM alloys is the complexity of the multinary systems being explored and often in the presence of metastable phases. With the prospect of electrical PCM applications on the research horizon, the new insights gained into a deeper theoretical understanding [8.54] of phase change materials hold great promise for insightful development of optimized PCM alloys and the fine-tuning of their properties.

Acknowledgements. The XAFS measurements were carried out at SPring-8 under approval of JASRI proposals 2001B-0099-NP-nx, 2001B0099-BX, and long-term proposal 2005A0004-LX-np.

References

[8.1] Kolomiets, B. T.: Vitreous semiconductors (I). Phys. Stat. Sol. **B 7**, 359-372 (1964)

[8.2] Ovshinsky, S.: Reversible electrical switching phenomena in disordered structures. Phys. Rev. Lett. **21**, 1450-1453 (1968)

[8.3] Wuttig, M, Yamada, N.: Phase-change materials for rewriteable data storage. Nature Mater. **6**, 824-832, (2007)

[8.4] Wuttig, M., Lüsebrink, D., Wamwangi, D., Welnic, W., Gillessen, M., Dronskowski, R.: The role of vacancies and local distortions in the design of new phase-change materials. Nature Mater. **6**, 122-128 (2007)

[8.5] Abrikosov, N. K., Danilova-Dobryakova, G.: An investigation of the structural diagram of Sb_2Te_3 - GeTe. Izvestiya Akademii Nauk SSSR, Neorganicheskie Materialy **1**, 204-207 (1965)

[8.6] Agafonov, V., Rodier, N., Céolin, R., Bellissent, R., Bergman, C., Gaspard, J.: Structure of Sb_2Te. Crystal Structure Communications **47**, 1141-1143 (1991)

[8.7] Petrov, I., Imamov, R., Pinsker, Z.: Electronographic determination of the structures of $Ge_2Sb_2Te_5$ and $GeSb_4Te_7$. Sov. Phys. Cryst. **13**, 339-344 (1968)

[8.8] Goldak, J., Barrett, C., Innes, D., Youdelis, W.: Structure of alpha GeTe. J. Chem. Phys. **44**, 3323 (1966)

[8.9] Gaspard, J. P., Pellegatti, A., Marinelli, F., Bichara, C.: Peierls instability in covalent structures I. electronic structure, cohesion and the Z=8-N rule. Phil. Mag. B **77**, 727-744 (1998)

[8.10] Chattopadhyay, T., Boucherle, J., Vonschnering, H.: Neutron-diffraction study on the structural phase-transition in GeTe. J. Phys. C **20**, 1431-1440 (1987)

[8.11] Steigmeier, E.: Soft phonon mode and ferroelectricity in GeTe. Solid State Comm. **8**, 1275-1279 (1970)

[8.12] Balde, L., Legendre, B., Balkhi, A.: Etude du diagramme d'equilibre entre phases du systeme ternaire germanium-etain-tellure. J. Alloys Compd. **216**, 285-293 (1995)

[8.13] Chen, M., Rubin, K. A., Barton, R. W.: Compound materials for reversible, phase-change optical data storage. Appl. Phys. Lett. **49**, 502-504 (1986)

[8.14] Ovshinsky, S.: Optically induced phase changes in amorphous materials. J. Non-Cryst. Solids **141**, 200-203 (1992)

[8.15] Raty, J., Godlevsky, V., Ghosez, P., Bichara, C., Gaspard, J. P., Chelikowsky, J. R.: Evidence of a reentrant Peierls distortion in liquid GeTe. Phys. Rev. Lett. **85**, 1950-1953 (2000)

[8.16] Onodera, A., Sakamoto, I., Fujii, Y., Mori, N., Sugai, S.: Structural and electrical properties of GeSe and GeTe at high pressure. Phys. Rev. B **56**, 7935-7941 (1997)

[8.17] Lankhorst, M. H. R., Ketelaars, B. W. S. M. M., Wolters, R. A. M.: Low-cost and nanoscale non-volatile memory concept for future silicon chips. Nature Mater. **4**, 347-352 (2005)

[8.18] Bordas, S., Clavaguera-Mora, M., Legendre, B., Chhay,: Phase diagram of the ternary system Ge-Sb-Te. ii. The subternary Ge-GeTe-Sb_2Te_3-Sb. Thermochim. Acta **107**, 239-265 (1986)

[8.19] Yamada, N., Ohno, E., Nishiuchi, K., Akahira, N., Takao, M.: Rapid phase-transitions of GeTe-Sb_2Te_3 pseudobinary amorphous thin-films for an optical disk memory. J. Appl. Phys. **69**, 2849-2856 (1991)

[8.20] Yamada, N., Ohno, E., Akahira, N., Nishiuchi, K., Nagata, K., Takeo, M.: High speed overwritable phase change optical disk material. Jpn. J. Appl. Phys. **26**, 61-66 (1987)

[8.21] Welnic, W., Pamungkas, A., Detemple, R., Steimer, C., Blügel, S., Wuttig, M.: Unravelling the interplay of local structure and physical properties in phase-change materials. Nature Mater. **5**, 56-62 (2005)

[8.22] Matsunaga, T., Morita, H., Kojima, R., Yamada, N., Kifune, K., Kubota, Y., Tabata, Y., Kim, J. J., Kobata, M., Ikenaga, E., Kobayashi, K.: Structural characteristics of GeTe-rich GeTe-Sb_2Te_3 pseudobinary metastable crystals. J. Appl. Phys. **103**, 093511 (2008)

[8.23] Matsunaga, T., Umetani, Y., Yamada, N.: Structural study of a $Ag_{3.4}In_{3.7}Sb_{76.4}Te_{16.5}$ quadruple compound utilized for phase-change optical disks. Phys. Rev. B **64**, 184116 (2001)

[8.24] Wang, K., Wamwangi, D., Ziegler, S., Steimer, C., Wuttig, M.: Influence of Bi doping upon the phase change characteristics of $Ge_2Sb_2Te_5$. J. Appl. Phys. **96**, 5557-5562 (2004)

[8.25] Wang, K., Wamwangi, D., Ziegler, S., Steimer, C., Kang, M., Choi, S., Wuttig, M.: Influence of Sn doping upon the phase change characteristics of $Ge_2Sb_2Te_5$. Phys. Stat. Sol. A **201**, 3087-3095, (2004)

[8.26] Kolobov, A., Fons, P., Frenkel, A. I.., Ankudinov, A.L.: Understanding the phase-change mechanism of rewritable optical media. Nature Mater. **3**, 703-708 (2004)

[8.27] Kolobov, A. V., Haines, J., Pradel, A., Ribes, M., Fons, P., Tominaga, J., Katayama, Y., Hammouda, T., Uruga, T.: Pressure-induced site-selective disordering of $Ge_2Sb_2Te_5$: A new insight into phase-change optical recording. Phys. Rev. Lett. **97**, 035701 (2006)

[8.28] Fons, P., Kolobov, A. V., Tominaga, J., Katayama, Y.: High-pressure induced structural changes in metastable $Ge_2Sb_2Te_5$ thin films: An x-ray absorption study. Nucl. Inst. Meth. B **238**, 160-162 (2005)

[8.29] Kolobov, A. V., Haines, J., Pradel, A., Ribes, M., Fons, P., Tominaga, J., Steimer, C., Aquilanti, G., Pascarelli, S.: Pressure-induced amorphization of quasibinary GeTe-Sb_2Te_3: The role of vacancies. Appl. Phys. Lett. **91**, 021911 (2007)

[8.30] Kuypers, S., van Tendeloo, G., van Landuyt, J., Amelinckx, S.: Electron microscopic study of the homologous series of mixed layer compounds $R_2Te_3(GeTe)_n$(R= Sb, Bi). J. Sol. State Chem. **76**, 102-108 (1988)

[8.31] Karpinsky, O., Shelimova, L., Kretova, M., Fleurial, J.: An x-ray study of the mixed-layered compounds of $(GeTe)_n(Sb_2Te_3)_m$ homologous series. J. Alloys and Compounds **268**, 112-117 (1998)

[8.32] Shelimova, L., Karpinskii, O., Zemskov, V., Konstantinov, P.: Structural and electrical properties of layered tetradymite-like compounds in the (GeTe)-$(Bi_2Te_3)_m$ and GeTe-$(Sb_2Te_3)_m$ systems. Inorg. Mat. **36**, 235-242 (2000)
[8.33] Shelimova L, Karpinskii O, Konstantinov P, Kretova M, Avilov E, Zemskov V: Composition and properties of layered compounds in the (GeTe)-$(Sb_2Te_3)_m$ system. Inorg. Mat. **37**, 342--348 (2001)
[8.34] Karpinskii, O., Shelimova, L., Kretova, M., Fleurial, J.: Structural study of ternary layered compounds in the $(GeTe)_n \cdot (Bi_2Te_3)_m$ and $(GeTe)_n \cdot (Sb_2Te_3)_m$ homologous series. Inorg Mat **34**, 225-232 (1998)
[8.35] Shelimova, L., Karpinskii, O., Kosyakov, V., Shestakov, V., Zemskov, V., Kuznetsov, F.: Homologous series of layered tetradymite-like compounds in Bi-Te and (GeTe)-$(Bi_2Te_3)_m$ systems. J. Struct. Chem. **41**, 81-87 (2000)
[8.36] Shelimova, L., Karpinskii, O., Konstantinov, P., Avilov, E., Kretova, M., Zemskov, V.: Crystal structures and thermoelectric properties of layered compounds in the ATe-$(Bi_2Te_3)_m$ (A = Ge, Sn, Pb) systems. Inorg. Mat. **40**, 451-460 (2004)
[8.37] Iwasaki, H., Ide, Y., Harigaya, M., Kageyama, Y., Fujimura, I.: Completely erasable phase change optical disk. Jpn. J. Appl. Phys. **31** (2B Part 1), 461-465 (1992)
[8.38] Tominaga, J., Kikukawa, T., Takahashi, M., Kato, T., Aoi, T.: Optical phase change disc without bulk laser initialization and a quick bulk initialization structure. Jpn. J. Appl. Phys. **36**, 3598-3601 (1997)
[8.39] Matsunaga, T., Umetani, Y., Yamada, N. Structural study of a $Ag_{3.4}In_{3.7}Sb_{76.4}Te_{16.5}$ quadruple compound utilized for phase-change optical disks. Phys. Rev. B **64**, 184116 (2001)
[8.40] Tominaga, J., Kikukawa, T., Takahashi, M., Phillips, R. T.: Structure of the optical phase change memory alloy, Ag-V-In-Sb-Te, determined by optical spectroscopy and electron diffraction. J. Appl. Phys. **82**, 3214-3218 (1997)
[8.41] Ghosh, G., Lukas, H. L., Delaey, L.: A thermodynamic assessment of the Sb-Te system. Zeitschrift für Metallkunde **80**, 731-736 (1989)
[8.42] Kifune, K., Kubota, Y., Matsunaga, T., Yamada, N.: Extremely long period-stacking structure in the Sb-Te binary system. Acta. Cryst. B 61(Part 5), 492-497 (2005)
[8.43] Shamoto, S., Yamada, N., Matsunaga, T., Proffen, T., Richardson, J. W., Jr., Chung, J.-H., Egami, T.: Large displacement of germanium atoms in crystalline $Ge_2Sb_2Te_5$. Appl. Phys. Lett. **86**, 081904-081906 (2005)
[8.44] Rehr, J., Albers, R.: Theoretical approaches to x-ray absorption fine structure. Rev. Mod. Phys. **72**, 621–654 (2000)
[8.45] Kolobov, A. V., Tominaga, J., Fons, P., Uruga, T.: Local structure of crystallized GeTe films. Appl. Phys. Lett. **82**, 382-384 (2003)
[8.46] Kolobov, A., Fons, P., Tominaga, J., Ankudinov, A., Yannopoulos, S., Andrikopoulos, K.: Crystallization-induced short-range order changes in amorphous GeTe. J. Phys. Cond. Mat. **16**, S5103-S5108 (2004)
[8.47] Andrikopoulos, K. S., Yannopoulos, S. N., Voyiatzis, G. A., Kolobov, A. V., Ribes, M., Tominaga, J.: Raman scattering study of the a-GeTe structure and possible mechanism for the amorphous to crystal transition. J. Phys.-Cond. Mat. **18**, 965-979 (2006)
[8.48] Tani, K., Yiwata, N., Harigaya, M., Emura, S., Nakata,Y.: EXAFS study of Sb-Te alloy films. J. Synchrotron Rad. **8**, 749-751 (2001)
[8.50] Raoux, S., Rettner, C. T., Jordan-Sweet, J. L., Kellock, A. J., Topuria, T., Rice, P. M,, Miller, D. C.: Direct observation of amorphous to crystalline phase transitions in nanoparticle arrays of phase change materials. J. Appl. Phys. **102**, 094305 (2007)
[8.51] Raoux, S., Zhang, Y., Milliron, D., Cha, J., Caldwell, M., Rettner, C., Jordan-Sweet, J., Wong, H.: X-ray diffraction studies of the crystallization of phase change nanoparticles produced by self-assembly-based techniques. Proc. Europ. Symp. On Phase Change and Ovonic Science, September (2007)

[8.52] Fons, P., Brewe, D., Stern, E., Kolobov, A., Fukaya, T., Suzuki, M., Uruga, T., Kawamura, N., Takagaki, M., Ohsawa,, Tanida, H., Tominaga, J..: Sub-nanosecond laser-induced structural changes in the phase change material $Ge_2Sb_2Te_5$ measured by an optical pump/x-ray probe technique. In: Proceedings of the 9th European Phase Change and Ovonics Symposium, Zermatt Switzerland (2007)

[8.52] Fons, P., Kolobov, A. V., Fukaya, T., Suzuki, M., Uruga, T., Kawamura, N., Takagaki, M., Ohsawa, H., Tanida, H., Tominaga, J.: Sub-nanosecond time-resolved structural measurements of the phase-change alloy $Ge_2Sb_2Te_5$. Jpn J Appl Phys Part I **46**, 3711-3714 (2007)

[8.54] K. Shportko, S. Kremers, M. Woda, D. Lencer, J. Robertson, and M. Wuttig: Resonant bonding in crystalline phase-change materials. Nat Mater, advanced online publication:, 2008.

9. Optical and Electrical Properties of Phase Change Materials

Bong-Sub Lee and Stephen G. Bishop

Abstract Optical and electrical properties of the phase change material $Ge_2Sb_2Te_5$ are reviewed for its three phases. Implications of these data for the energy distribution of the density of electron states in the vicinity of the band edges are described. Near-band edge optical data obtained from ellipsometry and optical absorption spectra have been fitted with dispersion equations to determine values of ~ 0.7 eV for the optical bandgap of the amorphous phase, and ~ 0.5 eV for the highly conducting fcc and hexagonal phases. Infrared absorption spectra exhibit the effects of free carrier absorption in both crystalline phases. Conductivity at ~ 300 K is dominated by holes in all three phases; hole concentrations and mobilities are 10^{20} cm^{-3} and ~ 30 cm^2/V-s in hexagonal, and 10^{20} cm^{-3} and ~ 1 cm^2/V-s in fcc. Temperature dependence of the sheet resistance of thin film $Ge_2Sb_2Te_5$ documents the dramatic decreases in electrical resistance at the amorphous-to-fcc and fcc-to-hexagonal phase transitions. Comparison of the temperature dependence of the electrical properties in each of the three phases of $Ge_2Sb_2Te_5$ provides additional insights concerning their conduction mechanisms.

9.1 Introduction

The performance of phase change memory devices is based on the significant change of optical or electrical properties upon phase transformation, which must be carefully controlled. Therefore, the understanding of optical and electrical properties, as well as the electronic band structure, is essential for the efficient development of the technologies. While the technological development has been achieved primarily by heuristic approaches, recent fundamental studies have provided deeper understanding of these properties. In this chapter, we will discuss the optical and electrical properties of the prototypical phase change material $Ge_2Sb_2Te_5$ in its amorphous, metastable distorted fcc (face-centered cubic), and hexagonal forms. In addition, the dependence of these properties upon materials composition and synthesis technique will be considered.

S. Raoux, M. Wuttig (eds.), *Phase Change Materials*,
DOI 10.1007/978-0-387-84874-7_9,

9.2 Optical Constants and Optical Bandgap

A rewriteable CD or DVD comprises a multilayered structure that includes a thin layer of phase change alloy, sandwiched between dielectric layers, deposited on a substrate and capped with a protective layer (see Chapter 12 for details). The actual reflection from a disc is determined by the optical constants (index of refraction - n, coefficient of extinction - k) and thickness of each layer, as well as surface and interface roughness, etc. In other words, if we know the optical constants of the phase change material in each phase (and those of the other layers) at the wavelengths of interest, we can calculate the reflection spectrum from the designed combination of layers [9.1]. (Note that optical constants depend on the wavelength, or photon energy, of incident light. They are not really "constants.")

The absorption coefficient α of a material is defined by the relation

$$I = I_o \exp(-\alpha x) \,. \qquad (9.1)$$

If light of initial intensity I_0 propagates a distance x in the material, its intensity decreases from I_0 to I. (Note that the initial intensity I_0 of the light within the medium is the intensity of the fraction of the light that is incident upon the layer in question that is not reflected or scattered at its surface.) The value of α determined from Eq. (9.1) is related to the optical constant k by the expression

$$\alpha = 4\,\pi\,k \,/\, \lambda \,, \qquad (9.2)$$

where λ is the wavelength of incident light. The measurement of α as a function of wavelength for a semiconducting material (its optical absorption spectrum) is a very important and widely applied method of measuring the material's energy bandgap and several other features of the band structure, as discussed below.

9.2.1 Determination of the Optical Constants and Absorption Coefficient

Several measurements of the optical constants of phase change materials by spectroscopic ellipsometry (e.g., [9.2] and references therein) and by optical transmission / reflection measurements (e.g., [9.3]) have been reported. In addition, more specialized techniques such as photo-thermal deflection spectroscopy (PDS) [9.4] have been used to obtain the constants of these thin film materials in the spectral range corresponding to low absorption coefficients [9.5].

9.2.1.1 Transmission and Reflection Measurements

Considering Eq. (9.1), direct measurement of the transmission and reflection may appear to be a straightforward way to determine the absorption coefficient, but it is not always so simple. It is relatively simple for the case of a bulk sample with smooth surfaces; measurement of the transmission spectrum alone can provide a good estimate if the reflection and scattering at the surface are not so important. In the phase change materials, bulk samples can usually be obtained in the stable crystalline form, e.g., the hexagonal phase of $Ge_2Sb_2Te_5$. However, the amorphous and metastable crystalline phases, which have greater technological importance, are available only in the form of a thin film on a substrate. In these cases, all of the internal reflection and interference must be considered, so the accurate analysis of data requires a more sophisticated approach as done in Ref. [9.3].

9.2.1.2 Ellipsometry

Ellipsometry measures the change in polarization of light upon reflection from a sample [9.6]. The raw data are the amplitude ratio upon reflection (tan Ψ) and the phase shift Δ of the ratio between the normalized s and p components of the reflected light (the s component is oscillating perpendicular to the plane of incidence and parallel to the sample surface, and the p component is oscillating parallel to the plane of incidence). Ψ and Δ are sensitive to the optical constants, so they can be obtained by proper mathematical treatments. The data analysis is usually performed by a computer program provided by the manufacturer of the instrument. The user first chooses a simple model to represent the layer structure under investigation – for example, a substrate, a phase change layer, and a surface layer that can represent the roughness and/or surface oxide. Such a surface layer may be modeled with 50% void and 50% phase change layer in the Bruggemann effective medium approximation (EMA). The optical constants and/or thickness can then be obtained by numerical fitting, starting from initial estimates that the user provides. It is possible to obtain the optical constants at each wavelength for which data are recorded, which can amount to hundreds of data points for a broad spectral analysis. However, it is also useful to employ an appropriate dispersion relationship or equation that relates the optical constants and wavelengths with just a few fitting parameters. Several theoretical or empirical dispersion equations based on different conditions are available. For example, the Tauc-Lorentz relationship has been used successfully to fit the optical data obtained for amorphous $Ge_2Sb_2Te_5$, and the Lorentz oscillator model has been applied to the data for fcc and hexagonal $Ge_2Sb_2Te_5$ [9.2]. A computer program adjusts the values of the parameters to achieve an optimal fit of the chosen dispersion relation to the data points. In fact, ellipsometry is highly sensitive to many characteristics of multilayer structures and their constituent materials, including the thickness, roughness, composition, anisotropy, etc. Such characteristics can be obtained if properly modeled and analyzed.

9.2.1.3 Optical Contrast between Amorphous and Crystalline Phases

Typical values of the optical constants and absorption coefficient obtained from thin samples of $Ge_2Sb_2Te_5$ are shown in Fig. 9.1. Note that the values for the crystalline fcc and hexagonal phases are similar, but significant differences between the optical constants of the amorphous and crystalline phases are evident in the visible and near-infrared range. This striking difference is the enabling feature of the phase change optical recording. It has been reported that significant changes in the local order and the number of bonds that occur very rapidly upon amorphization are responsible for the contrast in the optical properties of the crystalline and amorphous states [9.7, 9.8] (see also Chapter 2).

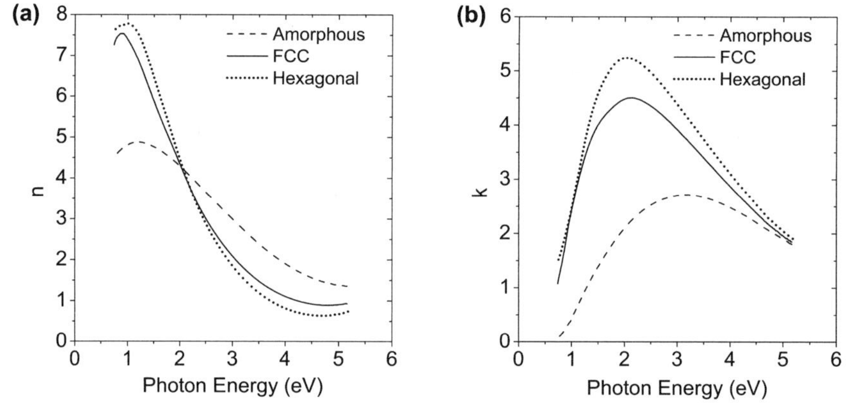

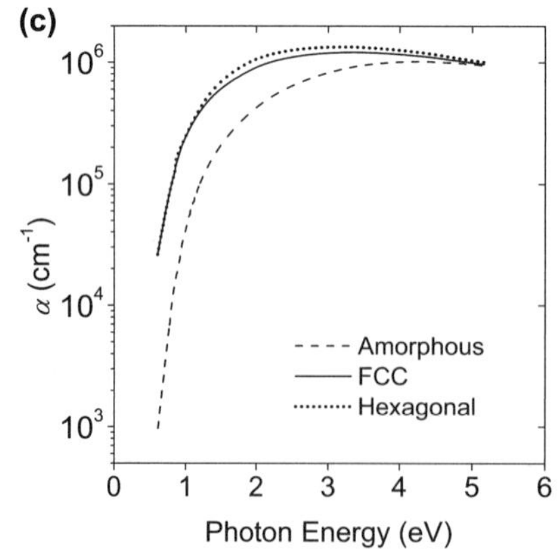

Fig. 9.1. Typical values of the optical constants (a) n, (b) k, and (c) the absorption coefficient α of the amorphous, fcc, and hexagonal phases of $Ge_2Sb_2Te_5$. (c) Reprinted with permission from [9.2]. Copyright 2005. American Institute of Physics.

This large optical contrast can be utilized to study the kinetics of the phase change process, often *in-situ*. In a pump-probe laser technique, an intense laser beam (pump beam) supplies the heat for phase transformation, and another beam (probe beam) monitors the change of reflectivity after or during the transformation, thereby providing insights concerning the kinetics of nucleation and growth [9.9, 9.10]. Sophisticated *in-situ* ellipsometry measurements have also been performed to study the phase change process [9.11, 9.12].

9.2.2 Optical Bandgap

The optical bandgap of a semiconductor is determined by the onset of optical absorption that occurs when the photon energy $h\nu$ of incident light just equals the energy separation of the highest occupied electron states in the valence band (VB) and the lowest empty electron states in the conduction band (CB) of the semiconductor. Photons with energy greater than this threshold are absorbed by the excitation of electrons from the filled VB states to the empty CB states. The magnitude of this interband optical absorption increases rapidly with photon energy as the density of filled electron states in the VB and the density of empty electron states in the CB both increase with energy. If $h\nu$ is smaller than this bandgap energy, no photons can be absorbed in an ideal semiconductor, i.e. one with a perfect periodic lattice containing no defects or impurities. In Fig. 9.1 (c), the absorption coefficient α indeed rapidly increases above the bandgap energy. However, the value of the bandgap energy cannot be inferred readily from these semi-logarithmic plots. There are defect states within the bandgap, enabling nonzero absorption at energies less than the bandgap that obscures any obvious onset of absorption at the bandgap energy.

It is known that the absorption coefficient α and the bandgap E_g^{opt} of semiconductors generally follow a relationship of the form

$$\alpha h\nu = (\text{constant})\,(h\nu - E_g^{opt})^r \tag{9.3}$$

with appropriate selection of the exponent r, when α is greater than 10^4 cm^{-1}. Here, we used the symbol E_g^{opt} because it is the optically determined bandgap, or optical bandgap. If this relationship holds, then E_g^{opt} is found by linearly extrapolating the plot of $(\alpha h\nu)^{1/r}$ vs. $h\nu$ and finding the intersection with the abscissa [9.13]. For crystalline semiconductors with parabolic bands, the exponent r becomes 1/2 for a direct gap and 2 for an indirect gap (both for allowed transitions). For amorphous semiconductors, Tauc et al. showed that r = 2 holds assuming parabolic bands [9.14]. It is often confusing that the Tauc relationship is similar to that of an indirect gap crystalline semiconductor. Although the transition in an amorphous semiconductor is sometimes called "direct" because it does not need phonon assistance, it still occurs without the *k*-space conservation rule as in an indirect transition.

The optical bandgap of $Ge_2Sb_2Te_5$ in each phase has been determined by the method described above [9.2, 9.15, 9.16], as illustrated in Fig. 9.2.

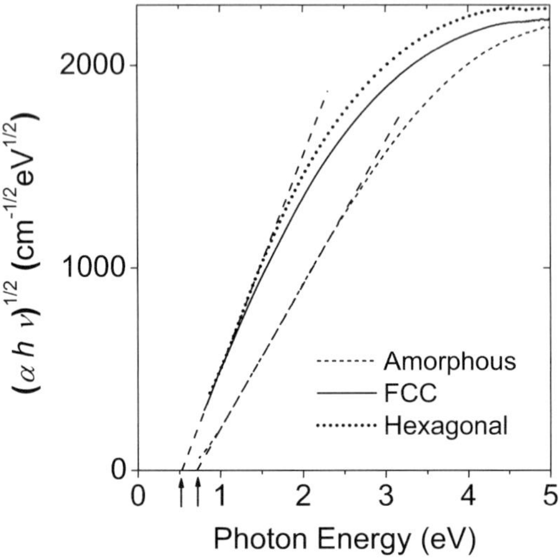

Fig. 9.2. Plots of $(\alpha h\nu)^{1/r}$ vs. photon energy for the amorphous, fcc, and hexagonal phases of $Ge_2Sb_2Te_5$ (redrawn with the data in [9.2]). The relationship with $r = 2$ is appropriate for all phases; the dashed straight line extrapolations to zero (as indicated by arrows) give an optical bandgap of 0.7 eV for the amorphous phase (Tauc gap) and 0.5 eV for the crystalline phases (indirect gap). Reprinted with permission from [9.2]. Copyright 2005. American Institute of Physics.

The relation $r = 2$ for indirect transitions is found appropriate for the fcc and hexagonal phases, giving the optical bandgap of ~ 0.5 eV. The Tauc relationship, also with $r = 2$, is appropriate for the amorphous phase and gives a bandgap energy ~ 0.7 eV, which is independent of the specific method used to obtain the optical constants. (It shows that the Tauc-Lorentz dispersion model can be successfully employed in the analysis, in which the bandgap is included as a fitting parameter.) Again, the difference in bandgap between the amorphous and crystalline phases is quite significant, considering the ease and speed with which the reversible phase transformations can be induced.

A frequently employed experimental method to estimate the optical bandgap of a semiconductor (especially for amorphous semiconductors) is simply to determine the photon energy at which the absorption coefficient is 10^4 cm^{-1} [9.17]. This value is often called E04. For example, in each of the three absorption spectra plotted in Fig. 9.1 (c), the value of the photon energy for which α is ~ 10^4 cm^{-1} corresponds closely to the value of the optical bandgap determined by fitting the absorption spectrum to Eq. (9.3) for each phase under the assumption that $r = 2$. Some previously reported studies of the optical bandgap of Ge-Sb-Te made the unfortunate choice of using $r = 1/2$ for every phase, and reported values of the bandgaps in the range 1.5 - 2 eV, as we have noted in [9.2]. These values are much larger than the apparent E04 values, and are inconsistent with other physical properties of $Ge_2Sb_2Te_5$ such as the temperature dependence of the electrical conductivity or Hall coefficient (Sections 9.4 - 9.5). Furthermore, the crystal structures of the $Ge_2Sb_2Te_5$ alloy have been shown to incorporate a large concentration of vacancies which are apparently responsible for the very high concentrations of free carriers (~ 10^{20} cm^{-3}) observed in the crystalline phases (Section 9.5). In semiconductors exhibiting such properties, the determination of the bandgap is likely to be complicated by phenomena such as the tailing of band states into the gap, or the widening of the optical bandgap by the Burstein-Moss shift that can occur in degenerate semiconductors [9.18].

9.2.3 Infrared Absorption: Band Tails and Free Carrier Absorption

Equation (9.3) describes the energy dependence of the absorption coefficient in the vicinity of the bandgap energy, but not over the entire range of the optical absorption spectrum. The density of states is no longer parabolic at energies far from the band edges, which causes a deviation from Eq. (9.3) at high photon energies as shown in Fig. 9.2. Deviation also occurs at lower energies (not shown in Fig. 9.2) due to defect-related absorption, free carrier absorption, etc.

9.2.3.1 Urbach Edge

In amorphous semiconductors the optical absorption edge is characterized by an exponential dependence of the absorption coefficient on photon energy, as shown in Fig. 9.3 for $Ge_2Sb_2Te_5$. This exponentially rising absorption spectrum, which is referred to as the Urbach edge [9.13], can be described by the empirical relation $\alpha \propto \exp(h\nu/\Delta E)$, where ΔE is the inverse slope of the Urbach edge, illustrated by the dashed line in Fig. 9.3.

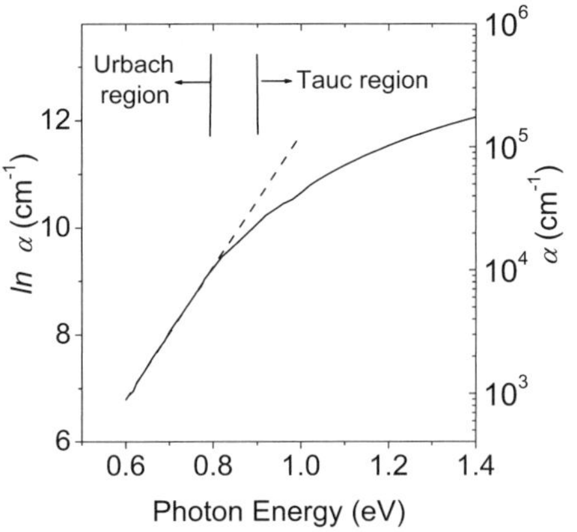

Fig. 9.3. Optical absorption of amorphous $Ge_2Sb_2Te_5$ near the bandgap energy [9.2]. The exponential, straight line segment of the spectrum is called the Urbach edge. It indicates the presence of exponentially decaying band tails. The inverse slope of the Urbach edge *(ΔE)*, which is referred to as the width of the Urbach edge or tail, is highly dependent upon sample preparation conditions. Reprinted with permission from [9.2]. Copyright 2005. American Institute of Physics.

In prototypical chalcogenide glasses such as As_2S_3 and As_2Se_3, which can be obtained in highly stable bulk form by cooling from the melt, this parameter is highly reproducible in slowly cooled or annealed samples. However, for the metastable amorphous phase change chalcogenide alloys that can only be obtained by thin film deposition, the magnitude of ΔE is a sensitive function of preparation conditions. For example, various workers have reported [9.2, 9.5, 9.15, 9.16] values of the inverse slope ΔE of the Urbach edge of $Ge_2Sb_2Te_5$ ranging from approximately 50 meV to 200 meV, while they all obtained values of the optical bandgap close to 0.7 eV. In the study reported in [9.5], it was demonstrated that ΔE depends upon film growth rate and oxygen content.

9.2.3.2 Free Carrier Absorption

The crystalline phases of $Ge_2Sb_2Te_5$ are known to contain large concentrations of free carriers even at room temperature. The effect of these free carriers (in this case, holes) on the infrared absorption spectra has been measured for the fcc and hexagonal crystalline phases, not for the resistive amorphous phase [9.19, 9.20]. The optical properties of metals can be approximated by the Drude dispersion model, in which the free carriers are treated as Lorentz oscillators without restoring force. The infrared transmission spectra of these crystalline phases could be fitted to the Drude model relatively well; the optical constants from the fitting are shown in Fig. 9.4 [9.20]. Note the high values of k at very low energies for the crystalline phases. We observed in Fig. 9.1 (b) that k decreases as the photon energy decreases, but it again sharply increases at very low energies due to the free carrier absorption in the crystalline phases.

Since this infrared absorption is attributed to the charge carriers, there is a link between the optical and electrical properties. Because the Drude model includes the conductivity as a fitting parameter, it is possible to obtain a measure of the ac electrical conductivity. The value of electrical conductivity determined from Drude model fitting in [9.20] is 2230 S/cm for the hexagonal phase, which is consistent with the measured bulk dc conductivity of 2400 S/cm [9.21]. The dc conductivity of a thin-film polycrystalline fcc sample [9.2, 9.22, 9.23], where grain boundary scattering significantly inhibits the current flow, is much lower than the ac value determined from the Drude model.

9.2.4 Effects of Composition and Preparation Conditions

It has been reported that the optical properties of various Ge-Sb-Te alloys depend on such factors as thickness [9.24, 9.25], temperature [9.26], film microstructure

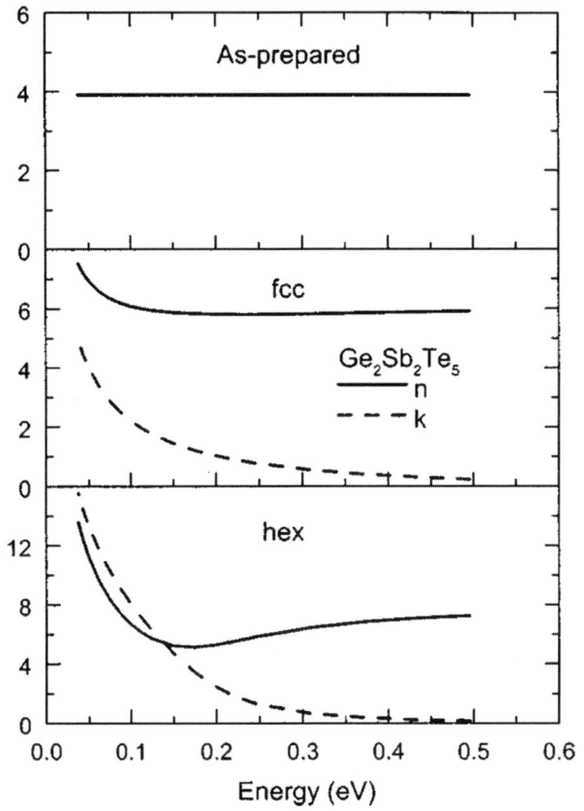

Fig. 9.4. Spectral distribution of the optical constants of the amorphous, fcc, and hexagonal phases of $Ge_2Sb_2Te_5$, obtained by fitting the infrared absorption spectra to the Drude dispersion model. High values of k at very low energies are caused by the free carriers in the highly conductive crystalline phases, which are not observed for the resistive amorphous phase. Reprinted with permission from [9.20]. Copyright 2000. American Institute of Physics.

[9.27] and, of course, compositional variation by incorporating additional elements. The addition of nitrogen [9.28] or oxygen [9.5] generally causes a slight increase of the optical bandgap, which is reasonable because the oxides or nitrides of constituent elements have larger bandgaps. The optical constants change accordingly.

Note that those factors may be changed not only by the user's intention but also by the differences in preparation conditions. For example, although higher growth rate may produce films with broader Urbach edges, the concomitant reduction in growth time can lower the concentration of oxygen incorporated during growth [9.5] which could also affect the properties of the sample. In addition, the increased growth rate might increase the actual growth temperature due to the heat of condensation. Many other conditions, which are often uncontrollable, may affect the properties. Therefore it is usually difficult, or impossible, to produce thin films of chalcogenide glasses having identical properties at different laboratories.

Although we have confined our discussions to measurements of the properties of $Ge_2Sb_2Te_5$, similar procedures can be applied to analyze the optical properties of other compositions in the Ge-Sb-Te system, and also the Ag-In-Sb-Te and Ge-Sb alloys. The bandgap energies are usually in the range of 0.3 ~ 0.9 eV. Some of the reported values of optical bandgap are incorrect because the wrong value of r was chosen when fitting Eq. (9.3) to the data, or the data were obtained in a photon energy range for which the analysis was inappropriate, i.e. too far above the band gap. Equation (9.3) is valid only for energies in the vicinity of the bandgap, where the absorption coefficient usually becomes ~ 10^4 cm^{-1}. On the other hand, special compositions like KSb_5S_8 [9.29] and Ag-Sb-S [9.30] have significantly higher bandgap of near 1.7 eV. It was argued that the wider bandgaps of these sulfides make them more responsive to ultraviolet light, which is needed for higher resolution and recording density [9.29]. However, the industrial applicability of these compositions has yet to be confirmed.

9.3 Photo-induced Effects

Optical illumination can induce reversible and irreversible changes in the structural, optical, and electronic properties of amorphous chalcogenides. From the standpoint of practical applications, the most important photo-induced effect is the laser pulse-induced reversible amorphous-to-crystalline phase transformation in the chalcogenide-based phase change materials. This effect is essentially of thermal origin, although some workers have proposed that bond rupture is at least partially attributable to electronic excitation [9.31]. In the case of $Ge_2Sb_2Te_5$, it has been shown that optical illumination can also increase the charge carrier concentration and even enhance the susceptibility to oxidation without inducing a phase change.

9.3.1 Photo-induced Current and Optical Nonlinearity

The photoexcitation of electron-hole pairs can give rise to an increased electrical conductivity under illumination that is greater than the dark conductivity under the same electric field. The origin of this photo-induced current may be either photoconductive or photo-thermal. If the absorbed photons excite electron-hole pairs that contribute directly to the increased conduction, the effect is called photoconductivity. On the other hand, if the photo-excited carriers recombine non-radiatively before reaching the electrical contacts, thereby increasing the sample temperature and causing a thermally activated increase in the conductivity, the effect is called photo-thermal conductivity (i.e. a bolometric effect). In either case, the photocurrent is proportional to the sample's optical absorption, so it sharply increases for photon energies greater than the bandgap energy. An optically induced increase in electrical conductance at photon energies corresponding to the band edge absorption has been observed in amorphous $Ge_2Sb_2Te_5$. These experiments, carried out using phase-sensitive detection under mechanical chopping, were originally interpreted in terms of photoconductivity [9.2]. Subsequent experiments carried out under continuous illumination (no chopping) clearly demonstrated a dominant bolometric contribution to the effect in $Ge_2Sb_2Te_5$ [9.32]. While the existence of a photoconductive effect is possible [9.33], its relative contribution is not known. There are indications that the extremely short lifetime of photoexcited carriers severely limits the photoconductive contribution [9.32].

The photo-generation of carriers has also been proposed as the possible origin of the nonlinear optical characteristics of crystalline $Ge_2Sb_2Te_5$, although this suggestion has not yet been confirmed. The absorption coefficient decreases as the power of incident light increases, possibly due to the band filling by thermally generated carriers under illumination [9.34]. It demonstrates the possibility of a super-resolution readout, which is discussed in Chapter 13.

9.3.2 Photo-Oxidation

A rich variety of non-thermal, photo-induced effects have been documented in the more conventional, wider bandgap chalcogenide glasses (e.g. As_2S_3 and As_2Se_3), including photo-oxidation, photodarkening (decrease of bandgap), photo-bleaching, photo-induced optical anisotropy, etc. [9.35]. In the case of amorphous $Ge_2Sb_2Te_5$, photo-oxidation was confirmed: illumination at intensities $\sim 110\ \text{mW/cm}^2$ for tens of hours produced a surface oxide layer of several nanometers [9.36]. This oxide can be significantly etched by hydrofluoric acid, or even by water in an ultrasonic bath. Of course, the presence of this oxide layer changes the transmission and reflection. However, when the $Ge_2Sb_2Te_5$ sample was covered by a transparent silicon nitride layer, illumination did not make any change in the optical response; there was no observable change in optical constants or bandgap. Although there has been a report of photodarkening in the nonstoichiometric $Ge_xSb_{20-x}Te_{80}$ system [9.37], the possibility of photo-oxidation was not considered in its analysis; this result should be re-visited. The observation of negligible photodarkening is consistent with previous work [9.35]; the sensitivity to photodarkening is generally lower in tellurides than in selenides or sulfides, and the introduction of highly coordinated constituent atoms like Ge minimizes photostructural effects.

9.4 Conductivity and Phase Transformation

9.4.1 Temperature-dependence of Resistivity

The temperature dependence of the sheet resistance of a thin film of $Ge_2Sb_2Te_5$ is shown in Fig. 9.5 [9.22]. The most striking feature is the 2-3 orders of magnitude drop in resistivity that coincides with the amorphous-fcc transformation at T_1, which is near 150 °C. This extraordinary contrast in resistivity is the key feature for Phase Chance Random Access Memory (PCRAM): the transformation of a very small volume produces sufficient resistance contrast to enable digital readout. This metastable fcc phase transforms into the stable hexagonal phase at a higher temperature T_2, which varies somewhat depending on the sample characteristics. The two plots of Fig. 9.5 actually show the sheet resistance of an initially 80 nm film, not the resistivity, because the film thickness changes during heating [9.22]; the biggest thickness reduction (~ 5 %) occurs at T_1 due to the density difference between the amorphous and fcc phases [9.38].

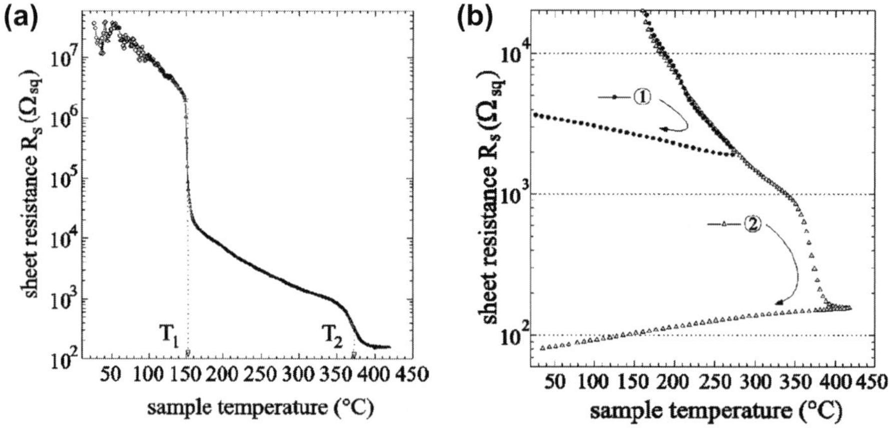

Fig. 9.5. The sheet resistance of an initially 80 nm thin $Ge_2Sb_5Te_5$ film vs. temperature. (a) Sheet resistance measured with increasing temperature at the heating rate 1.6 K/min. Sharp drops accompany the amorphous-fcc transformation at T_1 and the fcc-hexagonal transformation at T_2. (b) Sheet resistance upon heating (1.2 K/min for Curve 1 and 1.15K/min for Curve 2) and subsequent cooling. The resistance of the fcc phase (Curve 1) increases with cooling, whereas that of the hexagonal phase (Curve 2) decreases. Reprinted with permission from [9.22]. Copyright 2000. American Institute of Physics.

Fig. 9.5(a) also shows the continuous decrease of resistance in each phase, but the origin is different for each. The resistivity of amorphous $Ge_2Sb_2Te_5$ exponentially decreases as more carriers are excited at higher temperature. If it is cooled again, the resistivity essentially returns to its original value. However, the resistivity decrease of a crystalline phase is mainly due to the increase of mobility rather than carrier concentration [9.23]; crystal grains grow during heating and the scattering by grain boundaries decreases, increasing the mobility. This decrease in resistivity is irreversible, as shown in Fig. 9.5(b), and the lower resistance persists as the sample is cooled. During cooling, the fcc phase becomes slightly more resistive (curve 1) while the hexagonal phase becomes more conductive (curve 2) [9.22]. We will discuss these behaviors in more detail in Section 9.5.

The heat of crystallization is generated by electrical Joule heating in a PCRAM cell, and it actually involves a very interesting *electronic switching* phenomenon. As the voltage applied across an amorphous region increases, the current also increases but the magnitude is small due to the large resistivity. However, when the voltage exceeds a certain threshold voltage (V_{th}), the material suddenly exhibits a negative differential resistance, allowing a high current. Consequently, the Joule heat from this high current can crystallize the amorphous region. This critically important phenomenon is discussed in greater detail in Chapter 14.

9.4.2 Intermediate States: Percolation and Multilevel Recording

As the temperature of amorphous $Ge_2Sb_2Te_5$ is increased from room temperature and approaches the phase transformation temperature T_1, small crystalline nuclei are formed in the amorphous matrix (nucleation), and these nuclei grow in size to eventually make the material fully crystalline (growth). The relative speeds of nucleation and growth vary greatly as a function of composition. In any case, between the initiation and completion of crystallization, the material exists as a mixture of two different phases. Note that the conductivity of this mixture is *not* proportional to the volume fraction of crystalline phase, whereas the variation of the optical reflectivity is closely proportional to the volume fraction of crystalline material. Imagine a constant voltage is applied to a volume of amorphous $Ge_2Sb_2Te_5$ under heating. As crystalline nuclei are growing, if they are not yet in physical contact such that they form a low resistance current path through the sample, the increase of current is small. At the moment at which multiple crystallites meet each other and form a continuous crystalline path between the two electrodes, the current immediately becomes much larger. The fraction of crystalline phase need not be very large to support this percolation phenomenon. (The role of percolation in modeling the performance of PCRAM devices is discussed in Chapter 14.) In other words, a huge increase of conductivity may occur even though the crystalline fraction is relatively small. Indeed it has been observed that the increase in electrical conductivity is initiated significantly earlier in the heat pulse cycle than the increase in optical reflectivity. This dynamic behavior of the conductivity and reflectivity of phase change materials during electrically or optically induced transformation is a strong function of material composition and device design [9.39]. It has also been suggested that such nonlinear electrical behavior can be utilized to realize a "cognitive" information processing system [9.40].

The temperature dependence of sheet resistance curves in Fig. 9.5 show that the phase change material traverses an intermediate range of resistance between the highly resistive state of the fully amorphous phase and the more conductive state of the fully crystalline phase. This suggests the possibility of using this intermediate range of resistivity as the basis for multi-level (multi-state) memory cells that would increase memory or recording density. Pulsing schemes that can achieve mixed-phase states of reproducible intermediate resistance are now being investigated [9.40-42].

9.4.3 Effects of Composition and Processing Conditions

The conductivity of phase change materials is strongly dependent upon composition, processing conditions, and the resulting microstructure. We already mentioned that annealing causes grain growth in a polycrystalline sample to reduce the resistivity. As apparent in Fig. 9.5(b), a variation in annealing temperature and/or time can change the resistivity of an fcc sample by an order of magnitude. Speaking in terms of device characteristics, dramatically different electrical properties may be induced by different power/length pulse profiles or by changes in device design that modify thermal confinement. Although annealing does not produce significant changes in the carrier concentration of a single phase sample of a phase change material [9.23], a large number of repeated phase transformations in a PCRAM cell may cause a significant change in microstructure and chemical homogeneity [9.43], thus changing the overall electrical properties.

Incorporation of additional elements may also affect the electrical properties, e.g., by inhibiting grain growth. However, the electrical conductivity of chalcogenide glasses in general is remarkably insensitive to the introduction of impurity elements at concentrations corresponding to doping levels in conventional crystalline semiconductors (e.g. 10^{15} - 10^{20} cm^{-3}). When the term "doping" is used in the field of phase change materials, it usually refers to the incorporation of a few atomic percent of other/additive elements. Perhaps the most widely used additive to $Ge_2Sb_2Te_5$ is nitrogen. Nitrogen-doping of $Ge_2Sb_2Te_5$ induces strain fields in the grains and causes much finer grain size [9.44], which accordingly increases the resistivity and transformation temperature [9.28]. It has been reported that at low doping concentrations nitrogen atoms reside inside the grains and increase the resistivity, but with increasing doping levels excess nitrogen atoms aggregate at grain boundaries [9.45]. High resistivity of the crystalline phase is beneficial because a smaller RESET current (current to switch the cell from the crystalline to the amorphous state by melt-quenching) can generate the required Joule heat. The incorporation of oxygen also increases the resistivity and, if the oxygen concentration is above 10 %, it can increase even by an order of magnitude due to the oxide at grain boundaries [9.46]. These are just a few examples of on-going studies of the possible use of compositional variation to reduce the RESET current and enhance stability, recycle lifetime/reliability, switching speed, etc.

9.5 Electronic Transport Properties and Band Structure

In the previous sections, we have discussed in qualitative terms the dependence of the electrical resistance of phase change materials upon composition, preparation conditions, phase changes, grain size and grain boundaries, temperature, and optical illumination. However, the fundamental, quantitatively measurable characteristics that govern the electric current density of the phase change materials are the concentration and mobility of charge carriers (electrons and/or holes) in the materials. The study of these electronic transport properties, together with the measured optical properties, gives valuable information on the electronic band structure of each phase. Specifically, insights can be gained regarding the energy distribution of the density of electron states in the vicinity of the band edges. We will briefly review the characterization techniques, and discuss the transport properties in each phase of $Ge_2Sb_2Te_5$.

9.5.1 Characterization of Transport Properties

9.5.1.1 Hall Measurements

This is perhaps the most widely used technique to study the majority carrier concentration and mobility. If a magnetic field is applied to a material where an electric current is flowing, the charge carriers are forced to one side by the Lorentz force (for example, see [9.47]). Consequently, the spatial distribution of charge carriers becomes asymmetric, creating an electric potential difference (Hall voltage) across the sample in the dimension perpendicular to both the direction of current flow and the magnetic field. The sign of the Hall voltage V_H depends on the type of majority carrier.

$$R_H = V_H t / (IB) = 1 / (q p) \tag{9.4}$$

Equation (9.4) shows the Hall coefficient R_H when holes are dominating the conduction, where I is the current, t is the thickness of sample, B is the applied magnetic field, q is the absolute value of the charge of an electron, and p is the hole concentration (if electrons are dominating, -*n* represents the electron concentration and the sign of the charge carriers). The conductivity is given by

$$\sigma = q\,(p\,\mu_p + n\,\mu_n) \tag{9.5}$$

where μ_n and μ_p are the mobilities of electrons and holes, respectively. Therefore, if holes are dominating, the mobility (Hall mobility) is σR_H. If the conduction is not dominated by one type of carrier (mixed conduction), two-carrier analysis is needed. In the usual low magnetic field case, and ignoring the effect of mean free time between scattering events, the Hall coefficient is expressed by the following [9.48]

$$R_H = \frac{p - (\mu_n/\mu_p)^2 n}{q(p + (\mu_n/\mu_p)n)^2}\,. \tag{9.6}$$

The standard procedure for Hall effect measurements is described in [9.49]. The sample should be symmetric, and the size of electric contacts on a thin film sample should be minimized; a more sophisticated shape (e.g., clover shape) is beneficial to reduce the error due to finite contact size. Typically, a magnetic field of ~ 0.5 Tesla is used. Measurements on resistive samples usually incur larger experimental errors; the use of a higher magnetic field can mitigate this problem for resistive samples, but care must be taken to ensure that the dependence on magnetic field is linear.

9.5.1.2 Thermoelectric Effect

When there is a temperature gradient in a material, the charge carriers diffuse to the colder side. Accordingly, a voltage builds up because of the asymmetric distribution of charge. The Seebeck coefficient *S* (also called thermopower) of a material is defined as the ratio of the voltage difference ΔV and the temperature difference ΔT:

$$S = \Delta V / \Delta T \tag{9.6}$$

The sign of the Seebeck coefficient depends on the majority carrier type. More detailed theory and experimental techniques are described in [9.50]; a pair of heaters induce a temperature difference, and copper-constantan thermocouples measure the temperature at each side. The copper branches of thermocouples are used to measure the voltage difference. Thin-film thermocouples may be used as appropriate [9.16].

It is interesting that the thermoelectric effect may be useful for practical purposes, not only for analysis. In another kind of thermoelectric effect, the Thomson effect, a conductor emits or absorbs heat when it is subjected to a temperature gradient and electric current flows through it. It has been reported that, in addition to Joule heat, the Thomson effect in line-type PCRAM cells can reduce RESET currents by ~ 5 % [9.51].

9.5.2 Hexagonal $Ge_2Sb_2Te_5$

The hexagonal state of $Ge_2Sb_2Te_5$ is not of immediate technological interest, but it is the stable phase and is obtainable in either bulk or thin film form, which is advantageous for various kinds of fundamental experiments. In fact, several Ge-Sb-Te bulk alloys have been studied by Hall experiments since the early 1970s. In these early studies it was determined that these alloys are p-type and have very high hole concentrations of over 10^{20} cm^{-3} [9.52-54]. It has been reported that hexagonal $Ge_2Sb_2Te_5$ has a hole concentration of ~ 3 x 10^{20} cm^{-3}, and a mobility of ~ 30 cm^2/V-s in thin film form [9.2, 9.23] and ~ 50 cm^2/V-s in bulk form [9.21] at room temperature. Again, the mobility is limited by scattering, and is therefore dependent upon the microstructure and preparation conditions. The p-type behavior is consistent with the existence of numerous Ge and/or Sb vacancies in the lattice structure [9.15, 9.21, 9.54].

Fig. 9.6 shows the result of conductivity and Hall effect measurements at various temperatures.

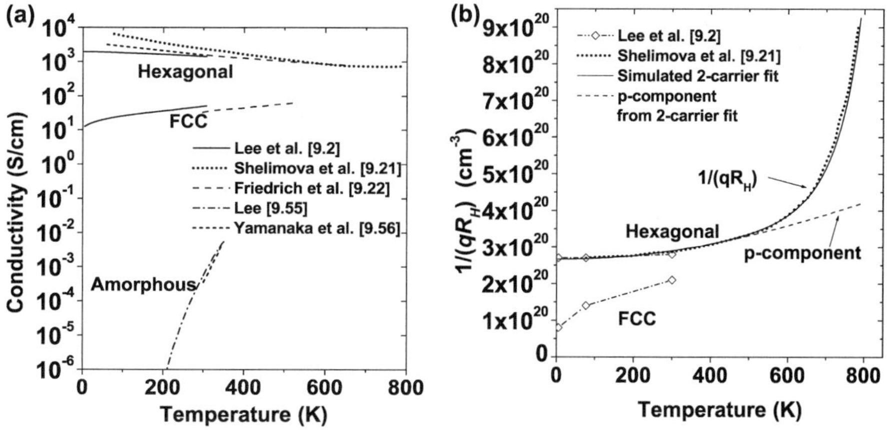

Fig. 9.6. The temperature-dependence of (a) conductivity and (b) $1/(qR_H)$ of $Ge_2Sb_2Te_5$ [9.2, 9.21, 9.22, 9.55, 9.56]. The quantity $1/(qR_H)$, where R_H is the Hall coefficient, represents the hole concentration when the effect of electrons is negligible. The high-temperature behavior of $1/(qR_H)$ in hexagonal $Ge_2Sb_2Te_5$ shows a significant electron contribution. Reprinted with permission from [9.2]. Copyright 2005. American Institute of Physics.

If the electron contribution is negligible, which is expected at room or lower temperature, the quantity $1/(qR_H)$ represents the hole concentration and σR_H represents the Hall mobility from Eqs. (9.4) and (9.5). One important feature in Fig. 9.6 is that, even at ~ 5 K, the conductivity remains very high due to the huge concentration of holes (over 10^{20} cm^{-3}); there is no freeze-out of carriers. The implication of this result is that the Fermi level lies within the valence band. Otherwise, there would be almost no free carriers at temperatures near 0 K, considering the Fermi-Dirac statistics. Therefore, it is reasonable to conclude that hexagonal $Ge_2Sb_2Te_5$ is a narrow-gap degenerate semiconductor, where the Fermi level is within the valence band [9.2, 9.16, 9.21]. This is not at all surprising considering the very large carrier concentration. In fact, because of the numerous vacancies in the crystal structure, there must be a defect band near the valence band, which may overlap and distort the valence band edge [9.57]. Another possible phenomenon is the Burstein-Moss shift [9.18]. As shown in Fig. 9.7, optical excitation requires that the photon energy should be higher than the electronic bandgap E_g, so the measured optical bandgap E_g^{opt} (0.5 eV in this case) is larger than E_g.

The Fermi-Dirac statistics also show that, when the Fermi level is relatively deep inside the valence band, an increase in temperature causes only a slight increase in the majority carrier concentration, which is apparent for temperatures below 200 K in Fig. 9.6(b). Fig. 9.6(a) shows that the hexagonal phase becomes slightly more conductive at lower temperature, which can be attributed to an increased mobility because of lower phonon scattering. Now, the high-temperature behavior in Fig. 9.6 may seem strange because $1/(qR_H)$ increases sharply at temperatures above ~ 500 K. It is explained by the contribution of electrons in this narrow-gap semiconductor; if the temperature is very high, we can expect from Fig. 9.7 that electrons may be excited to the conduction band (intrinsic conduction), so Eq. (9.6) should be used instead of (9.4).

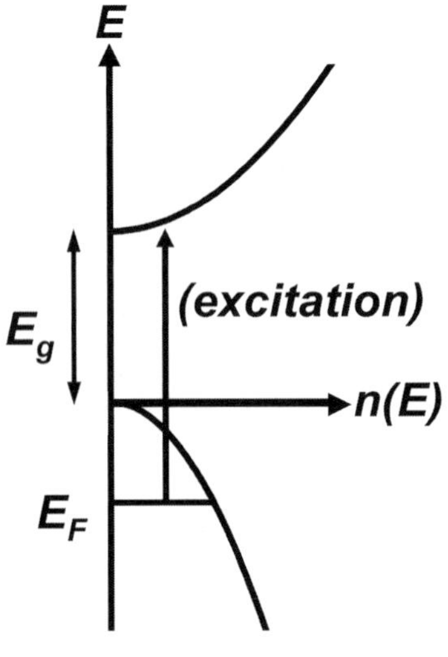

Fig. 9.7. The schematic band diagram of a p-type degenerate semiconductor. The onset of interband optical excitation occurs at photon energy significantly larger than the electronic bandgap E_g. The position of E_F deep in the valence band indicates a large concentration of free holes attributable to defects. Such a highly defective semiconductor is expected to have a defect band, near or overlapping the valence band edge.

This experimental result [9.21] was fitted by a simple calculation assuming parabolic bands [9.2]: a 2-carrier analysis using Eq. (9.6) leads to satisfactory model fits ("Simulated 2-carrier fit" in Fig. 9.6), while (9.4) could not fit the entire curve. The "p-component" curve in Fig. 9.6 represents the calculated hole concentration, which is almost identical to $1/(qR_H)$ at low temperatures but significantly different at high temperatures. Of course, the hole mobility should be obtained from the measured conductivity and this p-component; otherwise, the mobility would appear to decrease as temperature increases in an unphysically rapid manner.

9.5.3 Face-centered-cubic $Ge_2Sb_2Te_5$

The fcc phase is p-type and has a hole concentration of ~ 10^{20} cm^{-3} and a mobility of ~ 1 cm^2/V-s at room temperature [9.2, 9.23], which depends strongly upon preparation conditions. Both properties are smaller than those of the hexagonal phase, which explains the increase of conductivity that accompanies the fcc-hexagonal transformation. From Fig. 9.6, we find that the hole concentration remains very large (~ 8 x 10^{19} cm^{-3}) at ~ 5 K. The huge concentration of free carriers and the absence of freeze-out are characteristics of a degenerate semiconductor; these results suggest that the Fermi level of the fcc phase must be inside a band.

On the other hand, there are conflicting results that suggest the fcc phase is non-degenerate. In a recent hard x-ray photoemission spectroscopy study [9.58], which reported the valence band spectra of Ge-Sb-Te alloys, the data from fcc $Ge_2Sb_2Te_5$ have been interpreted in terms of a Fermi level ~ 0.15 eV higher than the valence band edge. This non-degenerate scheme was also suggested [9.16] on the basis of the temperature-dependence of conductivity near room temperature (Fig. 9.6(a)), which appears to have an activation energy of ~ 0.15 eV. However, if this phase were assumed as an ideal non-degenerate semiconductor having a Fermi level ~ 0.15 eV above the valence band, with an effective hole mass similar to the rest mass, the hole concentration would be only 10^{16} ~ 10^{17} cm^{-3} (as indicated in [9.16]); it would require an unreasonably high mobility of 10^3 ~ 10^4 cm^2/V-s in order to be consistent with the measured value of conductivity. In fact, the increase of conductivity with temperature can also be attributed to the gradual increase of carrier concentration in a degenerate semiconductor (combined with a subtle change of mobility). Therefore, the degenerate scheme appears to provide more satisfactory explanation of the electrical properties of the fcc phase, although it is not yet fully confirmed.

It should also be noted that this phase is highly defective. Due to the numerous vacancies, the existence of trap states near the valence band has been predicted [9.15], which at sufficient concentration might constitute a defect band.

9.5.4 Amorphous $Ge_2Sb_2Te_5$

The dark conductivity of amorphous $Ge_2Sb_2Te_5$ shows thermally activated behavior. The activation energy varies between 0.3 eV and 0.45 eV among published reports [9.2, 9.16, 9.55, 9.56, 9.59, 9.60]. If the Fermi level is pinned around the center of the bandgap, these values indicate a gap of 0.6 ~ 0.9 eV, which is consistent with the optical bandgap of 0.7 eV. The activation energy changes slightly with temperature: a thin film sample showed activation energies of 0.41 eV at 270 - 350 K, and 0.35 eV at 140 - 230 K [9.55]. It also depends on the applied voltage: 0.33 eV at 0.4 V and 0.29 eV at 1 V from a PCRAM cell [9.60]. Chapter 14 presents a theoretical model to describe the conduction in the amorphous phase, including this voltage dependence of the measured activation energy.

Although several studies of the transport properties of this disordered and resistive phase have been published [9.15, 9.16, 9.59], significant ambiguities persist because of the experimental and theoretical difficulties. The amorphous phase of $Ge_2Sb_2Te_5$ is known to be p-type, and it is expected to have a band tail originating from localized Te lone pairs (manifesting as an Urbach edge), and donor- and acceptor-like defect levels originating from the defects along Te-Te chains [9.15]. Seebeck coefficient measurements confirmed the p-type conduction, although detailed data and interpretation do not agree between reports [9.16, 9.59]. The Hall data from amorphous samples should be interpreted cautiously, since the conduction mechanism is very different from crystalline materials. Although reported Hall measurements were plagued by noisy data even at a magnetic field of ~ 7 Tesla due to the high resistivity, they nevertheless exhibited some important features [9.59]. Amorphous $Ge_2Sb_2Te_5$ showed an n-type Hall effect, but this result does not necessarily contradict the p-type result obtained from thermoelectric measurements [9.16, 9.59]. It has been suggested that this is an indication of thermally assisted hopping; in this case Hall mobility is often anomalously signed, is much larger than the conductivity mobility, and is relatively temperature-independent. The measured Hall mobility is 0.07 cm^2/V, but the actual conductivity mobility is expected to be lower. For example, a mobility of ~ 10^{-3} cm^2/V (similar to As_2Te_3) would correspond to a hole concentration of ~ 10^{18} cm^{-3} [9.16, 9.59].

9.6 Perspective for the Future

Since its initial discovery in the 1960s [9.61], the phase change phenomenon in chalcogenide alloys has been developed into today's robust phase change memory technologies (rewriteable CD/DVDs and PCRAMs) by largely heuristic means. This development has taken the better part of forty years. During those four decades, the conventional, wider-band gap chalcogenide glasses have been the subject of intensive fundamental investigation. Unfortunately, the insights and understanding gained from these investigations contributed only indirectly to the advancement of the phase change material based technologies. However, over the past ten years the narrow-band gap phase change alloys have attracted the enthusiastic attention of many researchers who are studying their basic structural, optical and electronic properties in order to gain a more complete understanding of the phase change switching mechanism(s). This understanding can motivate and guide the selection of promising new materials for PCRAMs. As further performance improvements and new functionalities are aggressively pursued, these developmental efforts can be supported and accelerated by parallel fundamental investigations of the properties of the phase change materials that have high technological relevance.

Acknowledgement. This material is based in part upon work supported by the National Science Foundation under Awards No. DMR 04-12939 and No. DMR 07-06267.

References

[9.1] Heavens, O.S.: Optical Properties of Thin Solid Films. Dover Publications, New York (1965)

[9.2] Lee, B.-S., Abelson, J.R., Bishop, S.G., Kang, D.-H., Cheong, B., Kim, K.-B.: Investigation of the optical and electronic properties of $Ge_2Sb_2Te_5$ phase change material in its amorphous, cubic, and hexagonal phases. J. Appl. Phys. **97,** 093509 (2005)

[9.3] Tsu, D.V.: Obtaining optical constants of thin $Ge_xSb_yTe_z$ films from measurements of reflection and transmission. J. Vac. Sci. Technol., A **17,** 1854-1860 (1999)

[9.4] Jackson, W.B., Amer, N.M., Boccara, A.C., Fournier, D.: Photothermal deflection spectroscopy and detection. Appl. Opt. **20,** 1333-1344 (1981)

[9.5] Olson, J.K., Li, H., Ju, T., Viner, J.M., Taylor, P.C.: Optical properties of amorphous GeTe, Sb_2Te_3, and $Ge_2Sb_2Te_5$: The role of oxygen. J. Appl. Phys. **99,** 103508 (2006)

[9.6] Tompkins, H.G., McGahan, W.A.: Spectroscopic Ellipsometry and Reflectometry: A User's Guide. Wiley, New York (1999)

[9.7] Welnic, W., Pamungkas, A., Detemple, R., Steimer, C., Blugel, S., Wuttig, M.: Unravelling the interplay of local structure and physical properties in phase-change materials. Nature Mater. **5,** 56-62 (2006)

[9.8] Welnic, W., Botti, S., Reining, L., Wuttig, M.: Origin of the optical contrast in phase-change materials. Phys. Rev. Lett. **98** (2007)

[9.9] Coombs, J.H., Jongenelis, A.P.J.M., van Es-Spiekman, W., Jacobs, B.A.J.: Laser-induced crystallization phenomena in GeTe-based alloys. I. Characterization of nucleation and growth. J. Appl. Phys. **78,** 4906-4917 (1995)

[9.10] Milliron, D.J., Raoux, S., Shelby, R., Jordan-Sweet, J.: Solution-phase deposition and nanopatterning of GeSbSe phase-change materials. Nature Mater. **6,** 352-356 (2007)

[9.11] Jeong, T.H., Kim, M.R., Seo, H., Kim, S.J., Kim, S.Y.: Crystallization behavior of sputter-deposited amorphous $Ge_2Sb_2Te_5$ thin films. J. Appl. Phys. **86,** 774-778 (1999)

[9.12] Kim, Y., Kim, S.J., Kim, S.Y., An, S.H., Suh, D.-S., Noh, J.-S., Lee, S.M., Kim, K.H.P., Shin, W.-C., Khang, Y.: Experimental setup for in situ investigation of phase changing behavior in phase-change random-access memory medium by microfocusing nanosecond-time-resolved ellipsometry. Jpn. J. Appl. Phys. Part 1 **45,** 6452-6454 (2006)

[9.13] Mott, N.F., Davis, E.A.: Electronic Processes in Non-crystalline Materials. 2nd ed. Clarendon Press; Oxford University Press, Oxford, New York (1979)

[9.14] Tauc, J., Grigorovici, R., Vancu, A.: Optical properties and electronic structure of amorphous germanium. Physica Status Solidi **15,** 627-637 (1966)

[9.15] Pirovano, A., Lacaita, A.L., Benvenuti, A., Pellizzer, F., Bez, R.: Electronic switching in phase-change memories. IEEE Trans. Electron Devices **51,** 452-459 (2004)

[9.16] Kato, T., Tanaka, K.: Electronic properties of amorphous and crystalline $Ge_2Sb_2Te_5$ films. Jpn. J. Appl. Phys. Part 1 **44,** 7340-7344 (2005)

[9.17] Boer, K.W.: Survey of Semiconductor Physics: Electrons and other Particles in Bulk Semiconductors. Van Nostrand Reinhold, New York (1990)

[9.18.] Moss, T.S.: Optical Properties of Semi-Conductors. Butterworths Scientific Publications, London (1959)

[9.19] Gonzalez-Hernandez, J., Lopez-Cruz, E., Yanez-Limon, M., Strand, D., Chao, B.B., Ovshinsky, S.R.: Free carrier absorption in the Ge:Sb:Te system. Solid State Commun. **95,** 593-596 (1995)

[9.20] Mendoza-Galvan, A., Gonzalez-Hernandez, J.: Drude-like behavior of Ge:Sb:Te alloys in the infrared. J. Appl. Phys. **87,** 760-765 (2000)

[9.21] Shelimova, L.E., Karpinskii, O.G., Konstantinov, P.P., Kretova, M.A., Avilov, E.S., Zemskov, V.S.: Composition and properties of layered compounds in the GeTe-Sb_2Te_3 system. Inorg. Mater. **37,** 342-348 (2001)

[9.22] Friedrich, I., Weidenhof, V., Njoroge, W., Franz, P., Wuttig, M.: Structural transformations of $Ge_2Sb_2Te_5$ films studied by electrical resistance measurements. J. Appl. Phys. **87,** 4130-4134 (2000)

[9.23] Lyeo, H.-K., Cahill, D.G., Lee, B.-S., Abelson, J.R., Kwon, M.-H., Kim, K.-B., Bishop, S.G., Cheong, B.-k.: Thermal conductivity of phase-change material $Ge_2Sb_2Te_5$. Appl. Phys. Lett. **89,** 151904 (2006)

[9.24] Miao, X.S., Chong, T.C., Huang, Y.M., Lim, K.G., Tan, P.K., Shi, L.P.: Dependence of optical constants on film thickness of phase-change media. Jpn. J. Appl. Phys. Part 1 **38,** 1638-1641 (1999)

[9.25] Liang, R.G., Peng, C.B., Nagata, K., Daly-Flynn, K., Mansuripur, M.: Optical characterization of multilayer stacks used as phase-change media of optical disk data storage. Appl. Opt. **41,** 370-378 (2002)

[9.26] Yamada, N., Ohno, E., Nishiuchi, K., Akahira, N., Takao, M.: Rapid-phase transitions of GeTe-Sb_2Te_3 pseudobinary amorphous thin-films for an optical disk memory. J. Appl. Phys. **69,** 2849-2856 (1991)

[9.27] Kim, J.-H., Kim, M.R.: Effects of microstructure on optical properties of $Ge_2Sb_2Te_5$ thin films. Jpn. J. Appl. Phys. Part 1 **37,** 2116-2117 (1998)

[9.28] Lai, Y.F., Qiao, B.W., Feng, J., Le, Y., La, L.Z., Lin, Y.Y., Tang, T.A., Cai, B.C., Chen, B.M.: Nitrogen-doped $Ge_2Sb_2Te_5$ films for nonvolatile memory. J. Electron. Mater. **34,** 176-181 (2005)

[9.29] Kyratsi, T., Chrissafis, K., Wachter, J., Paraskevopoulos, K.M., Kanatzidis, M.G.: KSb_5S_8: A wide bandgap phase-change material for ultra high density rewritable information storage. Adv. Mater. **15,** 1428-1431 (2003)

[9.30] Gutwirth, J., Wagner, T., Bezdicka, P., Vlcek, M., Kasap, S.O., Frumar, M.: Influence of silver concentration in $Ag_x(Sb_{0.33}S_{0.67})_{100-x}$ thin amorphous films on photoinduced crystallization. J. Non-Cryst. Solids **353,** 1431-1436 (2007)

[9.31] Kolobov, A.V., Fons, P., Frenkel, A.I., Ankudinov, A.L., Tominaga, J., Uruga, T.: Understanding the phase-change mechanism of rewritable optical media. Nature Mater. **3,** 703-708 (2004)

[9.32] Tanaka, K.: Comment on "Investigation of the optical and electronic properties of $Ge_2Sb_2Te_5$ phase change material in its amorphous, cubic, and hexagonal phases" [J. Appl. Phys. 97, 093509 (2005)]. J. Appl. Phys. **101,** 026111 (2007)

[9.33] Lee, B.-S., Abelson, J.R., Bishop, S.G., Kang, D.-H., Cheong, B.-K., Kim, K.-B.: Response to "Comment on 'Investigation of the optical and electronic properties of $Ge_2Sb_2Te_5$ phase change material in its amorphous, cubic, and hexagonal phases'" [J. Appl. Phys. 97, 093509 (2005)]. J. Appl. Phys. **101,** 026112 (2007)

[9.34] Lee, H.S., Cheong, B.-K., Lee, T.S., Jeong, J.-H., Lee, S., Kim, W.M., Kim, D.: Origin of nonlinear optical characteristics of crystalline Ge-Sb-Te thin films for possible super-resolution effects. Jpn. J. Appl. Phys. Part 2 **46,** 277-279 (2007)

[9.35] Shimakawa, K., Kolobov, A., Elliott, S.R.: Photoinduced effects and metastability in amorphous semiconductors and insulators. Adv. Phys. **44,** 475-588 (1995)

[9.36] Lee, B.-S., Xiao, Y., Bishop, S.G., Abelson, J.R., Raoux, S., Deline, V.R., Kwon, M.-H., Kim, K.-B., Cheong, B., Li, H., Taylor, P.C.: Photo-oxidation and the absence of photodarkening in $Ge_2Sb_2Te_5$ phase change material, In: Edwards, A.H., Taylor, P.C., Maimon, J.Kolobov, A. (eds.) Chalcogenide Alloys for Reconfigurable Electronics, MRS Proc., Vol. 918, p. 0918-H0902-0904, San Francisco, CA (2006)

[9.37] Pamukchieva, V., Szekeres, A.: Influence of illumination on the optical bandgap energy value of $Ge_xSb_{20-x}Te_{80}$ films. J. Optoelectron. Adv. Mater. **7,** 1277-1280 (2005)

[9.38] Kalb, J., Spaepen, F., Wuttig, M.: Atomic force microscopy measurements of crystal nucleation and growth rates in thin films of amorphous Te alloys. Appl. Phys. Lett. **84,** 5240-5242 (2004)

[9.39] Kim, D.H., Merget, F., Laurenzis, M., Bolivar, P.H., Kurz, H.: Electrical percolation characteristics of $Ge_2Sb_2Te_5$ and Sn doped $Ge_2Sb_2Te_5$ thin films during the amorphous to crystalline phase transition. J. Appl. Phys. **97** (2005)

[9.40] Ovshinsky, S.R.: Optical cognitive information processing - A new field. Jpn. J. Appl. Phys. Part 1 **43,** 4695-4699 (2004)

[9.41] Ovshinsky, S.R., Czubatyj, W.: New developments in optical phase change memory, In: Proceedings of SPIE - The International Society for Optical Engineering, vol. 4085, p. 15-22, Society of Photo-Optical Instrumentation Engineers, Shanghai (2001)

[9.42] Ventrice, D., Fantini, P., Redaelli, A., Pirovano, A., Benvenuti, A., Pellizzer, F.: A phase change memory compact model for multilevel applications. IEEE Electron Device Lett. **28,** 973-975 (2007)

[9.43] Ryu, S.O., Yoon, S.M., Choi, K.J., Lee, N.Y., Park, Y.S., Lee, S.Y., Yu, B.G., Park, J.B., Shin, W.C.: Crystallization behavior and physical properties of Sb-excess $Ge_2Sb_{2+x}Te_5$ thin films for phase change memory (PCM) devices. J. Electrochem. Soc. **153,** G234-G237 (2006)

[9.44] Jeong, T.H., Kim, M.R., Seo, H., Park, J.W., Yeon, C.: Crystal structure and microstructure of nitrogen-doped $Ge_2Sb_2Te_5$ thin film. Jpn. J. Appl. Phys. Part 1 **39,** 2775-2779 (2000)

[9.45] Kim, Y., Hwang, U., Cho, Y.J., Park, H.M., Cho, M.H., Cho, P.-S., Lee, J.-H.: Change in electrical resistance and thermal stability of nitrogen incorporated $Ge_2Sb_2Te_5$ films. Appl. Phys. Lett. **90,** 021908 (2007)

[9.46] Gonzalez-Hernandez, J., Herrera-Fierro, P., Chao, B., Kovalenko, Y., Morales-Sanchez, E., Prokhorov, E.: Structure of oxygen-doped Ge:Sb:Te films. Thin Solid Films **503,** 13-17 (2006)

[9.47] Streetman, B.G., Banerjee, S.: Solid State Electronic Devices. 5th ed. Prentice Hall, Upper Saddle River, N.J. (2000)

[9.48] Smith, R.A.: Semiconductors. 2d ed. Cambridge University Press, Cambridge ; New York (1978)

[9.49] Measuring resistivity and Hall coefficient and determining Hall mobility in single-crystal semiconductors (ASTM Standard) (1986)

[9.50] Nolas, G.S., Sharp, J., Goldsmid, H.J.: Thermoelectrics: Basic Principles and New Materials Developments. Springer, Berlin (2001)

[9.51] Castro, D.T., Goux, L., Hurkx, G.A.M., Attenborough, K., Delhougne, R., Lisoni, J., Jedema, F.J., Zandt, M.A.A.i.t.A., Wolters, R.A.M., Gravesteijn, D.J., Verheijen, M.A., Kaiser, M., Weemaes, R.G.R., Wouters, D.J.: Evidence of the thermo-electric Thomson effect and influence on the program conditions and cell optimization in phase-change memory cells, In: Goux, L. (ed.) Electron Devices Meeting, IEEE International, p. 315-318, (2007)

[9.52] Frumar, M., Tichy, L., Klikorka, J., Horak, J.: Preparation and some physical properties of semiconducting $GeSb_2Te_4$ crystals. Mater. Res. Bull. **7,** 1075-1085 (1972)

[9.53] Frumar, M., Tichy, L., Matyas, M., Zelizko, J.: Some physical-properties of semiconducting $GeSb_4Te_7$ crystals. Phys. Status Solidi A **22,** 535-541 (1974)

[9.54] Yanez-Limon, J.M., Gonzalez-Hernandez, J., Alvarado-Gil, J.J., Delgadillo, I., Vargas, H.: Thermal and electrical properties of the Ge:Sb:Te system by photoacoustic and Hall measurements. Phys. Rev. B: Condens. Matter **52,** 16321-16324 (1995)

[9.55] Lee, B.-S.: Optical and Electronic Properties, Nanoscale Structural Order, and Transformation Kinetics of Phase Change Materials. Ph. D. thesis, University of Illinois at Urbana-Champaign (2006)

[9.56] Yamanaka, S., Ogawa, S., Morimoto, I., Ueshima, Y.: Electronic structures and optical properties of GeTe and $Ge_2Sb_2Te_5$. Jpn. J. Appl. Phys. Part 1 **37,** 3327-3333 (1998)

[9.57] Blakemore, J.S.: Semiconductor Statistics. Dover, New York (1987)

[9.58] Kim, J.J., Kobayashi, K., Ikenaga, E., Kobata, M., Ueda, S., Matsunaga, T., Kifune, K., Kojima, R., Yamada, N.: Electronic structure of amorphous and crystalline $(GeTe)_{(1-x)}(Sb2Te3)_{(x)}$ investigated using hard x-ray photoemission spectroscopy. Phys. Rev. B **76,** 115124 (2007)

[9.59] Baily, S.A., Emin, D., Li, H.: Hall mobility of amorphous $Ge_2Sb_2Te_5$. Solid State Commun. **139,** 161-164 (2006)

[9.60] Ielmini, D., Zhang, Y.G.: Evidence for trap-limited transport in the subthreshold conduction regime of chalcogenide glasses. Appl. Phys. Lett. **90**, 192102 (2007)

[9.61] Ovshinsky, S.R.: Reversible electrical switching phenomena in disordered structures. Phys. Rev. Lett. **21,** 1450 (1968)

10. Development of Materials for Third Generation Optical Storage Media

Noboru Yamada

Abstract In 1987, the breakthrough discovery of the GeTe-Sb_2Te_3 pseudo-binary alloys removed the difficulties that had inhibited phase-change materials from being used in practical applications. The keys issues that had to be solved were related to i) the material composition and ii) the conceptual change in designing materials for phase-change devices. The prominent features of GeTe-Sb_2Te_3 pseudo-binary alloys enabled a very fast crystallization process and high cyclability at once, and the conceptual change from "how to crystallize the essentially amorphous material" to "how to amorphize the essentially crystalline material" enabled the application of single phase materials with high crystallization speed and high melting temperature, T_m. In this chapter, it will be mainly described how such phase-change alloys were obtained (history), how the compositions are superior (results), and what are the essential points of them (secrets) especially in the optical uses.

10.1 Introduction

In order to apply the reversible phase-change phenomenon to optical memory, recording marks in the amorphous state should be stable at room temperature conditions for a long period of at least several 10 years. On the other hand, the amorphous marks should be crystallized by a short laser pulse with a duration of less than several tens of nanoseconds for data rewriting. This implies that phase-change optical memory materials must respond quite differently under static and dynamic conditions. The difference between these response times reaches the order of 10^{17} (see Fig. 10.1). If phase-change memory is applied to computer uses, the required data rewriting cycle times will be at least 10^5 for replacing flash memory, and can be as high as 10^{15}-10^{18} for replacing dynamic random access memory (DRAM). What are the material characteristics to satisfy these surprising requirements? In this chapter we will investigate which factors control the recording speed, data retention time and cyclability.

Chalcogenide semiconductor thin films historically have been used as phase-change memory materials [10.1]. In particular, single phase compositions have been actively utilized in recent years. Why should these materials be chalcogenides and why should they be single phase compositions? In this chapter, the necessary conditions for selecting phase-change materials will be explained and an outlook will be given for obtaining next generation phase-change materials [10.2, 10.3].

S. Raoux, M. Wuttig (eds.), *Phase Change Materials*,
DOI 10.1007/978-0-387-84874-7_10,

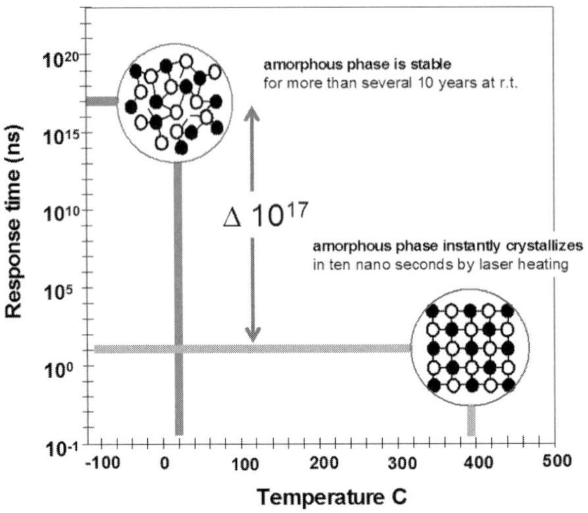

Fig. 10.1. Response time difference between the laser heating time required for crystallization and the amorphous stability storage time at room temperature. The time difference reaches 17 orders of magnitude.

10.2 Requirements for a Phase-change Material

Theoretically, every material has the potential to undergo reversible structure changes between amorphous and crystalline phases. For example, even the formation of amorphous Ar was reported by Nishio et al. even though it was just in a computer simulation [10.4]. However, not all materials easily undergo reversible phase-changes and not all existing phase-change materials will be applied for practical optical memories. Many of these materials do not satisfy the requirements for phase-change memory materials. The requirements for an optical memory material are:

1. appropriate laser light absorption to produce the laser-induced structural changes
2. a large change in the refractive index or absorbance accompanied by the structural changes
3. existence of both, the amorphous and crystalline phases, stable for a long period of time at the using or operating environmental conditions.

No metals or dielectrics but chalcogenide semiconductors have been utilized as phase-change memory materials just for these reasons. The first laser-induced phase-change demonstration was reported by Feinleib et al. (Fig. 10.2) [10.5].

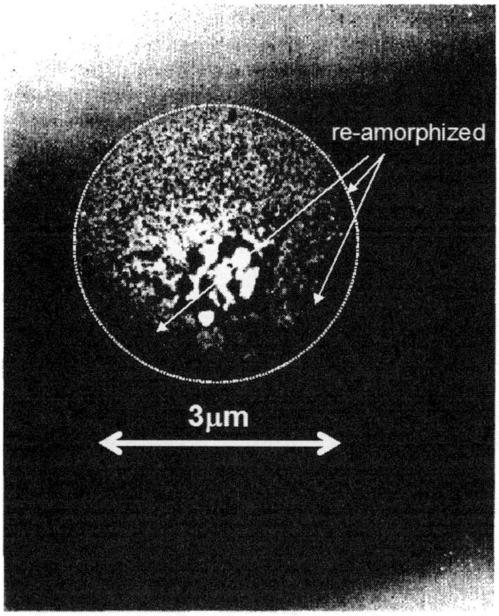

Fig. 10.2. The first demonstration of laser-induced reversible phase-changes on chalcogenide semiconductor amorphous film of $Te_{81}Ge_{15}Sb_2S_2$ by Feinleib et al. [10.5]. Some corners of the crystallized part in the white circle (added by the author) were partially re-amorphized. Reprinted with permission from [10.5]. Copyright (1971), American Institute of Physics.

Table 10.1 shows merits and demerits of various inorganic materials in terms of their usefulness as phase-change materials.

Table 10.1. Merits and demerits of metals, semiconductors and dielectrics. T_m is the melting point. *Italic* indicates the optical and thermal properties desirable for phase-change materials.

Material	Merit	Demerit
Metals		Include many free electrons Produce little change in optical properties between amorphous and crystalline phase
Semiconductors	*Proper optical absorption at typical laser wavelength, phase change detectable*	
Tetrahedral (Si, Ge)		Rigid structure and high T_m (difficult to melt)
Chalcogenide (Se, Te)	*Flexible structure and low T_m (easy to melt and amorphize)*	
Dielectric		Transparent at typical laser wavelength, not possible to change phase with laser

Why are metals not suitable phase-change materials? In the case of metals, the second condition (large change in optical constants) is not satisfied. Many free electrons absorb the light over a wide wavelength range and metals are basically opaque for visible light. Since the material structures are basically not different before and after the phase-change, no prominent optical changes required for phase-change materials are observed. Then, why are dielectrics not suitable, either?

In the case of dielectrics, they are not suitable for a phase-change optical memory since they lack the first condition (absorption). For laser light absorption it is necessary to excite electrons from the valence band to the conduction band; however, the wide band gap of the dielectrics suppresses the excitation by the usual exposure of visible laser light. In other words, dielectrics are essentially transparent for visible light in both, the amorphous and crystalline states.

Hence, they are not used for a phase-change film but as a protecting film in the phase-change optical discs.

Finally, semiconductors have electronic properties between those of metals and dielectrics. Their energy gaps are rather narrow compared to dielectrics, and the valence band electrons can be easily excited to the conduction band by irradiation using visible laser lights. The electronic properties are rather different between the amorphous and crystalline phases, and the structural changes become optically detectable. For the semiconductors, two types are known: tetrahedral semiconductors such as Si and Ge and chalcogenide semiconductors such as Te, Se and S, and their alloys. Why did we not select tetrahedral semiconductors but chalcogenides? That is the theme of the next section.

10.3 Why Chalcogenide Semiconductors for Optical Memory?

The atomic configuration in crystalline Si or Ge is quite different from that in Te or Se. The former exhibits the rigid diamond type structure and each atom forms tight covalent bonds with four surrounding atoms. This results in high melting temperatures T_m of 952 °C for Ge and 1414 °C for Si, and high viscosity in the molten state. These properties cause some difficulties for their practical application as phase-change materials. It means that crystalline (c-) Si or c-Ge are difficult to amorphize due to the high T_m and simultaneously amorphous (a-) Si or a-Ge are difficult to crystallize due to their high viscosity (i.e., slow atomic diffusion).

On the contrary, Te and Se are essentially two-fold coordinated forming coil like structures; e.g., Te atoms form strong covalent bonds with two adjacent Te atoms resulting in a -Te-Te-Te- coil like structure. Every -Te-Te-Te- coil is bound by a weak van der Waals force to another to form a hexagonal crystal. This is the reason why Te has a very low melting temperature (T_m = 450 °C). It was reported that Te was two-fold coordinated as usual but partially three-fold coordinated in the molten state [10.6]. From these facts, the ability of Te to form easily the amorphous phase can be interpreted as follows (Fig. 10.3).

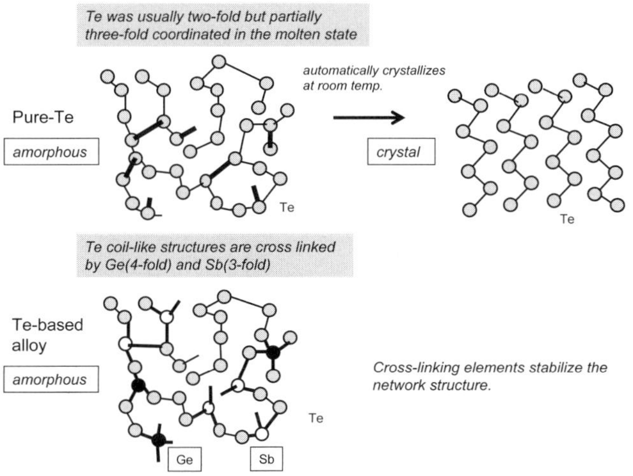

Fig. 10.3. Basic atomic configuration of Te: The three-fold formation of an amorphous network structure is not stable at room temperature. By adding some cross-linking elements, the network structure is stabilized. Reprinted with permission from [10.7].

1) Once Te is heated up above T_m, it will easily melt by cutting van der Waals bonds between -Te-Te-Te- coils. The -Te-Te-Te- coils remain in the melt.
2) It forms a random network structure since Te has partially 3 bonds in the molten state.
3) By rapid quenching, the network structure is frozen to form the amorphous solid.

Because the three-fold coordination in the amorphous Te is not stable at room temperature, the amorphous Te rapidly and automatically crystallizes by removing the cross-linking bonds.

This fragile, amorphous pure Te was the starting point from which the pioneers in the field of fast phase-change materials discovered and developed the Te-based eutectic materials. The above steps teach us that stable amorphous materials will be obtained by adding some cross-linking elements into Te. Here, cross-linking elements should have three-fold or four-fold coordination with Te, such as P, As and Sb (three-fold) and Si, Ge and Sn (four-fold) etc. (Fig. 10.4). By adoption of these cross-linking elements, it became possible to obtain Te-based eutectic alloys such as Te-Ge and Te-Sb with very low T_m. These were the first generation optical phase-change materials.

10.4 Merits and Demerits of the Te Based Eutectic Alloys

From the early 1970s through the first half of the 1980s Te-based eutectic compositions with low T_m were actively studied. Based on the typical eutectic systems such as Te-Ge and Te-Sb, various studies were carried out as illustrated in Fig. 10.4. However, these attempts did not reach the final goal for practical optical memories because the Te based eutectic alloys exhibited some basic and essential demerits in addition to some merits.

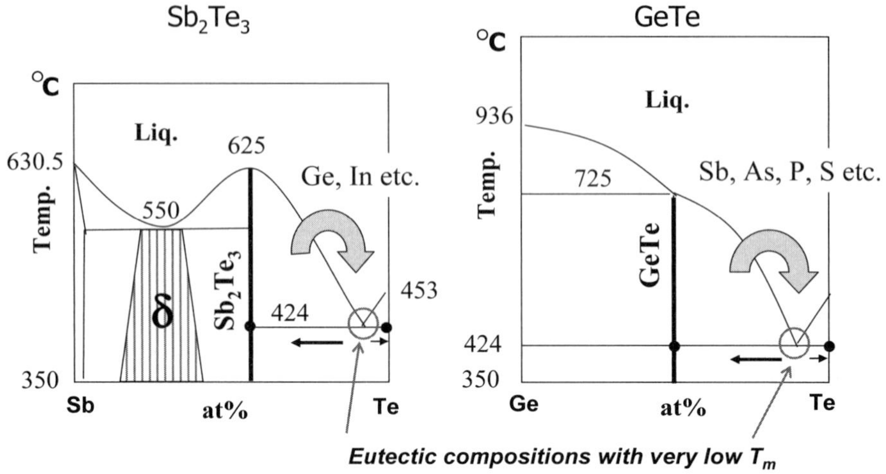

Fig. 10.4. Early 1970s trials for obtaining Te-based eutectic phase-change materials; various combinations of additives were studied as the third element in order to realize high crystallization rate and amorphous stability at once. Reprinted with permission from [10.7].

As shown in Table 10.2, Te-based eutectic compositions intrinsically possess the ability to easily form the amorphous phase as a large merit. This merit is naturally brought about by their very low melting temperatures and their flexible coil like structures.

Table 10.2. Merits and demerits of Te-based eutectic alloys. T_x is the crystallization temperature. ΔE is the free energy difference that is proportional to the driving motive force for crystallization, see Fig. 10.5.

	Merits	Demerits
Optical properties	Proper absorption at typical laser wavelengths in the visible light spectrum	
Thermal properties	Low T_m - easy to melt	Low T_x – unstable amorphous state (T_x = 1/3 T_m – 2/3 T_m)
Crystallization Process	Easy amorphization due to small ΔE	Low crystallization speed due to small ΔE, requires long atomic diffusion, small rewrite cycle number due to compositional deviation

However, they have three significant insufficiencies: slow crystallization speed, low thermal stability of the amorphous state and insufficient cyclability.

Let us here clarify the essential issues of the eutectic materials using a schematic phase diagram. Figure 10.5 shows such a schematic binary phase diagram of a eutectic system consisting of the elements A and B.

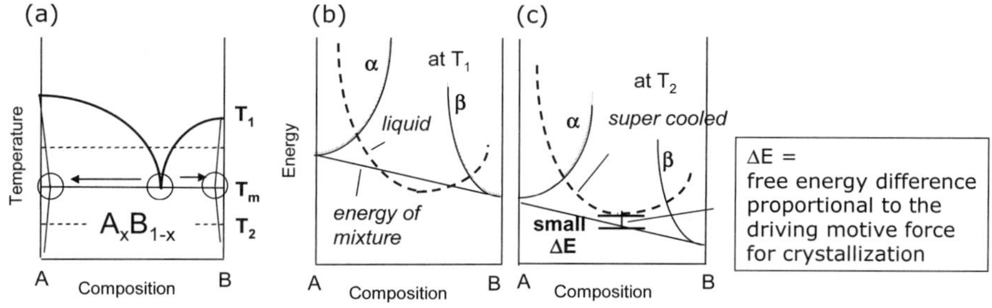

Fig. 10.5. A schematic binary phase diagram of the A-B eutectic system (a) and the relations between the compositions and free energy. (b) and (c) indicate the cases at temperature $T > T_m$ and $T < T_m$, respectively. Reprinted with permission from [10.8].

As a particular property of the eutectic binary systems, the liquidus curve has a minimum at the eutectic point, where the T_m of the eutectic point, denoted as $T_{m(eu)}$, sometimes becomes much lower than T_m of either, A or B. Figures 10.5b and 10.5c show the relationship between the compositions and the free energies. In these two figures, curves α and β indicate the free energies assuming that every composition is a single phase. Both values drastically rise for compositions other than pure A and pure B. On the other hand, straight lines that are drawn as the mutual tangent line to the curves α and β indicate the free energies of the solid phase wherein the α phase and β phase co-exist as a mixture.

The dotted curves show the free energy of the liquid phase (single phase). Generally, the mixed liquids or mixed gases A+B have lower total free energy than that for only A or only B. Hence, the free energy curve tends to fall at the center.

We will first discuss Fig. 10.5b. When the temperature T of the material system is higher than T_m, the free energy of the liquid phase becomes lower than that of the two phases co-existing at the eutectic composition. It merely indicates that the liquid phase is the thermal equilibrium phase at $T > T_m$. Here, assuming that the temperature drops to below T_m, the free energy of the two co-existing phases largely decreases and is reduced below the free energy of the liquid phase. That means, the thermal equilibrium phase moves to the solid phase of two co-existing phases, and accordingly, phase separation and crystallization should occur if the system is kept at this temperature. However, in the case of the eutectic compositions, the crystallization does not proceed so fast by the following two reasons. One is that the driving motive force for nucleation that is proportional to the difference of the free energies between the liquid and two co-existing phases is small, and the other is that the atomic mobility is very small at the eutectic point since the temperature of the liquid is quite low at the eutectic point. Thus, the eutectic materials become intrinsically amorphous phase formation materials.

The above discussion means that the eutectic compositions involve significant demerits for a phase-change material since nucleation rate and crystal growth speed will be limited for the above mentioned two reasons. In addition, phase separation causes poor cyclability. If the α and β phases have largely different thermal properties such as melting points and viscosities, it can lead to spatial phase-segregation after repetitious melt-quenching processes. Local deviation of composition from the eutectic will cause significant variation of recording power conditions, increased optical noise, and limited cycle numbers. The final demerit is their low crystallization temperature. According to Sakka and Mackenzie, the glass transition temperature T_g of materials corresponds to 1/3 - 2/3 of their T_m in Kelvin [10.9]. It means that T_g (which is a lower limit for T_x) of these materials is essentially low; therefore, their thermal stability of the amorphous phase will be insufficient.

In this section, we explained that the historical Te-based eutectic compositions have essential demerits and they could not be the final solution for a phase-change optical memory material. In the next section, the importance of the single phase compositions will be described.

10.5 Merits and Demerits of the Te-based Single Phase Materials

In order to compare Te-based single phase materials to the eutectic materials, their merits and demerits are listed in Table 10.3. Reproduced by permission of the MRS Bulletin from [10.2].

Table 10.3. Merits and demerits of Te-based single phase materials.

	Merits	Demerits
Optical properties	Proper absorption at typical laser wavelengths in the visible light spectrum	
Thermal properties	High T_x - stable amorphous state ($T_x > T_g = 1/3\ T_m - 2/3\ T_m$)	High T_m – difficult to melt
Crystallization Process	Rapid crystallization due to large ΔE, good cyclability – no phase separation	

Single phase materials, typically stoichiometric compounds, possess some advantages such as rapid crystallization while they have only one demerit. The characteristics will be thermodynamically explained below. Figure 10.6 shows a schematic binary phase diagram of the materials A and B having a compound phase (γ- phase).

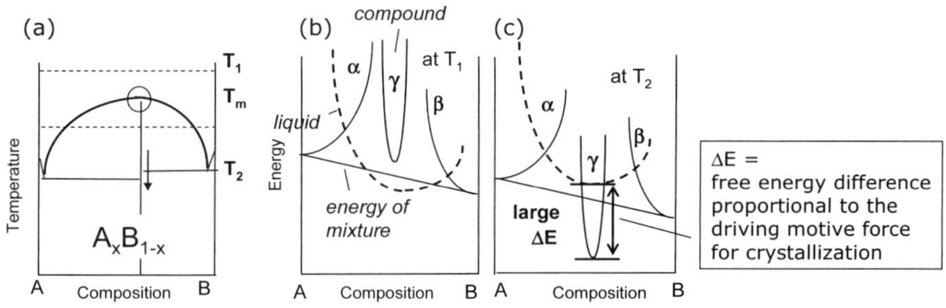

Fig. 10.6. A schematic binary phase diagram of the A-B compound system (a) and the relations between the compositions and free energy. (b) and (c) indicate the cases at $T > T_m$ and $T < T_m$, respectively. Reprinted with permission from [10.8].

In this system, thermally stable solid phases are the edge compositions of α and β and the intermediate compound phase of γ, wherein the liquidus curve usually has a maximum at the compound point. When the temperature of the system is higher than T_m, the free energy of the liquid phase is lower than that of the γ phase. It means that the liquid phase is the thermal equilibrium phase at $T > T_m$. If the temperature falls below T_m, the free energy of the γ phase largely decreases and becomes lower than that of the liquid phase; i.e., the stable phase is now the γ phase at $T < T_m$. In the case of such a compound system, a faster crystallization process can be expected compared to the eutectic systems for the following reasons. At first, the driving motive force for nucleation will become very large since it will be proportional to the free energy difference between the liquid phase and the γ phase. Second, the kinetic energy of the atoms will be rather large since the temperature of the molten phase is generally high at the compound point. At last, the crystallization process does not require any phase separation.

Thus, it was known that the compound composition would be a promising candidate as a phase-change material. The amorphous substance having a compound composition will likely have at the same time a large crystallization speed, a good reversibility, and a high thermal stability due to high T_m. Even though the discussion above was made for a stable single phase compound, one can argue similarly for the case of the meta-stable single phase compound as will be proven in the next section. As listed in Table 10.4, the issue of difficult amorphization is inevitably produced despite the above merits. However, the problem of difficult amorphization can be solved by the design of the device structure, such as a multi-layered stacking structure, producing very large cooling rates of 10^9 to 10^{11} K/s [10.10, 10.11].

10.6 From Eutectic to Single Phase Compositions

Figure 10.7 shows the Ge-Te binary phase-diagram [10.12] and the experimental results for the laser heating time required for crystallization (left axis) and the crystallization temperature (right axis) of Ge-Te thin films measured by Chen et al. in 1986 [10.13].

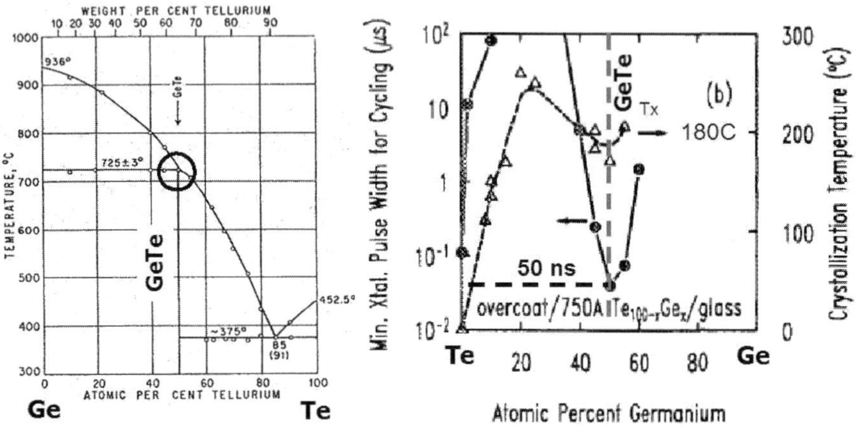

Fig. 10.7. Left - phase diagram of GeTe. Right - laser heating time required for crystallization (left axis) and the crystallization temperature (right axis) of the Ge-Te binary system. Reprinted with permission from [10.13]. Copyright (1986), American Institute of Physics.

It can be seen from the figure that both, the laser pulse time for crystallization and the crystallization temperature T_x, drastically change with film composition. They show a maximum at around 30 at. % Ge, a minimum at 50 at. % Ge, and increase again at higher Ge content. The experimental results vary between more than 100 µs crystallization time and 160 °C crystallization temperature for $Te_{85}Ge_{15}$ (eutectic composition) and about 50 ns and 180 °C for the GeTe stoichiometric compound. It should be noted that the laser heating time for crystallizing GeTe is three orders of magnitude shorter than that of $Te_{85}Ge_{15}$. This was the first data reporting that a stoichiometric compound can be a possible candidate for a phase-change memory film.

In 1986, another historically important result was reported on the $(Te_{80}Ge_5Sn_{15})_{100-x}Au_x$ system. Yamada et al. [10.14, 10.15] demonstrated that the existence of a metastable, single phase remarkably accelerated the crystallization process. Figure 10.8 shows the influence of the Au content on the laser heating time for crystallization, wherein the Au was added in the $Te_{80}Ge_5Sn_{15}$ composition as $(Te_{80}Ge_5Sn_{15})_{100-x}Au_x$ for shortening the -Te-Te-Te- coil-like structures.

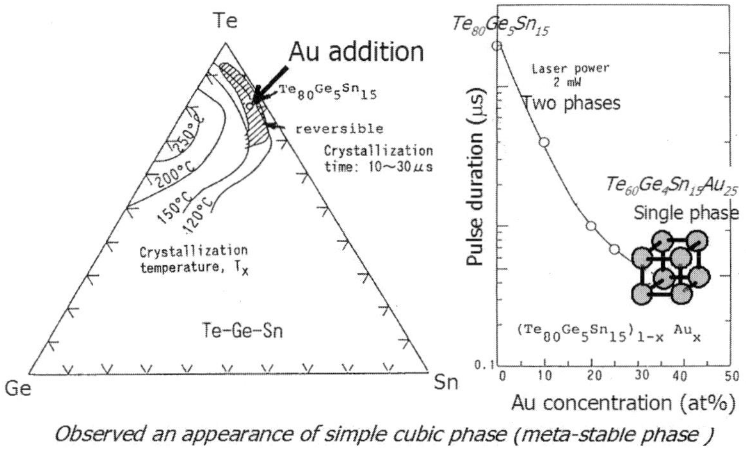

Fig. 10.8. Laser heating time required for crystallization of $(Te_{80}Ge_5Sn_{15})_{100-x}Au_x$. Reprinted with permission from SPIE from [10.14].

Au addition produced remarkable results; i.e., the laser heating time drastically decreased to 1/100 with Au addition from the 20 μs without Au to 300 ns for 30 at. % Au content. Added Au did not function merely as producing nucleation sites. By structural characterization using x-ray and electron diffraction analyses, it was clarified that a meta-stable single phase with simple cubic structure appeared for $(Te_{80}Ge_5Sn_{15})_{100-x}Au_x$ while $Te_{80}Ge_5Sn_{15}$ showed the usual multiple phases of Te and (Ge, Sn)Te as shown in Fig. 10.9. Thus, it was proven that the discussion for the single phase compound in the last section can be argued similarly for the case when the meta-stable single phase compound was produced. These two reports completely changed the directions of research and development in the field of phase-change memory materials. As shown in Table 10.4, the investigation of low-T_m, eutectic compositions that were mainly studied until 1985, shifted to single phase compositions starting in 1986 [10.2].

10.7 Discovery of the GeTe-Sb_2Te_3 Pseudo-binary System

The superior characteristics of the GeTe-Sb_2Te_3 pseudo-binary system were first demonstrated in 1987 by Yamada et al. [10.16]. This thin film material not only enabled rewritable optical memory technology such as DVD-RAM (Digital Versatile Disc – Random Access Memory) and Blu-ray discs (see also Chapter 12), but also revived the research and development activities for non-volatile electrical memory devices.

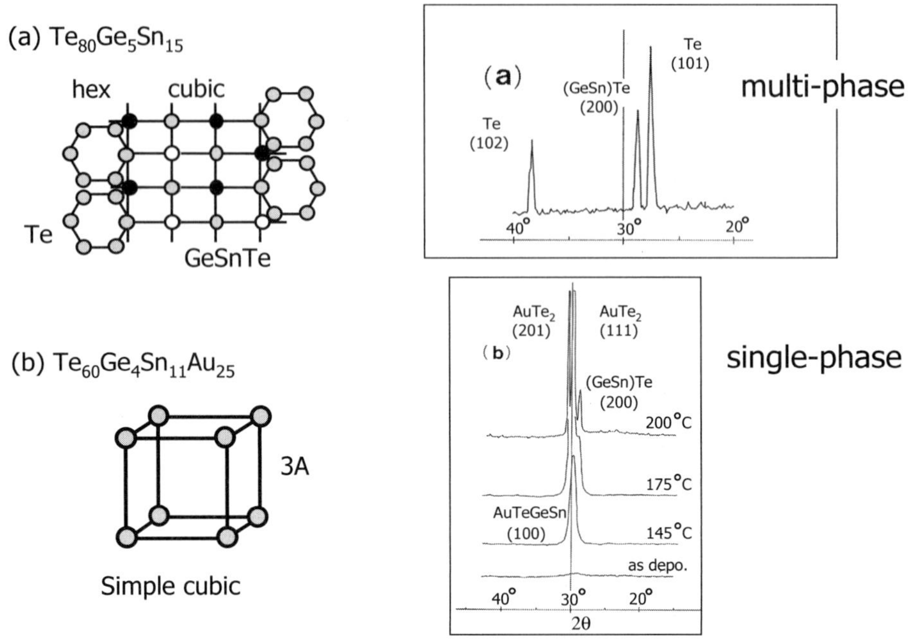

Fig. 10.9. Crystal structures and x-ray diffraction data of (a) $Te_{80}Ge_5Sn_{15}$ (reprinted with permission from [10.15]), and (b) $(Te_{80}Ge_5Sn_{15})_{0.75}Au_{0.25}$. The former has multiple phases of (Ge, Sn)Te and Te, and the latter has a single phase with simple cubic structure. Reprinted with permission from SPIE from [10.14].

Table 10.4. Research and development history of phase-change materials [10.2]. Eutectic (multiphase) materials are indicated in *italic*, compound or single-phase materials in **bold**. An apparent turning point is observed from eutectic alloys to compounds around 1986.

Year	Compositions
1971	*$Te_{81}Ge_{13}Sb_2S_2$*
1974	*$Te_{93}Ge_5As_2$*
1983	*$Te_{64}Ge_4Sn_{12}O_{20}$*
1985	*$Te_{80}Ge_5Sn_{15}$, Te-Sn-Se, Te-Se-Ga*
1986	**GeTe, In-Se, $Te_{60}Ge_4Sn_{11}Au_{25}$, Sb_2Se,** *Bi-Se-Sb, Te-Ge-Sn-Pd*
1987	**GeTe-Sb2Te3 (GeSb2Te4, Ge2Sb2Te5), In-Se-Te-Co**
1988	*In-Sb-Te,* **$In_3Sb_2Te_5$**
1989	**$GeTe-Sb_2Te_3-Sb$, Ge-Sb-Te-Pd, Ge-Sb-Te-Co, $Sb_2Te_3-Bi_2Se_3$**
1991	**Ge-Sb-Te-N, Ag-In-Sb-Te**
1998	**Ge-Sb-Te-Ag, Au-In-Sb-Te**
2000	**$(Sb_{70}Te_{30})$-Ge**

It is not an exaggeration to say that phase-change technology development was restarted after the discovery of $GeTe-Sb_2Te_3$ pseudo-binary compositions.

This material system naturally satisfies the above discussed merits of single phase materials. These are:

1. Quite large optical contrast over the whole laser wavelength range from infrared to violet (Fig. 10.10) [10.17],
2. Short laser heating times for crystallization of less than 30 ns,
3. Stable amorphous phase at room temperature of more than several tens of years,
4. Large cycle numbers of more than 50,000 cycles (Fig. 10.11) [10.18].

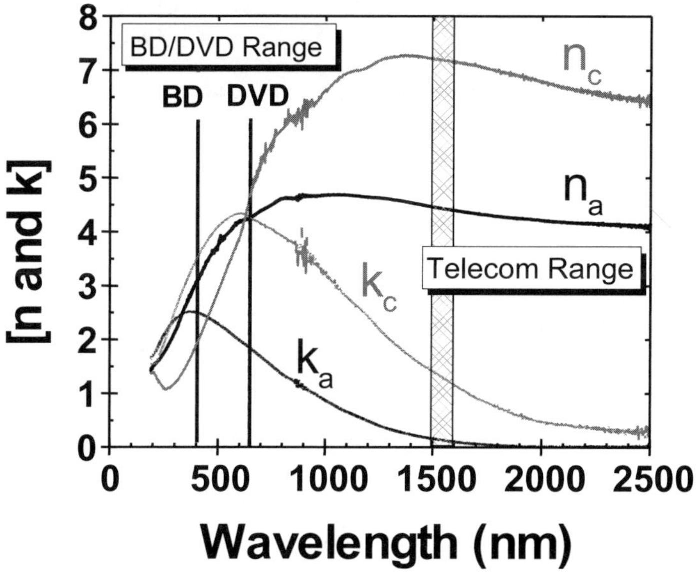

Fig. 10.10. Variations of optical constants *n* and *k* between the amorphous (a) and crystalline (c) states of $Ge_2Sb_2Te_5$ thin films. Reprinted with permission from EPCOS from [10.17].

Figure 10.12 summarizes the experimental results on the crystallization properties of Ge-Sb-Te material thin films. The two triangular figures show the dependency of the crystallization temperature T_x (left) and the laser heating time for crystallization τ_{cry} (right) on film composition [10.16, 10.19]. It can be seen from the figures that films in a wide compositional range have T_xs substantially above room temperature and very short τ_{cry}s of less than 100 ns.

It should be noted in these figures that both T_x and τ_{cry} show a continuous and gentle variation along the GeTe-Sb_2Te_3 tie-line even though there are more than three ternary compounds on this tie line.

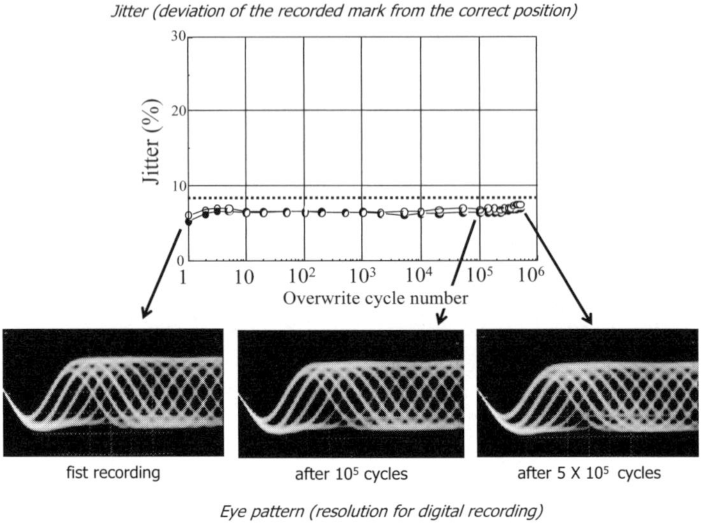

Fig. 10.11. Repetitious overwriting test results using a 4.7 GB capacity DVD-RAM optical disc. The top graph shows the dependence of jitter value on the overwrite cycle number when random signals are recorded on a same track. It is required to be less than 8.5 %. The bottom photos indicate the eye-pattern signals for evaluating the resolution power of digital signals, where it is required that the diamond–shaped spaces surrounded by curves will be clearly open after many times of overwritings. As can be seen in the photos, the eye patterns are very clear and unchanged after 500,000 cycles. Reprinted with permission from [10.18].

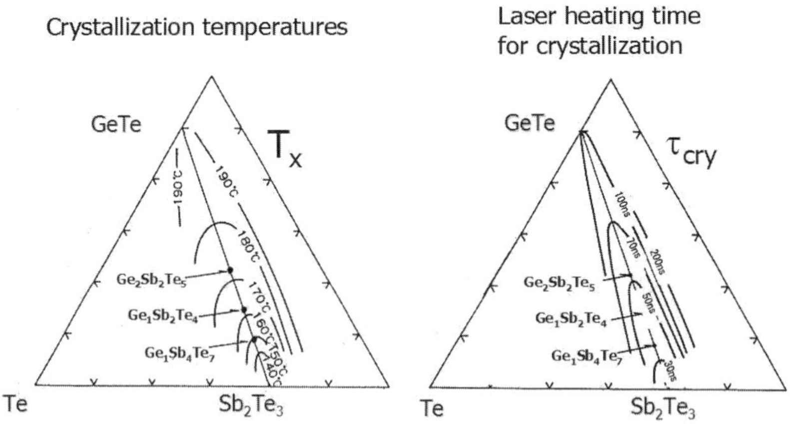

Fig. 10.12. Compositional dependencies of the crystallization temperature T_x (left) and the laser heating time for producing crystallization τ_{cry} (right) of Ge-Sb-Te thin films. The heating rate for T_x measurement is 10 K/min and the laser power for τ_{cry} measurement is 8 mW. It is very interesting to note that both T_x and τ_{cry} do not have particular values such as singularities at compound compositions. Reprinted from [10.14].

Practically, these properties are very useful in a real information recording system, where the recording speed can be sometimes very different. It is slow for audio data but high for visual data, and for professional uses the required speed becomes even higher. The material films applicable for various systems can be optimized by modifying the film compositions.

Here a few questions arise. Why do the stoichiometric compositions not show any particular characteristics though they are particular compositions as explained in section 10.5? Why does the GeTe-Sb_2Te_3 material system show such a good cyclability as described in Fig.10.11? It was reported that GeTe lacks in cyclability even though it is a single phase composition?

In the next section, various results on the analytical studies will be described.

10.8 Importance of the Cubic Structure and Vacancies

Figure 10.13 shows the differential scanning calorimetry (DSC) measurement results of various amorphous thin films: $Ge_{14}Sb_{29}Te_{57}$ ($GeSb_2T_4$), $Ge_{19}Sb_{25}Te_{56}$, $Ge_{22}Sb_{22}Te_{56}$ ($Ge_2Sb_2Te_5$), and $Ge_{25}Sb_{18}Te_{55}$ on the GeTe-Sb_2Te_3 tie line. As seen in the figure, the four samples are very similar; i.e., each curve shows two exothermic peaks and one endothermic peak.

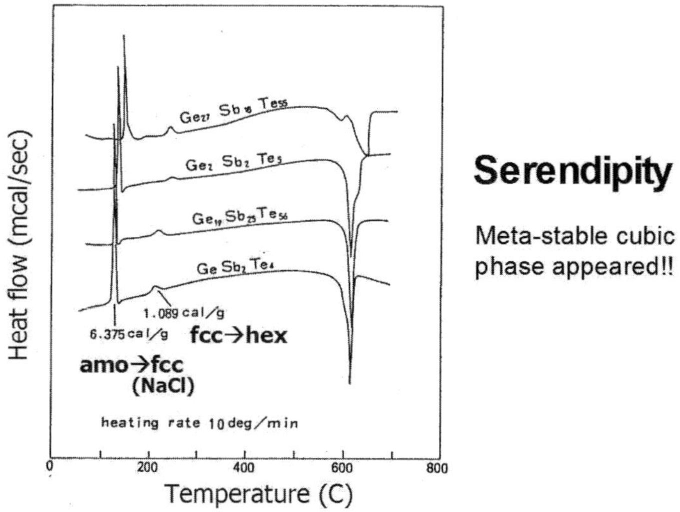

Fig. 10.13. DSC measurement results of GeTe-Sb_2Te_3 pseudo-binary amorphous films; the heating rate is 10 K/min for each. Every composition shows two steps in the crystallization process, from the amorphous to the metastable fcc phase and from the fcc phase to the stable hexagonal phase. Reprinted with permission from [10.19]. Copyright (1991), American Institute of Physics.

By x-ray diffraction analyses, the two exothermic peaks were identified to correlate to the two-step crystallization process, the transitions from the amorphous to the metastable cubic phase and from the cubic phase to the stable hexagonal phase. Importantly, the stable phase is hardly ever produced when films are crystallized by a short laser pulse. Hence, in the case of phase-change optical discs, we utilize the reversible phase changes between the amorphous and metastable cubic phases. In a sense, every amorphous film in this system behaves as if it is a stoichiometric compound [10.19]. Thus, we can obtain the explanation to the first question; this unique crystallization process is a reason why this material system shows such a wide compositional tolerance.

Next, the structure of GeTe-Sb_2Te_3 materials will be described. Figure 10.14 reveals the results for the metastable cubic phase for a typical composition, $Ge_2Sb_2Te_5$.

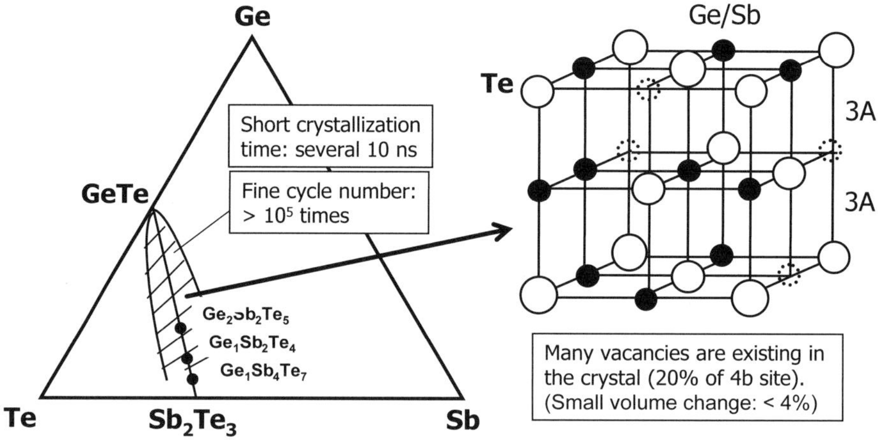

Fig. 10.14. Metastable NaCl type cubic phase observed along to the GeTe-Sb_2Te_3 tie line: 4a sites are occupied exclusively by Te atoms and 4b sites are randomly occupied by Ge atoms, Sb atoms and many vacancies. Reprinted with permission from [10.20]. Copyright (2000), American Institute of Physics.

The metastable phase has a rocksalt structure. In the metastable crystal, the 4a sites are occupied exclusively by Te atoms while the 4b sites are randomly occupied by Ge atoms, Sb atoms, and vacancies. It is very interesting that a large number of vacancies as high as 20 % are randomly distributed at the 4b sites [10.2, 10.19-10.21]. These observed structural characteristics make the crystal very symmetric and cause a rather small change in the volume between the amorphous and crystalline states. The amount of volume change is about 4 %, which is much smaller than the volume change of 10 % for GeTe.

The small volume change will reduce the mechanical stress in the layer stacks of optical discs, enabling a large number of repetitious structure changes. In the case of GeTe, it was reported that remarkable mechanical destruction occurs after several ten or hundred data rewritings. Here, we also obtained the answer to the second question, why this material reveals such a good cyclability.

In recent years, many experimental and simulation results have been reported on this material utilizing x-ray diffraction, extended x-ray absorption fine structure (EXAFS) spectroscopy, neutron diffraction, reverse Monte Carlo (RMC) simulation and so on [10.20-10.25] (see also Chapters 4 and 8). Summarizing these results, the metastable crystalline phase has a rather distorted rock salt structure, where all kinds of atoms, especially Ge atoms, deviate from the fcc lattice positions. Figure 10.15 shows the RMC simulation result of $Ge_2Sb_2Te_5$ in the meta-stable state [10.26]. It shows the structures at a given instant. The large atomic displacements are an important factor to realize the rapid crystallization of this material.

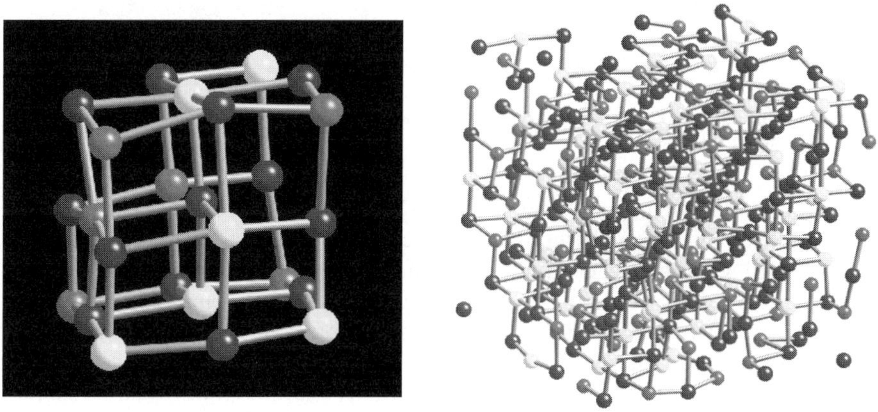

Fig. 10.15. Crystalline structure model obtained by RMC simulation: the left and right figures indicate a snap shot during a certain time of the calculation showing one cell (left) and about all 4000 atoms (right) [10.26].

10.9 Secrets of the Present Phase-change Materials I

In this section, secrets of the present phase-change materials will be discussed. Why do the present phase-change films show such high-speed crystallization processes while exhibiting at the same time a good thermal stability of the amorphous phase? It is not an easy question; therefore only some possibilities will be proposed here.

As described in the last section, GeTe-Sb_2Te_3 films have a high symmetry in the atomic distribution. Interestingly, this fact is common to good phase-change materials such as TeGeSnAu and AgInSbTe. As shown in Fig. 10.16, TeGeSnAu has a simple cubic structure with a lattice parameter of 0.3 nm wherein the four kinds of atoms randomly occupy the lattice sites [10.27].

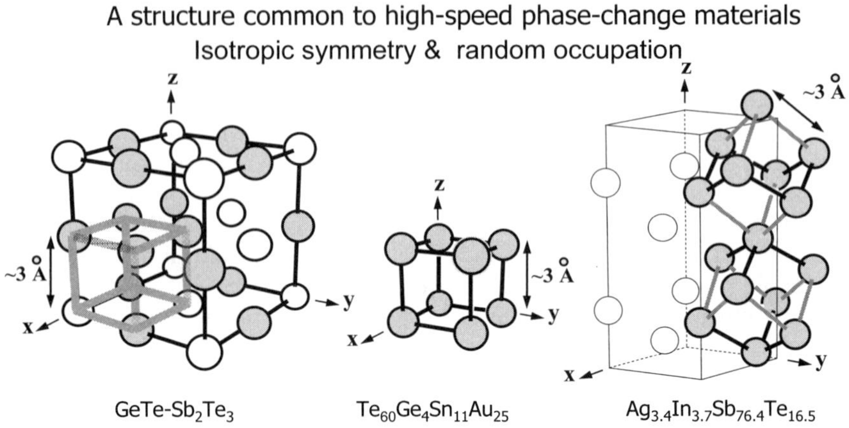

Fig. 10.16. Crystal structures of some superior phase-change materials. They have similar cubic structures with a lattice parameter of about 0.3 nm. Reprinted with permission from [10.27, 10.28]. Copyright (2001) by the American Physical Society from [10.28].

For AgInSbTe, it has a change at high temperature and transforms into a somewhat distorted simple cubic phase [10.28].

The obtained phase strongly resembles TeGeSnAu; i.e., it has a lattice parameter of 0.3 nm and every kind of atom randomly occupies the lattice sites [10.27]. Observing again GeTe-Sb_2Te_3 from the same view point, it can be seen that the rocksalt structure is constructed from simple cubic parts with a lattice parameter of 0.3 nm. It can be said that all of these superior phase-change materials have very high symmetry and their crystalline phases have rather random atomic distributions like liquids. Figures 10.17 and 10.18 reveal one more particular property common to superior phase-change materials.

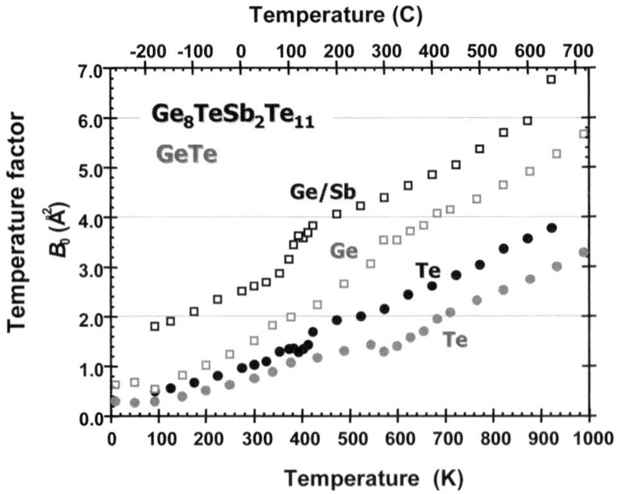

Fig. 10.17. The temperature dependencies of thermal vibrations B_0 of a typical phase-change material for GeTe-Sb_2Te_3 with GeTe as reference. Reprinted with permission from [10.29]. 4a site and 4b site are independently revealed for both compositions. The thermal vibration is obtained as a temperature factor of the Rietveld analysis [10.30].

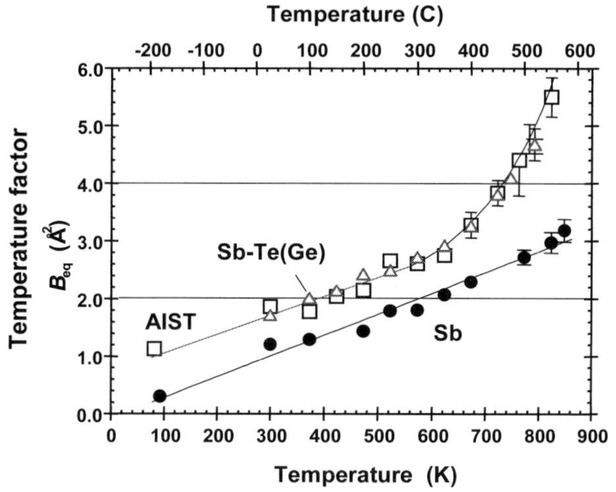

Fig. 10.18. Temperature dependencies of thermal vibrations B_0 for $Ag_{3.4}In_{3.5}Sb_{75}Te_{15}$ (AIST), $Ge_7Sb_{78}Te_{15}$ and pure Sb as reference. Reprinted with permission from [10.28]. Copyright (2001) by the American Physical Society.

It is demonstrated that GeTe-Sb_2Te_3 and Sb-Te based alloys show very large atomic vibrations at high temperature.

Figure 10.17 shows the results for $Ge_8Te_2Sb_{11}$ (a composition on the GeTe-Sb_2Te_3 tie line) and GeTe. It is apparent in the figure that thermal vibrations increase with increasing temperature for each case; however, the absolute values of the thermal vibrations are larger for $Ge_8Te_2Sb_{11}$ than GeTe for both, 4a and 4b sites. In particular, the 4b site of $Ge_8Te_2Sb_{11}$ has a large atomic displacement even at 0 K.

Figure 10.18 shows the results for Sb-Te based alloys such as $Ag_{3.4}In_{3.5}Sb_{75}Te_{15}$ (AIST) and Ge-doped Sb-Te. It can be seen in the figure that Sb-Te based alloys possess significant static distortions at 0 K and show a distinct over-linear temperature dependence. Generally, for a material without static distortions the thermal vibrations (temperature factor) form a straight line that passes through the origin.

Here, Eq. (10.1) [10.31] indicates the relation between the thermal vibration and the actual spatial atomic displacement from the lattice point.

$$<\Delta r^2> = B_0 / 8\pi^2 \qquad (10.1)$$

where Δr and B_0 denote the atomic displacement and the temperature factor, respectively. The maximum atomic displacement is obtained just below T_m. The maximum for the 4a site of $Ge_8Te_2Sb_1$ reaches values as high as 0.29 Å, and for AIST it is as high as 0.28 Å which is about 10 % of the nearest neighbor bond length of 3.0 Å for both cases.

These results indicate that the superior phase-change material has rather large distortions at low temperature, the displacement largely increases over-linearly with increased temperature, and finally the atomic displacements reach even up to 1/10 of the nearest neighbor bond length.

Consequently, we can very simply summarize the secret of the very rapid crystallization speeds as follows.

1) Superior phase-change materials have a very highly symmetric, cubic crystalline phase with rather large random atomic distributions such as simple cubic and rock-salt structure.
2) Their crystalline phases show very large thermal vibrations, in other words large atomic displacements, particularly at high temperature. It much increases the randomness of the atomic distributions in the crystalline phase; i.e., their crystal structures in a sense become rather similar to that of the liquid phase.
3) Their amorphous structure will naturally reflect the very isotropic atomic configuration in the liquid state.
4) Hence, the resemblance of the atomic distribution between the amorphous and crystalline phases at high temperature will reduce the barrier between the two states that will shorten the time for atomic rearrangement for crystallization.

Of course, it is indispensable to fully understand the amorphous structure for further discussion of the rapid crystallization processes. In recent years, various analytical studies have been actively carried out [10.22, 10.24, 10.32]. In the details, they describe rather different phase-change processes (chemical bonding); however, they have in common the point that crystallization will be completed by only small atomic rearrangements.

Figure 10.19 illustrates a model of Kohara et al. [10.24] for a phase-change process of a $Ge_2Sb_2Te_5$ material film simulated by RMC. The characteristics of the crystal structure such as i) even-numbered rings and ii) bond-angles centering around 90 degrees remain in the amorphous structure.

Since the required atomic rearrangement is so small, the crystallization will be completed by a very short laser heating.

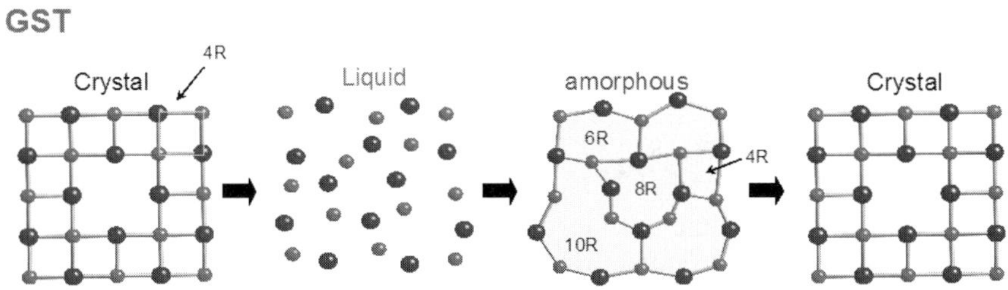

Fig. 10.19. A model of the crystal-amorphous-crystal phase-change process for $Ge_2Sb_2Te_5$ calculated by RMC simulation. It is characterized by even-numbered ring structures and bond angles that are centered at 90 degrees (not shown here). Reprinted with permission from EPCOS [10.24].

10.10 Materials for Blue Laser and Multi-layer Applications

In this section, we will trace again the research and development history of phase-change materials, especially focusing on practical applications, and will discuss future directions.

The ternary phase diagram in Fig.10.20 depicts various phase-change alloys with their year of discovery as a phase-change alloy and their use in different optical storage products.

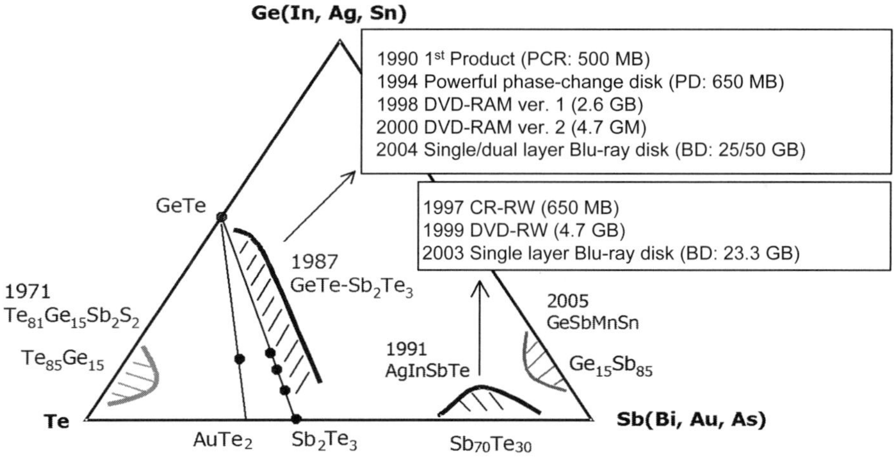

Fig. 10.20. Ternary phase diagram depicting different phase-change alloys, their year of discovery as a phase-change alloy and their use in different optical storage products. Reprinted by permission from Macmillan Publishers Ltd: Nature Materials [10.3], copyright (2007).

The first materials used were good glass formers such as Te-based eutectic alloys, represented by $Te_{85}Ge_{15}$, doped with elements such as Sb, S and P [10.5]. Although these materials already showed electrical switching that could be used for electronic storage, the time for crystallization was on the order of microseconds, partly because the first alloys did not crystallize in a single-phase material. The first materials to show fast re-crystallization and good optical contrast were GeTe [10.13] and $Ge_{11}Te_{60}Sn_4Au_{25}$ [10.14]. This triggered the discovery of pseudo-binary alloys along the GeTe–Sb_2Te_3 tie line, such as $Ge_1Sb_4Te_7$, $Ge_1Sb_2Te_4$ and $Ge_2Sb_2Te_5$ [10.16, 10.18]. Nowadays, $Ge_2Sb_2Te_5$ and related materials such as GeSbTeN, GeSnSbTe, GeBiSbTe, GeBiTe and GeInSbTe [10.33-10.36] have been tried, some of which are frequently used in commercial products. A second material family of doped Sb_2Te alloys was found later, in which dopants such as Ag, In and Ge were often used [10.37-10.39]. A composition close to $Ag_5In_5Sb_{60}Te_{30}$ (AIST) is frequently employed in rewriteable optical storage media such as DVD-RW (re-recordable DVD) and DVD+RW [10.40]. In recent years a third material family has been found [10.41, 10.42] that uses Ge-doped Sb.

It has been and will be indispensable for new phase-change alloys that they must respond to various changing requirements stemming from different aspects of the total system. The typical requirements are related to 1) various laser wavelengths, 2) multi-layer recording, and 3) higher recording speed.

The wavelength λ of the available laser diode has gotten shorter and shorter, which has contributed to an increase in recording density. It started from infrared lasers (λ = 830 nm) for the first products in 1990, changed to red lasers (λ = 780 nm and 650 nm) for CD-RW in 1997 and for DVD-RAM in 1998, and changed again to blue-violet lasers (λ = 405 nm) for BD in 2003.

Although the optical contrast of GeTe-Sb_2Te_3 alloys, for example, tends to degrade with shortening of the wavelength, these alloys have been utilized for phase-change optical disc systems continuously from infrared systems in 1991 to blue-violet systems in 2004 as listed in Fig. 10.20. This is because the GeTe-Sb_2Te_3 pseudo-binary films remain a single phase if the compositional ratio changes, if the constitutional component is substituted by another element, and if some elements are added beyond stoichiometry. In other words, optical properties of these GeTe-Sb_2Te_3 pseudo-binary alloy films can be finetuned by only small changes in the film composition.

Modification of optical properties for blue-violet wavelengths was achieved using the GeTe-Sb_2Te_3 system by changing the compositions as shown in Fig. 10.21.

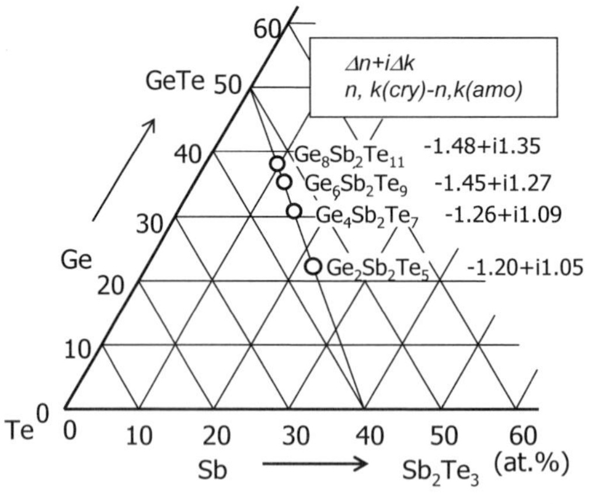

Fig. 10.21. Change of refractive index (*n*) and extinction coefficient (*k*) with stoichiometry for a wavelength of 405 nm. The contrast increases with increasing Ge content along the pseudo-binary line. Reprinted with permission from SPIE from [10.43].

The experimental results reveal the change of refractive index (*n*) and extinction coefficient (*k*) of the alloys when switched between the amorphous and crystalline states for a wavelength of 405 nm. The contrast increases with increasing Ge content along the pseudo-binary line [10.43]; thus, the Ge-rich compositions become applicable for blue-violet systems such as Blu-ray discs.

Multi-layer recording is one of the most effective ways for multiplying the recording capacity. Importantly, the phase-change film thickness has to be extremely reduced for making it highly transparent without sacrificing the crystallization speed and optical variations. Here it becomes an issue that crystallization speed decreases with reduced film thickness as shown in Fig. 10.22 [10.44].

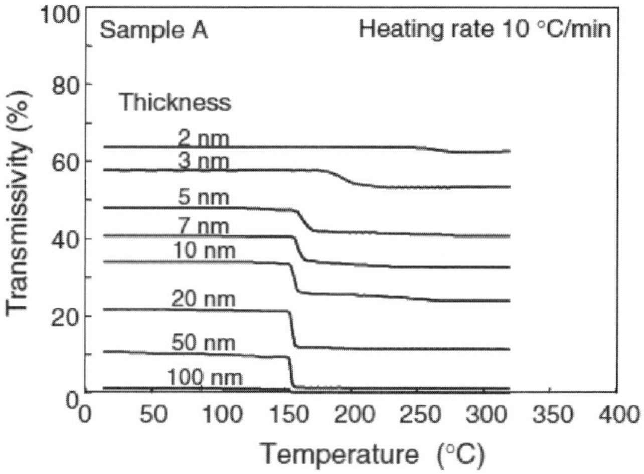

Fig. 10.22. Thermal transition curves of the transmissivity of amorphous $Ge_2Sb_2Te_5$ thin films with various thicknesses. The reduction of transmissivity corresponds to crystallization. The critical temperature abruptly begins to increase as the film thickness becomes lower than 10 nm. Reprinted with permission from [10.44].

As observed from the figure, the crystallization temperature is constant for films with a thickness of 10 nm or more; however, it abruptly begins to increase as the film thickness is reduced to less than 10 nm. It is to be noticed also that the steepness of the transmittance change becomes more and more gradual with reduced film thickness indicating a degradation of the crystallization speed.

As an effective method it was proven that substitution of Ge or Sb in the GeTe-Sb_2Te_3 alloys for heavy elements distinctly shortened the laser crystallization time. Figure 10.23 shows the case where Sn was partially substituted for Ge in $Ge_{4-x}Sn_xSb_2Te_7$ ternary amorphous alloy films [10.34]. Obviously, the degradation of crystallization time is improved by substituting Sn for Ge especially for very thin films with less than 6 nm in thickness. A similar effect was obtained for example by substituting Pb for Ge and Bi for Sb. It can be said that these results enabled the multi-layer recording for phase-change optical discs.

As described through this chapter, a dominantly important point for superior phase-change materials is that they crystallize into a single phase. Here, once more another important point should be added, namely the crystalline phase should have a wide compositional tolerance; in other words, the properties of the film should be easily modified by compositional changes or by some additives while maintaining the single phase.

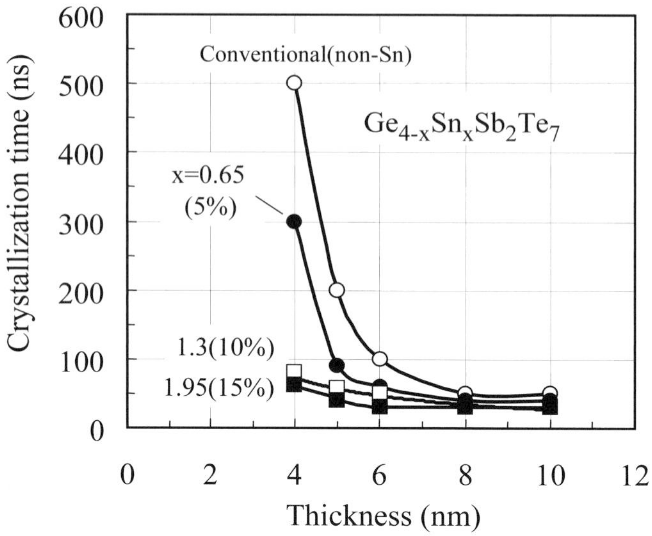

Fig. 10.23. Change in laser crystallization time vs. film thickness for $Ge_{4-x}Sn_xSb_2Te_7$ ternary amorphous films. The crystallization time for Sn poor films shows a distinct increase as the film thickness is reduced while that with Sn substitution for Ge shows rather constant tendency. Reprinted with permission from [10.34].

10.11 Secrets of Present Phase-change Materials II

The particular properties of superior phase-change alloys can also be discovered in their phase-diagrams. Through recent studies it was clarified that all of GeTe-Sb_2Te_3 alloys, GeTe-Bi_2Te_3 alloys, and Sb-Te based alloys have essentially similar phase diagrams, see Fig. 10.24.

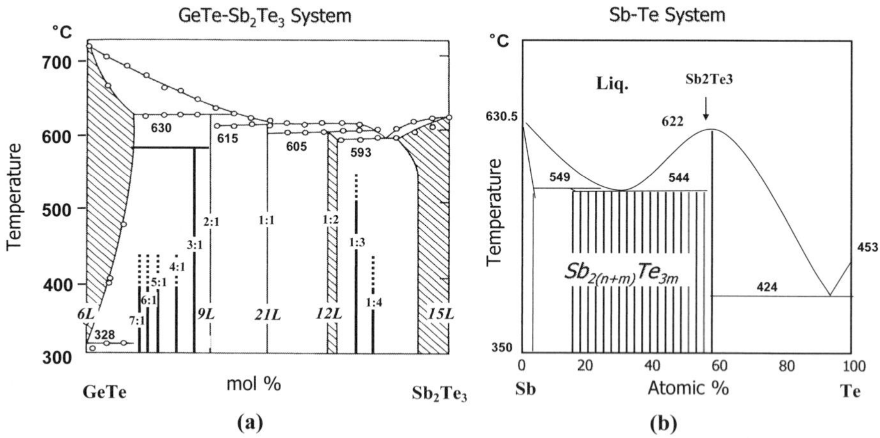

Fig. 10.24. Phase-diagrams for (a) GeTe-Sb_2Te_3 and (b) Sb-Te systems. (a) was summarized by Matsunaga et al. [10.45] by adding recent data [10.46] to the phase diagram of Abrikosov et al. [10.47] (b) was proposed by Kifune et al. [10.48]. Reprinted with permission from [10.7].

In the GeTe-Sb_2Te_3 pseudo-binary system, many ternary compounds denoted as $(GeTe)_n(Sb2Te3)_m$ exist. It is quite similar to the GeTe-Bi_2Te_3 pseudo-binary system (not shown here). These compounds have a similar structure with small differences in composition and generically are called homologous compounds. On the other hand, in the case of the Sb-Te system, numerous homologous series were reported by Kifune et al. [10.48], who revealed that the δ- phase in the Sb-Te system consists of many compounds denoted by $Sb_{2n}Te_3$ although this eutectic-like phase has been treated as an unknown phase.

In the above two systems, the reported stoichiometries are in the hexagonal phase with a long atomic periodicity; however, it is very noteworthy that metastable cubic phases appear over wide compositions when the films are heated by laser crystallizations. For the Ge-Sb system, such a homologous series has not been reported, yet; however, there remains a possibility if some third element will be added because metastable cubic phases can be observed for a rather wide compositional range also in this system when the film is formed by sputter deposition.

10.12 Conclusions

Research and development history of phase-change memory materials reaches back about 40 years since Ovshinsky first reported on phase-change materials in 1968 [10.1]. At the early stages, Te-based eutectic alloys were chiefly studied, but these activities did not reach the goal of fast switching.

In 1987, reports of the GeTe-Sb_2Te_3 pseudo-binary alloys by Yamada et al. [10.16] reversed the shrinking interest in the field of the phase-change materials, and it finally led to the first phase-change optical disc product in 1990 from Matsushita Electric (Panasonic Corporation). It also revived the research and development activities on electrical phase-change solid state memory that is continuing to present days.

In this chapter, the prominent characteristics of present phase-change alloys were described, in particular the GeTe-Sb_2Te_3 system. The "secrets" are revealed that make these alloys the most successful phase-change materials today. They were obtained through various developments of optical discs and their analytical study.

Knowing the research and development history and essential points of today's developments will lead us to new steps and new materials. Hopefully, this short article will be helpful for many students and beginners in the phase-change field.

References

[10.1] Ovshinsky, S. R.: Reversible Electrical Switching Phenomena in Disordered Structures. Phys. Rev. Lett. **21**, 1450 - 1453 (1968)

[10.2] Yamada, N.: Phase-chase-change optical materials. Mater. Res. Soc. Bulletin **21**, 48–50 (1996)

[10.3] Wuttig, M, Yamada, N.: Phase-change materials for rewriteable data storage. Nature Mater. **6**, 824-832 (2007)

[10.4] Nishio, K., Koga, J., Yamaguchi, T., Yonezawa, F.: Molecular dynamics study on freezing of Lennard-Jones argon in an open-ended cylindrical pore. J. Non-Cryst. Solids **345-346**, 694-697 (2004)

[10.5] Feinleib, J., deNeufville, J., Moss, S. C., Ovshinsky, S. R.: Rapid reversible light-induced crystallization of amorphous semiconductors, Appl. Phys. Lett. **18**, 254-257 (1971)

[10.6] Bichara, C, Raty, J.-Y., Gaspard, J.-P.: Structure and bonding in liquid tellurium. Phys. Rev. B **53**, 206 - 211 (1996)

[10.7] Yamada, N.: Phase Change materials in optical storage application. Mater. Res. Soc. Spring Meeting, San Francisco, March 2008, Phase Change Materials: Science and Applications, tutorial notes.

[10.8] Yamada, N.: Research relating to phase change materials and their applications. PhD thesis, Kyoto University, Japan, 2000

[10.9] Sakka, S., Mackenzie, J. D.: High pressure effects on glass, J. Non-Cryst. Solids **1**, 107-142 (1969)

[10.10] Ohta, T., Inoue, K., Uchida, M., Yoshioka, K., Akiyama, T., Furukawa, S., Nagata, K., Nakamura, S.: Phase change disk media having rapid cooling structure. Jpn. J. Appl. Phys. **28**, Supplement 28-3, 123-128 (1989)

[10.11] Nishiuchi, K., Yamada, N., Akahira, N., Takenaga, M., Akutagawa, R.: Laser diode beam exposure instrument for rapid quenching of thin-film material. Rev. Sci. Instrum. **63**, 3425–3430 (1992)

[10.12] Hansen, M., Anderko, K.: Constitution of Binary Alloys. McGraw-Hill, New York (1958)

[10.13] Chen, M., Rubin, K. A., Barton, R. W.: Compound materials for reversible, phase-change optical data storage. Appl. Phys. Lett. **49**, 502-504 (1986)

[10.14] Yamada, N., Takao, M., Takenaga, M.: Te-Ge-Sn-Au Phase change recording film for optical disk. Proc. SPIE **695**, Optical Data Storage II, San Diego, 79-85 (1986)

[10.15] Ohno, E., Yamada, N., Kurumizawa, T., Kimura, K., Takao, M.: TeGeSnAu alloys for phase change type optical disk memories. Jpn. J. Appl. Phys. **28**, 1235-1240 (1989)

[10.16] Yamada, N., Ohno, E., Akahira, N., Nishiuchi, K., Nagata, K., Takao, M.: Jpn. J. Appl. Phys. **26**, Suppl. 26–4, 61-66 (1987)

[10.17] Strand, D., Tsu, D.V., Miller, R., Hennessey, M., Jablonski, D.: Optical Routers Based on Ovonic Phase Change Materials. *Proc. E\PCOS06* (2006), Grenoble ; available at http://www.epcos.org>.

[10.18] Yamada, N., Otoba, M., Kawahara, K., Miyagawa, N., Ohta, H., Akahira, N., Matsunaga, T.: Phase-Change Optical Disk Having a Nitride Interface Layer, Jpn. J. Appl. Phys. **37**, 2104-2110 (1998)

[10.19] Yamada, N., Ohno, E., Nishiuchi, K., Akahira, N., Takao, M.: Rapid-phase transition of GeTe-Sb_2Te_3 pseudobinary amorphous thin films for an optical disk memory. J. Appl. Phys. **69**, 2849–2856 (1991)

[10.20] Yamada, N., Matsunaga, T.: Structure of laser-crystallized $Ge_2Sb_{2+x}Te_5$ sputtered thin films for use in optical memory. J. Appl. Phys. **88**, 7020-7028 (2000)

[10.21] Nonaka, T., Ohbayashi, G., Toriumi, Y., Mori, Y., Hashimoto, H.: Crystal structure of GeTe and Ge2Sb25 meta-stable phase. Thin Solid Films **370**, 258-261 (2000)

[10.22] Kolobov, A.V., Fons, P., Frenkel, A.I., Ankudinov, A.L., Tominaga J., Tomoya, U.: Understanding the phase-change mechanism of rewritable optical media. Nature Mater. **3**, (2004), 703–708 (2004)

[10.23] Shamoto, S., Yamada, N., Matsunaga, T.: Large displacement of germanium atoms in crystalline Ge2Sb2Te5. Appl. Phys. Lett. **86**, 081904 (2005)

[10.24] Kohara S., Kimura S., Tanada H., Yasuda N., Fukuyama Y., Murayama H., Kim J., Takata M., Kato K., Tanaka Y., Usuki T., Suzuyu K., Tanaka H., Moritomo Y., Matsunaga T., Kojima R. and Yamada N., Structural basis for fast phase change of DVD-RAM -Topological order in the amorphous phase. Proc. European Phase Change and Ovonic Science Symp., Zermatt, Switzerland, 2007

[10.25] Matsunaga, T., Yamada, N., Kubota, Y.: Structures of stable and metastable $Ge_2Sb_2Te_5$, an intermetallic compound in GeTe-Sb_2Te_3 pseudobinary systems. Acta Cryst. **B60**, 685-691 (2004)

[10.26] Private communication, Kohara, S. (JASRI) (2006)

[10.27] Matsunaga, T., Yamada, N.: A Study of highly symmetrical crystal structures, commonly seen in high-speed phase-change materials, using synchrotron radiation. Jpn. J. Appl. Phys. **41**, 1674-1678 (2002)

[10.28] Matsunaga, T., Umetani, Y., Yamada, N.: Structural study of a $Ag_{3.4}In_{3.7}Sb_{76.4}Te_{16.5}$ quadruple compound utilized for phase-change optical disks. Phys. Rev. **B 64**, 184116 (2001)

[10.29] Matsunaga. T, Yamada, N.: Crystal structure and bonding nature of Ge8Sb2Te11, a suitable matee material for high-speed, high-density phase-change recording. Proc. PCOS2004, 1-4(2004)

[10.30] H. M. Rietveld: A profile refinement method for nuclear and magnetic structures, J. Appl. Cryst. **2,** 65-71 (1969)

[10.31] Clegg, W., Blake, A.J., Gould, R.O., Main, P., Edited by Clegg, W.: Crystal Structure Analysis Principles and Practice. IUCr Texts on Crystallography **6**, International Union of Crystallography / Oxford University Press, 164 (2001)

[10.32] Baker, D. A., Paesler, M. A., Lucovsky, G., Agarwal, S. C., Taylor, P. C.: Application of bond constraint theory to the switchable optical memory material $Ge_2Sb_2Te_5$. Phys. Rev. Lett. **96,** 255501 (2006)

[10.33] Kojima, R., Okabayashi, S., Kashihara, T., Horai, K., Matsunaga, T., Ohno, E., Yamada, N., Ohta, T.: Nitrogen doping effect on phase change optical disks. Jpn. J. Appl. Phys. Part 1 **37,** 2098–2103 (1998)

[10.34] Kojima, R., Yamada, N: Acceleration of crystallization speed by Sn addition to Ge–Sb–Te phase-change recording material. Jpn. J. Appl. Phys. Part 1 **40,** 5930–5937 (2001)

[10.35] Yusu, K., Nakai, T., Ashida, S., Ohmachi, N., Morishita, N. & Nakamura, N.: Highspeed crystallization characteristics of Ge–Sb–Te–Bi materials used for next generation rewritable DVD with blue laser and NA = 0.65. Proc. E\PCOS05 (2005); available at http://www.epcos.org

[10.36] Kusada, H., Hosaka, T., Kojima R., Yamada, N.: Effect of excess Sb on GeTe– S b2Te3–Bi2Te3 recording films. Proc. 18th Symp. PCOS2005 32–35 (2006)

[10.37] Iwasaki, H., Ide, Y., Harigaya, M., Kageyama., Y., Fujimura, I. : Completely erasable phase-change optical disk. Jpn. J. Appl. Phys. Part 1 **31,** 461–465 (1992)

[10.38] Horie, M., Nobukuni, N., Kiyono, K. & Ohno, T. High-speed rewritable DVD up to 20 m/s with nucleation-free eutectic phase-change material of Ge(Sb70Te30)+Sb. Proc. SPIE **4090,** 135–143 (2000)

[10.39] Kato, T., Hirata, H., Komaki, T., Inoue, H., Shingai, H., Hayashida, N., Utsunomiya, H., The phase change optical disc with the data recording rate of 140 Mbps. Jpn. J. Appl. Phys. Part 1 **41,** 1664–1667 (2002)

[10.40] Iwasaki, H., Harigaya, M., Nonoyama, O., Kageyama, Y., Takahashi, M., Yamada, K., Deguchi, H., Ide, Y.: Completely erasable phase-change optical disc. II. Application of Ag–In–Sb–Te mixed-phase system for rewritable compact disc compatible with CD-velocity and double CD-velocity. Jpn. J. Appl. Phys. Part 1 **32**, 5241–5247 (1993)

[10.41] Afonso, C. N., Solis, J., Catalina, F. & Kalpouzos, C. Ultrafast reversible phase-change in GeSb films for erasable optical storage. Appl. Phys. Lett. **60**, 3123–3125 (1992)
[10.42] Yuzurihara, H., Iwasa, H. & Kaneko, Y. GeSbSnMn for high speed BD-RE media. Proc. 17th Symp. PCOS2005 19–22 (2005)
[10.43] Yamada, N., Kojima, R., Uno, M., Akiyama, T., Kitaura, H., Narumi, K., Nishiuchi, K.: Phase-change material for use in rewritable dual-layer optical disk. Proc. SPIE **4342,** 55–63 (2002)
[10.44] Nishiuchi, K., Yamada, N., Kawahara, K., Kojima, R.: Effect of dielectric material films on crystallization characteristics of $Ge_2Sb_2Te_5$ phase-change memory film. Jpn. J. Appl. Phys. 46, 7421-7423 (2007)
[10.45] Private communication, Matsunaga, T. (Panasonic), from EPCOS2007 Presentation (2007)
[10.46] Abrikosov, N. Kh., Danilova-Dobryakova, G. T.: Study of the Sb_2Te_3-GeTe phase diagram. Izv. Akademii Nauk SSSR, Neorg. Mater. **1**, 204-209 (1965)
[10.47] Shelimova, L.E., Karpinskii, O.G., Konstantinov, P.P., Kretova. M.A., Avilov, E.S., Zemskov, V.S.: Composition and properties of layered compounds in the GeTe–Sb_2Te_3 System. Inorganic Materials **37**, 342-348 (2001)
[10.48] Kifune, K., Kubota, Y., Matsunaga, T, Yamada. N.: Extremely long period-stacking structure in the Sb-Te binary system. Acta Cryst. B**61**, 492-497 (2005)

11. Novel Deposition Methods

Delia J. Milliron, Qiang Huang and Yu Zhu

Abstract Recently, alternatives to conventional sputter deposition have emerged as novel routes to phase change materials. Applying chemical vapor deposition (CVD), electrodeposition, or solution-phase deposition to these materials offers potential advantages in fabrication of phase change memory cells or optical discs, yet each method brings its own challenges. In this chapter, we review the basics of each strategy, emphasizing the advantages and the difficulties of applying them to phase change materials. In addition, we discuss the recent reports of chemically synthesized phase change nanowires and nanoparticles and more broadly explore the potential for nanomaterials to contribute to the development and understanding of phase change memory.

11.1 Chemical Vapor Phase Deposition

Most of the phase change thin films studied in the past decades were deposited by physical vapor deposition (PVD), such as sputtering, evaporation, or laser ablation; however, conformality of PVD films is poor. Chemical vapor deposition of phase change materials is of interest mainly for its conformal deposition profile and the potential application in phase change memory. Chemical vapor deposition (CVD) may be defined as the deposition of a solid on a heated surface from a chemical reaction in the vapor phase. Generally, the precursors and co-reactant vapors are introduced into a deposition chamber under vacuum, where the substrate can be heated to elevated temperatures. The chemical reaction takes place on the heated surface or in vapor phase above the substrate surface. The advantage of CVD over PVD is its conformal deposition profile, i.e. the film thickness in the field is the same or close to the thickness on the side wall in a trench or via structure. CVD can be used to fill deep recesses, holes, and other difficult three-dimensional configurations. The major disadvantage of CVD is the need for a relatively high deposition temperature. In addition, extensive engineering controls can be required due to the high toxicity and volatility of CVD precursors. Most thermal CVD processes are achieved at temperatures of 600 °C and above, and many substrate materials are not thermally stable at these temperatures. The deposition temperature can be reduced to below 400 °C, by introducing plasma in the

S. Raoux, M. Wuttig (eds.), *Phase Change Materials*,
DOI 10.1007/978-0-387-84874-7_11,

chamber to enhance the reaction (PECVD) or by using metal-organic precursors (MOCVD), which have lower reaction temperatures than their inorganic counterparts. Currently, CVD has been implemented in a lot of deposition processes for integrated circuit (IC) manufacturing. For instance, the deposition of poly-silicon, silicon dioxide, silicon nitride, metal nitrides, and tungsten are accomplished by CVD processes. The tradeoff is that CVD is generally a much more complex process than PVD, with more opportunities for introduction of contaminates both beneficial and detrimental [11.1, 11.2].

Phase Change Random Access Memory (PCRAM) is one of the most promising candidates for next-generation nonvolatile memory. The effort toward CVD of phase change materials is motivated by the development of PCRAM. One of the main development goals for PCRAM is to reduce the RESET current (the current required to melt the material in order to switch it from the crystalline to the amorphous phase), which can be achieved by confining the phase change material, $Ge_2Sb_2Te_5$ (GST) for example, in a small volume [11.3]. Reactive ion etching (RIE) or trimming the blanket GST to scale down the volume becomes problematic due to the damage layer from RIE [11.4]. Filling GST into small vias by PVD is difficult due to the nature of the non-conformal PVD deposition profile. Using CVD to fill small feature size, high aspect-ratio vias with GST can scale down the phase change material volume and is critical for fabrication of high-density phase change memory (see Chapters 15 – 17 for details on device designs and fabrication).

Chemical vapor deposition of GST can be achieved using metal-organic precursors. The germanium precursors which have been evaluated for CVD of GST include: tetra(allyl)germanium ($Ge(C_3H_5)_4$) [11.5], tetra(isobutyl)germanium ($Ge(i\text{-}C_4H_9)_4$) [11.6, 11.7], isobutylgermane ($H_3\text{-}Ge\text{-}(i\text{-}C_4H_9)$ [11.8], and tetra(dimethylamino)-germanium ($Ge(N(CH_3)_2)_4$ [11.9]. The antimony precursors for CVD of GST include: tri(isopropyl)antimony ($Sb(C_3H_7)_3$) [11.5-11.7] and tri(dimethylamino)antimony ($Sb(N(CH_3)_2)_3$) [11.9]. One of the commonly used tellurium precursors for CVD of GST is di(isopropyl)tellurium ($Te(C_3H_7)_2$). Some of these precursors have been studied for metal-organic chemical vapor deposition (MOCVD) of compound semiconductor materials previously. Generally, the metal organic precursors are less toxic and decompose at lower temperature in comparison with their hydride or halide counterparts, the inorganic precursors. This allows the CVD of GST to take place below 400 °C. Because of insufficient vapor pressure, carrier gas is used to assist the delivery of the precursor vapor into the deposition chamber. Some metal-organic precursors need be delivered through a vaporizer. The direct liquid source delivery or injection system can allow a wider range of chemicals than the precursors mentioned above for CVD of phase change materials. Hydrogen is often used as co-reactant in CVD of GST.

The composition of GST films can be altered by the process parameters. For instance, the germanium incorporation into GST films by MOCVD is controlled by the precursor bubbling temperature, carrier gas flow rate, deposition temperature and pressure. Generally, when deposition is dominated by surface reactions, film composition is sensitive to the deposition temperature and pressure. When deposition is dominated by mass transport, film composition is sensitive to the ratio of precursor vapor partial pressures. Different carrier gas flow rates or different precursors bubbling temperatures can be used to adjust the ratio of precursor partial pressures, and therefore adjust the GST film composition. This is demonstrated as an example in Fig. 11.1 which shows the x-ray diffraction (XRD) data on films deposited at different pressures [11.5].

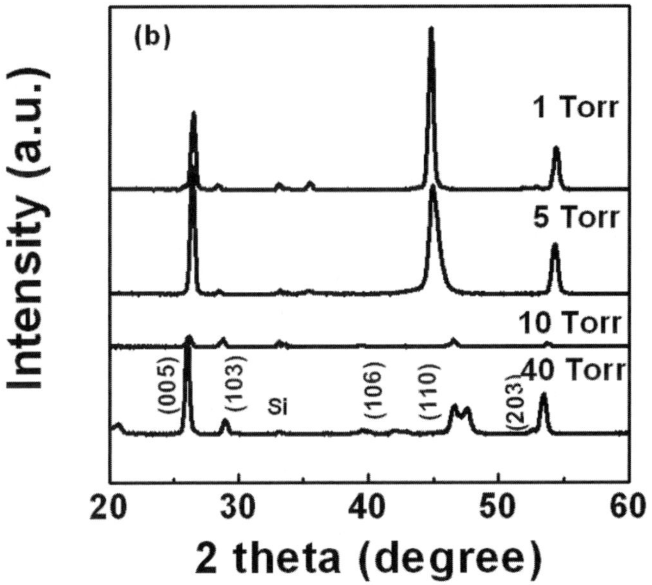

Fig. 11.1. XRD patterns of GST films deposited at different pressures and at 350 °C. Reprinted with permission from [11.5]. Copyright (2006), American Institute of Physics.

At low pressure the XRD patterns show the formation of Sb_2Te_3 while at higher pressure $Ge_2Sb_2Te_5$ is formed in the hexagonal structure ($Ge_2Sb_2Te_5$ also has a metastable fcc crystalline phase that is typically formed at around 150 ºC and transforms into the stable hexagonal phase around 310ºC for PVD deposited films).

In general, all of the deposition parameters mentioned above need to be tuned and optimized simultaneously to achieve GST films with the desired stoichiometry [11.5].

The choice of precursors is also critical to control the film stoichiometry. For instance, $Ge_2Sb_2Te_5$ films can be deposited using $Ge(allyl)_4$, $Sb(iPr)_3$ and $Te(iPr)_2$ under optimized deposition condition. When switching the Ge precursor to $Ge(iBu)_4$ and using the same Sb and Te precursors, the Ge amount in GST film is only 0.4%, despite the 63% partial pressure of Ge precursor during the deposition [11.7].

Atomic layer deposition (ALD) is an ultimate way to control the phase change material composition. ALD is a chemical gas phase thin film deposition method based on alternate saturated surface reactions [11.10]. Ideally, monolayers of different elements can be deposited, and by altering the number of layers deposited, as well as the sequence, the composition can be well controlled. Though true ALD type deposition of GST has not been demonstrated yet, CVD of GST in an ALD fashion has been reported [11.9]. In these studies, single component films of Ge, Sb and Te were deposited, and then by adjusting the number of pulses for each element, $Ge_{23}Sb_{21}Te_{55}$ could be achieved. It was also noticed, that depositions of Ge, Sb and Te are sensitive to the substrate surface. For instance, Ge can be deposited on TiN surfaces, but nucleation was difficult on SiO_2 surfaces when using certain precursors. For this reason, the sequence of the precursor pulses also must be optimized for stoichiometric composition [11.6, 11.9]. CVD of GST always shows large grains, with a rough surface, which could be problematic for filling in small vias. By separating the precursor flow scheme, ALD or pulsed CVD techniques can interfere with the large grain growth and minimize the grain size and surface roughness.

It is known that the resistivity and crystallization behavior of GST can be modified by doping with nitrogen [11.11], oxygen [11.12] or other dielectric materials. For PVD GST films, doping material can be mixed into the sputtering target, co-sputtered from a different target, or introduced by flowing nitrogen or oxygen gas during sputtering. Similarly, doping can be achieved in CVD of GST films. Separate precursors for different dopants can be added during deposition, or dopants can be introduced by flowing nitrogen, ammonia, or oxygen during deposition. Doping can also be done by engineering the precursor. For instance, by attaching amide groups in the precursor, nitrogen can be incorporated in the resulting film. By using tri(hexamethyldisilylamino)antimony for CVD of GST, Si doped amorphous GST can be achieved [11.9].

CVD phase change materials' morphology and composition can be obtained through conventional thin film characterization techniques such as x-ray diffraction, scanning electron microscopy, atomic force microscopy, Auger electron spectroscopy, and transmission electron microscopy. Film morphology is strongly dependent on the film composition. Generally, CVD GST films tend to be polycrystalline, with large grains and rough surfaces. Figure 11.2 shows a scanning electron microscope (SEM) image of a CVD $Ge_2Sb_2Te_5$ film [11.5] with grain sizes in the micrometer range. Sometimes, CVD GST films show inhomogeneous morphologies with incomplete lateral growth. These issues need to be improved before implementing CVD of GST for PCRAM, since large grains may clog up the top of vias and hinder the fill-in of high aspect ratio vias; and inhomogeneous composition could cause memory cell to cell variation.

Besides the structural characterization, the phase change properties need to be characterized as well. The amorphous to crystalline phase transition can be detected for thin-film phase change material in the amorphous phase, like most of the sputtered films, by annealing the thin-film above the crystallization temperature and monitoring the film sheet resistance decrease.

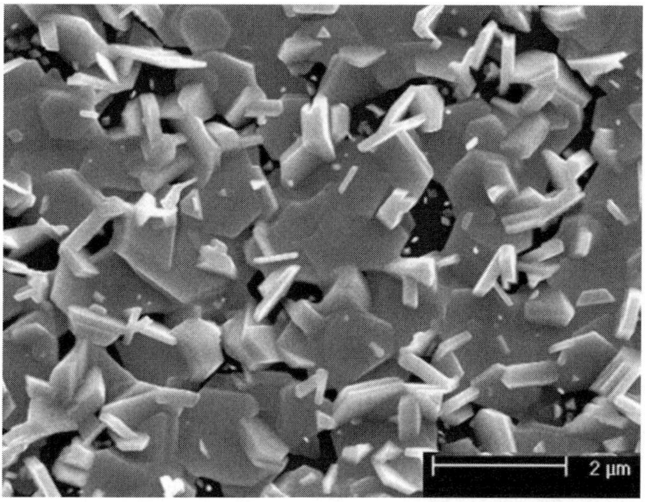

Fig. 11.2. SEM image of $Ge_2Sb_2Te_5$ CVD film deposited at 350ºC and 10 Torr for 30 min. Reprinted with permission from [11.5]. Copyright (2006), American Institute of Physics.

Thin films deposited by CVD are typically in the crystalline state, because most CVD processes are done at about 300 °C or higher, which is above the crystallization temperature. Laser pulses can melt and quench the phase change material, and are used to examine the switching from the crystalline to the amorphous phase [11.13].

Scanning probe microscopy (SPM) or conducting tip atomic force microscopy can focus the current through a sharp tip, inducing a nanoscale phase change in a crystalline GST film [11.14]. Although, the ultimate test will be integrating the CVD phase change material into a functional device, these types of characterization techniques are desired for screening different CVD GST films with quick feedback for PCRAM applications.

GST is the most popular material for PCRAM. CVD is compatible with semiconductor process technologies. The combination of these two, i.e. CVD of GST, is critical for developing high density PCRAM in terms of scaling down the GST volume. A research group from Samsung Electronics demonstrated the first highly scalable phase change memory with CVD GST for sub-50 nm lithography generation. A CVD film with an approximate composition of $Ge_2Sb_2Te_5$ was deposited using metal organic precursors and H_2 at 350 °C. Figure 11.3 shows a transmission electron microscope (TEM) image of these $Ge_2Sb_2Te_5$ devices fabricated by CVD with 50 nm contact diameter to the bottom electrode. One can see the excellent via filling capabilities of CVD that enable the cell design with a confined geometry for the phase change material. The figure also demonstrates the very good control over the materials composition as measured by EDX (Energy-Dispersive x-ray spectroscopy).

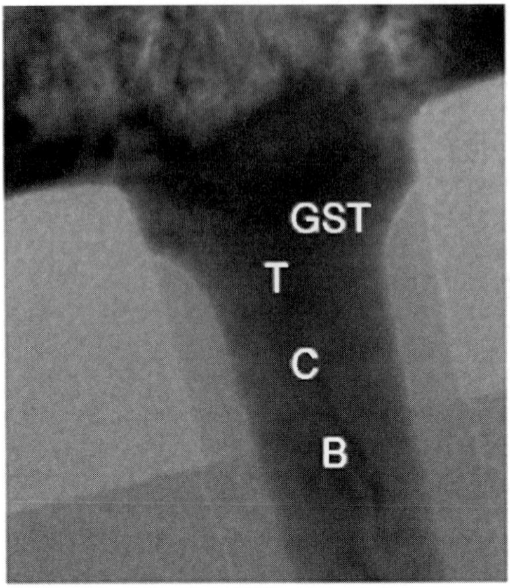

Position	Ge	Sb	Te
T	22.6	22.0	55.3
C	23.5	21.2	55.1
B	24.4	20.0	55.5

Fig. 11. 3. TEM image of $Ge_2Sb_2Te_5$ CVD PCRAM device (left) and composition of GST material at different locations within the via determined by EDX in atomic % (right) (© (2007) IEEE, from [11.15]).

This cell geometry with a better confinement of the phase change material compared to a typical mushroom cell (see Chapter 17) leads to a better thermal isolation of the material and thus to a reduction of the RESET current (about 50 % in [11.15]). Good data retention (no data loss after 48 h at 140 °C) and cyclability (up to 10^8 cycles) was demonstrated (Fig. 11.4).

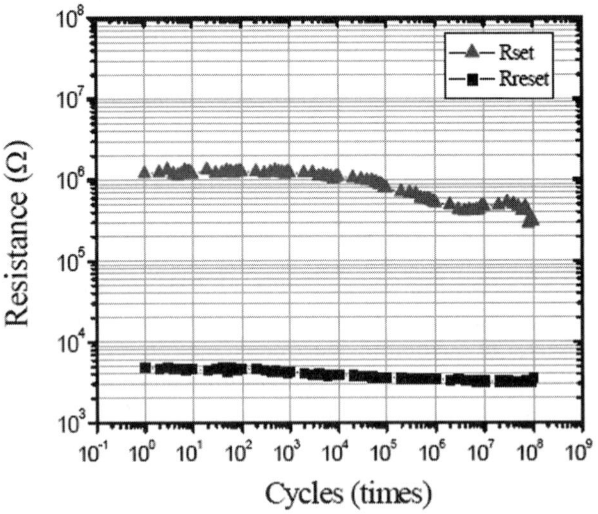

Fig. 11.4. Endurance characteristic of confined cell structure. Plotted are the resistances of the cell in the set (crystalline) and reset (amorphous) state after repeated switching (from [11.15], © (2007) IEEE).

In order to make the implementation of CVD GST successful, various challenges need to be addressed, for instance: the crystallization of the GST during deposition may hinder via fill-in, so minimizing the grain size or depositing amorphous films is needed; phase change material is sensitive to temperature, a narrow temperature window for process conditions can be expected; a dependency of the nucleation probability of GST on different surfaces has been observed, in a memory cell, obtaining the same growth rate of GST on dielectric and bottom electrode material may be important; device integration may require surface cleaning, such as wet clean, sputter clean or plasma treatment, the difference of the nucleation on different surfaces may be minimized or enhanced by these surface treatments; precisely controlling the composition within the vias and across the wafer surface is also critical. CVD of phase change material is a relatively new subject compared with the history of phase change material research, and will require extensive understanding of the processes as well as the deposited film properties for implementation of this technique in devices.

11.2 Electrodeposition

Electrodeposition (electroplating) refers to a method where the deposition occurs through an electrochemical reaction in a solution, in most cases, a reduction reaction of the dissolved precursor species in a liquid solution. An electric power source is generally used to provide the electrons and the driving force for the reaction. In addition to its traditional applications in the surface coating and metallurgy industries, electrodeposition has also been integrated into microelectronics and MEMS (micro-electro-mechanical system) fabrication [11.16, 11.17].

During a typical electrodeposition process, a cationic species in the solution is reduced to the elemental state by externally supplied electrons, forming a thin film or a bulk material on the substrate, the cathode. The reaction can be written as $M^{n+} + ne^- \longrightarrow M$, where e is a free electron, and M^{n+} and M stand for the electrochemical pair in the cationic and elemental form, respectively. While the detailed principles involved in electrodeposition are available in other references [11.18-11.20], the simplified characterization of this reaction includes three aspects: thermodynamics, reaction kinetics and transport.

For an electrochemical reaction, the reaction rate is usually referred to as the current density. Figure 11.5(a) illustrates the simplified relation between the applied potential and the current density, obtained with a technique called linear sweep voltammetry (LSV). The thermodynamics determines the reversible potential, $E^0_{M^{n+}/M}$, a macroscopic threshold potential for the reduction to occur. The reversible potential depends on all the components, their activities (approximately, their concentrations) and the temperature. On the other hand, the reaction kinetics determines how fast the deposition rate is, which depends on the nature of the reactants, the temperature and the applied potential. The current density increases with the applied potential until it becomes limited by the transport of the cationic reactant from bulk solution to the cathode surface. In this case, the reactant is depleted at the interface of the electrolyte and cathode. This transport-limited rate, the limiting current density, depends on the concentrations and diffusion coefficients of the reactants, and the agitation.

Also shown in Figure 11.5(a) is the alloy deposition of two non-interactive species, *M* and *M'*.

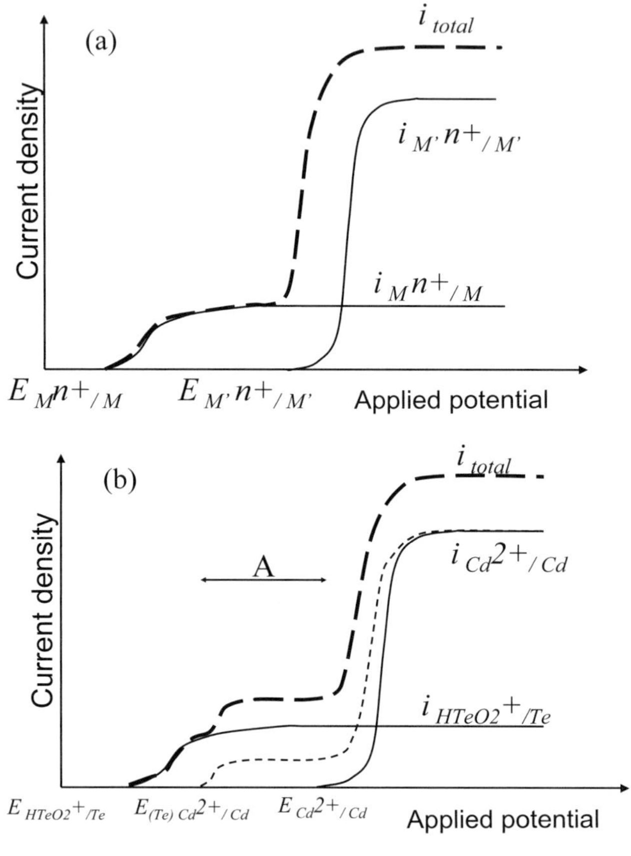

Fig. 11.5. LSV diagrams of elemental deposition (solid lines) and alloy deposition (dashed lines) for (a) two non-interactive species; (b) Cd and Te, sketched according to [11.21]. Reproduced from [11.21] by permission of The Electrochemical Society, Copyright 1978.

When a higher concentration of the less noble species, *M'*, is used, alloys rich in *M* and *M'* can be obtained at low and high potentials, respectively. The composition of the alloys can be easily calculated with the deposition rate of the individual depositions. However, the real practice is often more complicated than the ideal case illustrated therein, either due to the electrochemical reaction of the solvent or the interaction between the elements to be plated.

A great number of studies have been reported on the electrodeposition of tellurium compounds and other chalcogenides. Many elements were found to interact with tellurium when they are co-deposited. Fig. 11.5(b) shows the illustrative LSV diagram of the co-deposition of CdTe in an aqueous solution, sketched according to reference [11.21]. In that article, the authors found that the reduction of Cd^{2+} on Te was shifted to a less negative potential due to the formation energy of the CdTe, as shown in the region A in the diagram. In order to deposit the stoichiometric CdTe compound, an electrolyte with a high concentration of the less noble component, Cd, and a low concentration of the noble component, Te, was used. In the potential range A, Te deposition was diffusion controlled. Cd deposition only occurs on the Te surface, forming the compound CdTe. Elemental Cd was not deposited because the reversible potential of Cd^{2+}/Cd was not reached. Therefore, the deposition rate of Cd only depends on the $HTeO_2^+$ concentration in the solution, but not the Cd^{2+} concentration. Since the reduction of Cd^{2+} consumes two electrons and $HTeO_2^+$ four electrons, the deposition current of Cd is always half that of Te.

This phenomenon where the Cd deposition was shifted to a less negative potential due to the formation of a compound with the substrate, Te, is called under potential deposition (UPD). This UPD has also been observed in the electrodeposition of other Te compounds and other chalcogenides. An electrochemical atomic layer epitaxy (EC-ALE) method has been developed, based on the concept of UPD, to form single crystalline semiconductor compounds in a layer by layer fashion [11.22].

In addition to CdTe [11.21, 11.23-11.28], many other Te compounds and chalcogenides have been electrodeposited, mainly for optoelectronic [11.29-11.38] and thermoelectric [11.39-11.43] applications. Among them, the thermoelectric materials Bi_2Te_3, and $Bi_{0.5}Sb_{1.5}Te_3$ have dominated the activities in V-VI compounds, with very few studies dedicated to Sb_2Te_3.

Triggered by the interest in PCRAM and the requirement for good via filling capabilities for some cell designs and fabrications (see Chapters 15–17) electrodeposition of phase change materials was recently investigated. A study on the electrodeposition of SbTe phase change alloys was carried out at room temperature [11.44]. The LSVs of elemental and alloy deposition are shown in Fig. 11.6(a). The electrodeposition of Sb in the presence of Te starts at a lower potential than the deposition by itself, similar to the CdTe case. However, the Sb deposition rate was also found to be independent of the Te concentration in the solution, different from the CdTe case. A mechanism involving a proton-incorporated intermediate was proposed to explain this so-called induced deposition of Sb. While polycrystalline SbTe was electrodeposited at near boiling temperature [11.45], electrodeposition at room temperature resulted in amorphous deposits [11.43, 11.44]. The phase change properties of the electrodeposited SbTe alloys were characterized with time resolved x-ray diffraction, as shown in Fig. 11.6(b). Plotted is the intensity of diffracted x-rays over a 2θ range of 24°–38° as a function of temperature while the sample was heated in a purified He atmosphere at a rate of 1K/s. The x-ray energy was 6.9 keV. A transition from the amorphous state to crystalline Sb_2Te_3 was observed at about 120 °C for SbTe alloys with 37% to 57% Sb.

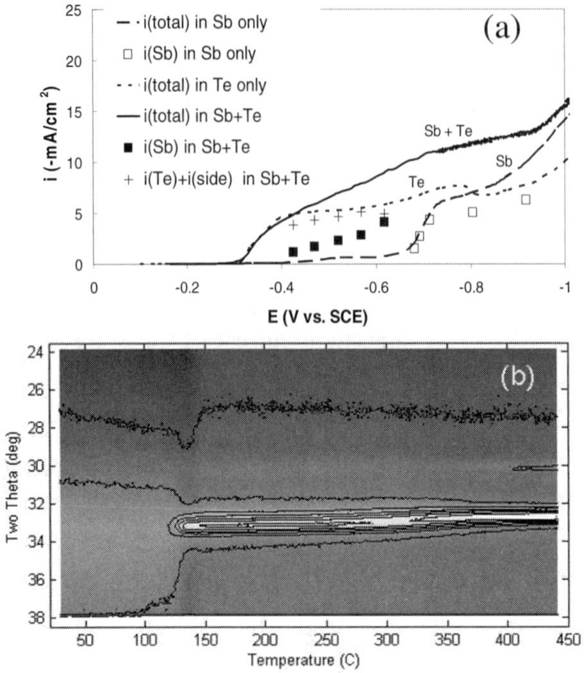

Fig. 11.6. (a) LSV in Sb, Te, and SbTe alloy deposition, where a shift of Sb deposition was observed in alloy deposition (solid squares as compared with open squares). (b) Time resolved x-ray diffraction pattern of the electrodeposited SbTe alloy with temperature ramping, showing crystallization at 120 °C. Reproduced from [11.44] by permission of The Electrochemical Society, Copyright 2008.

As the addition of some third elements are expected to improve the phase change properties of SbTe [11.46, 11.47], electrodeposition of ternary alloys were also carried out. However, no improvement in the transition temperature was observed so far for electrodeposited AgSbTe and InSbTe.

Electrodeposition of amorphous Ge films was achieved on Si from a non-aqueous solution. A re-crystallization into single crystal was observed upon annealing at 400 °C [11.48]. Despite the fact that Ge deposition has never been achieved from aqueous solution due to a very high reduction rate of water on Ge surfaces [11.49], GeSbTe alloys with up to 10% Ge have been electrodeposited from an aqueous solution [11.50]. However, the compositional uniformity, the incorporation of other elements such as oxygen from the solution, and the phase change properties of the films still remain unknown.

Electrodeposition is a solution based process with an external driving force. Metastable alloys and phases can be obtained due to the non-equilibrium nature of electrodeposition. Furthermore, the amount of the impurities is highly dependent on the chemistry and process. Therefore, the electrodeposited films are expected to behave differently from the films deposited from vacuum processes.

Different from many other deposition methods, electrodeposition requires an electrically conductive substrate to carry the electrons for the electrochemical reaction to occur. This limitation necessitates extra steps to remove the seed layer in the fabrication of phase change memory devices. But even before the development of such an integration scheme the properties of the electrodeposited materials have to be well understood as blanket films. However, the seed layer used for electrodeposition, which conducts electrons and heat, complicates or even prevents the electrical and thermal tests for the films formed thereon.

Another difficulty of electrodepositing phase change materials lies in the electrodeposition of non-metallic elements, such as N and P, which may be used to improve the properties of the alloys. Elemental N and P are highly reactive in ambient conditions and the electrodeposition of these elements has never been achieved in aqueous solutions unless being co-deposited with a transition metal, such as Ni. Even the more metallic elements like Ge have never been successfully deposited from aqueous solution. Organic solvents, molten salts and ionic liquids may find application here.

Despite the many unknowns and limitations, the electrodeposition method is extremely attractive for phase change memory devices due to its unique capability of forming microstructures or nanostructures onto irregular surfaces and recessed substrates. High aspect-ratio nanowires of different Te compounds [11.51-11.56], including Sb_2Te_3 nanowires [11.57] have been demonstrated with electrodeposition. Figure 11.7 shows SbTe via structures electrodeposited in a patterned SiO_2 layer, with an aspect ratio of 8.

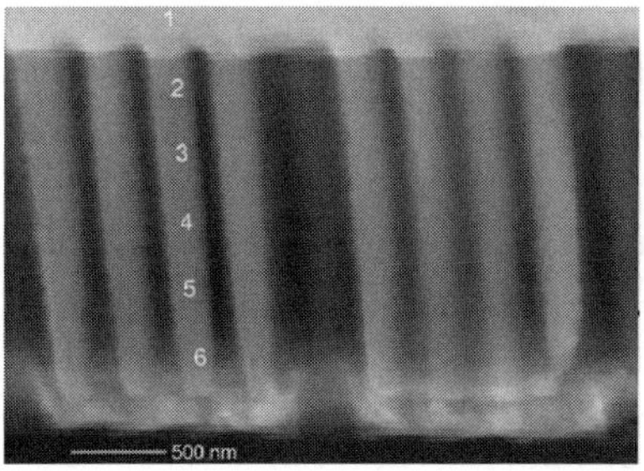

Point	At % Sb	At % Te	At % O	Sb/Te
1	25	50	25	0.50
2	22	46	32	0.48
3	17	44	38	0.39
4	18	39	43	0.46
5	19	40	41	0.48
6	19	39	43	0.49

Fig. 11.7. Electrodeposited via structures of SbTe alloy in patterned SiO_2 layer and the composition along its length, analyzed with Auger electron spectroscopy. The bottom layer under the vias is Cu. The vias are 200 nm in diameter and 1.6 μm long.

The Auger electron spectroscopy analysis showed a uniform $SbTe_2$ composition throughout the length of via. In phase change memory devices, the phase change material is surrounded by a dielectric material and contacted on both sides by a conducting electrode. Such a medium aspect-ratio nanostructure would be a perfect application for electrodeposition.

In addition to the ease of filling vias with phase change material, electrodeposition offers another advantage by using the EC-ALD method. The phase change material formed with this method can be single crystalline. Therefore, the volume shrinkage upon the initial crystallization (needed for as-deposited amorphous material) is avoided. In fact, single crystalline Sb_2Te_3 and some ternary alloys have been electrodeposited with this method [11.58, 11.59].

In summary, electrodeposition finds its niche in phase change device fabrication due to its unique capability in forming nanostructures and single crystal materials. However, knowledge barriers, including the incorporation of desired elements, the properties of the electrodeposited material, and the effects and distributions of impurities in the deposits, have to be overcome before this method is integrated into the device fabrication. These challenges provide great opportunities for the collaboration among electrochemists, material scientists and electrical engineers.

11.3 Solution-phase Deposition

Solution-phase deposition of thin films is accomplished by coating the substrate using spin coating, dip coating, spraying, printing, etc. Besides their technological simplicity, such methods offer excellent conformal coating of substrate features as long as the solution wets the substrate material. In this way, these techniques offer another route to filling high aspect ratio vias with phase change material in order to reduce the reset current of PCRAM devices. Solution deposition is most easily accomplished when the material to be deposited is itself soluble, so that the film is formed directly upon drying. Solution deposited polymer films such as photoresist are commonly deposited this way, but this approach has limited utility for inorganic materials such as phase change materials, which tend not to be directly soluble at concentrations sufficient for coating. Even so, some metal chalcogenides are soluble in amine-containing solvents and this direct approach has been used to coat films of amorphous $As_2S_{3-x}Se_x$, $GeS_{2-x}Se_x$, $GeSe_{2-x}Te_x$, and even As_2Te_3 [11.60-11.64]. The range of chalcogenide materials accessible by solution phase deposition is greatly expanded if a less direct, precursor-based approach is used.

In this case, a soluble precursor is coated, then thermally decomposed to form the metal chalcogenide film. For thin films, like those typical for phase change applications, it is sometimes possible to deposit the precursor and dry the film before decomposing it all at once in a subsequent thermal anneal. For thicker films, this approach can trap decomposition products within the film, so it is preferred to heat the substrate during deposition so that the precursor decomposes as the film forms, releasing byproducts continuously. This approach is related to CVD and has been explored, for example, in the spray deposition of $CuIn_{1-x}Ga_xSe_2$ for photovoltaic applications [11.65, 11.66].

Recently, Mitzi et al. [11.67-11.69] described a new class of soluble precursors for metal chalcogenide thin films which are easily prepared, reduce the chance of contamination, and exhibit minimal shrinkage upon thermal decomposition. This class of precursors consists of salts or neutral complexes of metal chalcogenide clusters with the reducing inorganic solvent hydrazine. First applied in the fabrication of thin film transistors, their application has now been extended to include the phase change materials KSb_5S_8 [11.70] and Ge-doped Sb_2Se_3 [11.71]. In the latter case, the precursor decomposition temperature is sufficiently low that amorphous films may be deposited in order to characterize the thermal crystallization of these novel materials. Then, their properties may be evaluated by similar methods to those used for conventional sputtered materials.

X-ray diffraction during thermal ramping provides a clear indication of the crystallization temperature of a phase change thin film. Using this method, solution deposited KSb_5S_8, crystalline after precursor decomposition was shown to melt around 440 °C [11.70]. Because this material is very slow to crystallize, it could be quenched to the amorphous state simply by allowing the film to cool after melting. The amorphous film could then be ramped again until diffraction peaks reappeared at the crystallization temperature, around 280 °C for a ramp rate of 1 K/s.

Ge-doped Sb_2Se_3 films were deposited by Milliron et al. [11.71] by mixing a Ge-Se and a Sb-Se precursor in solution, spin coating, and thermally annealing. In this case, the crystallization temperature could be tuned by adjusting the Ge content so that amorphous films were formed after annealing to decompose the precursors. Followed by x-ray diffraction, the crystallization temperature was tuned in the range of 200-250 °C by adjusting the Ge content between 2 and 6% (Fig. 11.8).

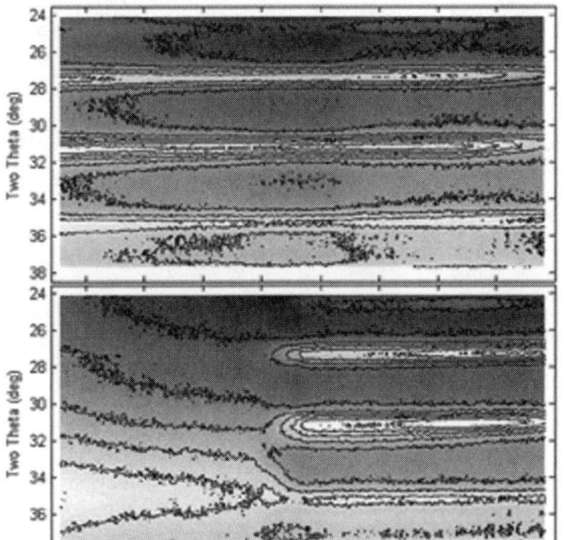

Fig. 11.8. Intensity of diffracted x-ray peaks as a function of temperature for spin-on $Sb_1Se_{1.7}$ (top) and $Ge_{2.4}(Sb_1Se_{1.8})_{97.6}$ (bottom) films. It is apparent that the addition of Ge has a great influence on the crystallization temperature. While Ge-free films are crystalline after deposition (which includes a bake at 160 ºC to remove the solvent) Ge-containing films crystallize at temperatures between 200 - 250 ºC depending on the composition. The film composition can be fine-tuned by adjusting the ratio of the precursors, and good control over the crystallization temperature can be achieved. Reprinted by permission from Macmillan Publishers Ltd: Nature Materials [11.71], copyright (2007).

For these films, crystallization was fairly rapid, so that laser pulse annealing was employed to investigate crystallization of the as-deposited and melt-quenched films. Comparable to some more conventional sputtered materials, melt-quenched spots in a crystalline matrix could be re-crystallized in as little as 100 ns. Resistivity contrast between the amorphous and crystalline phases was also observed by measuring film resistance during thermal ramping. While the properties of these GeSbSe films do not compare to the best sputtered films, their qualitatively similar functionality reveals the potential of solution deposition to yield viable phase change materials.

The potential advantage of solution phase deposition is the convenient filling of small and high aspect ratio structures to reduce reset current in PCRAM devices. It was recently demonstrated that small (ca. 20 nm) vias may be filled with metal chalcogenide materials using the coating methods described above [11.72]. When the surface of the material is hydrophilic, it is easily wet by the solution, and capillary forces drive the precursor to fill recessed vias and trenches. By controlling the solution viscosity (through concentration) and the drying rate (through spin coating rate), bottom up filling of vias occurs. Furthermore, these parameters can be adjusted to selectively fill the vias only partially, or to planarize the surface above completely filled vias as shown in Fig. 11.9. Such control is clearly advantageous for integration of metal chalcogenide materials into PCRAM devices.

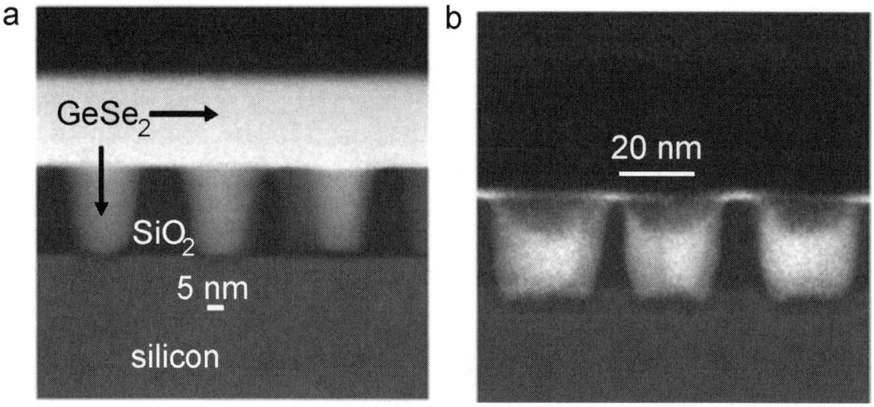

Fig. 11.9. A spin coating process was used to demonstrate filling of nanoscale vias with the metal chalcogenide material $GeSe_2$. By controlling the viscosity and drying time, the vias may be filled completely and the surface planarized (a) or partially filled from the bottom up (b) as these cross-sectional transmission electron microscope images show.

11.4 Nanomaterials

Semiconductor nanowires have been used to investigate and push the limits of scaling microelectronics devices. High quality, single crystalline nanowires have been grown of elemental semiconductors (such as silicon), compound semiconductors (such as GaAs), and heterostructures in various configurations using the vapor-liquid-solid (VLS) method [11.73, 11.74]. Very recently, this methodology has been brought to bear on phase change nanowires with intriguing initial results [11.75-11.77]. These materials can be grown in a CVD reactor from reactive precursors or by vapor transport of bulk material. The material is induced to grow as nanowires, rather than a continuous thin film through the use of metal nanoparticles which act as a catalyst. In the VLS mechanism for nanowire growth, reactants arrive by vapor transport and are dissolved into a metal particle. The particle becomes a liquid as the melting temperature is depressed by the dissolved components. When the dissolved material reaches supersaturation, the nanowire begins to form, growing from the drop of liquid metal. Nanowires have very recently been grown of phase change materials, including GeTe nanowires down to 50 nm [11.75, 11.76] and $Ge_2Sb_2Te_5$ nanowires as small as 30 nm in diameter [11.77], and several microns long. Figure 11.10 shows the SEM images of $Ge_2Sb_2Te_5$ nanowires with faceted structure [11.77].

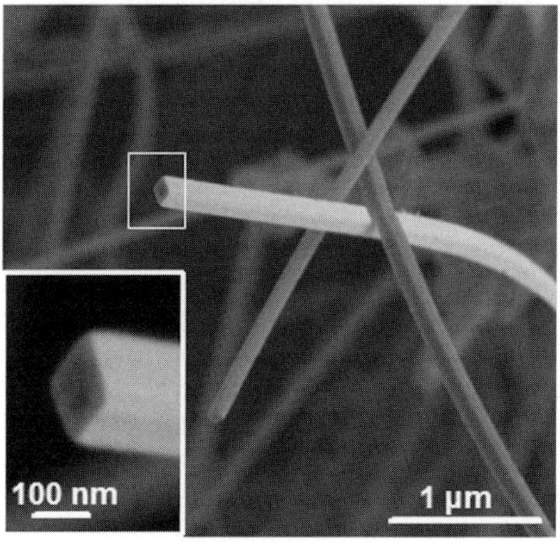

Fig. 11.10. SEM image of $Ge_2Sb_2Te_5$ nanowires. Reprinted by permission from Macmillan Publishers Ltd: Nature Nanotechnology [11.77], copyright (2007).

These nanowires are typically formed as single crystals and their phase switching behavior can be investigated electrically, following earlier demonstrations of semiconductor nanowire devices [11.74, 11.78]. Park et al. demonstrated reversible electrical switching of GeTe nanowire devices with average wire diameters around 65 nm [11.76]. Their results indicate that nanowires can function qualitatively like phase change devices fabricated by conventional methods. Agarwal et al. used $Ge_2Sb_2Te_5$ nanowires to investigate the dependence of electrical switching characteristics on wire diameter [11.77]. Very low power operation could be achieved in small, 30 nm diameter wires. The results show that data retention time is critically and predictably dependent on wire diameter, with 80 °C retention times of 3 years and 25 years for 30 and 60 nm diameter wires, respectively. Such correlations are crucial to predicting the ultimate scalability of phase change memory devices (see also Chapter 6).

Additional insight into scaling properties of phase change materials can be gained by patterning films into arrays of nanoscale islands. In this case, the dimensional extent is limited in all dimensions to determine the impact of reduced switching volume on fundamental characteristics such as crystallization temperature. Raoux et al. [11.79] utilized high resolution electron beam lithography to fabricate arrays of cone-shaped nanoscale islands (down to about 30 nm) of several different phase change materials. Comparing the crystallization of these arrays to that of thin films, they concluded that volume reduction effects are minor, but material-dependent. In the case of Sb_2Te a 40 °C increase in the crystallization temperature was observed, implying that the amorphous state stability improves for this material at small size scales. Such observations underscore the necessity of evaluating materials characteristics at small sizes early in the selection and integration process.

The properties of materials at the ultimate scaling limit, approaching molecular dimensions, can be probed by chemically synthesizing nanoparticles in the sub-10 nm range. For example, size-dependent melting point depression has been observed in metal [11.80, 11.81] and semiconductor [11.82] nanocrystals. In these extremely small particles, a significant fraction of the atoms lie at or very near the surface and their incomplete bonding network decreases the solid state stability of the particle as a whole. In a very small scale phase change memory device, the interface between the small volume of phase change material and the surrounding dielectric and contacts can be expected to strongly influence the depression of melting temperature. Colloidal nanocrystals have also been used to reveal fundamental physics of crystal-crystal phase transitions, notably the 4- to 6-coordinate transition in CdSe [11.83]. Remarkably altered kinetics of this phase transition were observed as a result of the single crystalline nature of the nanocrystals. Lacking high energy grain boundaries or other major crystal defects, the nucleation of the new phase is suppressed. Such effects may become significant at small sizes for phase change memories which might achieve multiple bits through the transition to various crystal phases. Finally, phase switching in vapor deposited Ga metal and GST nanoparticles has recently been demonstrated [11.84-11.86] as discussed more fully in Chapter 6. While these particles are far larger than colloidal particles which have been used to study phase transitions, and the results are preliminary, this approach shows promise for probing amorphous-crystalline transitions in nanoparticles.

Very recently, colloidal nanocrystals of GeTe have been synthesized [11.87]. These particles are faceted single crystals with dimensions down to ca. 4 nm (Fig. 11.11).

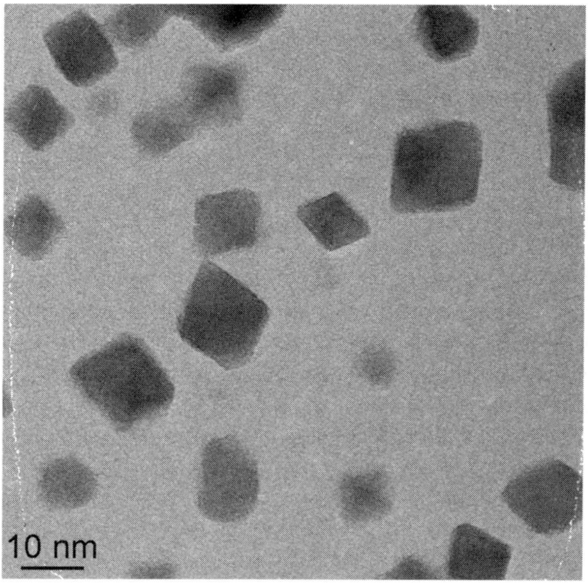

Fig 11.11. Low resolution TEM of GeTe nanocrystals [11.87].

Size-dependent trends in transition temperatures and phase change kinetics of these particles may establish new material or device design rules for scaling PCRAM to the ultimate high density limit.

11.5 Conclusions

The challenge of bringing electronic phase change memory successfully to market has inspired the application of several non-conventional methods for depositing thin films of phase change materials. CVD, electrodeposition, and solution-phase deposition all offer greatly improved filling of via structures versus physical vapor phase deposition (PVD). This advantage may be critical to reducing the RESET current of phase change memories to an acceptable level. Each method has been previously applied to the deposition of materials related to phase change materials; for example metal chalcogenides with applications in thermoelectrics, photovoltaics, etc. Very recently, each has also been applied to phase change materials, though significant challenges remain. A major hurdle involves compositional control since reactive driving forces exert a much greater impact on composition in each of these methods than in conventional PVD. The incorporation of non-metallic dopants such as N and O, which can strongly modulate phase change properties, can also be difficult to control in electrodeposition and solution phase coating. Finally, deposition morphology is a significant challenge since each of these methods often produce polycrystalline films with rough surface texture. Early results demonstrating the deposition of homogeneous amorphous films must be extended as these methods develop.

To succeed in impacting phase change memory technology, a new deposition method must offer advantages, and also must be integrated into the overall processing of the integrated circuit (IC). Of the methods discussed here, CVD has the most established position of integration into IC processing, and electrodeposition of metals is also now used in some fabrication processes. Solution deposition is used for sacrificial materials (e.g. photoresist), but has so far not been integrated for active device components. This is not a fundamental limitation, but does pose additional challenges for the application of this method to phase change memory.

Finally, chemically synthesized nanomaterials offer a chance to examine the scaling physics, and ultimate scaling limits, of phase change material properties. Phase change nanowires grown by the VLS mechanism are single crystals with diameters down to 30 nm or so. Devices fabricated from these wires have already begun to elucidate fundamental scaling properties of phase change memory at small dimensions. Nanopatterned thin films can illuminate the properties of phase change material confined to small volumes – a model for the memory devices of the future. Ultimately, sub-5 nm diameter nanoparticles of phase change materials may reveal the physics of the amorphous-crystalline phase transformation in these materials in the small-volume limit and enable predicting of changing device behavior due to scaling. Coupling such predictions with the growing understanding of how materials characteristics control device properties will allow phase change memory to be successfully scaled to the extremely high densities required in decades to come.

References

[11.1] Dobkin, D.M., Zuraw, M.K.: Principles of chemical vapor deposition. Kluwer, Dordrecht (2003)

[11.2] Pierson, H.O.: Handbook of chemical vapor phase deposition (CVD). Noyes, Park Ridge (1992)

[11.3] Cho, S.L., Yi, J.H., Ha, Y.H., Kuh, B.J., Lee, C.M., Park, J.H., Nam, S.D., Horii, H., Cho, B.K., Ryoo, K.C., Park, S.O., Kim, H.S., U-In Chung Moon, J.T., Ryu, B.I.: Highly scalable on-axis confined cell structure for high density PRAM beyond 256mb. Symp. VLSI Tech Dig. **2005,** 96-97 (2005)

[11.4] Joseph, E.A., Happ, T. D., Chen, S.-H., Raoux, S., Chen, C.-F., Breitwisch, M., Schrott, A. G., Zaidi, S., Dasaka, R., Yee, B., Zhu, Y., Bergmann, R., Lung, H.-L., Lam, C.: Patterning of N:$Ge_2Sb_2Te_5$ films and the characterization of etch induced modification for non-volatile phase change memory applications, in VLSI TSA. 2008.

[11.5] Kim, R.-Y., Kim, H.-G., Yoon, S.-G.: Structural properties of $Ge_2Sb_2Te_5$ thin film by metal organic chemical vapor deposition for phase change memory applications. Appl. Phys. Lett. **89,** 102107 (2006)

[11.6] Choi, B.J., Choi, S., Shin, Y.C., Hwang, C.S., Lee, J.W., Jeong, J., Kim, Y.J., Hwang, S.-Y., Hong, S.Y.: Cyclic PECVD of $Ge_2Sb_2Te_5$ films using metallorganic sources. J. Electrochem. Soc. **154,** H318-H324 (2007)

[11.7] Kim, R.-Y., Kim, H.-G., Yoon, S.-G.: Growth of Ge-doped Sb_2Te_3 thin films by metal-organic chemical vapor deposition. Integr. Ferroelectr. **90,** 80-87 (2007)

[11.8] Ovshinsky, S.R, and Kamepalli, S., US Patent No. 2006/0172067 A1

[11.9] Lee, J., Choi, S., Lee, C., Kang, C., Kim, D.: GeSbTe deposition for the PRAM application. Appl. Surf. Sci. **253,** 3969-3967 (2007)

[11.10] Ritala, M., Leskela, M.: Atomic layer deposition. In Nalwa, H.S. (ed.) Handbook of thin film materials, pp. (2002)

[11.11] Do, K., Sohn, H., Ko, D.-H.: Phase transformation behavior of N-doped $Ge_2Sb_{2+x}Te_5$ thin films (x = 0, 0.2) for phase change memory. J. Electrochem. Soc. **154,** H867-H870 (2007)

[11.12] Ebina, A., Hirasaka, M., Nakatani, K.: Oxygen doping effect on Ge–Sb–Te phase change optical disks. J. Vac. Sci. Technol. A. **17,** 3463-3466 (1999)

[11.13] Chen, Y.C., Rettner, C.T., Raoux, S., Burr, G.W., Chen, S.H., Shelby, R.M., Salinga, M., Risk, W.P., Happ, T.D., McClelland, G.M., Breitwisch, M., Schrott, A., Philipp, J.B., Lee, M.H., Cheek, R., Nirschl, T., Lamorey, M., Chen, C.F., Joseph, E., Zaidi, S., Yee, B., Lung, H.L., Bergmann, R., Lam, C.: Ultra-thin phase-change bridge memory device using GeSb. Tech. Dig. - Int. Electron Devices Meet. **30,** 1-4 (2006)

[11.14] Satoh, H., Sugawara, K., Tanaka, K.: Nanoscale phase change in crystalline $Ge_2Sb_2Te_5$ films using scanning probe microscopes. J. Appl. Phys. **99,** 024306 (2006)

[11.15] Lee, J.I., Park, H., Cho, S.L., Park, Y.L., Bae, B.J., Park, J.H., Park, J.S., An, H.G., Bae, J.S., Ahn, D.H., Kim, Y.T., Horii, H., Song, S.A., Shin, J.C., Park, S.O., Kim, H.S., Chung, U.-I., Moon, J.T., Ryu, B.I.: Highly scalable phase change memory with CVD GeSbTe for sub 50nm generation. Sym. VLSI Tech. Digest **102,** 102-103 (2007)

[11.16] Romankiw, L.T.: A path: From electroplating through lithographic masks in electronics to liga in mems. Electrochim. Acta. **42,** 2985-3005 (1998)

[11.17] Andricacos, P.C., Uzoh, C., Dukovic, J.O., Horkans, J., Deligianni, H.: Damascene copper electroplating for chip interconnections. IBM J. Res. Dev. **42,** 567-574 (1998)

[11.18] Bard, A.J., Faulkner, L.R.: Elecrochemical methods: Fundamentals and applications. John Wiley and Sons Inc, New York (2002)

[11.19] Newman, J.S., Thomas-Alyea, K.E.: Electrochemical systems. John Wiley and Sons Inc, New York (2004)

[11.20] Schlesinger, M., Paunovic, M.: Fundamentals of electrochemical deposition. John Wiley and Sons Inc, New York (2006)

[11.21] Panicker, M.P.R., Knaster, M., Kroger, F.A.: Cathodic deposition of CdTe from aqueous electrolytes. J. Electrochem. Soc. **125,** 566-572 (1978)

[11.22] Stickney, J.L.: Electrochemical atomic layer epitaxy (EC-ALE): Nanoscale control in the electrodeposition of compound semiconductors. In Alkire, R.C., Kolb, D.M. (eds.) Advances in electrochemical science and engineering, pp. Wiley-VCH, Verlag GmbH (2001)

[11.23] Bonilla, S., Dalchiele, E.A.: Electrochemical deposition and characterization of cdte polycrystalline thin films. Thin Solid Films. **204,** 397-403 (1991)

[11.24] Saraby-Reintjes, A., Peter, L.M., Özsan, M.E., Dennison, S., Webster, S.: On the mechanism of the cathodic electrodeposition of cadmium telluride. J. Electrochem. Soc. **140,** 2880-2888 (1993)

[11.25] Das, S.K., Morris, G.C.: Preparation and properties of CdS/CdTe thin film solar cell produced by periodic pulse electrodeposition technique. Sol. Ener. Mater. Sol. Cell. **30,** 107-118 (1993)

[11.26] Hayden, B.E., Nandhakumar, I.S.: In situ STM study of CdTe ECALE bilayers on gold. J. Phys. Chem. B. **102,** 4897-4905 (1998)

[11.27] Varazo, K., Lay, M.D., Sorenson, T.A., Stickney, J.L.: Formation of the first monolayers of CdTe on Au (111) by electrochemical atomic layer epitaxy (EC-ALE): Studied by LEED, Auger, XPS, and in-situ STM. J. Electroanal. Chem. **522,** 104-114 (2002)
[11.28] Venkatasamy, V., Jayaraju, N., Cox, S.M., Thambidurai, C., Happek, U., Stickney, J.L.: Optimization of CdTe nanofilm formation by electrochemical atomic layer epitaxy (EC-ALE). J. Appl. Electrochem. **36,** 1223-1229 (2006)
[11.29] Colyer, C.L., Cocivera, M.: Thin-film cadmium mercury telluride prepared by nonaqueous electrodeposition. J. Electrochem. Soc. **139,** 406-409 (1992)
[11.30] Rajeshwar, K.: Electrosynthesized thin films of group II-VI compound semiconductors, alloys and superstructures. Adv. Mater. **4,** 23-29 (1992)
[11.31] Kumaresan, R., Gopalakrishnan, R., Moorthy Babu, S., Ramasamy, P., Kruger, D., Zaumseil, P.: X-ray photoelectron spectroscopic studies of electrodeposited mercury cadmium telluride semiconductor thin films. J. Phys. Chem. Solid. **61,** 765-771 (2000)
[11.32] Lincot, D.: Electrodeposition of semiconductors. Thin Solid Films. **487,** 40-48 (2005)
[11.33] Vaidyanathan, R., Stickney, J.L., Happek, U.: Quantum confinement in PbSe thin films electrodeposited by electrochemical atomic layer epitaxy (EC-ALE). Electrochim. Acta. **49,** 1321-1326 (2004)
[11.34] Venkatasamy, V., Jayaraju, N., Cox, S.M., Thambidurai, C., Mathe, M., Stickney, J.L.: Deposition of HgTe by electrochemical atomic layer epitaxy (EC-ALE). J. Electroanal. Chem. **589,** 195-202 (2006)
[11.35] Venkatasamy, V., Jayaraju, N., Cox, S.M., Thambidurai, C., Stickney, J.L.: Studies of HgCdTe formation by electrochemical atomic layer deposition and investigations into bandgap engineering. J. Electrochem. Soc. **154,** H720-H725 (2007)
[11.36] Mahalingam, T., Kathalingam, A., Velumani, S., Lee, S., Moon, H., Kim, Y.D.: Electrosynthesis and studies on $Zn_{1-x}Hg_xTe$ thin films. J. New Mater. Electrochem. Sys. **10,** 21 - 25 (2007)
[11.37] Orts, J.L., Diaz, R., Herrasti, P., Rueda, F., Fatas, E.: $CuInTe_2$ and In-rich telluride chalcopyrites thin films obtained by electrodeposition techniques. Sol. Ener. Mater. Sol. Cell. **91,** 621 - 628 (2007)
[11.38] Taunier, S., Guimard, D., Lincot, D., Guillemoles, J.-F., Grand, P.-P., Method of producing thin films of compound i-iii-vi, promoting the incorporation of iii elements in the film. 2007: US2006015133A1.
[11.39] Magri, P., Boulanger, C., Lecuire, J.M.: Synthesis, properties and performances of electrodeposited bismuth telluride films. J. Mater. Chem. **6,** 773-779 (1996)
[11.40] Miyazaki, Y., Kajitani, T.: Preparation of Bi_2Te_3 films by electrodeposition. J. Crystal Growth. **229,** 542-546 (2001)
[11.41] Tittes, K., Bund, A., Plieth, W., Bentien, A., Paschen, S., Plötner, M., Gräfe, H., Fischer, W.J.: Electrochemical deposition of Bi_2Te_3 for thermoelectric microdevices. J. Solid State Electrochem. **7,** 714-723 (2003)
[11.42] Wang, W., Huang, Q.H., Jia, F.L., Zhang, Z.R.: Structure and property characterization of $Bi_{2-x}Sb_xTe_3$ thermoelectric films prepared by electrodeposition. Wuji Cailiao Xuebao(Journal of Inorganic Materials). **20,** 1234-1238 (2005)
[11.43] Del Frari, D., Diliberto, S., Stein, N., Boulanger, C., Lecuire, J.M.: Comparative study of the electrochemical preparation of Bi_2Te_3, Sb_2Te_3, and $(Bi_xSb_{1-x})_2Te_3$ films. Thin Solid Films. **483,** 44-49 (2005)
[11.44] Huang, Q., Kellock, A.J., Raoux, S.: Electrodeposition of SbTe phase-change alloys. J. Electrochem. Soc. **155,** D104-D109 (2008)
[11.45] Leimkühler, G., Kerkamm, I., Reineke-Koch, R.: Electrodeposition of antimony telluride. J. Electrochem. Soc. **149,** C474-C478 (2002)
[11.46] van Pieterson, L., Lankhorst, M.H.R., van Schijndel, M., Kuiper, A.E.T., Roosen, J.H.J.: Phase-change recording materials with a growth-dominated crystallization mechanism: A materials overview. J. Appl. Phys. **97,** 083520 (2005)

[11.47] Privitera, S., Bongiorno, C., Rimini, E., Zonca, R.: Crystal nucleation and growth processes in $Ge_2Sb_2Te_5$. Appl. Phys. Lett. **84,** 4448-4450 (2004)

[11.48] Huang, Q., Bedell, S.W., Saenger, K.L., Copel, M., Deligianni, H., Romankiw, L.T.: Single-crystalline germanium thin films by electrodeposition and solid-phase epitaxy. Electrochem. and Solid-State Lett. **10,** D124-D126 (2007)

[11.49] Fink, C.G., Dokras, V.M.: Electrodeposition and electrowinning of germanium. J. Electrochem. Soc. **95,** 80-97 (1949)

[11.50] Huang, Q., Kellock, A.J., Shao, X., Venkatasamy, V., A method of electrodepositing germanium compound materials on a substrate. 2007, International Business Machine: US patent application.

[11.51] Sapp, S.A., Lakshmi, B.B., Martin, C.R.: Template synthesis of bismuth telluride nanowires. Adv. Mater. **11,** 402-404 (1999)

[11.52] Chen, R., Xu, D., Gui, L.: Silver telluride nanowires prepared by dc electrodeposition in porous anodic alumina templates. J. Mater. Chem. **12,** 2435-2438 (2002)

[11.53] Sander, M.S., Prieto, A.L., Gronsky, R., Sands, T., Stacy, A.M.: Fabrication of high-density, high aspect ratio, large-area bismuth telluride nanowire arrays by electrodeposition into porous anodic alumina templates. Adv. Mater. **14,** 665-667 (2002)

[11.54] Chen, R., Xu, D., Guo, G., Tang, Y.: Electrodeposition of thin films and single-crystalline nanowires of Ag_7Te_4. Chem. Phys. Lett. **377,** 205-209 (2003)

[11.55] Gandhi, T., Raja, K.S., Misra, M.: Templated growth of cadmium zinc telluride (czt) nanowires using pulsed-potentials in hot non-aqueous solution. Electrochim. Acta. **51,** 5932-5942 (2006)

[11.56] Li, L., Yang, Y., Huang, X., Li, G., Zhang, L.: Pulsed electrodeposition of single-crystalline Bi_2Te_3 nanowire arrays. Nanotechnology. **17,** 1706-1712 (2006)

[11.57] Jin, C., Zhang, G., Qian, T., Li, X., Yao, Z.: Large-area Sb2Te3 nanowire arrays. J. Phys. Chem. B. **109,** 1430-1432 (2005)

[11.58] Yang, J., Zhu, W., Gao, X., Bao, S., Fan, X., Duan, X., Hou, J.: Formation and characterization of Sb_2Te_3 nanofilms on Pt by electrochemical atomic layer epitaxy. J. Phys. Chem. B. **110,** 4599-4604 (2006)

[11.59] Huang, Q., Shao, X., Stickney, J.L., Venkatasamy, V., Method of making phase change materials by electrochemical atomic layer deposition. 2007, International Business Machine: US patent application.

[11.60] Chern, G.C., Lauks, I.: Spin-coated amorphous-chalcogenide films. J. Appl. Phys. **53,** 6979-6982 (1982)

[11.61] Gutwirth, J., Wagner, T., Kohoutek, T., Vlcek, M., Schroeter, S., Kovanda, V., Vlcek, M., Frumar, M.: Physical properties and structure of amorphous $Ag_x(Sb_{0.33}S_{0.67})_{(100-x)}$ prepared by optically-induced diffusion and dissolution of silver into spin-coated amorphous $Sb_{33}S_{67}$ films and their application for optical recording. J. Optoelectron. Adv. Mater. **5,** 1139-1146 (2003)

[11.62] Hajto, E., Ewen, P.J.S., Belford, R., Hajto, J., Owen, A.E.: Optical-properties of spin-coated amorphous-chalcogenide thin-films. J. Non-Crystall. Solids. **97-8,** 1191-1194 (1987)

[11.63] Kohoutek, T., Wagner, T., Frumar, M., Vlcek, M.: Spin coated chalcogenide films of Ge-Se-Te: Physio-chemical properties. Phys. Chem. Glass.-Europ. J. Glass Sci. Tech. Part B. **47,** 250-253 (2006)

[11.64] Kohoutek, T., Wagner, T., Orava, J., Krbal, M., Fejfar, A., Mates, T., Kasap, S.O., Frumar, M.: Surface morphology of spin-coated As-S-Se chalcogenide films. J. Non-Crystall. Solids. **353,** 1437-1440 (2007)

[11.65] Shirakata, S., Kannaka, Y., Hasegawa, H., Kariya, T., Isomura, S.: Properties of Cu(In,Ga)Se2 thin films prepared by chemical spray pyrolysis. Jap. J. Appl. Phys. **38,** 4997-5002 (1999)

[11.66] Tembhurkar, Y.D., Hirde, J.P.: Structural, optical and electrical properties of spray pyrolytically deposited films of copper indium diselenide. Thin Solid Films. **215,** 65-70 (1992)
[11.67] Milliron, D.J., Mitzi, D.B., Copel, M., Murray, C.E.: Solution processed metal chalcogenide films for p-type transistors. Chem. Mater. **18,** 587-590 (2006)
[11.68] Mitzi, D.B., Copel, M., Murray, C.E.: High-mobility p-type transistor based on a spin-coated metal telluride semiconductor. Adv. Mater. **18,** 2448-2452 (2006)
[11.69] Mitzi, D.B., Kosbar, L.L., Murray, C.E., Copel, M., Afzali, A.: High-mobility ultrathin semiconducting films prepared by spin coating. Nature. **428,** 299-303 (2004)
[11.70] Mitzi, D.B., Raoux, S., Schrott, A.G., Copel, M., Kellock, A., Jordan-Sweet, J.: Solution-based processing of the phase-change material KSb_5S_8. Chem. Mater. **18,** 6278-6282 (2006)
[11.71] Milliron, D.J., Raoux, S., Shelby, R.M., Jordan-Sweet, J.: Solution-phase deposition and nanopatterning of GeSbSe phase change materials. Nature Mater. **6,** 352-357 (2007)
[11.72] Milliron, D.J., Caldwell, M.A., Wong, H.-S.P.: Synthesis of metal chalcogenide nanodot arrays using block copolymer-derived nanoreactors. Nano Lett. **7,** 3504-3507 (2007)
[11.73] Law, M., Goldberger, J., Yang, P.: Semiconductor nanowires and nanotubes. Ann. Rev. Mat. Res. **34,** 83-122 (2004)
[11.74] Lu, W., Lieber, C.M.: Semiconductor nanowires. J. Phys. D. **39,** R387-R406 (2006)
[11.75] Meister, S., Peng, H., McIlwrath, K., Jarausch, K., Zhang, X.F., Cui, Y.: Synthesis and characterization of phase-change nanowires. Nano Lett. **6,** 1514-1517 (2006)
[11.76] Yu, D., Wu, J., Gu, Q., Park, H.: Germanium telluride nanowires and nanohelices with memory-switching behavior. J. Am. Chem. Soc. **128,** 8148-8149 (2006)
[11.77] Lee, S.-H., Jung, Y., Agarwal, R.: Highly scalable non-volatile and ultra-low-power phase-change nanowire memory. Nature Nanotech. **2,** 626-630 (2007)
[11.78] Lu, W., Lieber, C.M.: Nanoelectronics from the bottom up. Nature Mater. **6,** 841-850 (2007)
[11.79] Raoux, S., Rettner, C. T., Jordan-Sweet, J. L., Kellock, A. J., Topuria, T., Rice, P. M. and Miller, D. C.: Direct observation of amorphous to crystalline phase transitions in nano-particle arrays of phase change materials. J. Appl. Phys. **102**, 94305 (2007)
[11.80] Buffat, P., Borel, J.-P.: Size effect on the melting temperature of gold particles. Phys. Rev. A. **13,** 2287-2298 (1976)
[11.81] Dick, K., Dhanasekaran, T., Zhang, Z., Meisel, D.: Size-dependent melting of silica-encapsulated gold nanoparticles. J. Am. Chem. Soc. **124,** 2312-2317 (2002)
[11.82] Goldstein, A.N., Echer, C.M., Alivisatos, A.P.: Melting in semiconductor nanocrystals. Science. **256,** 1425-1427 (1992)
[11.83] Tolbert, S.H., Alivisatos, A.P.: Size dependence of a first order solid-solid phase transition: The wurtzite to rock salt transformation in CdSe nanocrystals. Science. **265,** 373-376 (1994)
[11.84] Soares, B.F., Jonsson, F., Zheludev, N.I.: All-optical phase-change memory in a single gallium nanoparticle. Phys. Rev. Lett. **98,** 153905 (2007)
[11.85] Soares, B.F., MacDonald, K.F., Zheludev, N.I.: Resetting single nanoparticle structural phase with nanosecond pulses. Appl. Phys. Lett. **91,** 043115 (2007)
[11.86] Suh, D.-S., Lee, E., Kim, K.H.P., Noh, J.-S., Shin, W.-C., Kang, Y.-S., Kim, C., Khang, Y., Yoon, H.R., Jo, W.: Nonvolatile switching characteristics of laser-ablated $Ge_2Sb_2Te_5$ nanoparticles for phase-change memory applications. Appl. Phys. Lett. **90,** 023101 (2007)
[11.87] Caldwell, M.A., Raoux, S., Urban, J.J., Milliron, D.J., Wong, H.-S.P.: unpublished.

Part II:
Applications: Optical, Solid State Memory and Reconfigurable Logic

12. Optical Memory: From 1st to 3rd Generation and its Future

Luping Shi

Abstract In this section the development of 1st, 2nd, and 3rd generation optical discs is introduced with focus on phase-change rewritable optical discs. The 4th generation and the possible development trends of future optical discs are also discussed.

12.1 Introduction

Data storage technology is one of the key elements in information technology. Tremendous efforts have been put forward to search for data storage devices with low cost, high sensitivity, fast data transfer rate, high performance, high reliability and reusability. The rapid growth of the data storage market has been driven by the personal computer revolution followed by the multimedia revolution. Fueled by the development of portable electronics, e.g. mobile phones, and network applications, it is expected that the data storage market is going to expand even faster.

Generally speaking, there are three main types of data storage technologies: solid-state memory, magnetic storage and optical storage. Each of them has its own characteristics, and within each type there is also a wide variation in the capabilities of each product. So far, no practical universal storage medium exists, and all forms of storage have some drawbacks. Therefore a computer system usually contains several kinds of storage, each with an individual purpose. Solid-state memories, which have high-speed and compact size, are mainly used as the primary memories, while magnetic and optical storage devices are typically used as the secondary storage device for computer systems. With the rapid development of portable electronics, such as mobile phones and digital cameras, non-volatile solid state memory is becoming more and more important. Hard disc drives, which are the primary type of magnetic storage devices, are ubiquitous for high performance and high growth rate in areal density. They have been and still remain the choice for the secondary storage device in most of the computer systems due to the combination of high speed with low cost. Recently removable hard disc drives have become popular. They have also been applied in portable electronics, and inserted into Digital Versatile Disc (DVD) recorders. Optical storage offers a reliable and

S. Raoux, M. Wuttig (eds.), *Phase Change Materials*,
DOI 10.1007/978-0-387-84874-7_12,

removable storage medium with excellent robustness, long lifetime, low cost and non-contact data retrieval and provides read only, write once read many and rewritable functions. Optical discs have been widely used in multimedia to store digitized audio, video, animation and images.

12.2 Three Generations of Optical Media

Initial work on optical storage began in the 1950s and 1960s. Although the research had been conducted for several decades, the first popular system was the Compact Disc-Read Only Memory (CD-ROM) which was introduced in 1982, adapted to data storage (the CD-ROM format) with the 1985 Yellow Book, and re-adapted as the first mass market optical storage medium with write once read many capability as CD-R (Compact Disc–Recordable) in 1988 and with rewritable capability as CD-RW (Compact Disc-ReWritable) in 1997.

The Compact-Disc (CD) stores data in microscopic grooves running in a spiral around the disc. All types of CD drives use laser beams to scan these grooves. Minuscule reflective bumps (called lands) and non-reflective holes (called pits) aligned along the grooves represent the zeros and ones of digital information.

Until now three generations of optical media have been developed, CD, DVD (Digital Versatile Disc) and Blu-ray (BD) [12.1-12.3]. Every generation was developed following a main driving force from the market. Optical storage has been commercially successful in the form of CDs for audio and software distribution, DVDs for video distribution, and BD for high definition TV (HDTV). They have mainly emerged into the consumer market and the computer-based data storage market with developments in lasers and optical media. In order to develop every new generation of optical storage technology, the following main technological issues should be considered: density and capacity, data transfer rate, access time, performance such as overwriting cycle and lifetime, and cost.

The optical media can be grouped into three types, read only memory (ROM), write once read many or recordable (R) and rewritable (RW), also called Random Access Memory (RAM) optical disc. Read only discs such as CD-ROM and DVD-ROM are currently the most important optical media and have become the standard medium for distributing large quantities of information in a reliable package, such as data, video and audio. The write once and rewritable media will become more and more important because of the rapid growth of the internet which generates enormous amounts of data. In addition more data are transferred through the internet. Write-once optical discs are mainly used for archival purposes to store information permanently, such as medical records and legal documents. To meet the rewritable requirement, several types of rewritable optical discs such as CD-RW and DVD-RAM/±RW and BD-RE (recordable and erasable) have been developed. Because the data can be rewritten many times, rewritable optical discs can be used as the peripheral data storage devices in computer systems and optical recorders.

Optical storage is removable. That is the reason why standards are needed for every generation. When a new generation is developed, such as moving from CD to DVD, and DVD to BD, one very critical issue is the compatibility because the global standards for music, video and computer software are well accepted. Thus each generation of optical storage needs to be compatible with interchangeable media and backward standards. DVD players and DVD-ROM drives can play back or read CDs without substantial additional costs. BD drivers also play back or read DVDs and CDs. This compatibility facilitates the introduction of each new generation of technology in the market. But this comes at the costs of a time consuming process on standard generation for each new generation.

12.2.1 The First Generation: Compact Discs (CDs)

The CD was initially developed as a device for the storage and replay of digitally encoded audio. As computer software developers began writing larger computer applications that exceeded the storage capacity of the then commonly used floppy disc the CD became a standard exchange format of software distribution. CDs with 650 MB capacity apply laser diodes with a wavelength of 780 nm, a numerical aperture (*NA*) of the focusing lens of 0.45 and 1.6 μm track pitch. The history of the CD, the "mother" of all current mainstream optical storage formats, includes the following developments after the first introduction of CD-audio:

- CD-ROM for data applications
- Video CD for storing video on a CD
- CD-R for writing user content on discs
- CD-RW for adding the rewritable option
- Increasing transfer rates for fast reading (up to 72×) and writing (up to 52×).

The main specifications of CDs are listed in Table 12.1.

12.2.2 The Second Generation: Digital Versatile Disks (DVDs)

The development of DVD was initially driven by video distribution. The keyword is "versatile". DVDs are similar to CDs, but use a different type of recording technology and have a much larger data capacity. DVDs provide superb storage and access of video, audio and data -- all on one disc. They can hold up to 4.7 GB of data. DVDs use a smaller track pitch of 0.74 μm by applying a laser with wavelength of 650 nm, *NA* of 0.6 and new modulation and error correction methods. These technologies allow storing amounts of data that are seven times as large as the capacity of a CD. Currently there are three types of DVDs namely DVD-ROM for read only, DVD±R for write once and DVD-RAM, and DVD±RW for

Table 12.1. The main specifications of CDs.

Parameters	CD-ROM	CD-R	CD-RW
Disc thickness (mm)	1.2	1.2	1.2
Storage capacity (GB)	0.65	0.65	0.65
Data transfer rate (Mbps)	1.41(1×)	1.41(1×)	1.41(1×)
Recording method	In-groove	In-groove	In-groove
Embossed address information	-	no	no
Recording track pitch (μm)	1.6	1.6	1.6
Channel bit length (nm)	300	300	300
Laser wavelength (nm)	780	780	780
Numerical aperture	0.45	0.45	0.45

rewritable applications. The main specifications of DVDs are listed in Table 12.2.

Three application formats can be stored on a DVD-ROM disc. The first is DVD-Video for playback of pictures and sound. Offering superior picture and sound quality and versatile functions, DVD-Video makes interactive entertainment possible. The second is DVD-Audio which is mainly for music. Offering ultrahigh sound quality and long playback times, DVD-Audio delivers a vivid you-are-there audio experience. The third is DVD-ROM for computer applications which take full advantage of DVD's large capacity. A dual-layer, single-sided disc, which allows the data recorded on both sides to be read from a single side, has a large capacity of 8.5 GB. To enable a dual-layer DVD to be read from a single side, the reflection film of the first layer of the disc is semi-transparent. Moreover, the bonding agent is also transparent to enable instant access from the first layer to the second layer and seamless continuous playback from the first layer to the second layer. A dual-layer, double-sided disc is made by back-to-back bonding of two dual-layer, single-sided discs and has four layers in total.

In summary, a one-layer single-sided disc has a 4.7 GB capacity. A dual-layer single-sided disc has an 8.5 GB capacity. Back-to-back bonding of two such discs achieves a remarkably large capacity of 17 GB capacity with 8.5 GB per side. The structures of DVD 5, DVD 9, DVD 10 and DVD 18 GB capacity are shown in Fig. 12.1, respectively.

Table 12.2. The main specifications of DVDs.

Parameters	DVD-ROM	DVD-R	DVD+R	DVD-RW	DVD+RW	DVD-RAM
Disc thickness	0.6mm ×2	0.6mm×2	0.6mm ×2	0.6mm×2	0.6mm ×2	0.6mm×2
Storage capacity (GB)	4.7	4.7	4.7	4.7	4.7	4.7
Data transfer rate (Mbps)	11.08(1×)	11.08(1×)	11.08(1×)	11.08(1×)	11.08(1×)	22.16(2×)
Recording method	In groove	In-groove	In-groove	In-groove	In-groove	Land and groove
Embossed address information		yes	no	yes	no	yes
Recording track pitch (μm)	0.74	0.74	0.74	0.74	0.74	0.615
Channel bit length (nm)	133	133	133	133	133	140-146
Laser wavelength (nm)	650	635/650	650	635/650	650	650
Numerical aperture	0.6	0.6	0.65	0.6	0.65	0.6

The transfer rate of reading DVD-ROM has progressed to 16 × speed, where the number refers to the speed compared with the baseline speed of the format. Recordable (DVD±R) and rewritable (DVD±RW, DVD-RAM) DVDs are on the market for video and data recording. The industry has not been able to come up with a single standard. For recordable DVDs there are two competing formats: DVD+R and DVD-R. For rewritable DVDs, there are three formats: DVD+RW, DVD-RW, and DVD-RAM. This has lead to a longer drive qualification process, and hence a higher cost.

Within all the formats, only DVD-RAM and DVD+RW can randomly rewrite blocks of data on disc by using lossless linking technology for compatibility with the continuous spiral. For rewritable DVDs, higher speed upgrades have been developed for fast writing. Currently, the maximum speeds that have been commercialized are 20 × for DVD±R, 8 × for DVD±RW and 16 × for DVD-RAM.

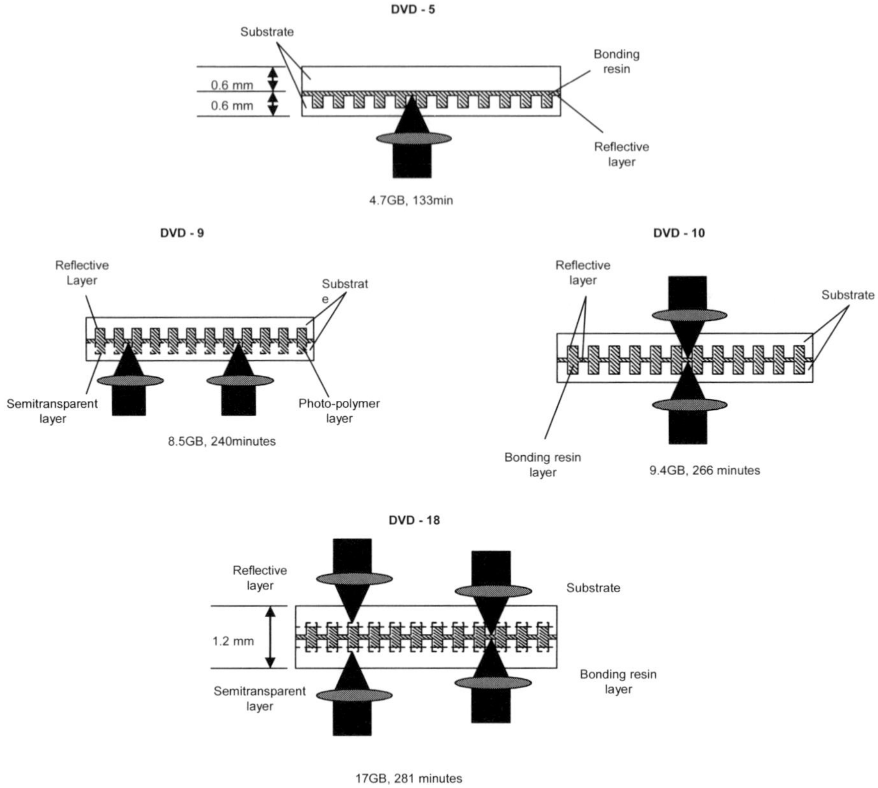

Fig. 12.1. The structures of DVD 5, DVD 9, DVD 10 and DVD 18, respectively.

12.2.3 The Third Generation: Blu-ray Discs (BDs)

With increased global availability of HDTV broadcasting, it is necessary to develop a new generation of optical storage technology with higher density and faster date transfer rate. As with the CD for music and DVD for standard definition video, the 3rd generation optical format enables higher quality for HD visual content as well as a convenient way to store and reproduce a large quantity of material without requiring the inconvenience of spanning it across multiple discs. The most prominent competing technology is the Blu-ray disc (BD).

12.2.3.1 Blu-ray Discs

The BD was introduced into the market in March 2003. It uses a 405 nm wavelength blue-violet laser technology, 0.85 *NA* and track pitch of 0.32 μm. The benefit of using a blue-violet laser (405 nm) is that it has a shorter wavelength than a red laser (650 nm), which makes it possible to focus the laser spot with even greater precision. This allows data to be packed more tightly and stored in less space so that it can even fit more data on the disc with the same physical size as a CD/DVD. Together with the increasing of *NA* to 0.85, it enables BDs to have higher density. The rewritable Blu-ray disc, with a data transfer rate of 36 Mbps ($1 \times$ speed) can hold up to 25 GB of data on a single-layer disc and 50 GB on a dual-layer disc. Combined with the use of advanced video and audio codes, it will offer consumers an unprecedented HD experience. BD's most significant application is the large scale publication and distribution of high definition motion pictures, games, music and other audio/video content. It is easy to make BDs backwards compatible with CDs and DVDs through the use of a BD/DVD/CD compatible optical pickup unit.

There are also three types of BD namely BD-ROM, BD-R and BD-RE for read only, write once and rewritable, respectively. Table 12.3 summarizes the main specifications of BDs.

12.3 The Basic Principle of Optical Recording

Optical storage systems consist of a drive unit and a storage medium in a rotating disc form. In general the discs are pre-formatted using grooves and lands (tracks) to enable the positioning of an optical pick-up and recording head to access the information on the disc. Under the influence of a focused laser beam emanating from the optical head, information is recorded on the media as a change in the material characteristics. The tiny pits in the tracks correspond to the zeros and ones of the digital information. The stored data is retrieved by detecting the reflection difference of the reflected laser beam as show in Fig. 12.2.

The diameter of the diffraction limited laser spot is directly proportional to the laser wavelength λ, and inversely proportional to the numerical aperture *NA* of the focus lens (Fig. 12.2). The area a of the spot is therefore proportional to

$$a \propto \left(\frac{\lambda}{NA}\right)^2 \tag{12.1}$$

and the resulting areal density D is then inversely proportional to this (spot) area a

$$D \propto \left(\frac{NA}{\lambda}\right)^2. \tag{12.2}$$

From above formula, it can be seen that in order to increase density a shorter wavelength or larger *NA* is needed.

Table 12. 3 The main specifications of BDs.

Parameters	BD-ROM	BD-R	BD-RE
Disc thickness	1.1+0.1	1.1+0.1	1.1+0.1
Storage capacity (GB) single layer	25	25	23/25/27
Data transfer rate (Mbps)	36(1×)	36(1×)	36(1×)
Recording method		In-groove	In-groove
Embossed address information			no
Recording track pitch (μm)	0.32	0.32	0.32
Data bit length (nm)			120/111.75/103.50
Laser wavelength(nm)	405	405	405
Numerical aperture	0.85	0.85	0.85

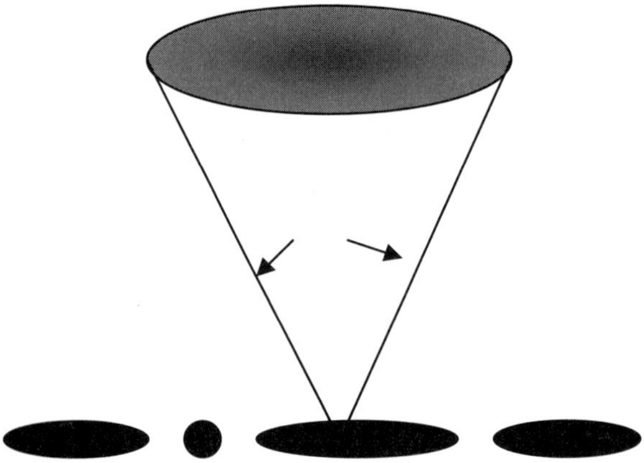

Fig. 12.2. Schematic drawing showing the optical system and media.

Thus smaller track pitch and smaller bit length can be achieved.

A DVD disc uses a laser with much shorter wavelength than that for CDs. The track pitch is about half of that on a CD so that tracks and pits are finer. To read and write the data a red laser is used. By focusing the laser light through a large-aperture lens, a narrower focus is obtained and high-density data reading is achieved. As a result, high-density data storage is achieved.

Nevertheless, because the laser is refracted considerably by the use of a large *NA* lens, it is difficult to obtain an accurate focus if the disc is not precisely horizontal in relation to the laser beam. This difficulty can be avoided by making the disc thinner. The thickness of 1.2 mm which is used for CDs would be too thick for a DVD disc and is not suitable for error-free reading of high-density data. Even a slight distortion can deteriorate the laser focus. The higher the data density, the greater is the risk of error during data reading. It can be expected that a thinner disc will reduce the impact of distortions on focus deterioration. But adoption of a thin disc makes it difficult to achieve CD compatibility. The solution compromising these issues is that the DVD disc consists of two discs. Each is 0.6 mm thick, and bonded back-to-back to achieve the strength and the same 1.2 mm thickness as CD's. Compared with a single 1.2 mm thick disc, the DVD configuration in which two discs are bonded is less prone to warping caused by heat or humidity. All external dimensions of the DVD are the same as those of a CD. Furthermore, both sides can be used. Based on the same reason, BDs use a substrate of 1.1 mm and a 0.1 mm cover layer.

The differences in capacity and data rate among the CD, DVD, and BD formats are mainly dependent on reducing laser beam spot size of the focus beam at the medium by reducing the wavelength and increasing the *NA*. Other factors were also found to affect the density improvement in DVDs and BDs, such as stronger modulation coding, signal processing, error correction, and tighter tolerances.

In the form of the read-only optical disc, such as CD-ROM, DVD-ROM, and BD-ROM, the data is pre-stored by the mastering process. Data are stored on the disc as a series of microscopic indentations ("pits", with the gaps between them referred to as “lands”). A laser is shone onto the reflective surface of the disc to read the pattern of pits and lands. Because the depth of the pits is approximately one-quarter to one-sixth of the wavelength of the laser light used to read the disc, the reflected beam's phase is shifted in relation to the incident beam, causing destructive interference and changing the reflected beam's intensity. This pattern of changing intensity of the reflected beam is converted into binary data.

Recordable optical discs or write once read many discs rely on using high power lasers to write marks that still can be read using a low power laser. The high power laser can locally modify the optical properties of the recording medium so as to provide a significant change of signal. On the contrary, the low power readout laser must not affect the properties of the media. Processes such as ablation caused by a dye layer are irreversible and thus are well suited for recordable optical discs.

Current rewritable discs such as CD-RW, DVD±RW, DVD-RAM, and BD-RE are all based on phase-change optical recording technology using phase-change materials for the recording layers.

12.4 Phase-change Optical Recording and Related Technologies

In phase-change optical discs, recording and erasing are achieved by the crystallographic structural changes of thin film phase-change materials heated by a laser pulse. The reproduction of recorded information takes advantage of the fact that phase-change materials have different optical indices in their crystalline and amorphous states, leading to different reflectivities [12.4].

In this section, the principle of phase-change optical storage and the widely used phase-change media will be discussed. The typical structures of phase-change discs and key performance parameters will also be presented. Then the related technologies of phase-change optical storage will be introduced.

Recording density, data transfer rate and overwrite cycle are the three important factors in phase-change optical storage. The recording density is related to the mark size. The smaller the mark size the higher the recoding density. There are many methods to increase the recording density such as using laser diodes with shorter wavelength, lenses with larger *NA*, mark edge recording, land and groove recording, dual layers recording, multilevel recording, near-field recording and super resolution near-field structure disc (Super-RENS, see also Chapter 13).

The maximum data transfer rate that can be achieved in phase-change optical storage is highly dependent on the phase transition speed of phase-change materials. By increasing the linear velocity of the disc and reducing the laser pulse duration, the dwell time of the laser spot decreases, leading to a shorter energy deposition time for phase transitions. Overwrite cycle is related to phase-change materials and disc structure. Repeated melting, crystallization and amorphization of phase-change media results in material segregation, stress buildup, micro-crack formation etc. These factors tend to reduce the data reliability and cyclability of phase-change media.

12.4.1 Phase-Change Optical Storage

12.4.1.1 Principle of Phase-Change Optical Storage

The principle of phase-change optical storage is based on the concept that some physical property of a microscopic area of the recording layer on the disc surface is altered due to crystallographic structure changes when the films are irradiated by laser pulses. The reproduction of the recorded information takes advantage of the difference in reflectivity due to the difference in refractive index and extinction coefficient between the two phases as shown in Fig. 12.3.

Although there are two possible types of phase change (one is between amorphous and crystalline phases and another is between two different crystalline phases) the one applied in rewritable phase-change optical discs is only the first type. Before recording data on the phase-change optical discs, the as-deposited

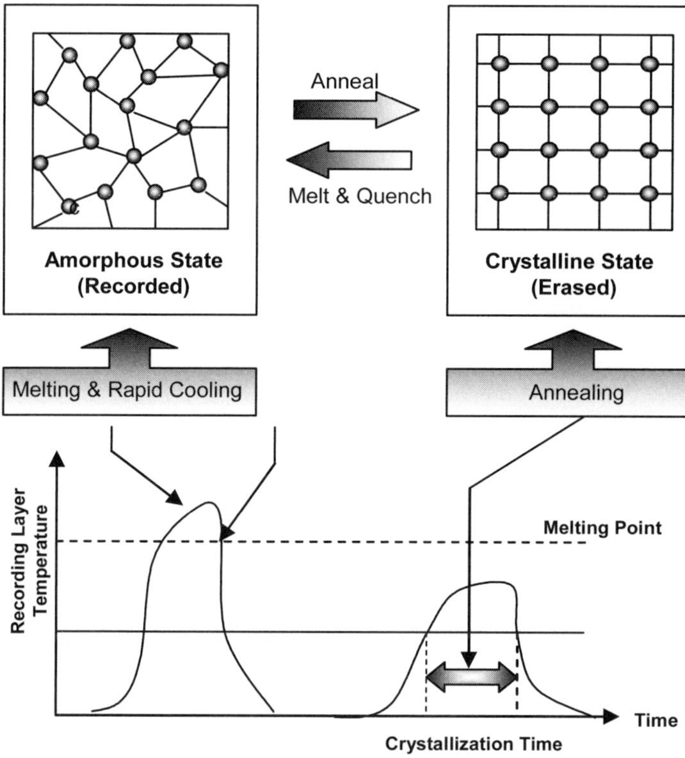

Fig. 12.3. Principle of phase-change recording and temperature profile of the recording layer for the writing and erasing process.

amorphous films are "initialized", i.e. transformed into the crystalline state. In the writing process, the amorphous state is achieved by heating the phase-change thin films with sufficient laser power above its melting point and then rapidly quenching it to room temperature. Because the atoms of the phase-change material after melting are in a disordered state and the cooling rate of the area irradiated by laser pulses is very high, the time is not sufficient for the atoms to be arranged into the ordered crystalline structure. Thus amorphous marks are formed. The absolute minimum quenching rates required for amorphization are different for various materials, ranging from 10^6 to 10^{11} K/s.

In the erasing process as shown in Fig. 12.3, the crystalline phase is realized by annealing the phase-change films at the temperature between the crystallization temperature T_c and the melting point T_m with a medium power laser irradiation. During the irradiation period, the atoms of the phase-change media are rearranged into an ordered structure; thus the amorphous region can be changed to the crystalline state. The phase changes in the phase-change optical discs are accomplished by using the irradiation of laser light which typically has a diameter on the order of 1 μm. When a laser beam having a 1 μm diameter moves on the recording thin films at a linear velocity of 10 m/s, as shown in Fig. 12.4, the irradiation time of a point on the film is only 100 ns. Hence, the available time for the energy deposition is also within this irradiation time duration.

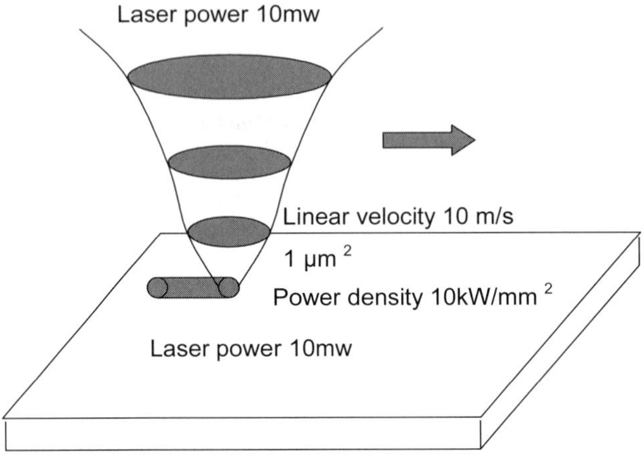

Fig. 12.4. Schematic representation of optical recording.

Assuming that the laser power is 10 mW, the power density of the light spot is on the order of 10 kW/mm^2. Since melting of a solid is virtually instantaneous, amorphization can be achieved by very short laser pulses, provided that the melting temperature is exceeded upon laser illumination. However because the atoms or molecules are re-arranged during crystallization, the time duration required for the crystallization process is determined by the physical characteristics of phase-change materials. In other words, each material has its own crystallization speed. Consequently, the materials for phase-change optical discs are required to have not only high thermal stability in the amorphous state, but also a high crystallization speed to enable that the atomic rearrangement process can be realized within the energy deposition time of 100 ns. The crystallization time of a material can be influenced by the adjacent layers and by the thermal properties of the layer stack.

The direct overwriting is a common performance in magnetic recording. However, it is an issue for optical recording due to the heating mode employed in current optical recording. If a thin film phase-change material has sufficiently high crystallization speed and the atoms can be rearranged within a short duration time of the laser beam, the direct overwriting can be accomplished by laser power modulation between a peak recording power level and a bias erasing level as shown in Fig. 12.5.

Before overwriting, there are some amorphous or recorded spots on the track. When peak laser power is applied, it raises the temperature above the melting point and quenches rapidly so that an amorphous mark is written on the same track. These amorphous marks are formed on the original spots of either an amorphous mark or the crystalline phase. When the bias power laser is applied, it heats up the phase-change material to a temperature between the crystallization

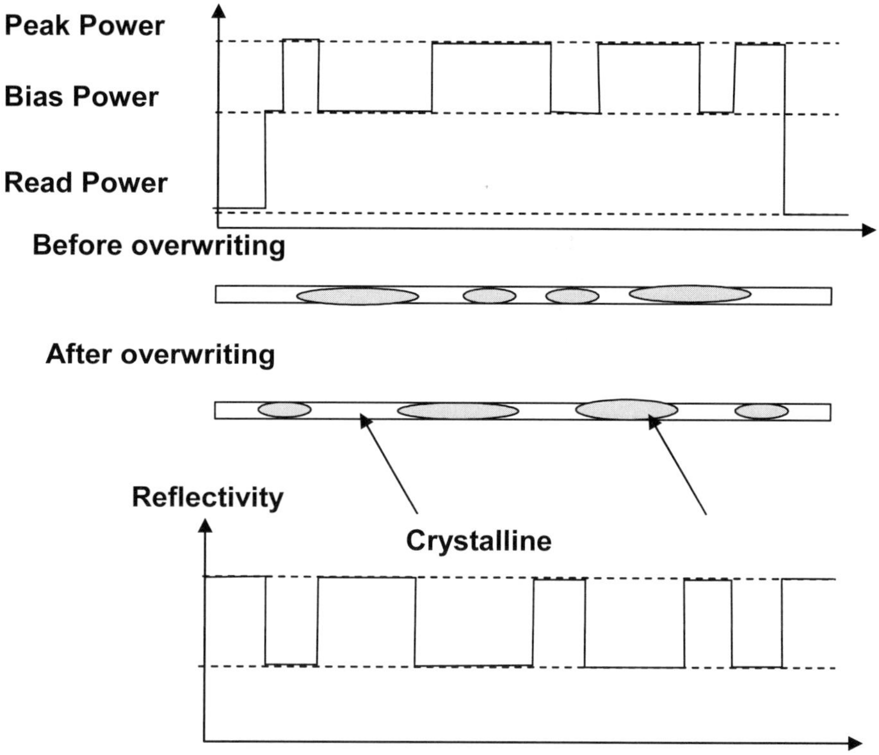

Fig. 12.5. Overwriting methods of phase-change optical recording.

temperature and the melting point so that a crystalline phase is formed. This overwrite method shows that no matter which phase the phase-change material is in before the overwriting, the films irradiated with the peak power become amorphous, and those irradiated with the bias power change to the crystalline phase.

12.4.1.2 Phase-Change Materials

Many materials, such as In-Sb, Ag-Zn, In-Sb,Ge-Sb-Te, Ge-Te-Sn, Sb-Se-Te, Ga-SeTe, In-Sb-Te, Ag-In-Sb-Te etc. have been reported to be potential candidates for phase-change optical data storage. Among all kinds of phase-change materials, stoichiometric compositions along the GeTe-Sb_2Te_3 pseudo-binary line (here referred to as GeSbTe) and quaternary AgInSbTe alloys are widely used in phase-change optical storage.

There is a well established classical crystallization theory (see also Chapter 7). According to this theory, crystallization is dependent on two distinct processes, the nucleation of small crystallites, followed by the subsequent growth. Both the nucleation rate and growth rate are functions of temperature. Coombs et al. have studied the crystallization behavior in phase-change materials using a static tester [12.5, 12.6]. They have demonstrated that this method can quantitatively characterize the crystallization of a phase-change thin film and separate the contribution from nucleation and growth. Phase-change materials have been classified into two types, so called nucleation-dominated materials and growth-dominated materials by distinguishing the relative contributions of nucleation and subsequent growth to the overall crystallization process [12.7]. If the nucleation rate of a material is higher than its growth rate, this material is called nucleation-dominated material. If the nucleation rate of a material is lower than its growth rate, this material is called growth-dominated material. Figure 12.6 shows the principle of nucleation-dominated and growth-dominated phase-change materials.

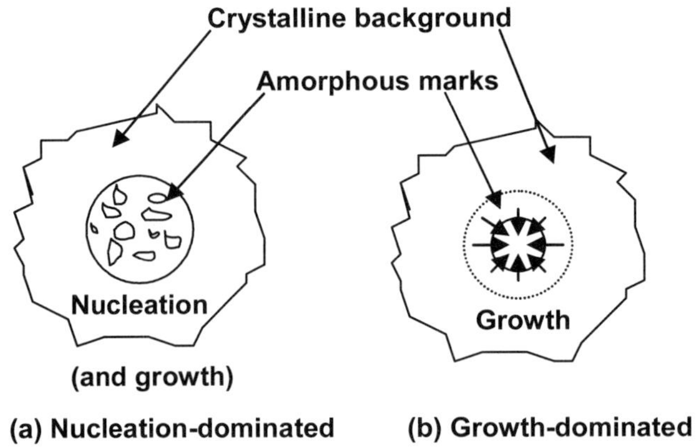

Fig. 12.6. Principle of nucleation dominated and growth dominated phase-change materials.

GeSbTe and AgInSbTe are two main phase-change materials used in current phase-change optical discs. Both phase-change materials have been characterized by measuring the complete crystallizing time (CCT) of an amorphous mark also called the complete erasing time (CET). CET is defined as the minimum duration of the erasing pulse for complete re-crystallization. The CET of AgInSbTe increases strongly with increasing mark size. Therefore, it is called a growth-dominated material. On the contrary, the CET of GeSbTe, which is a nucleation-dominated material, is only slightly dependent on the increase of the mark size.

12.4.1.3 Development of Phase-Change Optical Storage Media

The phase-change optical disc is one important type of rewritable optical disc available nowadays. When selecting a suitable material for the erasable phase-change recording layer, there are several important factors that must be considered:

1) Read/write cyclability. The material needs to be transformed between the amorphous and crystalline phases many times. Usually, materials without phase separation during the reversible phase transitions have good read/write cyclability.

2) Crystallization speed. The faster the speed at which the phase-change material crystallizes, the shorter the erasing time. To achieve rapid crystallization, the materials should have: (a) large atomic mobility in the amorphous and super-cooled states and (b) short atomic diffusion distance from the atomic location in the amorphous state to the lattice sites of the crystalline state. Atomic mobility is controlled by the viscosity of the super-cooled liquid. Generally speaking, a weak bond means a low viscosity force among atoms, which increases atom mobility and crystallization speed.

3) Melting point. Because the material must be molten by laser power, the melting point cannot be too high. However, if it is too low, the crystallization temperature will also be low and self-crystallization may occur resulting in the amorphous phase being unstable at room temperature. Hence, the materials are limited to those with melting points in the range of 500 ~ 1000 °C and glass transition temperatures about 1/2 to 2/3 of the corresponding melting point.

4) Thermal stability. The retention of the disc is dependent on thermal stability. High stability is needed for long life time of the disc at room temperature.

5) Optical constants. The material must be chosen such that it has enough absorption that shifts in the visible or near-infrared region with the phase transitions. Hence, metals and insulators are eliminated which leaves only semiconductors as promising materials. Amorphous semiconductors [12.8] which have limited long-range periodic order possess an optical behavior that is far different from their crystalline counterparts.

Research on phase-change optical storage media began many years ago. In 1968, S. R. Ovshinsky [12.9] discovered a rapid and reversible transition between a highly resistive (disordered structure) and a conductive state (ordered structure) in chalcogenide materials due to the reversible phase transition between amorphous and crystalline phases induced by an electric field. Soon after, a laser induced optical memory phenomenon in chalcogenide materials was observed by Feinleib et al. [12.10]. High speed and reversible phase transitions between an amorphous and a crystalline phase could be triggered by short laser pulses in a material with the composition of $Te_{81}Ge_{15}Sb_2S_2$, which led to a large change in optical reflection and transmission because of different refractive indices of the amorphous and crystalline phases. Significant progress was made in 1983 by Clemens [12.11]. By using low-doped Te films ($Te_{96.8}As_{3.0}Ge_{0.2}$) with thick capping layers, he realized reversible optical storage with over $4x10^4$ possible write/erase cycles.

In the same year, Takenaga et al. [12.12] claimed 10^6 write/erase cycles on a disc with 55 dB carrier-to-noise ratio (CNR) using a tellurium-oxide-based active layer. The amorphization was easily achieved and the data stability was longer than 1 year. However, the erasure time was longer than 1 μs and the observed optical property changes were mainly due to the segregation of Te from the TeO_2 matrix which had an adverse effect on the reversibility of the TeO_x based optical recording media. Furthermore, Te segregation from the TeO_2 matrix caused nucleation and crystal growth, resulting in an undesirable effect on the read back signal after recording. In 1985, Chen et al. [12.13] demonstrated for the first time that as-deposited amorphous $Te_{87}Ge_8Sn_5$ films could be optically switched between the crystalline and amorphous states more than 10^6 times. The reversibility was not limited by phase segregation, but by ablation. The medium had a high crystallization temperature and hence long data retention time. But for optical data storage applications, the minimum erasure time needed to be reduced and the crystallization temperature further increased. One year later, Chen et al. [12.14] investigated the laser induced and heating induced crystallization of $Te_{1-x}Ge_x$. They found that films with compound compositions, Te and GeTe, can be crystallized using laser pulses of less than 100 ns duration.

In 1991, Yamada et al. [12.15] found that stoichiometric compositions on the $GeTe\text{-}Sb_2Te_3$ pseudo-binary line, $GeSb_2Te_4$ and $Ge_2Sb_2Te_5$, are good candidates for phase-change media in optical storage. These materials have large optical contrast between the amorphous and crystalline phases. When they are sandwiched between heat-conductive ZnS layers, these materials can be transformed rapidly and reversibly between the amorphous and crystalline phases by laser irradiation with very short duration in less than 50 ns. The fast amorphization is due to the extremely high cooling speed of the sandwiched films of $\sim 10^6$ K/s, which permits the molten material to solidify while keeping the atomic distribution of the liquid state. The fast crystallization is attributed to their two-step crystallization processes [12.16-12.18]. When the amorphous phase of stoichiometric compositions on the $GeTe\text{-}Sb_2Te_3$ pseudo-binary line is crystallized it first transforms into a metastable face-centered cubic (fcc) structure at lower temperature, and then into a stable hexagonal (hex) structure at higher temperature. The metastable fcc structure has a high symmetric isotropic structure similar to that of the amorphous structure. Crystallization from the amorphous to the fcc structure occurs without phase separation of the stoichiometric compositions. Atoms need to travel only a short distance to transform the amorphous phase into the fcc crystalline lattice. This allows the fast crystallization of stoichiometric compositions on the $GeTe\text{-}Sb_2Te_3$ pseudobinary line from the amorphous to the fcc crystalline phase.

The AgInSbTe system was first proposed in 1992 by Iwasaki et al. [12.19] as a completely erasable phase-change material. This material has many advantages, especially in pulse width modulation recording, such as high erasability and high controllability of mark length. They suggested that the narrow widths of the written marks and no large crystalline grains in the periphery of the written marks are the major causes for the high erasability. Thus AgInSbTe based phase-change media could also be used in high density rewritable disc systems [12.20].

Other significant developments in phase-change media include In-Se-Te media having a short crystallization time of 0.2 μs reported by T. Nishida et al. in 1987 [12.21], and In_3SbTe_2 with reversible cycles above 10^5 presented by Maeda et al. in 1988 [12.22]. All of these media are good candidates for phase-change optical storage with high-speed erasing and long-term data stability. Furthermore, all these media can be overwritten directly with a single laser beam.

12.4.1.4 Disc Structure of Phase-Change Optical Disc

Figure 12.7 shows the typical structure of a phase-change optical disc of quadra-layered thin films on the polycarbonate substrate.

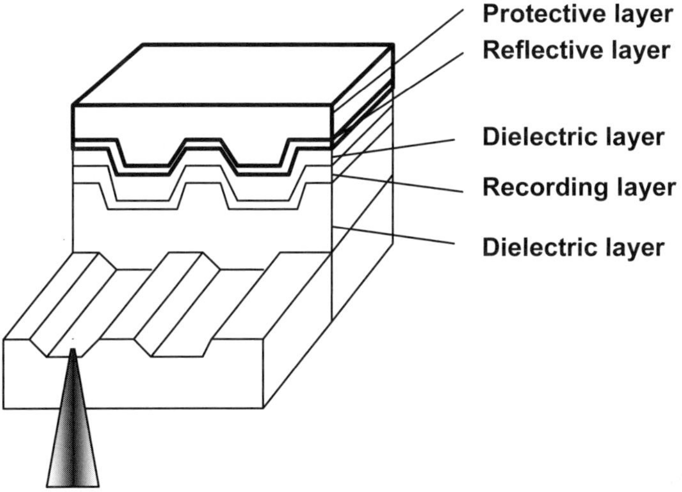

Fig. 12.7. The structure of a typical phase-change optical disc.

The phase-change layer is sandwiched between two dielectric protective layers made of ZnS-SiO_2, and a reflective layer made of Al alloy. The design of the individual layer thickness and the choice of material used are very important in the manufacturing of phase-change optical discs due to the following reasons:

- Optically, the layers are required to have large absorption efficiency at the wavelength of the laser light and large signal amplitude corresponding to the reflectivity difference between the amorphous and crystalline states.
- Thermally, the heating efficiency and the rapid quenching condition for amorphization have to be balanced and met by the disc structure design.
- Mechanically, the disc should withstand the thermal stress caused by the repeated heating and quenching cycles [12.23].

The protective dielectric layers and reflective layer have the following functions:

- Prevention of thermal damage to the substrate and mechanical protection against humidity.
- Optical modulation and enhancement of reflectivity difference between amorphous and crystalline phases.
- Controlling thermal performance during the writing and erasing processing.

For the reflective layer, the material should have properties that allow it to act as a mechanically protective layer. On the other hand as a heat sink, it prevents the thermal damage of the substrate as well as promotes rapid cooling of the phase-change layer by quenching it into the amorphous state during writing cycles. It also acts as a reflector of the laser light so as to achieve high sensitivity for measurements to attain the necessary carrier-to-noise ratio value.

The dielectric layer is made of the ZnS-SiO_2 compound. ZnS has a large refractive index of 2.4 which permits better laser spot size resolution while its high melting point of 1700 °C ensures that it is not melted by the laser heat. SiO_2 is added into ZnS to make an amorphous like structure with smaller grain size, to decrease its internal stress and reduce degradation during heating cycles of the phase-change recording process. ZnS-SiO_2 does not show grain growth phenomena even after 700 °C annealing and is therefore a thermally stable protective layer for phase-change optical discs, which allows millions of read/write cycles.

The lower dielectric layer is designed to be relatively thick to impede heat diffusion from the phase-change layer to the substrate, because the heat dissipated from the recording layer must not be allowed to damage the substrate which has a lower tolerance for heat. In addition, this layer provides anti-reflection and also functions to couple more laser light into the active layer to ensure a sufficient amount of heat in the recording spot of the films during writing.

The upper dielectric layer is designed to be relatively thin to allow the heat generated from the recording layer to dissipate quickly to the top metallic layer in order to achieve a rapid cooling effect. Thus, a thin layer improves the writing characteristics as the temperature of the molten recording spot can be decreased rapidly after writing, which allows the written spot to amorphize quickly.

A rapid quenching structure has been proposed to withstand the thermal stress caused by the repeated heating and quenching cycles [12.24]. In this rapid quenching structure, the phase-change, as well as the dielectric layer between the reflective and phase-change layer are deliberately made thin. This will allow the thermal energy produced in the recording phase-change layer to be diffused rapidly, leading to less damage to the other layers. As a result, it was reported that a million cycles of overwrite have been achieved [12.25]. The life spans of the phase-change optical discs have also been investigated and have been found to be sufficiently long for practical use. From an accelerated aging test, the life spans were estimated to be more than 60 years in the environment of 32 °C and 80 % relative humidity.

12.4.1.5 Models of Phase-Change Induced by Moving Laser Beam

A number of methods and models have been used to study the phase change induced by the moving laser beam. For example, the Kissinger equation and Johnson-Mehl-Avrami kinetic analysis are usually adopted to estimate the activation energy for crystallization and the reaction order of the crystallization process, respectively (see also Chapter 5). Ortiz et al. and Solis et al. have described the processes involved in laser-induced crystallization of thin films [12.26, 12.27]. At high temperature, the driving force for nucleation is small and the transformation proceeds slowly. At low temperature, the driving force for nucleation is large, but the atomic mobility is low. Peng et al. developed an empirical model based on their experimental results [12.28].

All of these models are based on the homogeneous nucleation and subsequent growth. By considering the contribution of nucleation and subsequent growth, phase-change materials have been characterized into two types, so called nucleation-dominated material and growth-dominated material. The characterization of both types of phase-change materials is related to the static phase change. From the material point of view this classification is correct. However, if a moving laser beam scans across the boundary between crystalline and amorphous regions, edge growth from the boundary between the amorphous area and the crystalline surrounding plays an important role during the crystallization process. Figure 12.8 clearly reveals the difference between static and dynamic irradiation of laser beam. In the static case, the laser beam does not move, while in the dynamic case the laser beam moves along the track. The dark areas are in the amorphous state induced by the laser beam.

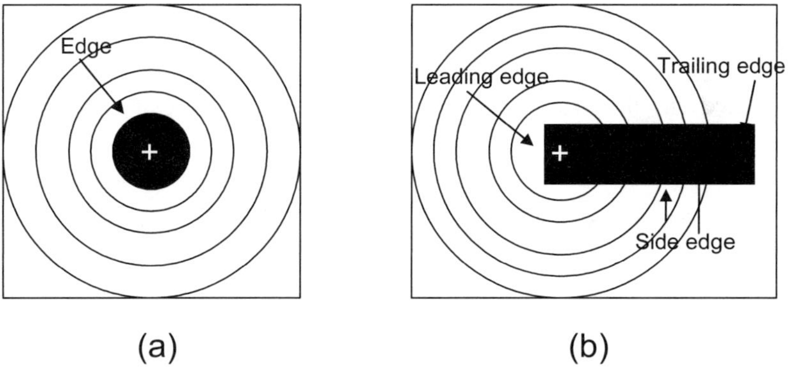

Fig. 12.8. Difference between static (a) and dynamic (b) irradiation of the laser beam

The crystallization in static and dynamic states is discussed in the following, respectively. For growth-dominated materials, the amorphous area edge growth is the main reason. For nucleation-dominated materials such as GeSbTe, the amorphous area edge growth also plays an important role. The relation between erasing of marks and mark length has been studied [12.29]. For the long amorphous areas the crystallization was a mixed case in which both the nucleation and growth and the growth from the boundary contributed to the phase change process. Usually the edge growth contributed at the initial stage while the nucleation and growth of the nucleus in the amorphous area center made the major contribution in the latter crystallization process. For small amorphous regions, the crystallization process was mostly dominated by the growth mechanism. The contribution of the nucleation and growth in the amorphous area center was very small. Based on these results, an edge growth-dominated nucleation assisted model was proposed to describe the dynamic crystallization mechanism of a moving laser induced phase change [12.30]. It can be summarized as follows: for the moving laser beam induced phase change, four factors are contributing to the crystallization, which are named leading edge growth, trailing edge growth, side edge growth, and nucleation and growth in the amorphous area center as shown in Fig. 12.8(b). These four factors are all related to the maximum temperature and the amorphous area size. The experimental and simulation results show that the growth from the edge plays an important role during the dynamic crystallization process. For a longer amorphous area, both the growth from the leading, sides and trailing edges and the nucleation and subsequent growth in the amorphous area centers contribute to the crystallization process. The growth from the leading edge is faster than that at the side edge. For a short amorphous area, crystallization is a growth process from the boundary between the amorphous area and the crystalline surrounding.

According to this model it can be deduced that both the nucleation-dominated model and the growth-dominated model can be viewed as two extreme cases of the edge growth-dominated nucleation assisted model. From this model we can also see that the size of the amorphous area plays a very important role. If the radius decreases, especially at the nanometer scale, the edge growth will contribute more to the dynamic crystallization process. The amorphous area edge growth and the nucleation-growth in the amorphous area center strongly depend on the amorphous area width and length, which are dependent on the laser wavelength, *NA* of the objective lens, the laser pulse length and the laser beam moving speed.

12.4.2 Techniques for Phase-Change Optical Storage

Numerous technologies have been developed to improve phase-change optical storage: (1) to achieve higher recording density; (2) to increase the data transfer rate; (3) to achieve better overwrite cyclability [12.31]. Advancement has been made in the area of the optical system, the coding/modulation, the disc structure, and the signal processing schemes. Many different methods have been proposed, such as land/groove recording, shorter laser wavelength recording, near-field phase-change optical recording, multilevel phase-change recording, super-resolution near-field phase-change optical recording, dual layer recording etc. The most promising techniques will be discussed in the following sections.

12.4.2.1 Short Wavelength Laser Diodes

From Eqs. (12.1) and (12.2) it can be easily understood that the obvious way to increase the storage density is to use a shorter wavelength laser diode. The earliest optical storage products adopted a laser diode with an infrared wavelength of 830 nm down to 780 nm, simply because these were the only available laser diodes with the required reliability, output and affordable cost in the 1980s. Moving from CD to DVD, the simplest way to increase density was to use a red laser with shorter wavelength available in the 1990s. In the 2000s blue laser diodes of 405 nm were used for BD.

Continued advances to even shorter wavelength have been made with novel diode laser materials and concepts. GaN lasers with wavelengths as short as 375 nm have been explored. However, for further reducing the wavelength of the laser, such as using a deep UV laser, almost all of the current components would need to be changed. This will result in a significant cost increase which is not very practical.

12.4.2.2 Large Numerical Aperture (*NA*)

Besides reducing the laser wavelength, an increase in *NA* of the objective lens can also reduce the spot size and increase areal density. Indeed, during the transition from CDs ($NA = 0.45$) to DVDs ($NA = 0.6$), and to BDs ($NA = 0.85$), the increase in storage density is partially attributed to the increase in *NA*. There are, however, practical limits to the increase in *NA*, such as manufacturing tolerances for diffraction limited optics. A very critical issue is the depth of focus. The depth of focus δ is directly proportional to the wavelength λ, but inversely proportional to the square of *NA*:

$$\delta \propto \left(\frac{\lambda}{NA^2}\right) \tag{12.3}$$

In order to increase the density, the recording media layer should be as close as possible to the focused diffraction-limited readout beam. However it results in a smaller manufacturing tolerance for the driver and requires a fast focusing servo. These issues become more critical and difficult as the optical depth of focus decreases.

12.4.2.3 Land/Groove Recording

In the early rewritable phase-change discs, amorphous marks were recorded only on the grooves, whereas the lands served as the guides for tracking and the suppression of heat flow from the adjacent groove tracks. Recording on both land and groove area is one of the simplest methods for doubling the track density, as shown in Fig. 12.9.

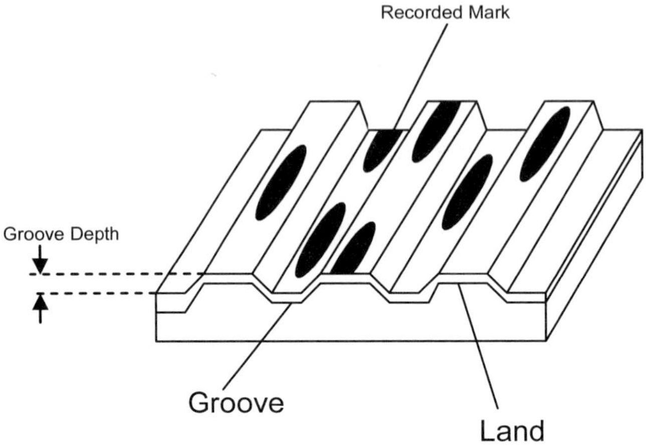

Fig. 12.9. Schematic drawing of land and groove recording method.

Both land and groove have similar widths and are available for recording. The advantages of this recording method are as follows:

- Doubling track density with conventional optical heads and drivers
- Compatible to conventional disc fabrication process.

However, this method of track density doubling leads to the problem that interference between adjacent tracks occurs, namely optical cross-talk and thermal cross-erasure. The former is caused by the interference of light reflected from both land and groove areas during reading. The latter arises due to excessive heat flowing from the laser-irradiated area, for example, from the groove area to the land area, and results in an accidental erase of the existing spot there.

12.4.2.4 Write Strategy

The accuracy of recording marks on an optical disc is related to many different physical parameters such as environmental conditions in a drive, beam profiles of pickup heads and thermal properties of the recording layer. Precise control of mark shapes maintains a low bit error rate (BER) of the reproduced signal. To record good mark shapes, it is necessary to adjust the parameters of a recording light pulse pattern which is called write strategy.

As an example we use DVD-RAM to introduce the write strategy. Figure 12.10 shows the wave form of the write strategy of DVD-RAM. The parameters used to define the waveform are pre-stored on the disc.

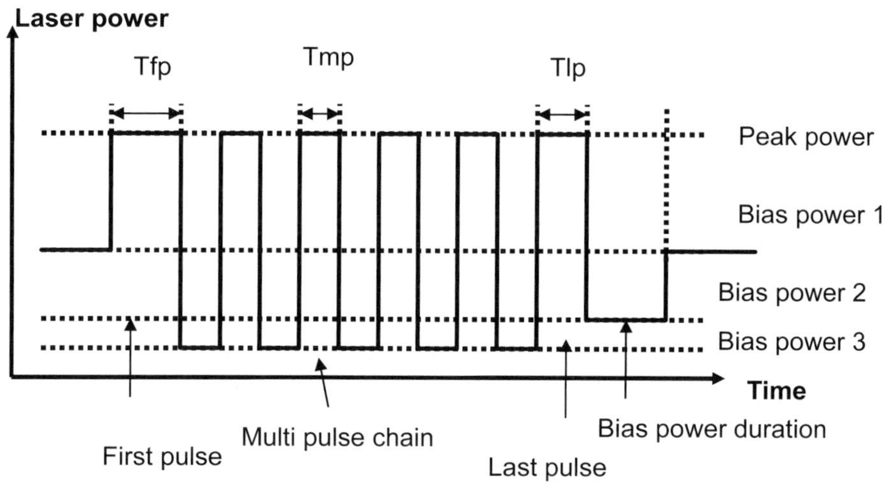

Fig. 12.10. The wave form of the write strategy of DVD-RAM. T_{fp} , T_{mp} and T_{lp} is the duration of the first pulse, multi pulse chain, and the last pulse, respectively.

In order to achieve high density, many methods have been used to develop new write strategies. Sequenced amplitude margin (SAM) is a procedure for evaluating the error rate performance of Partial-Response Maximum Likelihood (PRML) systems. Several SAM-based write strategy compensation methods have been proposed. However, the conventional methods are only for adjusting the recording pulse, the parameters of which are the timing and the length. Adachi et al. [12.32] have proposed a method for adjusting the write strategy using SAM (AWSAM). AWSAM is based on novel indices calculated from the sequenced amplitude margin. In AWSAM, the recording power adjustment is achieved by utilizing the equality between the indices corresponding to the long marks and the long spaces. The recording pulse adjustment is accomplished by utilizing the closeness of the indices classified by the mark lengths. These processes were implemented by detecting the amounts and directions of deviations in recording parameters. Experimental results showed that AWSAM enabled the determination of both optimum recording power and optimum recording pulse of each mark length, resulting in good recorded marks. A low bit error rate and a wide recording power margin were realized.

12.4.2.5 Cross Talk

In order to achieve higher capacity and super-high density optical discs, it is indispensable to narrow the track pitch. The track pitch determines the distance between the recorded marks on a track and the adjacent track. To narrow the track pitch, the following two technological issues should be considered. The first one is the interference signal from the adjacent track, which is called "optical cross talk".

Another one is the erasure of the adjacent track, which is called “thermal cross talk”.

Milster et al. [12.33] have analyzed cross talk in the read out signal resulting from narrow track pitches with Babinet decomposition of the light reflected from the disc surface. The analysis takes the inter-modulation of the scan-track mark pattern and the adjacent track mark pattern into account and shows its significance. The analysis presents a fundamentally new way to conceptualize cross talk. In this analysis, effects due to different terms of the Babinet decomposition can be studied separately or combined in meaning signal groups that relate directly to measurements. Quantification of the cross talk effects were presented with respect to shifts of signal transitions.

Morita et al. [12.34] have developed a super-high density optical disc using the deep groove method. Optical discs with a deeper-than-usual groove were manufactured. It was shown that for deeper groove less cross-erasure occurs. By adjusting the groove depth, cross erasure is suppressed.

BD with a capacity of 50 GB/layer has been reported by reducing track pitch to 0.24 μm, while that of the conventional BD is 0.32 μm. The cross talk becomes serious due to high track density. Therefore, Kudo et al. [12.35] have proposed a cross talk cancellation method to minimize the error between the cross talk canceled signal and the target signal. The main characteristic of the proposed method is that the target signal, which has no cross talk effect, is generated from the binary output of an adaptive-PRML decoder, and is used to evaluate the cross talk error in the readout signal. A series of simulations revealed that this method of cross talk cancellation is effective down to a track pitch of 0.2 μm. The experimental results obtained using this method showed that the BER is reduced to 1/100. A high-density recording of a 50 GB/layer was demonstrated using this method [12.35].

12.4.2.6 Super Resolution

The resolution of the conventional optical disc playback process, i.e., reading a conventional optical disc with a conventional optical head, is limited by diffraction. To overcome this resolution limit, one way is to minimize the point spread function of the optical pickup; this technique is called super resolution, where amplitude or phase modulation is usually applied to the field in the aperture of the pickup [12.36], see also Chapter 13.

Yamanaka et al. reported on a high density optical disc using the super resolution technique [12.37]. The focus spot size on the recording media is reduced to 80 % of that of a conventional optical recording media by placing a double rhomb prism in the collimating beam, which improves the light power loss. The readout signal degradation affected by focused beam side-lobes is suppressed spatially by extracting the main lobe region from the reflected light using a slit. The linear density has been improved by a factor of 1.2 compared to a convenient optical disc.

Recently, Wang et al. proposed to generate a super resolution beam with non-diffraction characteristics, [12.38]. This beam has a sub-wavelength scale and uniform axial intensity within an arbitrary limited region. Thus the defocus and the spherical aberrations of the imaging system are eliminated. This was achieved by placing a multi-belt pure phase type binary optical element on the pupil. The method mentioned in this paper is applicable to both paraxial and high numerical aperture optical systems.

12.4.2.7 Multilevel Phase-Change Recording

Multilevel data storage is very attractive due to its increased disc density as well as data transfer rate. Many attempts have been made to realize multilevel recording in optical disc recording. With multilevel recording technology, the feasibility to obtain 34 GB on the existing high-*NA* blue laser systems has been demonstrated [12.39]. Preliminary experiments on a 12-level multilevel code [12.40] have shown the potential to achieve over 40 GB on a single sided, single layer 12 cm disc.

By controlling the pulse width and power amplitude, multilevel reflection modulation recording can be realized. Different levels of input power give rise to different reflection levels. In order to study the mechanism for forming the multilevel reflection, partial crystallization effect and size effect were studied [12.41]. The partial crystallization effect was studied using differential scanning calorimeter (DSC) measurements and X-ray diffractometer (XRD) measurements. The size effect was studied by simulation and measurement. Based on the simulation and experimental results the combination of size effects and partial crystallization effects was proposed as the main reason to cause the multilevel reflection for rewritable phase-change optical discs.

12.4.2.8 Dual Layer Phase-change Optical Recording

Optical storage technologies, where individual bits are stored as distinct optical changes on the surface of a recording medium, are approaching physical limits where individual bits may be too small or too difficult to store and retrieve. Storing information throughout the volume of a medium instead of just on its surface offers an alternative way to achieve high capacity. Three-dimensional optical storage opens up another direction to increase the capacity of storage media. Trade-offs between density and data rate make it possible, at least in principle, to forgo some of this density to obtain very high data rates.

Several factors in volumetric optical storage are important, such as focusing on the stored layer and tracing along the track. These can be distinguished by the method used to address the stored data. The techniques include adjusting focus to access data on a particular layer; and using an interferometer (sensing differences in path length) to address a layer.

A complementary approach to obtain high storage capacities is to increase the number of storage layers. An important aspect of such discs is that all recording layers should be accessed from the same side of the disc. Bruneau et al. proposed solutions for the design of a dual layer phase-change rewritable optical disc [12.42]. The feasibility of such media was examined from different angles, such as recording possibility, materials aspect and readability. The intermodulation of the read information on two layers was suppressed using sum and differential detection schemes.

In 1999, a rewritable phase-change optical disc with dual layer structure was demonstrated [12.43]. In order to obtain equivalent signal amplitude and recording sensitivity from the first medium located on the beam incident side and the second medium, a high transmittance first medium with a very thin phase-change film of 7 nm without reflection layer, and a high recording sensitivity with 10 nm phase-change film and high reflectivity second medium with a semi-transparent front layer and a reflection layer were used. A recording density corresponding to 8.5 GB on a 120 mm disc with DVD substrates has been achieved.

One of the challenges for developing dual layer phase-change media is the requirement of semitransparent recording stacks. The transmission of the recording stack can be greatly enhanced by reducing the thickness of the reflector/heat-sink layer. However, to maintain the sufficiently high cooling rates that are necessary for the formation of melt-quenched amorphous marks, a thin silver heat sink (10 nm) was found to be indispensable [12.44]. For recording stacks utilizing a thin Ag layer, a substantial fraction of the incident light is absorbed in the phase-change layer, which has a relatively high imaginary part of the complex refractive index. Therefore, a further gain in transmission can be achieved by reducing the thickness of the phase-change layer as well.

Recently rewritable dual layer phase-change optical discs applying a blue laser of 405 nm wavelength were developed and demonstrated a maximum capacity of 50 GB [12.45].

12.4.2.9 Superlattice-like Phase-change Optical Disc

Phase-change materials are the key for rewritable phase-change optical discs. In order to optimize the disc performance, various factors need to be considered regarding the phase-change media. Unfortunately, these factors are interrelated and partly contradictory. Therefore, although in nature thousands of phase-change materials are available, only two or three types are basically suitable for practical applications. In order to increase the recording speed, a material with a high crystallization speed, which will result in a high phase-change speed, is needed. However, such a material is generally not stable in the amorphous phase against crystallization. In applications, the usual method to solve this contradiction is to sacrifice the speed in order to assure the stability.

Recently a "superlattice-like" (SLL) structure has been proposed to develop phase-change media with an artificial multilayer structure as shown in Fig. 12.11 [12.46]. The basic idea is to alternatively deposit two phase-change materials, one with a high crystallization speed and the other with a relatively low crystallization speed but a high thermal stability, to form the SLL structure.

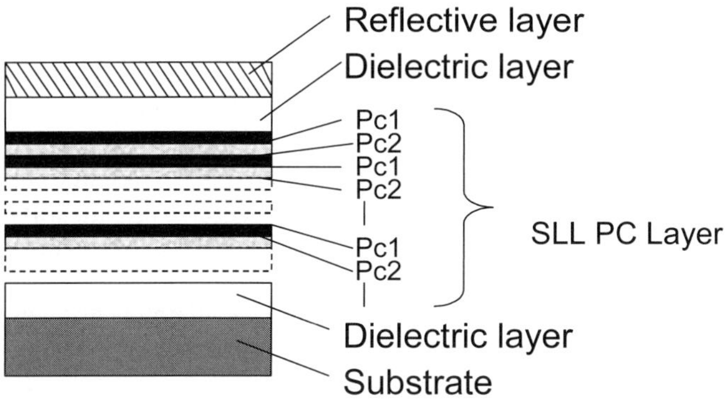

Fig. 12.11. The structure of SLL media. The structure uses two phase change materials, Pc1 and Pc2 to form SLL structure.

Thermal properties of superlattice structures have recently attracted increasing attention due to their excellent properties such as thermal management properties.

Theoretical and experimental studies have revealed that thermal conductivities along both in-plane and cross-plane directions deviate significantly from the corresponding bulk materials due to the interface phonon scattering and phonon confinement effects [12.47-12.49]. This reduction of thermal conductivity in the SLL structure can be used to confine the heat within the phase-change layer of SLL disc.

The SLL structure has demonstrated both a high speed and a good stability. The optical disc with this structure can operate at high speed due to the first material; meanwhile, it can maintain good stability due to the second material. More importantly, the new structure can incorporate the unique properties of two non-promising materials and transform them to achieve outstanding disc performance. With this structure, a new way was found to explore phase-change materials for optical disc.

12.4.2.10 Initialization Free Phase-change Optical Disc

One of the most important advantages of optical discs is their low costs. Thus the manufacturing costs of the optical disc are a very important issue. The as-deposited phase-change recording layer fabricated by a sputtering system is in the amorphous state, but the background of the phase-change optical discs should be in the crystalline state requiring pre-crystallization of the disc. The main reason is that if the amorphous state is used as the background state of the phase-change optical disc, in order to realize direct overwriting, high bias power P_m (melting power) and medium mark writing power P_w (crystallization power) have to be used. Otherwise the direct overwriting cannot be achieved and a mark ring will be left after erasing the recorded information because of the Gaussian distribution of the laser beam energy. As a result the laser power consumption is high.

This crystallization process from the as-deposited amorphous state to the crystalline state is called "initialization". In order to initialize the phase-change disc, several technologies were developed. The most commonly used method is laser beam irradiation. A big spot of a high power laser beam is used to irradiate the as deposited disc and many tracks can be crystallized at the same time. However, the initialization time is much longer than that of other manufacturing processes, such as bonding and sputtering. Normally 6 to10 initializers are required for one production line in order to reduce the manufacture time. The initialization process thus becomes one of the bottlenecks for phase-change disc mass manufacturing.

In order to reduce manufacture cost, an initialization free technology has been developed [12.50]. The basic concept is to add additional layers adjacent to the phase-change layer to induce phase-change during the sputtering process as shown in Fig. 12.12 (a). An additional top layer is also added to further enhance crystallization as shown in Fig. 12.12(b) [12.51].

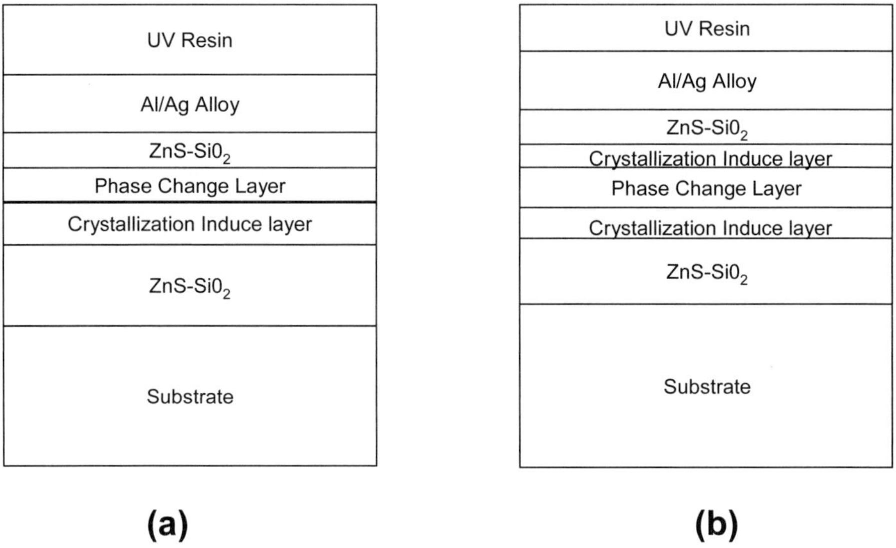

Fig. 12.12. Disc structure of the initialization-free phase-change optical disc (a) with one crystallization inducing layer (b) with two crystallization inducing layers.

The additional layers lead to the effect that the sputtered films are now crystallized during the sputtering process itself. Consequently it is unnecessary to use initializers. As a crystallization enhancement layer, the materials should meet the following requirements: (1) low crystallization temperature, (2) high crystallization speed, (3) lattice constant similar to the phase-change media, (4) similar crystalline structure. The first and second condition leads to the crystallization of the additional layer during sputtering. The third and fourth requirements ensure surface crystallization of the phase-change recording layer due to small interfacial free energy and small lattice miss-match.

For nucleation-dominated GeSbTe materials and growth-dominated AgInSbTe materials, Sb_2Te_3 and Sb were selected as the additional layer materials, respectively [12.51, 12.52]. The combination of surface crystallization induced by the additional layer(s) and a temperature increase of the substrate during the sputtering process is believed to be the main reason to induce crystallization during the sputtering process. Initialization-free DVD-RAM discs with Sb_2Te_3 and BD discs with additional Sb layers were successfully fabricated that can fully meet the industry standard [12.52].

12.4.2.11 Near-field Phase-Change Optical Storage

Near-field optical technique is one of the promising candidates showing the possibility for realizing ultra-high density beyond the diffraction limit. This method is based on the usage of an optical near-field generated from a sub-wavelength aperture. The near-field optical system, which makes use of the solid immersion lens (SIL), is very attractive for ultra-density optical data storage.

Another approach which goes beyond the optical diffraction limit and realizes ultra high-density optical data storage is the super-resolution near-field disc. This will be described in more detail in Chapter 13.

12.5 The Future of Optical Storage

In the era of the internet, massive amounts of information and multimedia can be easily accessed in every corner of the world. The demand for larger storage capacities of smaller devices, lower cost and faster storage media is ever increasing.

Current phase-change optical data storage technologies offer unparalleled capacity and reliability for long-term data storage and archival applications. Today, the rewritable phase-change optical discs in the form of the DVD-RAM and DVD±RW have become very popular. Recently, the 3rd generation BD have come to the market.

The pursuit of optical discs with larger storage capacities and faster data transfer rate will continue. NHK has proposed and demonstrated ultra-high-definition vision or UHDV, UHDTV and UHD, with a resolution of 7,680 × 4,320 pixels (16 times the pixel resolution compared to existing HDTV). The main driving force for 4th generation optical storage is super-high definition TV.

To further increase the storage capacity, data should be recorded in even smaller marks. Currently the working unit area of the data storage media is on the order of nano-scale. At the nano-scale, extreme dimensional and nano-structural constraints and the greater and greater influence of interfaces will cause the deviation of the phase-change behavior from that of bulk materials. Recently Shi et al. have discussed phase-change behavior of nanoscaled materials and the effect on data storage applications [12.53].

For decades, the major driving force for optical disc development has been to increase the storage density by reducing the spot size, moving to shorter wavelengths and larger *NA*s. Current BDs with 25 GB/side use a wavelength of 405 nm and an *NA* of 0.85. To increase the disc capacity, possible approaches are to reduce the wavelength of the laser diode and increase the *NA* of the objective lens. However, for 4th generation optical storage, it is not practical to further increase the density by using shorter wavelengths because almost all of the components will need to be changed if UV light is used, even though such light sources have not even been developed yet.

Researchers are seeking alternative options for the next generation optical storage. Until now, a few solutions have been proposed and listed in the INSIC optical roadmap 2006 [12.54] and ISOM optical disc roadmap 2006 [12.55] including near field recording, volumetric recording, holographic recording, and super resolution near field optical recording. Each technology has its advantages and drawbacks.

To explore the possible solutions of future optical recording, it is worthwhile and important to rethink the challenge from a fundamental point of view. The electric field of light can be described as

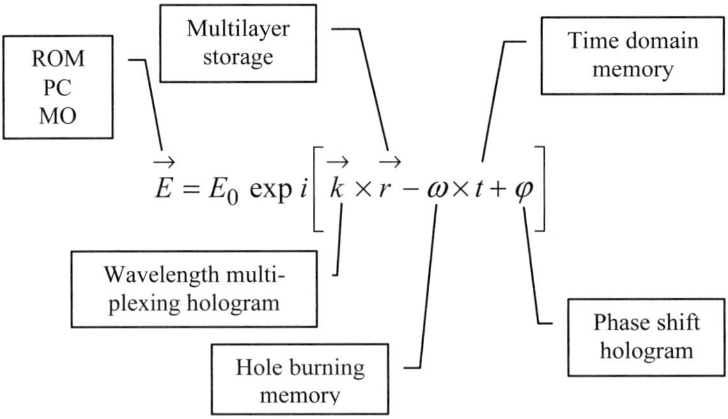

$$\vec{E} = E_0 \exp i\left[\vec{k} \times \vec{r} - \omega \times t + \varphi\right] \tag{12.4}$$

In fact optical recording technologies were developed by using different light parameters as shown in Eq. (12.4). From Eq. 12.4, it can be seen that with light having multiple parameters distinguishes it from other memory technologies. This special feature should be utilized. One possible solution is multi-dimensional multilevel recording [12.41]. Current multilevel recording technology normally uses one parameter. But the detection signal can show different levels at the same focused laser spot, for instance, multilevel reflection modulation optical recording. But even in this kind of technology, only one parameter of light is used and the level and consequently density increase will be limited. It can be expected that expanding one-dimensional multilevel recording to multi-dimensional multilevel recording can further increase the density significantly. The key of the new concept proposed in this work is to make use of different parameters (such as modulation amplitude, polarization, frequency, reflection, refraction, time, mark width, mark length and so on) of light to detect multi-dimensional multilevel signals. Data are recorded in the media in the form of marks, which are characterized by at least two physical properties of recording materials of the media. The idea of two dimensional multilevel recording is to create another group of multilevel signals based on the pre-multilevel signal. This idea can be further expended to multi-dimensional multi-level recording. The principle is shown in Fig. 12.13.

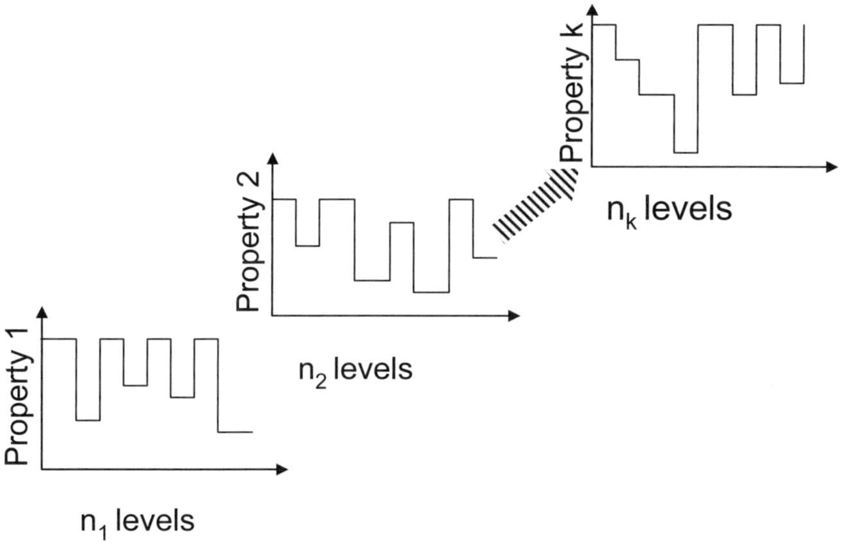

Figure 12.13. The principle of multi-dimension multilevel recording.

Different light parameters 1, 2, ..., k can be used, where k is the number of parameters used. For each parameter, by controlling the laser beam parameters, a multilevel signal can be formed as n_1, n_2, ..., and n_k. Each parameter can be viewed as one dimension. Thus the total signal level number is $n_1 \times n_2 \times \ldots \times n_k$ for every spot. By introducing multi-dimensional multilevels, the number of signals for every spot can be significantly increased by $n_1 \times n_2 \times \ldots\ldots \times n_k$ times.

In summary, the development of 1st, 2nd, and 3rd generation optical discs was introduced with the focus on phase-change rewritable optical discs. The 4th generation and the possible development trends for future optical discs were also discussed.

References

[12.1] Blu-ray standard

[12.2] CD standard

[12.3] DVD standard

[12.4] Wuttig, M. and Yamada, N., Phase-change materials for rewritable data storage. Nature Mater. **16**, 824-832 (2007)

[12.5] Coombs, J. H., Jongenelis, A. P. J. M., Es-Spiekman, W.V. and Jacobs, B.A.J.: Laser-induced crystallization phenomena in GeTe-based alloys. J. Appl. Phys. **78**, 4906-5917 (1995)

[12.6] Coombs, J. H., Jongenelis, A. P. J. M., Es-Spiekman, W.V. and Jacobs, B.A.J.: Composition dependence of nucleation and growth. J. Appl. Phys. **78**, 4918- 4928 (1995)

[12.7] Zhou, G.F., Borg, H.J., Rijpers, J.C.N. and Lankhorst, M.: Crystallization behavior of phase change materials: comparison between nucleation- and growth-dominated crystallization, Tech. Digest of Optical Data Storage 2000, 74-76 (2000)

[12.8] Tauc, J.: Amorphous and Liquid Semiconductors. London, New York, Plenum (1974)

[12.9] Ovshinsky, S. R., Reversible electrical switching phenomena in disordered structures. Phys. Rev. Lett. **21**, 1450-1453 (1968)

[12.10] Feinleib, J., Neufville, J.D., Moss and S.C., Ovshinsky, S.R.: Rapid reversible light-induced crystallization of amorphous semiconductors. Appl. Phys. Lett. **18**, 254-257 (1971)

[12.11] Clemens,P. C.: Reversible optical storage on a low-doped Te-based chalcogenide film with a capping layer. Appl. Opt. **22**, 3165- 3168 (1983)

[12.12] Takenaga, M., Yamada, N., Nishiuchi, K., Akahira, N., Ohta, T., Nakamura, S. and Yamashita, T.: TeOx thin films for an optical disc memory. J. Appl. Phys. **54**, 5376-5380 (1983)

[12.13] Chen, M., Rubin, K.A., Marello, V., Gerber U.G. and Jipson, V.B.: Reversibility and stability of tellurium alloys for optical data storage applications. Appl. Phys. Lett. **46**, 734-736 (1985)

[12.14] Chen, M., Rubin, K. and Barton, R.: Compound materials for reversible, phase-change optical data storage. Appl. Phys. Lett. **49**, 502-504 (1986)

[12.15] Yamada, N., Ohno, E., Nishiuchi, K., Akahira, N. and Takao, M. : Rapid-phase transitions of GeTe-Sb_2Te_3 pseudobinary amorphous thin films for an optical disk memory. J. Appl. Phys. **69**, 2849-2856 (1991)

[12.16] Mao, Z. L., Chen H. and Jung A.L. : The structure and crystallization characteristics of phase change optical disk material $Ge_1Sb_2Te_4$. J. Appl. Phys. **78**, 2338-2342 (1995)

[12.17] Matsunaga, T. and Yamada, N.: Structural investigation of $GeSb_2Te_4$: A high-speed phase-change material. Phys. Rev. B, Vol. **69**, 104111 1-8 (2004)

[12.18] Naito, M., Ishimaru, M., Hirotsu, Y. and Takashima, M.: Local structure analysis of Ge-Sb-Te phase change materials using high-resolution electron microscopy and nanobeam diffraction. J. Appl. Phys. **95**, 8130-8135 (2004)

[12.19] Iwasaki, H., Ide, Y., Harigaya, M., Kageyama and Y., Fujimura, I.: Completely erasable phase change optical disk. Jpn. J. Appl. Phys. **31**, 461-465 (1992)

[12.20] Shinotsuka, M., Shibaguchi, T., Abe, M. and Ide, Y.: Potentiality of the Ag–In–Sb–Te phase change recording material for high density erasable optical discs. Jpn. J. Appl. Phys. **36**, 536-538 (1993)

[12.21] Nishida, T., Terao, M., Miyauchi, Y., Horigome, S., Kaku, T. and Ohtaet, N.: Single-beam overwrite experiment using In-Se based phase-change optical media. Appl. Phys. Lett. **50**, 667-669 (1987)

[12.22] Maeda, Y., Andoh, H., Ikuta, I. and Minemura, H.: Reversible phase-change optical data storage in InSbTe alloy films. J. Appl. Phys. **64**, 1715-1719 (1988)

[12.23] Jong, C.A, Weileung, F., Lee, C.M. and Chin, T.S.: Mechanical properties of phase-change recording media: GeSbTe films. Jpn. J. Appl. Phys. **40**, 3320-3325 (2001)

[12.24] Ohta,T., Inoue, K., Uchida, M., Yoshioka K., Akiyama, T., Furukawa, S., Nagata K. and Nakamura, S.: Phase change disk media having rapid cooling structure. Jpn. J. Appl. Phys. **28** Suppl. 28-3, 123-128 (1989)

[12.25] Ohta, T., Uchida, M., Yoshioka, K., Inoue, K., Akiyama, T., Furukawa, S., Kotera, K. and Nakamura, S.: Million cycle overwritable phase change optical disk media. SPIE Proc. **1078**, 27-29 (1989)

[12.26] Ortiz, C. and Blatter. A.: Laser irradiation of amorphous thin films. Thin Solid Films **218**, 209-218 (1992)

[12.27] Solis, J., Rubin, K.A. and Ortiz C. : Structural and optical transformations by laser irradiation of InSb-based thin films , J. Mater. Res. **5**, 190-201 (1990)

[12.28] Peng, C., Cheng, L. and Mansuripur, M.: Experimental and theoretical investigations of laser-induced crystallization and amorphization in phase-change optical recording media. J. Appl. Phys. **82**, 4183-4191 (1997)

[12.29] Shi, L.P, Chong, T.C., Hu, X. and Yao, H.B.: Study of the dynamic crystallization behavior of GeSbTe phase change optical disk. Jpn. J. Appl. Phys. **42**, 841-847 (2003)

[12.30] Shi, L.P., Chong, T.C., Hu, X., Miao, X.S, and Li, J.M.:, : Investigation on dynamic erasing mechanism on phase change optical disk. Digest of Optical Data Storage 2004, 284 (2004).

[12.31] Ohta,T., Nishiuchi, K., Narumi, K., Kitaoka, Y., Ishibashi, H., Yamada, N. and Kozaki, T. : Overview and the future of phase-change optical disk technology. Jpn. J. Appl. Phys. **39**, 770-774 (2000)

[12.32] Adachi, Y., Etoh, A., Ishii, M., Maeda, S. and Kojima, K.: New method for adjusting write strategy using sequenced amplitude margin. Jpn. J.Appl. Phys. **45**, 1061-1065 (2006)

[12.33] Milster, T.D. and Upton, R. S.: Fundamental principles of crosstalk in optical data storage. Jpn, J. Appl. Phys. **38**, 1608-1613 (1999)

[12.34] Morita, S., Nishiyama, M. and Ueda, T. : Super-high-density optical disk using deep groove method. Jpn, J. Appl. Phys. **36**, 444-449 (1997)

[12.35] Kudo, H., Minemura, H., Miyamoto, H., Tamura, R. and Adachi, K. : Crosstalk cancellation for 50-GB/Layer optical recording. Jpn.J. Appl. Phys. **44**, 3445-3448 (2005)

[12.36] Francia, G. T. D. and Ronchi, A. F. G.: Super-gain antennas and optical resolving power, Nuovo Cimento Suppl. **9**, 426-428 (1952)

[12.37] Yamanaka, Y., Hirose, Y. and Fuji, H. : High density optical recording by super-resolution. Jpn, J. Appl. Phys, **28** Supplement 28-3, 197-200 (1989)

[12.38] Wang, H.F., Shi, L.P., Yuan, G., Tan, W. and Chong, T.C. : Subwavelength and super-resolution nondiffraction beam. Appl. Phys. Lett. **89**, 171102 (2006)

[12.39] Hieslmair, H., Stinebaugh, J., Wong, T., Neill, M., Kuijper, M. and Langereis, G.: 34 GB multilevel-enabled rewritable system using blue laser and high-numeric aperture optics. Jpn. J. Appl. Phys. **42**, 1074-1075 (2003)

[12.40] McLaughlin, S., Lo, Y. and Pepin, C., WarlanTech, D.: MultiLevel DVD: coding beyond 3 bits/data-cell. Dig. Int. Symp. Optical Memeory/Optical Data Storage 2002, 380-382 (2002)

[12.41] Shi, L.P., Chong, T.C., Tan, P.K., Miao, X.S., Ho, J.J. and Wu, Y.J.: Study of the multilevel reflection modulation recording for phase change optical disks. Jpn. J. Appl. Phys. **39**, 733-736 (2000)

[12.42] Bruneau, J.M, Bachevet, B. and Germain, C.: Optical design for a double level rewritable phase change disc. Jpn.J.Appl. Phys. **37**, 2168-2175 (1998)

[12.43] Nagata, K., Yamada, N. and Nishiuchi, K.: Rewritable dual-layer phase change optical disk. Jpn. J. Appl. Phys. **38**, 1679-1686 (1999)

[12.44] Mijiritskii, A., Hellmig, J. and Borg, H.: Development of recording stacks for a rewritable dual-layer optical disc. Jpn. J. Appl. Phys. **41**, 1668-1673 (2002)

[12.45] Hellmig, J., Mijiritskii, A.V., Herman J.B., Musialková, K. and Vromans P.: Dual-Layer Blu-ray disc based on fast-growth phase-change materials. Jpn.J.Appl.Phys. **42**, 848-851 (2003)

[12.46] Chong,T.C., Shi, L.P., Qiang, W., Tan, P.K., Miao X.S. and Xu, H.: Superlattice-like structure for phase-change optical recording. J. Appl. Phys, **91**, 3981-3987 (2002).
[12.47] Ren, S.Y. and Dow, J.D.: Thermal conductivity of superlattices. Phys. Rev. B **25**, 3750-3755 (1982)
[12.48] Chen, G. and Neagu, M.: Conductivity and heat conduction in supperlattices. Appl. Phys. Lett. **71**, 2761-2764 (1997)
[12.49] Fagas,G., Kozorezov, A.G., Lambert, C.J., Wigmore, J.K., Peacock, A., Poelaert, A. and Hartog, R. D.: Lattice dynamics of a disordered solid-solid interface. Phys. Rev B. **60**, 6459 -6464 (1999)
[12.50] Ogawa, S., Takeguchi, K. and Morimoto, I. : New layer structure of phase-change optical disk free from the initialization process. Proc. SPIE **3401**, 244-251 (1998)
[12.51] Miao, X.S., Chong, T.C., Shi, L.P., Tan, P.K. and Li, F.: New additional layer to realize initialization-free function for digital versatile disk-random access memory disk. Jpn. J. Appl. Phys. **39**, 729-732 (2000)
[12.52] Miao, X.S., Shi, L.P., Tan, P.K., Xu, W., Li, J.M., Lim, K.G. and Chong, T.C.: Initialization-free blue-laser optical disc. Jpn. J. Appl. Phys. **44**, 3612-3614 (2005)
[12.53] Shi, L.P. and Chong, T.C.: Nanophase change for data storage applications. J. Nanoscience & Nanotechnology, **7**, 65-93 (2007)
[12.54] INSIC optical storage roadmap 2006
[12.55] ISOM optical storage roadmap 2006

13. 4th Generation Optical Memories Based on Super-resolution Near-field structure (Super-RENS) and Near-field Optics

Junji Tominaga

Abstract Phase-change materials have additional potential for future nanotechnological devices besides their applications in optical disks and solid-state memory devices. When comparing the crystalline state and amorphous state, phase-change materials usually have two distinctly different refractive indices. Especially, one phase shows a positive dielectric constant, while the other one shows a negative constant. Once these phases exist together with a boundary on the nanometer length scale, the optical contrast can be used to produce localized plasmons there. Such a condition can be generated by focusing a laser beam on thin multilayer stacks. This is called super-resolution near-field structure (super-RENS). Since its invention in 1998, ultra-high density optical storage system based on this principle has been developed. In this chapter, the basic concept of plasmons and near-field optics, multilayer design, and the physical background this technology is based on in phase-change films are discussed.

13.1 Introduction

Optical memory is now widely available worldwide. Compact Disks (CD) and Digital Versatile Disks (DVD) are the typical and successful products in the market. For early products, no recordable or erasable disks were available because recording was not based on phase change materials (see Chapter 12). However, users gradually started to demand these kinds of recording media, and finally such disks were developed as CD-recordable (CD-R), CD-re-writable (CD-RW), DVD-R, and DVD-RW (-RAM) in the 1990s. The recordable and re-writable CDs or DVDs need additional layers compared to the older versions of CDs or DVDs that required only a reflective layer, namely layers with recording and erasing capabilities and layers with protective functions against oxidation or mechanical deformations are required. There are two kinds of layers for recording: one is based on breaking unsaturated chemical bonds and another is based on inducing phase-transitions (crystalline to amorphous, or amorphous to crystalline) in inorganic

S. Raoux, M. Wuttig (eds.), *Phase Change Materials*,
DOI 10.1007/978-0-387-84874-7_13,

materials. In the former, the focused, pulsed laser beam breaks unsaturated chemical bonds, as a result, the refractive index is substantially changed, because the unsaturated or conjugated bond length of the molecule determines the absorption wavelength. As the chemical bonds are randomly broken, the bonding of these molecules can be modified only once and the bond-breaking is irreversible. In the latter, on the other hand, the phase-transition material allows recording and erasing data repeatedly. After the invention of chalcogenide switching in the 1960s, a variety of chalcogenide alloys have been developed in order to achieve a large change in refractive index, a high switching speed and a large number of switching cycles [13.1]. One of the most successful and most studied materials is $Ge_2Sb_2Te_5$. It is a typical chalcogenide GeSbTe alloy that exhibits a large index change during the phase transition of 2.0 in the imaginary part of the complex refractive index, and more than 1 million times re-writable performance has been demonstrated [13.2]. In parallel, optical modules for recording and readout have also been developed and their performance has been advanced. The light wavelength of the solid-state lasers used to read and write CDs and DVDs has been shortened from 780 nm to 405 nm over the last 20 years leading to an increase in data storage density. The objective lens to focus the laser beam has also been improved from a numerical aperture (*NA*) of 0.45 for CDs to 0.85 for blu-ray disks (BD), and the beam spot available on the disk surface was reduced from more than 1 micron to 300 nm. In BDs, the smallest bit size reached is 140 nm. Even on such a nanometer level, chalcogenide thin films still have the capability to reliably store the tiny marks or bits at high speed. The question arises how small of an amorphous mark or bit can actually be recoded in chalcogenides. This question cannot be answered using an optical system because the laser beam cannot be focused beyond the diffraction limit of the optics. For example, the laser spot in a BD cannot be focused to a spot smaller than ~ 300 nm. Some experiments have been performed using the current through an Atomic Force Microscope tip to switch smaller regions (see Chapter 6).

13.2 Diffraction Limit and Near-Field Optics

Propagating electromagnetic waves cannot be focused into one singular point, and this is due to the diffraction theory of waves. Ernst Abbe derived an equation of the diffraction limit ζ of light with a wavelength λ and a numerical aperture NA of an objective lens [13.3] as

$$\zeta = \lambda/(2n \cdot \sin\theta). \tag{13.1}$$

Here, n and θ are the refractive index and beam incident angle, respectively, and the numerical aperture is defined as $NA = n \cdot \sin\theta$. In air, where $n = 1$, NA's maximum value becomes 1, where the lens bottom just touches the surface of the object. At present, the most advanced optical pick-up head of a BD has an *NA* of 0.85. The beam with the wavelength of 405 nm is focused to 405 nm/(2·0.85) = 238 nm. The size of two marks that still can be resolved is just a half of this value, 119 nm. This is the theoretical limit of optical resolution for two different marks. In BDs, the minimum mark size was strictly limited to 140 nm. The storage capacity of a BD disk cannot exceed 25 GB due to this diffraction limit. If an increase of the storage capacity is desirable or mandatory, alternative concepts have to be applied that circumvent the diffraction limit.

One of the alternatives is called near-field optics, which is a special field of optics dealing with non-propagating light [13.4]. Let us consider a grating that is prepared with a pitch of η and a monochromatic light beam is incident on the surface at normal incidence [13.5]. If the pitch η is large, the beam is reflected back towards the incident direction. If η is reduced to smaller and smaller values, the beam is reflected at some specific diffraction angles. For example, if one looks at the surface of a CD under white light one can see a beautiful rainbow of colors on the surface. In DVDs, as the grating pitch (track pitch) is shorter than that of CDs, one can see a more yellowish color. In BDs, the yellowish color is not anymore reflected from the surface, instead, a bluish color is more intense than the other colors because the pitch is much shorter than for the DVDs. So what happened to the light of other frequencies? In order to preserve energy, the energies of the other wavelengths must be preserved or dissipated into heat. A part of the higher order diffracted beam is actually dissipated into heat on the grating surface through *plasmons*, which are non-propagating light modes [13.6]. Plasmons are special light waves generated at an interface between a metal and dielectric material. The waves are generated by a collective motion of electrons on the metal surface. Under special circumstances, *plasmons* can be trapped at a local point! As shown in Fig. 13.1, light waves usually propagate in the upper region labeled A in ω-k space where ω is the angular frequency and k the wave vector.

Light waves usually propagate on a straight line A because $\omega = c \cdot k/n$. Here, n and c are a refractive index of a medium and the speed of light in vacuum, respectively. Curve B is the dispersion relation for light propagation in a metal such as Ag or Au. The line A and curve B have no intersection, thus, if two media are put together, no surface plasmons are generated. However, once the surface of the metal is corrugated with a grating period η, a new wave vector, $\Delta k = 2\pi/\eta$ is added in x direction, k_x. If a light strikes the metal surface with the grating at an incident angle θ, k_x is described as

$$k_x = \omega/c \cdot \sin\theta \pm \nu \cdot 2\pi/\eta. \qquad (13.2)$$

Here, ν is an integer. Therefore, as the pitch η becomes smaller, k_x *is* increased and finally crosses the light line and reaches plasmon line B. Surface plasmons are then generated on the grating.

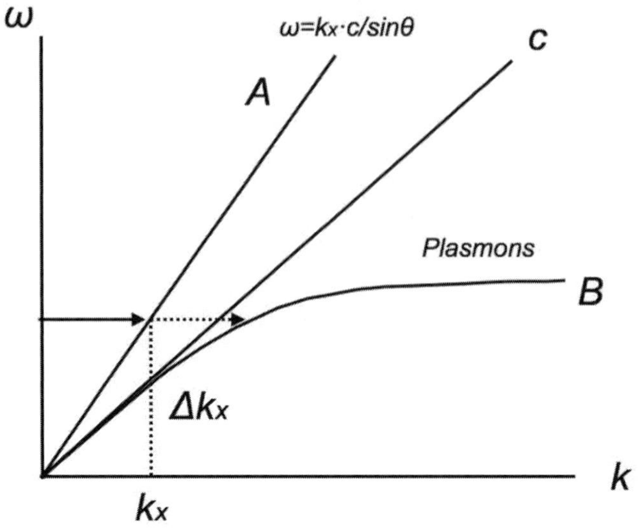

Fig. 13.1. Plasmon generation with a grating. Propagating light always exists above the straight-line *c* in a medium. However, a grating assists to reach k_x towards plasmon generation.

One of the tremendous advantages of these local photons (*plasmons*) is that they do not obey far field optical laws. Local photons are not constrained by the diffraction limit. Similar to a grating, local photons exist at all surfaces with small or large surface irregularities when light is incident on these surfaces. However, in most cases almost all of the light is transmitted or reflected at the surface, and only a very small fraction of the energy is transferred to local photons. To apply local photons to ultra-high density optical storage one must design a devise to accumulate and amplify local photons by trapping them in a laser spot for recording and readout. We will investigate in the next chapter how this can be realized.

13.3 Small Aperture and Non-propagating Photons

One possible solution towards this objective was provided by the use of a chalcogenide thin film in 1998 [13.7]. As already mentioned, some chalcogenides exhibit a great difference of refractive index between the amorphous and crystalline state. Especially, $Ge_2Sb_2Te_5$ and other compounds in the Ge-Sb-Te family show a large change of the imaginary index around 2. The index of the amorphous phase has a lower value than that of the crystalline phase. By radiating a laser beam with a Gaussian intensity profile on the chalcogenide thin film in the crystalline state a temperature profile is generated in the laser spot. As the laser beam intensity is increased, the temperature rises due to the strong light absorption caused by the large imaginary index. The distribution of the temperature is determined by the profile of the light intensity. As the temperature exceeds a phase transition point (melting or solid-solid phase transition), the refractive index, especially, the imaginary part, suddenly decreases to a lower value compared to the surrounding crystalline region. In this situation the molten region plays the role of a small aperture generated in the center of the laser spot. This aperture can be smaller than the diffraction limit ζ and can be used to control the behavior of the incident light. We will investigate the situation by simple Fourier optics analysis [13.8].

Let us consider the following case: Light is incident normally on a screen with a small aperture. The light source is placed at a far distance >> $\lambda/2$. We assume that the z-axis is along the incidence direction of the light. The electric field $E(x, y, z)$ can be expressed as follows with superposition of the three directional components k_x, k_y, and k_z

$$E(x,y,z) = \frac{E_0}{(2\pi)^3} \iiint F^*(k_x,k_y,k_z) \cdot \mathrm{e}^{-i\omega t} \cdot \mathrm{e}^{-i(k_x x + k_y y + k_z z)}\, dk_x dk_y dk_z \,. \qquad (13.2)$$

Here $F^*(k_x, k_y, k_z)$ is a Fourier transformation. In addition, the light wave is a propagating mode, that means, the mode must satisfy the condition [13.6]

$$\omega^2/c^2 = k_x{}^2 + k_y{}^2 + k_z{}^2. \qquad (13.3)$$

If the aperture size is about λ, k_z becomes 0. Therefore, the light cannot penetrate the aperture. In this case, the wave energy is transferred to k_x and k_y which are parallel to the screen. When the aperture size is reduced further, k_z becomes imaginary. The light wave in this case is described by

$$E = E_0 \cdot e^{-i\omega t} \cdot e^{-i(k_x x + k_y y)} \cdot e^{-k_z \delta} \qquad (13.4)$$

leading to

$$\delta = \sqrt{k_x^2 + k_y^2 - \omega^2 / c^2}\,. \qquad (13.5)$$

Therefore, as the aperture size is reduced, the light penetration becomes more and more difficult because of a strong exponential decay due to Eqs. (13.4) and (13.5). This is illustrated in Fig. 13.2.

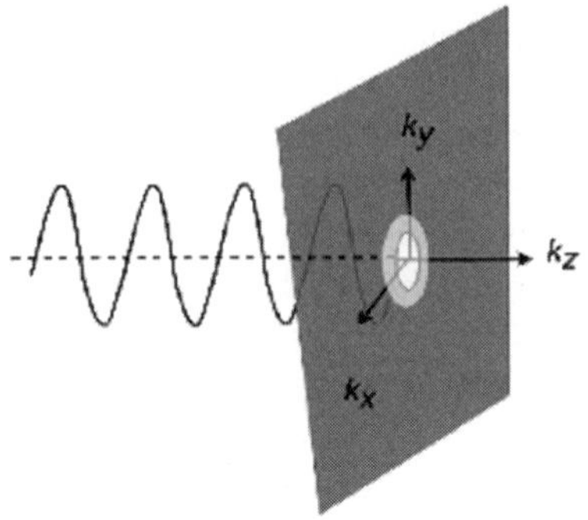

Fig. 13.2. Local photons trapped at an aperture smaller than the diffraction limit on a screen.

13.4 Super-resolution Near-field Structure (Super-RENS) Principle to Retrieve Non-propagating Light

From the discussion presented above we saw that an optical aperture smaller than ξ can generate non-propagating photons in a region around it. If the screen is very thin, the plasmon intensity normal to the screen decreases exponentially, and the region is smaller than the aperture size.

If a chalcogenide film is considered to be used as a screen to generate a small optical aperture, the thickness has to be less than 100 nm. This is not a technical problem because films thinner than 10 nm have already been applied successfully to BDs. Even in DVD-RW or DVD-RAM, the chalcogenide film thickness is only 15 - 20 nm, and these films are easily mass-produced using sputter deposition systems. To reproduce optical signals from marks or pits that are smaller than the diffraction limit, these marks must be very closely placed on the disk. In such a case, a multilayer stack can be applied with the following structure: *Ch/Di/Rec*. Here, *Ch, Di*, and *Rec* are the chalcogenide layer, dielectric layer, and recording layer, respectively. Note that in this case the first chalcogenide layer is *not* the recording layer but used as the layer in which the small aperture is formed, while the recording layer itself can also be a chalcogenide material. Because non-propagating photons from the aperture cannot propagate beyond 100 nm, the transparent dielectric layer thickness is also limited to 100 nm or less. The thinner the film, the more non-propagating light intensity is retrieved as a signal. The first experiment to demonstrate this effect was carried out in 1998 [13.7] using a structure with the following film sequence: *Di-1/Ch/Di-2/Rec/Di-3*. Here, *Di-1, -2*, and *-3* were made from SiN. Sb and $Ge_2Sb_2Te_5$ were used as *Ch* and *Rec*, respectively. The multilayer stack of *Ch/Di/Rec* with films thicknesses that lead to local photon scattering is called "*super resolution near-field structure*" or "*super-RENS*." Several types of *super-RENS* optical storage disks have been invented and improved since then, and the most up-to-date design is depicted in Fig. 13.3 [13.9, 13.10, 13.11].

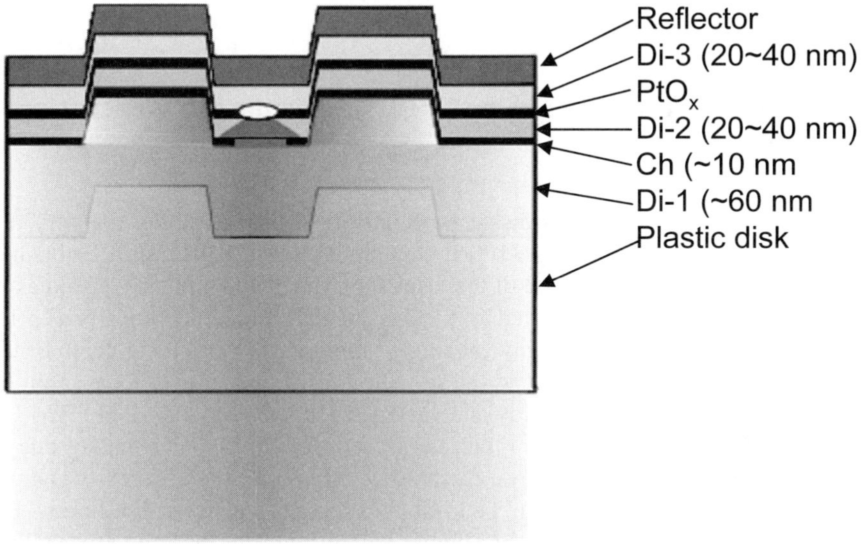

Fig. 13.3. 4th-generation super-RENS optical disk structure. The laser beam coming from the bottom is focused on a 0.6 mm thick plastic disk surface through a 0.60 or 0.45 *NA* lens. After the beam reaches the *Ch* layer, it generates a small optical window in the *Ch* film by inducing a phase transition leading to higher transparency. The window generates local photons (plasmons), which record a pit or are scattered by a PtO_x recording layer.

In the super-RENS disk, platinum oxide (PtO_x) is used as *Rec*, which decomposes into Pt nano-particles and an oxygen gas bubble. The bubble plays the role of the data pit. The super-RENS disk can clearly record and retrieve 37.5 nm pit tracks at a speed of 6.0 m/s!

Before the invention of super-RENS disks, recording was attempted by placing an optical head located above a recording medium to generate local photos. In this approach, high spatial control on the nanometer length scale between the head and the media at a high speed was required, and it was found to be too difficult to record and retrieve signals stably because the signal intensity fluctuated exponentially with head-media distance. Even by 2007, this barrier had not been overcome at an averaged DVD rotation speed of 6.0 m/s except for super-RENS disks. In super-RENS disks, the active region is buried under a protective transparent dielectric film, and head crashes do not easily destroy the data.

The key to increase the scattering signal from such a pit that is smaller than the diffraction limit depends on the chalcogenide material. In the early stage of development, the signal intensity from 100 nm pit tracks was less than 30 dB (dB = $20 \cdot log_{10}\, S/N$ where S is the signal and N is the noise intensity) [13.9]. Currently the signal intensity has been improved to ~ 50 dB for 100 nm pits [13.12, 13.13]. In practical applications, the signal intensity needs to exceed 40 dB. The 4th generation disk can now perform using 37.5 nm pits with more than 40 dB [13.40], even though the pit area is only 1/100 of the laser spot!

13.5 Origin of the Strong Scattered Signals for 4th Generation Super-RENS Disks

The properties of super-RENS disks, especially the 4th generation, are very unusual because more than 40 dB (100 times) signal to the noise ratio is obtained from an area which is only 1/100 of the laser spot. Since 2003, intensive studies of the mechanism have been carried out experimentally and theoretically [13.15, 13.16, 13.17]. In experimental investigations, the intensity and focal depth of the evanescent waves generated around the so-called "optical aperture" were carefully estimated by replacing the *Ch* alloy and varying the thickness of *Di-2* in Fig. 13.3. Figure 13.4 (a), (b) and (c) show the comparison of the signal intensity due to varying Te fraction of Sb-Te alloys. It can be easily understood that the signal intensity is correlated to the ratio of Te to Sb, but the dependency is not linear. Especially, Sb_2Te shows a remarkably higher signal intensity than the other alloys [13.18]. Also, the thickness dependency on *Di-2* shows that the highest signal intensity is obtained for the thinnest films. The signal intensity and focusing depth of the local plasmonic effect are very attractive in comparison with other Sb-Te alloys. Both features must be directly related to the refractive indices and electrons embedded in the films because resonant motions of electrons generate high-spatial frequencies and are the main sources for increasing the signal and resolution.

To understand the details, *ab-initio* computer simulations are a powerful and helpful tool. We developed a model of SbTe-9P (Te: 33 at%) on a computer. It is known that the unit cell of the alloy consists of an Sb block with 4 atoms and a Sb_2Te_3 block in a *R3-m* symmetric lattice. Under *ab-initio* local density approximation (LDA) using the CASTEP code [13.17] with ultrasoft potentials and a cut-off energy of 240 eV, the energy minimum was calculated. The result is shown in Figs. 13.5 (a) and (b).

The primitive lattice has the R3-m structure with lattice parameters $a = b =$ 4.25Å and $c =$ 16.91Å. As shown in Fig. 13. 5 (b), electrons cover all the atoms. The alloy has no band gap, that means it behaves like a metal. The alloy can still maintain the stable lattice for about 10 % volume expansion. For a larger volume expansion (more than this critical volume) the cell was expanded further by breaking the bonding (Te-Sb) between the Sb and Sb_2Te_3 blocks, see Fig. 13. 5 (c) and (d). Interestingly, the two blocks are freely movable like sheets and this bond breaking slightly expands the band gap.

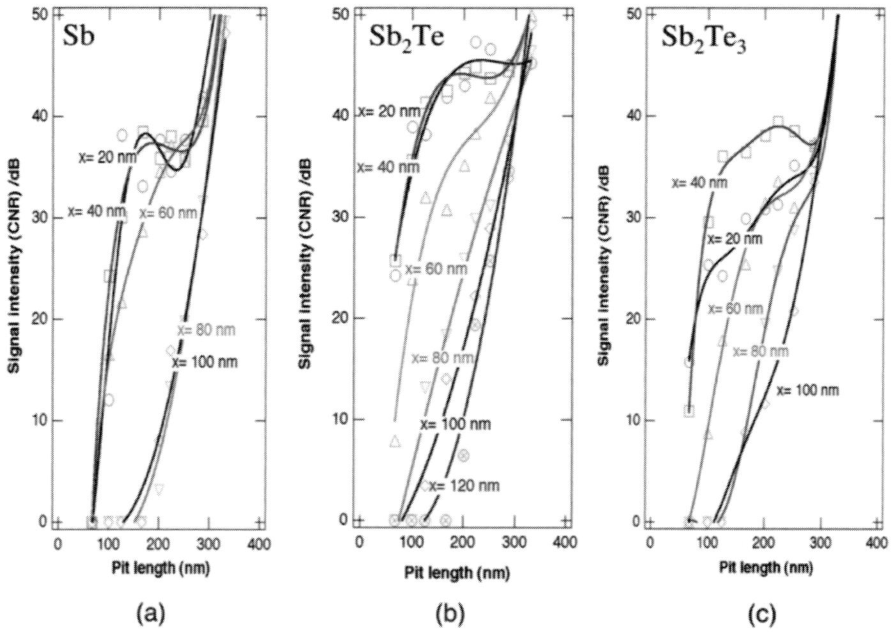

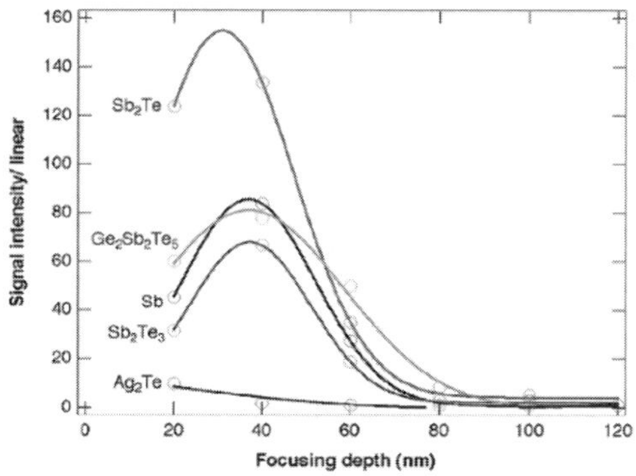

Fig. 13.4. Signal intensity for Sb, Sb_2Te and Sb_2Te_3 thin films as a function of intermediate film thickness (***x*** = 20, 40, 60, 80, 100, 120 nm) between the chalcogenide film and the PtO_x projection layer for laser powers above the threshold laser power. (a) Sb, (b) Sb_2Te, and (c) Sb_2Te_3. Focus depth for the chalcogenide film studied here as estimated by a Gaussian approximation for a 166 nm pit train. Reprinted with permission from the MRS, Warrendale, PA, from [13.19].

Figure 13.6 shows the complex refractive index versus volume change, which was calculated by a dipole approximation for the electron bands. It should be noted that all the simulations were carried out at a temperature of 0 K, not at room temperature. At increasing (compressive) pressure, the real index n is gradually reduced (the material becomes more metallic) while the imaginary index k steeply increases. In our simulation, the alloy seems to have a solid-solid phase transition at a volume of 0.8 V_0 where V_0 is the volume at a pressure of 0 GPa. At decreasing pressure (tensile) however n is slightly increased and saturates at around 4, while k is decreased to 2. n and k cross each other at around 1.1 V_0.

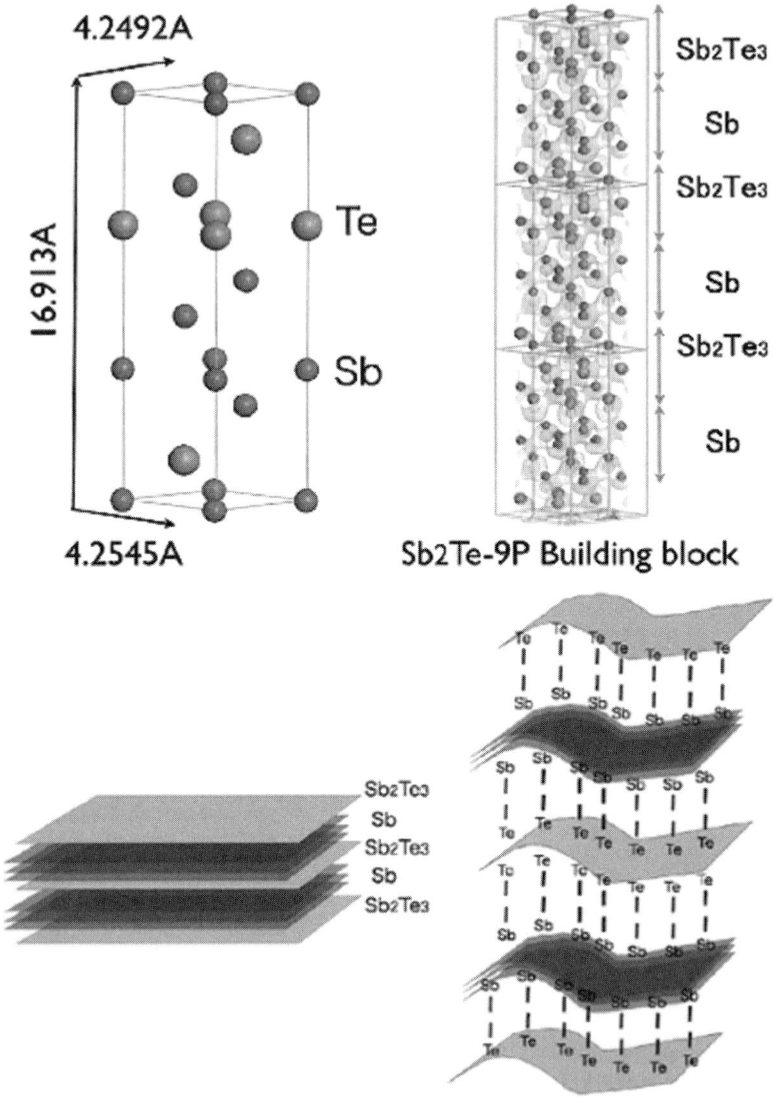

Fig. 13.5. Resolution SbTe-9P (Te:33at%) building block calculated by local density approximation (a) and electron density superimposed on two unit stacks (b). The alloy is composed of two different sheets alternatively stacked like (c), and the Te-Sb bonds connecting each sheet split at the melting point but the sheets do not melt at the same melting point (d). Therefore, the sheets can easily slide in two dimensions. Once cooled, the sheets recombine quickly to build the crystal.

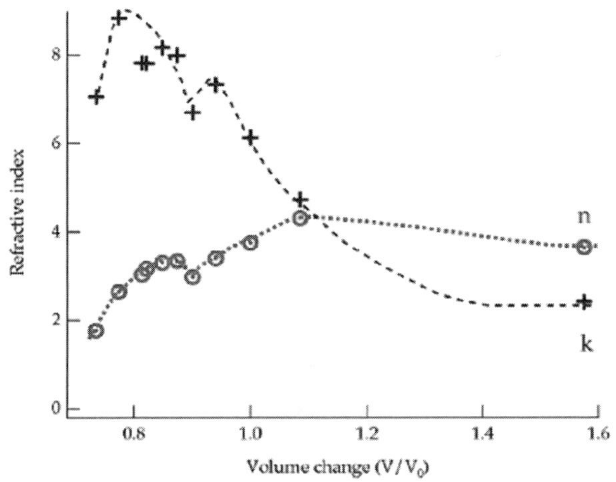

Fig. 13.6. Refractive index change of SbTe-9P structure vs. volume ratio. V_0 is a volume at 0 GPa, and V is varied by increasing or decreasing pressure between -5 GPa and 24 GPa in the simulation.

From this result, we do not inevitably need to assume an "umbrella-flip" for the Ge atoms to switch optical and electrical properties in phase change memory [13.18]. Rather, the electrons donated from Te atoms and contributing to the band structure are important, resulting in the large change in optical indices and electrical conductivity.

In the laser spot in a super-RENS disk two different states are produced as the laser power is increased. In the central area of the spot, a molten state containing sheet blocks (in Sb-Te alloys) or a high temperature state (in GeSbTe alloys) is generated when the power overcomes the threshold point. In the outer region surrounding the central part the crystal state occurs at temperatures less than the transition temperature. Therefore, two different phases (metal-semiconductor, or semiconductor – semiconductor) are adjacent similar to a *p-n* junction. At the boundary, the refractive index is miss-matched and discontinuous. In addition, the energy levels must be connected to each other to have equal Fermi levels. In the outer area, hot electron carriers are generated by the absorption of the laser, and dispersed rapidly with the moving laser spot. The electron carriers however are not easily eliminated within a uniform Gaussian distribution because of carrier traps generated at the boundary. The boundary probably accumulates electron carriers like a trap with a ring or crescent shape. Therefore, the very narrow boundary reflects the beam more than the other regions. If the electric field in this condition is reconstructed by Fourier optics, many high-spatial frequency components of Eq. (13.2) are required, especially at the boundary. The components inevitably exceed far beyond an inverse of the diffraction limit. Thus, a great number of localized evanescent waves are trapped at the boundary as a result. We computer-simulated such a model by a finite-differential time-domain (FDTD) method, inserting a thin silver ring at the boundary. The model succeeded to reproduce the experimental results with an evanescent focal depth of ~ 30 nm, while it was about 90 nm without the ring. This is shown in Fig. 13.7 [13.19].

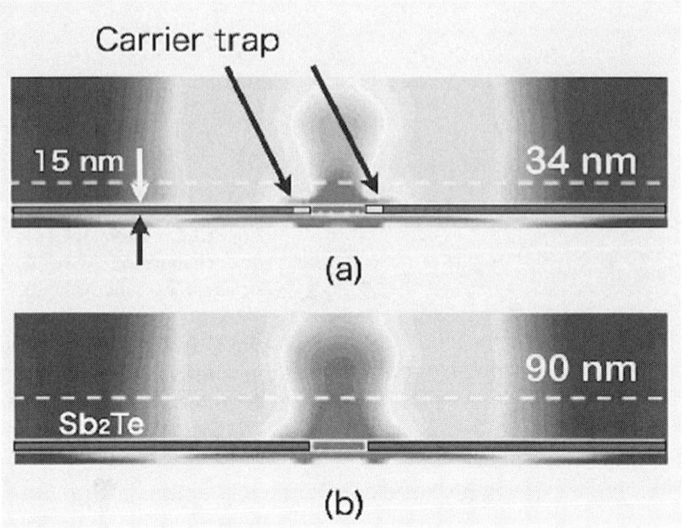

Fig. 13.7. Electrical field intensities at a small aperture generated in super-RENS disk (2D image) with a localized evanescent wave trap (a) and without a trap (b). The trap plays a role in confining evanescent waves around the aperture. Aperture size is set to 100 nm.

The boundary generation of two different phase-change states in adjacent areas is a very unique feature of the combination of chalcogenide alloys and laser heating in a small spot. The unique property of the localized evanescent light focusing appears due to the fact that a dielectric constant (n) in one phase is positive while a dielectric constant in the other phase is negative. The super-RENS disk is the first device where that the unique physical property is applied and actively controlled.

13.6 Beyond Super-RENS

As described so far, super-RENS has a lot of attractive physics and optics hidden in the nano-world of phase-change thin films. However, almost all the features were never generated by the phase change film itself, but occurred due to additional films that sandwiched the phase-change film and that were required because of oxidation and volume expansion due to the transition. Especially, the protecting layers consisting of ZnS-SiO_2 (with a ratio of 85:15) are the most suitable ones for these objectives. The multilayer structure can thus withstand pressures of a few GPa or more if the transition area is confined to a few tens of nanometers. The internal stress may become a driving force for the phase transition not only in optical disks but also in solid state phase-change RAM (PCRAM). The combination of the PCRAM structure and the internal stress forces do not set a limit to the size of a data although device fabrication methods in practice limit the size to manufacturable structures.

Over the last 40 years of studying of phase-change materials, we have gradually accumulated the knowledge how to use phase-change and which material provides a new face of physics and optics at the nano scale. The nano-world physics of phase-change films is very new research and has just started.

Acknowledgments. The author is grateful to Professor Alexander Kolobov and Dr. Paul Fons of AIST for valuable discussions and comments on phase change models and computer simulation.

References

[13.1] Ovshinsky S.R.: Reversible Electrical Switching Phenomena in Disordered Structures, Phys. Rev. Lett. **21**, 1450- 1453 (1968)
[13.2] Yamada N.Ohno E., Nishiuchi K., Akahira N. and Takao M.: Rapid-phase transitions of GeTe-Sb2Te3 pseudobinary amorphous thin-films for an optical disk memory. J. Appl. Phys. **69**, 2849-2856 (1991)
[13.3] Born B. and Wolf E. :Principles of Optics, Cambridge (1999)
[13.4] Paesler M. and Moyer P.: Near-field optics- Theory, Instrumentation, and Applications, John Willey & Sons (1996)
[13.5] Shalaev V. M. and Kawata S.: Nanophotonics with Surface plasmons, Elsevier (2007)
[13.6] Raether H.: Surface Plasmons – on smooth and rough surfaces and on gratings. Springer (1988)
[13.7] Tominaga J., Nakano T. and Atoda N.:An approach for recording and readout beyond the diffraction limit with an Sb thin film. Appl. Phys. Lett. **73**, 2078-2080 (1998)
[13.8] Tominaga J. and Nakano T.: Optical near-field recording – science and technology. Springer (2005)
[13.9] Tominaga J., Nakano T., Atoda N., Fuji H. and Sato A.: The characteristics and the potential of super-resolution near-field structure. Jpn. J. Appl. Phys. **39**, 957-961 (2000)
[13.10] Fuji H., Katayama H., Tominaga J., Men L. Nakano T. and Atoda N.: A near-field recording and readout technology using a metallic probe in an optical disc. Jpn. J. Appl. Phys. **39**, 980- 981(2000)
[13.11] Kikukawa T., Nanano T., Shima T. and Tominaga J.: Rigid bubble pit formation and huge signal enhancement in super-resolution near-field structure disc with platinum-oxide layer. Appl. Phys. Lett. **81**, 4697-4699 (2002)
[13.12] Shima T., Nakano T. and Tominaga J.: Effect of SiO_2 addition to PtOx recording layer of super-resolution near-field structure disc. Jpn. J. Appl. Phys. **46**, 3912-3916 (2007)
[13.13] Kim J., Bae J., Hwang I., Lee J., Park H., Chung C., Kim h., Park I., and Tominaga J.: Error rate reduction of super-resolution near-field structure disc. Jpn. J. Appl. Phys. **46**, 3933-3935 (2007)
[13.14] Fuji H., Kikukawa T. and Tominaga J.: Bit-by-bit detection on suoer-resolution near-field structure disk with platinum oxide layer. Jpn. J. Appl. Phys., **42**, L589-591 (2003)
[13.15] Tominaga J., Shima T., Kuwahara M., Fukaya T., Kolobov A. and Nakano T., Ferroelectric catastrophe: beyond nanometer-scale optical resolution. Nanotechnology **15**, 411-415 (2004)
[13.16] Kuwahara M., Shima T., Fons P., Fukaya T. and Tominaga J.: On a thermally induced readout mechanism in super-resolution optical disks. J. Appl. Phys. **100**, 043106 (2006)
[13.17] CASTEP code: Materials Studio 4.2. Accelrys. Com.

[13.18] Kolobov. A., Fons P., Frenkel A., Ankudinov A., Tominaga J. and Uruga T.: Understanding the phase-change mechanism of rewritable optical media. Nature Materials **3**, 703-708 (2004)

[13.19] Tominaga J., Fons P., Shima T., Kurihara K., Nakano T., Kolobov A. and Petit S.: Localized light focusing and super-resolution readout via chalcogenide thin film. Mater. Rec. Soc. Symp. Proc. **918**, 41-51 (2006)

14. Phase Change Memory Device Modeling

Daniele Ielmini

Abstract This chapter provides a review of the most important modeling issues for the simulation of the phase change memory (PCM) operation and reliability. After a short description of the PCM operation principle, the chapter will focus on electrical conduction characteristics for the amorphous and the crystalline phase of the chalcogenide material. The modeling of the programming operation and the methodology for calculating the programming current for phase transition will be described. Modeling-based methods for optimizing and scaling down the reset current will also be discussed. Cell reliability will finally be addressed, showing physics-based modeling approaches for crystallization and structural relaxation processes, which affect the stability of the amorphous phase and PCM data retention.

14.1 Introduction

The PCM, as many other new nano-electronic devices entering the arena of semiconductor industry, offers formidable challenges in terms of understanding and modeling of the physical and electrical operation. This is mainly due to the complicated physics of PCM cell operation and reliability, involving pure electronic effects (drift, diffusion, hopping and switching processes), thermal effects (Joule heating, heat conduction) and structural transitions (melting, crystallization, structural relaxation). In addition, the transport and physical properties of the chalcogenide compounds used as active materials in the PCM cell are still to be completely understood.

The purpose of this chapter is to provide an overview of the most significant modeling issues for PCM cells, offering a consistent approach to simulate the thermal, electrical and structural processes according to the most recent literature on this topic. Modeling will focus on PCM cells with $Ge_2Sb_2Te_5$ (GST) chalcogenide, which is the most widely used material for this technology. Electrical conduction will be analyzed first, discussing the transport and switching mechanisms in the amorphous phase and drift-diffusion mechanisms in the crystalline phase. The electrical modeling of the cell, together with thermal models for heat generation and conduction in the cell, provides the basis for the numerical modeling of

S. Raoux, M. Wuttig (eds.), *Phase Change Materials*,
DOI 10.1007/978-0-387-84874-7_14,

the *programming characteristics*, in particular of the current-demanding reset operation to form the amorphous phase by melting and quenching processes. The model capability to provide guidelines to optimize and scale down the programming current will be discussed. The modeling of *reliability characteristics* will then be studied, focusing on the crystallization kinetics, which controls both the high-temperature stability of the amorphous phase and the set operation of the cell. Structural relaxation effects which influence the stability of the electrical properties of the cell in the amorphous phase will be finally considered.

14.2 Device Operation

The PCM cell is a two-terminal device consisting of a bottom electrode, a chalcogenide layer and a top electrode. Fig. 14.1 displays a sketch of the cell layout for a very general vertical geometry and for the two logic states of the cell [14.1, 14.2].

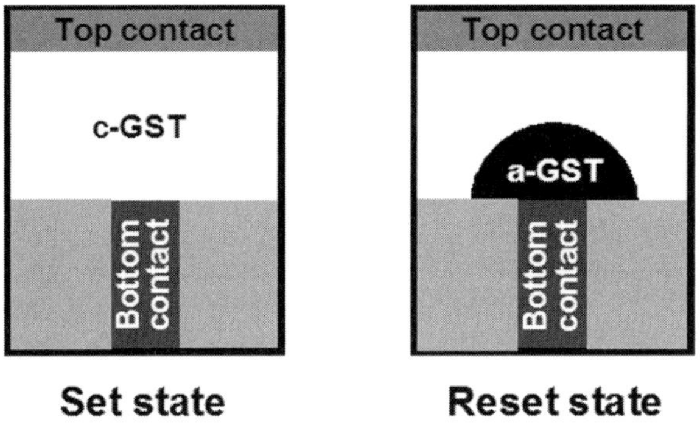

Fig. 14.1. Schematic for a PCM cell for the set state (left) and the reset state (right), where the chalcogenide material is in a crystalline (c-GST) or amorphous (a-GST) phase. The bottom contact, the chalcogenide layer and the top contact can be seen.

The cell operation is based on the thermally-induced change of phase in the active chalcogenide layer. The amorphous phase of the chalcogenide material is characterized by a large resistivity, while the resistivity in the crystalline phase is about 3 orders of magnitude lower than in the amorphous phase [14.3, 14.4]. As a result, the chalcogenide phase can be easily recognized by a voltage or current sensing of the cell. The phase in the chalcogenide material can be changed by the application of electrical pulses: to transform the crystalline phase into amorphous (*reset operation*), the pulse current must deliver enough Joule heating to raise the temperature above the melting point, thus allowing the transformation in the liquid phase during the pulse [14.4]. The liquid phase is then quenched in a disordered amorphous phase. The crystalline phase can be recovered by the *set operation*, namely with the application of a current pulse where Joule heating raises the temperature below the melting point, but high enough to allow for a fast crystallization of the amorphous structure, in the 100 ns – 1 μs range [14.4]. In the set state (Fig. 14.1, left), obtained by the set operation, usually the whole chalcogenide layer is in a crystalline phase, while in the reset state (Fig. 14.1, right), obtained by the reset operation, the amorphous phase occupies a typical dome-shaped volume. This is the case for the so-called mushroom cells (depicted in Fig. 14.1). Other cell designs are discussed in Chapters 15–17.

For efficient Joule heating in the cell, a strong confinement of heat and electrical currents is needed. This is usually made possible by a narrow bottom electrode, also known as the 'heater'. The heater must dissipate a large power and sustain a large temperature gradient, thus high electrical resistivity and low thermal conductivity are required.

Fig. 14.2 shows typical current-voltage (*I-V*) curves for the set and the reset state.

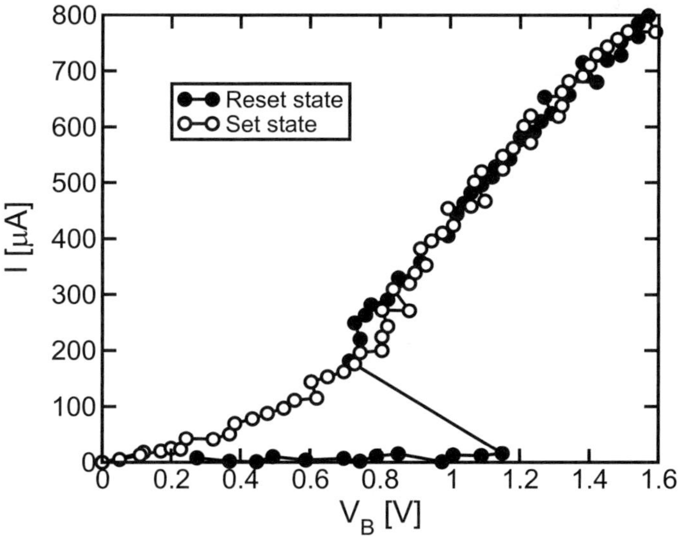

Fig. 14.2. Measured *I-V* curves for a PCM cell in the set (open symbols) or reset (filled symbols) states. The latter displays the threshold switching effect at about V_T = 1.2 V. (© (2007) IEEE, from [14.46]).

The set state features a non-linear *I-V* curve with a low-current resistance R_{set} of about 5-10 kΩ, and a high-current (ON state) resistance R_{ON} of about 1-3 kΩ. The reset state is characterized by a considerably larger resistance R_{reset}, usually in the range of few MΩ. However, the *I-V* curve suddenly changes at the so-called threshold voltage V_T (about 1.2 V in the figure) where the conductivity of the cell becomes comparable to that of the set state. This transition is known as *threshold switching* (see Section 14.4). The snap back in the figure from the switching point at V_T is controlled by a load resistance (R_L = 3 kΩ for data shown in the figure) which is usually put in series to the cell during the measurement to limit the current after threshold switching.

14.3 Modeling of Electrical Conduction in the Amorphous Phase

To study the conduction mechanism in the amorphous phase, we start considering the sub-threshold characteristic, i.e. below the threshold switching point. Fig. 14.3a shows experimental *I-V* curves for a PCM cell in the reset state, at temperatures T of 25, 45, 65 and 85 °C.

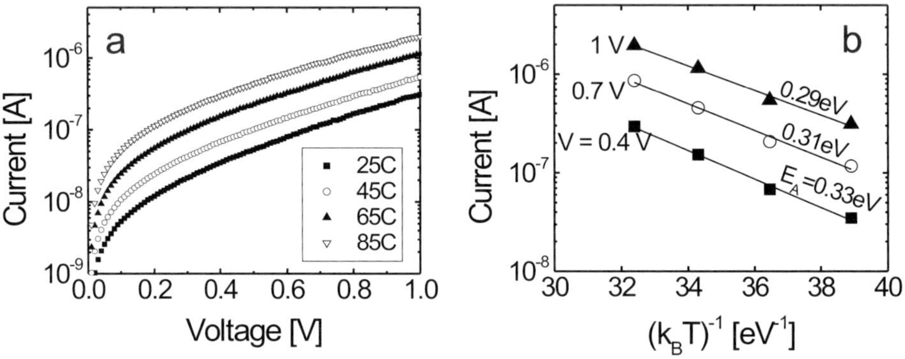

Fig. 14.3. (a) Measured *I-V* characteristics in the sub-threshold regime for amorphous GST at increasing temperature and (b) Arrhenius plot of the measured current for increasing voltage V = 0.4, 0.7 and 1 V. Reprinted with permission from [14.8]. Copyright [2007], American Institute of Physics.

The shapes of the *I-V* curves indicate a linear behavior for very small voltage, and an exponential increase of the current at relatively large voltage. To gain more insight into the physical mechanism responsible for this voltage dependence, the temperature dependence of the current was analyzed in Fig. 14.3b.

The Arrhenius plot reveals a relatively large activation energy $E_A = d\log I/d(1/k_BT)$ of about 0.3 eV (k_B is the Boltzmann constant). This is usually observed for most chalcogenide glasses, and is attributed to the Fermi level E_F being pinned at about mid gap as a result of the huge concentration of traps due to the disordered structure [14.7]. The value E_A = 0.3 eV is consistent with a midgap level within an energy gap of about 0.7 eV for amorphous GST [14.3]. Most interestingly, the activation energy in the figure is shown to decrease from 0.33 to 0.29 eV as the voltage increases from 0.4 to 1 V [14.8]. This feature is a signature for hopping transport of carriers by thermal emission over a potential barrier, which is generally lowered by the applied field [14.9].

Based on this indication, thermally activated hopping transport can be modeled as schematically shown in Fig. 14.4. It is assumed that the current is due to a Poole-Frenkel (PF) transport of electrons through traps, located at a distance Δz from each other. Electrons have to overcome a potential barrier equal to $\Delta\phi(0)$ at zero voltage (Fig. 14.4a), which is lowered to $\Delta\phi(V)$ for an applied voltage V. As a result, the

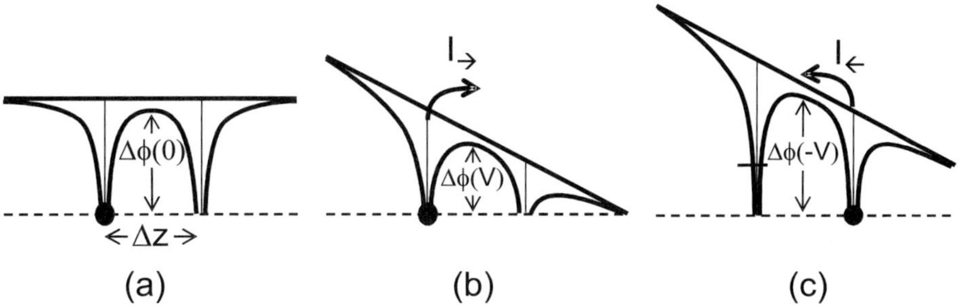

Fig. 14.4. Schematic for the analytical model for sub-threshold transport. The current is due to electrons being emitted from one trapped state to another, over a potential barrier. The potential barrier is equal to $\Delta\phi(0)$ when no voltage is applied (a), and drops to $\Delta\phi(V)$ as a voltage V is applied, resulting in an exponential enhancement of the forward current $I_{\rightarrow}$ (b). The reverse contribution to the current $I_{\leftarrow}$ is exponentially lowered by the increase of the potential barrier $\Delta\phi(-V)$ (c). Reprinted with permission from [14.10]. Copyright [2007], American Institute of Physics.

forward current $I_{\rightarrow}$ (Fig. 14.4b), defined as the current due to electrons flowing in the same direction as the electrostatic force, reads:

$$I_{\rightarrow} = qAN_T \frac{\Delta z}{\tau_0} \exp\left(-\frac{E_C - E_F - qV \dfrac{\Delta z}{2u_a}}{k_B T} \right), \tag{14.1}$$

where q is the electron charge, A is the area of contact of the active volume in the PCM cell, N_T is the total concentration of traps contributing to the electron current (i.e. from the Fermi level to the mobility edge), τ_0 is the characteristic attempt-to-escape time of the electron, E_C is the mobility edge at the conduction band and u_a is the thickness of the amorphous layer. Equation (14.1) results from the integration of all current contributions between E_F and E_C within the mobility gap [14.10]. Note in Eq. (14.1) the linear decrease of the potential barrier with voltage, which applies for a small distance Δz between traps [14.11]. For larger Δz, the conventional PF formula, where the potential barrier decreases with the square root of the voltage, is recovered [14.10].

In addition to the forward current in Eq. (14.1), also the reverse contribution should be taken into account, especially at low voltages (Fig. 14.4c). This is due to electrons being thermally emitted in the opposite direction of the electrostatic force. In this case, the potential barrier is increased by an amount $qV\Delta z/2u_a$, i.e. equal to the lowering for the forward current. Thus the reverse current can be written as:

$$I_{\leftarrow} = qAN_T \frac{\Delta z}{\tau_0} \exp\left(-\frac{E_C - E_F + qV \frac{\Delta z}{2u_a}}{k_B T} \right), \tag{14.2}$$

the only difference from Eq. (14.1) being the reversed sign of the potential barrier change in the exponent. The total current can be calculated from the difference between $I_{\rightarrow}$ [Eq. (14.1)] and $I_{\leftarrow}$ [Eq. (14.2)], yielding:

$$I = 2qAN_T \frac{\Delta z}{\tau_0} \exp\left(-\frac{E_C - E_F}{k_B T} \right) \sinh\left(\frac{qV}{k_B T} \frac{\Delta z}{2u_a} \right), \tag{14.3}$$

where the sinh function results from the difference between the two equal exponential terms, with opposite signs. The sinh-type voltage dependence in Eq. (14.3) can account for the linear and exponential increase of the current for low and high voltage, respectively, in Fig. 14.2. From Eq. (14.1), describing the exponential regime of Eq. (14.3), the activation energy can be obtained as:

$$E_A = \frac{d \log I}{d1/(\mathrm{k_B T})} = E_C - E_F - qV \frac{\Delta z}{2u_a}, \tag{14.4}$$

and the sub-threshold slope (STS) can be defined and calculated by:

$$\mathrm{STS} = \frac{d \log I}{dV} = \frac{q}{k_B T} \frac{\Delta z}{2u_a}. \tag{14.5}$$

Both E_A and STS in Eqs. (14.4) and (14.5) depend on the microscopic parameters Δz and u_a, namely on the ratio between the average trap distance and the thickness of the amorphous layer. To confirm Eqs. (14.4) and (14.5), E_A and STS were extracted for increasing voltage and temperature, respectively, from experimental *I-V* curves in Fig. 14.2, and are displayed in Fig. 14.5. Calculated results from Eqs. (14.4) and (14.5) are also shown in the figure, demonstrating a good agreement with data for $\Delta z / u_a = 0.17$.

A more general expression for the activation energy, including both the linear and the exponential regions of conduction, can be given calculating the derivative of Eq. (14.3), which yields:

$$E_A = E_C - E_F - qV_A \coth\left(\frac{qV_A}{k_B T} \frac{\Delta z}{2u_a} \right). \tag{14. 6}$$

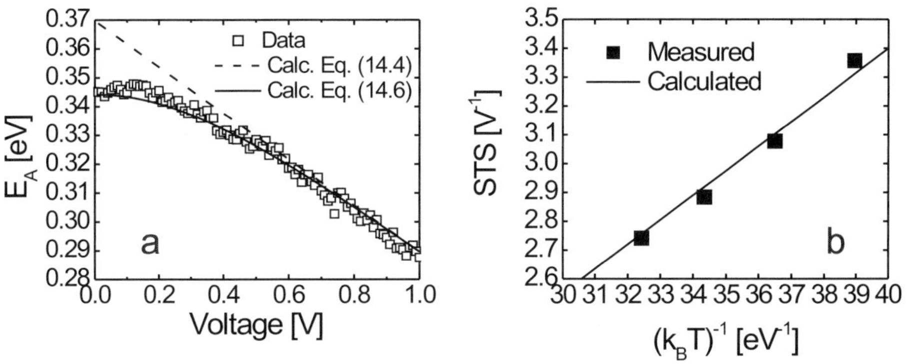

Fig. 14.5. (a) Activation energy E_A, as a function of voltage, and (b) STS as a function of $1/k_BT$, both obtained from experimental *I-V* curves at variable temperature. Both figures also display calculations according to Eq. (14.4) and Eq. (14.5), assuming Δz / $u_a = 0.17$. Also shown in (a) are the simulation results from the full analytical expression for E_A, from Eq. (14.6). Reprinted with permission from [14.8] and [14.10], both copyright [2007], American Institute of Physics.

Calculations according to Eq. (14.6) are shown in Fig. 14.5a, displaying a good agreement in both low-voltage and high voltage regimes. In particular, the model is able to reproduce the saturation of E_A at low voltage to the value $E_C - E_F - k_BT$ [14.10], which is due to the competition between forward and reverse currents. For large voltages, the linear decrease of E_A is recovered by Eq. (14.6), as already observed in Fig. 14.2.

Fig. 14.6 shows experimental and calculated *I-V* curves for increasing temperature. The good agreement in the figure demonstrates that the model correctly captures the essential physical nature of the sub-threshold transport in our samples.

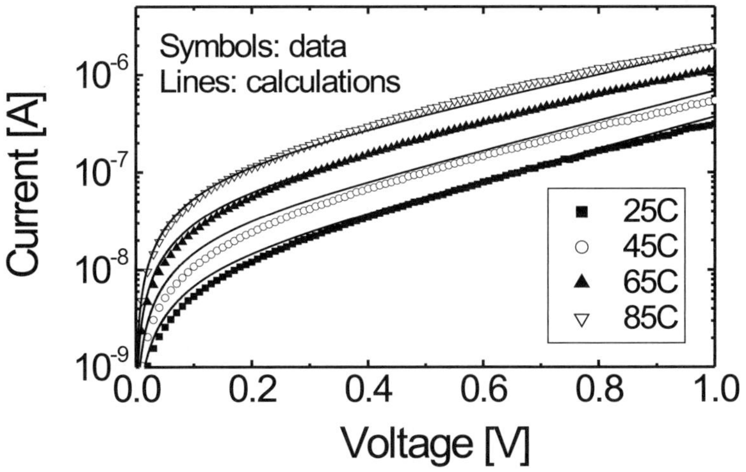

Fig. 14.6. Measured and calculated *I-V* curves for increasing temperature in the sub-threshold regime. Reprinted with permission from [14.10]. Copyright [2007], American Institute of Physics.

14.4 Threshold Switching in the Amorphous Chalcogenide

Threshold switching is the typical increase of conductivity observed for many chalcogenide materials in their amorphous phase, and results from a negative differential resistance (NDR) behavior. In some chalcogenide glasses, including GST, the threshold switching usually results in a transition from the amorphous to the crystalline phase, while for other materials the switching process leaves the phase unaltered [14.12]. This difference can be explained by the Joule heating resulting from the large current increase at switching, which, for sufficiently long electrical pulses, can contribute to the transition to the crystalline phase for glasses with a low crystallization point [14.13].

From the application point of view, threshold switching plays an essential role in the operation and performance of the PCM cell: in fact, the threshold voltage V_T defines the boundary between the voltage ranges for reading and writing of the memory cell, thus its predictability is vital for a controlled operation of the cell. On the other hand, the switching time, estimated to be in the range below 1 ns [14.14], represents the ultimate limit for the programming speed in the PCM and for the switching speed in threshold switches based on the same principle [14.15].

Threshold switching was attributed to a voltage-current instability due to electronic excitation at high fields [14.14, 14.16-19]. The conduction instability mechanism at the basis of threshold switching can be explained by the high-field PF process, schematically shown in Fig. 14.7.

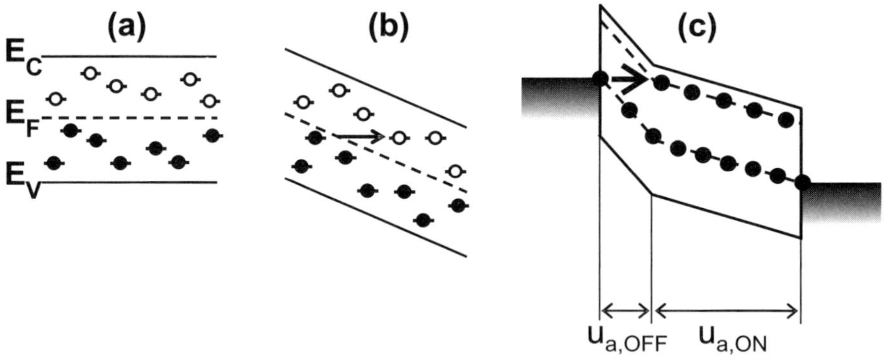

Fig. 14.7. Schematic for the injection process taking place at high-field and responsible for threshold switching in the amorphous phase of the chalcogenide material. At low electric field, carriers are at equilibrium (a), while at high field energy-gain processes lead to a non-equilibrium electron distribution (b). Due to the presence of a dead space for electron energy gain $u_{a,\mathrm{OFF}}$, electrons are non-equilibrium distributed only in a portion $u_{a,\mathrm{ON}}$ of the amorphous layer, resulting in a significant field non-uniformity in the amorphous chalcogenide (c). Reprinted with permission from [14.10]. Copyright [2007], American Institute of Physics.

For low voltage, the current is due to PF transport under equilibrium energy distribution of electrons, namely a Fermi-Dirac distribution controlled by E_F

(Fig. 14.7a). For higher electric fields, trapped electrons can gain a significant energy as a result of the electric field [14.10, 14.20, 14.21]. This can be modeled as a transition of electrons from low-energy to high-energy trap states, as shown in Fig. 14.7b. An increase of electron energy corresponds to a huge increase of PF 'mobility' for electrons, as a result of the exponential dependence of current on the potential barrier in Eq. (14.1). However, note that a finite distance (OFF space $u_{a,\mathrm{OFF}}$) is needed for electrons to gain sufficient energy under the applied field, as shown in Fig. 14.7c. The resulting change in conductivity along the chalcogenide thickness leads to a non-uniform field to sustain the continuous PF current. The electric field has to be large in the low-conductivity, OFF space $u_{a,\mathrm{OFF}}$, where the energy distribution of electrons is close to equilibrium, and small in the high-conductivity ON layer $u_{a,\mathrm{ON}}$, where electrons have a large average energy. The field collapse in the ON region results in the negative differential resistance (NDR) effect, appearing as a voltage snap back along the measured *I-V* curve [14.10].

Based on this physical picture, an analytical model was developed in [14.10] for threshold switching. The energy gain process was modeled as a tunnelling injection mechanism from deep states, close to E_F, to high-energy (shallow) states close to the conduction band E_C, and the current was computed as the sum of the PF contributions due to deep and shallow levels. Fig. 14.8 shows measured and calculated *I-V* curves, spanning from the sub-threshold regime to the switching region. The good agreement supports the physical basis of the proposed model.

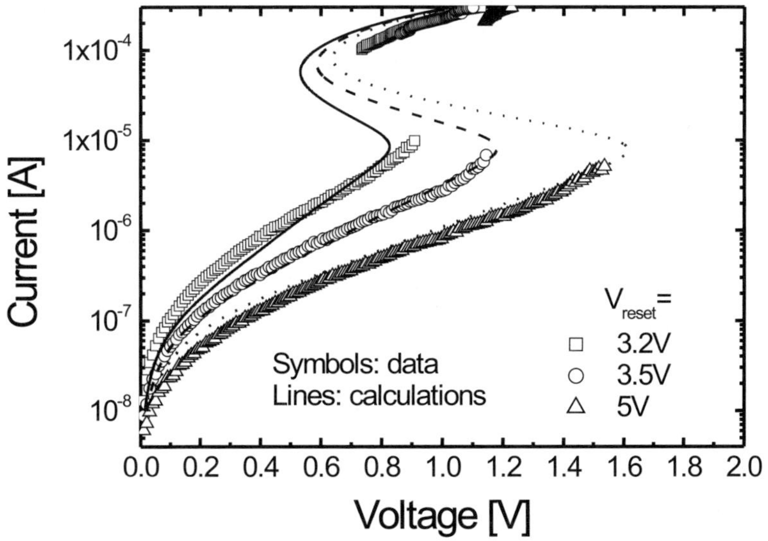

Fig. 14.8. Measured and calculated *I-V* curves for a PCM cell programmed at different voltages V_{reset}, corresponding to increasing thickness of the amorphous region in the cell. Reprinted with permission from [14.10]. Copyright [2007], American Institute of Physics.

14.5 Modeling the Electrical Conduction in the Crystalline Chalcogenide

The programming operation of the PCM takes place in the high current regime of the set or reset state. Referring to Fig. 14.2, it is clear that the amorphous and crystalline *I-V* characteristics are almost indistinguishable. This indicates that, while the amorphous conduction behaviour is essential to predict the read properties of the reset state (e.g. resistance stability discussed in Section 14.7 and 14.8), conduction in the crystalline phase is critical for modeling the programming operation.

Electrical experiments indicate that the chalcogenide in the crystalline phase behaves as a p-doped semiconductor [14.3], with relatively small activation energy for conduction, in the range of few tens of meV [14.22]. Optical measurements have shown that the energy gap is about 0.5 eV for crystalline GST. According to these observations, a band diagram for the crystalline phase as in Fig. 14.9a can be assumed, where the acceptor centres are attributed to equilibrium vacancies in the crystalline structure [14.3].

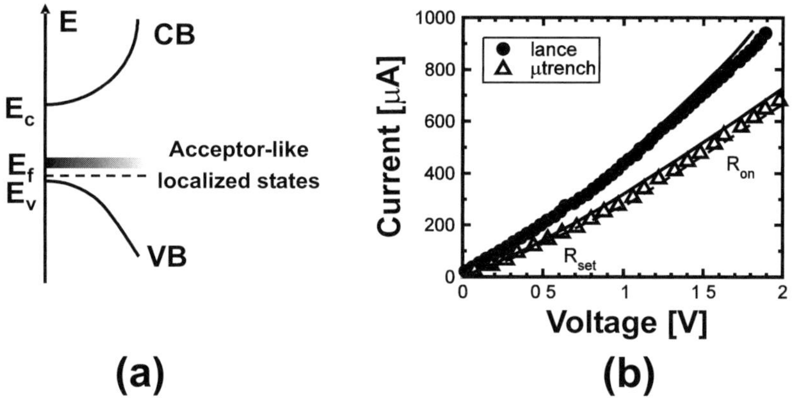

Fig. 14.9. Band diagram assumed for the crystalline phase of GST (a) and comparison of measured (symbols) and calculated (lines) *I-V* characteristics for PCM cells in lance and μtrench cells (b). (© (2008) IEEE, from [14.23]).

As a result, a drift-diffusion model can be used to describe conduction in the crystalline phase, in analogy with standard semiconductor-device simulations. Fig. 14.9b shows measured and calculated *I-V* characteristics for two different cell architectures, namely the *lance* cell, characterized by a cylindrical bottom electrode [14.1, 14.2], and the μtrench cell, where a thin-film bottom electrode is in contact with the chalcogenide layer through a sub-lithographic trench [14.1, [14.6]. Both measured *I-V* curves feature the same non-linear behaviour as in Fig. 14.2, with a transition from a low-current resistance R_{set}, to a high-current resistance R_{on}: The non-linear effect can be explained by the cooperation of thermal effects, increasing the concentration of carriers via thermal generation in the small band-gap semiconductor (about 0.5 eV for the crystalline phase of GST [14.3]) and field effects, increasing the concentration and mobility of carriers by impact ionization and PF mechanisms. These non-linear effects result in a drop of the chalcogenide resistance at high current, with a remaining R_{on} to be attributed to the bottom electrode.

14.6 Electro-thermal Modeling of the Programming Characteristics

Modeling of the programming operation requires the development of an electro-thermal numerical tool, able to solve electrical conduction, thermal generation and thermal conduction and to deal with any arbitrary geometry of the cell. An electro-thermal model with such features was demonstrated in [14.22]: the model was based on a drift-diffusion simulator for semiconductor devices and was able to simulate conduction in both the crystalline (set state) and the amorphous phases (reset state). For the amorphous phase, a crystalline-like band structure was assumed, to allow simple implementation in the semiconductor device simulator.

To simulate the reset operation, that is the transition from the crystalline to the amorphous phase, the temperature profile in the cell was calculated solving self-consistently the electrical conduction (drift diffusion) and the thermal conduction (Fourier) equations in the cell structure. Transient simulations are necessary to consider the effects of electrical delays, associated to cell and parasitic capacitances, and thermal delays, resulting from the finite thermal capacitance of the cell. The temperature within the chalcogenide volume was compared to the GST melting point (T_m = 888 K), and mesh elements with temperature exceeding T_m were switched to the liquid phase. During the quenching edge of the electrical/thermal pulse, the molten chalcogenide elements were switched from the liquid to the amorphous phase. For particularly large quenching rates, the transition from the amorphous to the crystalline phase was considered during the temperature decay. This can be achieved by a Monte Carlo model for nucleation and growth (N/G) of the crystalline phase within the amorphous environment, which will be discussed in Section 14.7. This procedure allows to predict the phase distribution after the application of the pulse, while the corresponding resistance can be calculated simply applying the drift diffusion code at low voltage.

Fig. 14.10 shows the calculated *R-I* characteristics for the two cell architectures considered in Fig. 14.9b. The measured *R-I* characteristics collect the values of the resistance *R* measured on a cell after the application of a pulse with current amplitude *I*. Prior to the application of this programming pulse, the cell was initialized in the set state (low resistance).

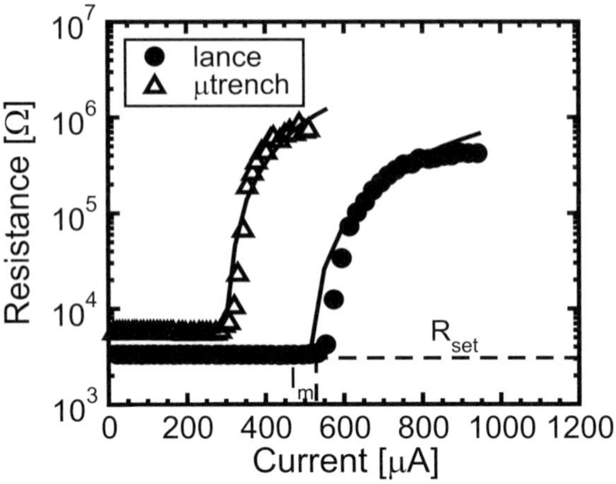

Fig. 14.10. Measured and calculated *R-I* curves, i.e. programming characteristics, for lance and μtrench PCM cells. (© (2008) IEEE, from [14.23]).

The rise of the resistance from low to high values marks the initiation of melting in the chalcogenide layer and of the consequent formation of an amorphous volume. From Fig. 14.10, the melting current I_m, corresponding to the minimum condition for the melting of the chalcogenide material all across the bottom contact area, can be identified.

Simulation results in Fig. 14.10 indicate a good agreement with data, and demonstrate the capability to predict the programming characteristics for the transition from the set to the reset state for any cell geometry [14.23]. It is worth recalling that the reset operation is the most demanding in terms of current consumption, since a temperature above the melting point must be achieved for significant amorphization in the cell [14.24, 14.25]. As a result, means to minimize the programming current either by careful optimization of the cell geometry and materials, or by scaling the cell structure are strongly needed [14.23].

The reset current can be optimized by finding the cell geometry, in terms of length of the bottom electrode L_h and of the chalcogenide layer L_c, that maximizes the temperature in the chalcogenide layer close to the bottom electrode for a given value of the programming current. In turn, the optimum cell minimizes the programming current needed to achieve the melting point (I_m). This is shown in Fig. 14.11 for a lance structure: the figure compares three cells with different geometries but the same contact diameter ϕ and the same read resistance R_{set}. To maintain the same R_{set}, L_c is decreased for increasing L_h. The conditions of equal R_{set} and ϕ were chosen in order to limit the degrees of freedom in the optimization process [14.23]. The temperature profile shown in the figure was calculated at the melting condition. For cells (a) and (c), the programmable region was very close to the bottom and top thermal sinks, respectively, where the temperature is kept at ambient temperature due to the presence of highly conductive metallic lines or vias. As a result, the current I_m was relatively high.

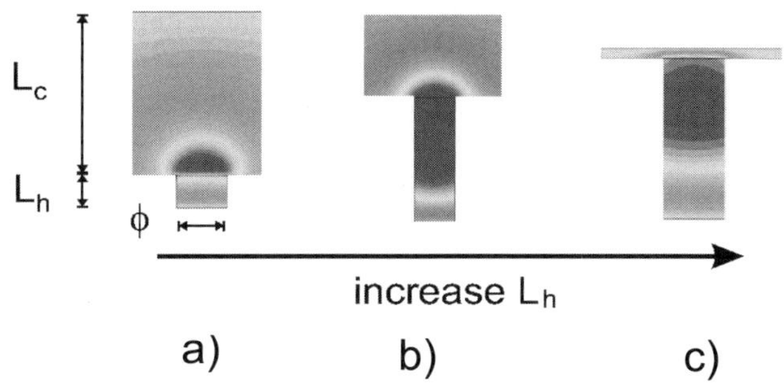

Fig. 14. 11. Calculated temperature profiles for three lance cells with different length of the bottom electrode (L_h), chalcogenide thickness (L_c) and constant R_{set} and ϕ. Cell (b) is optimum, in that it requires the minimum current for the melting condition. (© (2008) IEEE, from [14.23]).

An optimum case was found for cell (b), where the programmable region is centric with respect to the overall cell structure, thus keeping a relatively high thermal resistance toward the top and bottom thermal sinks and allowing for a minimum I_m. Fig. 14.12 shows the calculated I_m as a function of L_h and for four different R_{set} = 2, 3, 4 and 5 kΩ.

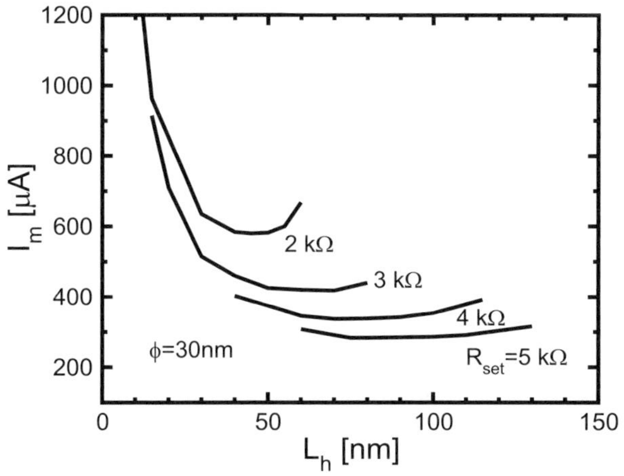

Fig. 14.12. Calculated I_m as a function of the heater length L_h, for increasing R_{set} and for a fixed heater diameter ϕ = 30 nm. (© (2008) IEEE, from [14.23]).

Each curve displays a characteristic U-shape, where the minimum corresponds to that geometry where the programmable region features comparable thermal resistances toward the bottom and top electrodes. Note also that as the resistance R_{set} is increased in Fig. 14.12, the minimum in the U-shaped curve moves to lower I_m values. This is because, for the same bottom electrode diameter, an increase of electrical resistance (hence of thermal resistance) corresponds to an increase of Joule power dissipation and heating for a given current, thus to a lower

I_m. In other words, the programming current can be reduced at the expense of an R_{set} increase, which may degrade the read-out time in the cell [14.23].

One of the most important strengths of the electro-thermal modeling of the cell is indeed to provide a scaling perspective for the PCM technology. In particular, scaling is an attractive way to reduce the programming current, thus cell modeling can be very useful in providing a comparison among different scaling strategies in terms of the calculated programming and readout performance. Fig. 14.13 shows the calculated I_m for lance cells as a function of F, that is the lithographic feature that sets the scale of the cell.

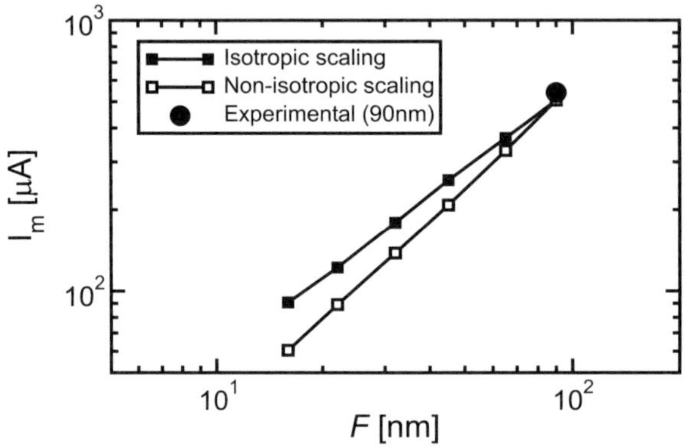

Fig. 14.13. Calculated I_m as a function of F, for isotropic and non-isotropic scaling. The symbol represents the experimental I_m from [14.1]. (© (2008) IEEE, from [14.23]).

As a reference, note that for $F = 90$ nm the bottom contact ϕ was assumed to be 56 nm, and the total height of the cell was about 120 nm (bottom electrode and chalcogenide layer), resulting in $R_{set} = 3$ kΩ. Two different scaling approaches were assumed, namely isotropic scaling, where all the transversal and vertical dimensions were proportionally reduced by the same scaling factor, and non-isotropic scaling, where only the bottom-contact diameter ϕ was scaled down [14.23, 14.26]. From the figure, the melting current for the isotropic scaling is inversely proportional to the lithographic feature F: This can be explained noting that the increase of temperature in the cell due to Joule heating can be approximated as $T_m - T_0 = \alpha R_{th} P$, where T_0 is the ambient temperature, R_{th} is the equivalent thermal resistance from the chalcogenide hot spot to the top and bottom thermal contacts of the cell, P is the dissipated Joule power and α is a constant.

Since the Joule power at high current is given by $R_{on} I^2$, the melting current can be obtained by:

$$I_m = \sqrt{\frac{T_m - T_0}{\alpha R_{on} R_{th}}}\,. \tag{14.7}$$

For isotropic scaling, both R_{th} and R_{on} increase for decreasing F ($R_{on} \propto F^{-1}$ and $R_{th} \propto F^{-1}$), thus yielding $I_m \propto F$, in agreement with the results in Fig. 14.13. For non-isotropic scaling, $R_{\text{on}} \propto F^{-2}$ since in this case ϕ is reduced, while the length is kept constant. R_{th} scales instead as F^{γ}, with γ between -1 and -2: R_{th} is in fact dominated by thermal loss through the chalcogenide layer, which is strongly affected by spreading effects from the small bottom contact to the top contact, leading to $R_{th} \propto F^{-1}$ and, from Eq. (14.7), to $I_m \propto F^{1.5}$. More detailed simulation results indicate that $R_{th} \propto F^{-0.7}$, as a result of parallel heat conduction through the insulating material surrounding the bottom electrode, thus leading to $I_m \propto F^{1.3}$, as shown in Fig. 14.13 [14.23].

The electro-thermal model can be used to compare the performance of the cell under programming (I_m) with the one under read operation (R_{set}). The latter dictates the cell current available during reading, thus it is critical for the careful design of the memory circuit. Fig. 14.14 shows calculated I_m and R_{set} for the lance cell, for technology nodes F = 90 to 16 nm along both isotropic and non-isotropic scaling paths.

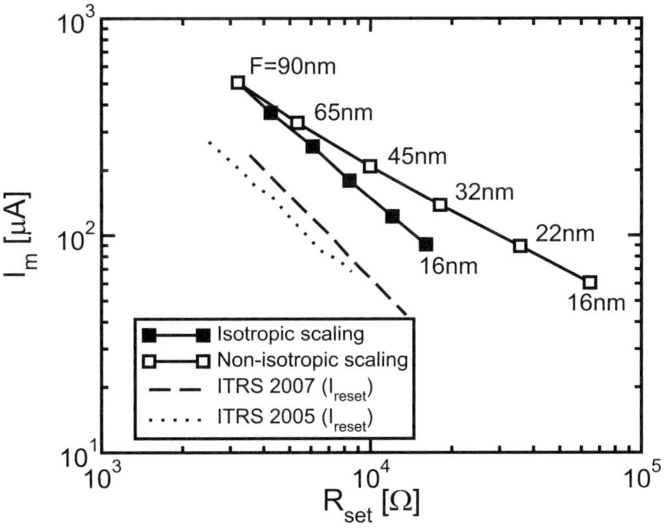

Fig. 14.14. Calculated I_m as a function of R_{set}, for isotropic and non-isotropic scaling in a lance cell. Also shown are values reported in the ITRS 2005 and 2007 Releases [14.25].

Note that $R_{\text{set}} \propto F^{-1}$ holds for isotropic scaling, while $R_{\text{set}} \propto F^{-2}$ holds for non-isotropic scaling. This can account for the results in Fig. 14.14, namely $I_m \propto F \propto R_{\text{set}}^{-1}$ for isotropic scaling, and $I_m \propto F^{1.3} \propto (R_{\text{set}}^{-1/2})^{1.3} = R_{\text{set}}^{-0.65}$ for non-isotropic scaling. The comparison in the figure shows that programming current reduction is faster for non-isotropic scaling, although the latter also leads to a larger increase in R_{set} as compared to isotropic scaling. Also shown in Fig. 14.14 are values reported in the International Technology Roadmap for Semiconductors for PCM scaling, from F = 72 to 22 nm [14.25]. The values shown are lower than the corresponding results from our simulations, for both the programming current and the set resistance. This can be explained assuming different material properties (hence electrical and thermal resistivities) for evaluating the programming current and cell resistance.

14.7 Modeling the Amorphous to Crystalline Phase Transformation

Phase transformation from the amorphous to the crystalline phase is essential for the proper operation of the PCM cell. On the one hand, thermally-activated crystallization can lead to unwanted phase change in the memory cell at relatively low temperature and without the application of any electrical pulse, thus possibly affecting data retention [14.27-29]. On the other hand, phase change controls the set operation, where crystallization occurs in the 100 ns - 1 μs timescale with the application of a large current pulse yet at temperatures below the melting point [14.4]. Thus a careful modeling of the phase transformation is indeed a central task for this technology.

Fig. 14.15 shows a typical resistance–time (*R*–*t*) curve for *T* = 190°C.

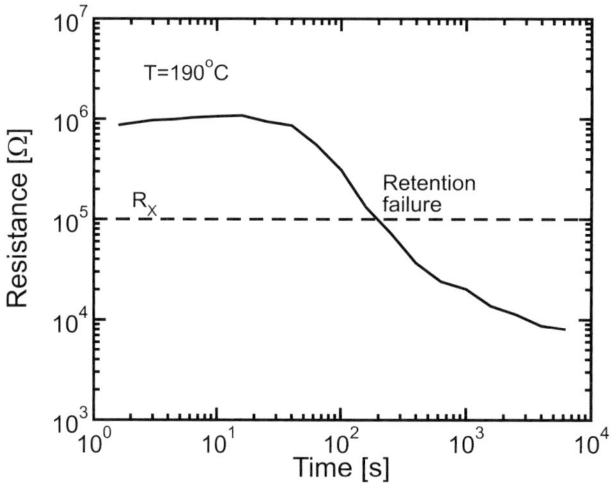

Fig. 14.15. Measured resistance of a cell initially programmed in the reset state, as a function of time during annealing at *T* = 190°C. (© (2007) IEEE, from [14.29]).

The measured *R* displays a transition from the amorphous (large *R*) to the crystalline phase (low *R* of few kΩs) in a time range between about 50 and 2000 s. Correspondingly, a retention lifetime can be defined as the time needed for the resistance to fall below a threshold value of 100 kΩ, i.e. close to the geometric mean value $(R_{reset}R_{set})^{1/2}$ in the middle of the resistance window of the cell.

Fig. 14.16 shows the cumulative distributions of crystallization lifetimes measured for the same PCM cell at temperatures of 180, 190, and 210 °C [14.28, 14.29]. The cumulative probability *f* is plotted on a Weibull scale: This is because crystallization in the amorphous phase affects the electrical resistance via a *percolation* process, which normally obeys Weibull statistics [14.31, 14.32]. Results indicate a relatively large spread of crystallization times, due to the random N/G mechanisms at the origin of the phase transformation [14.28, 14.30].

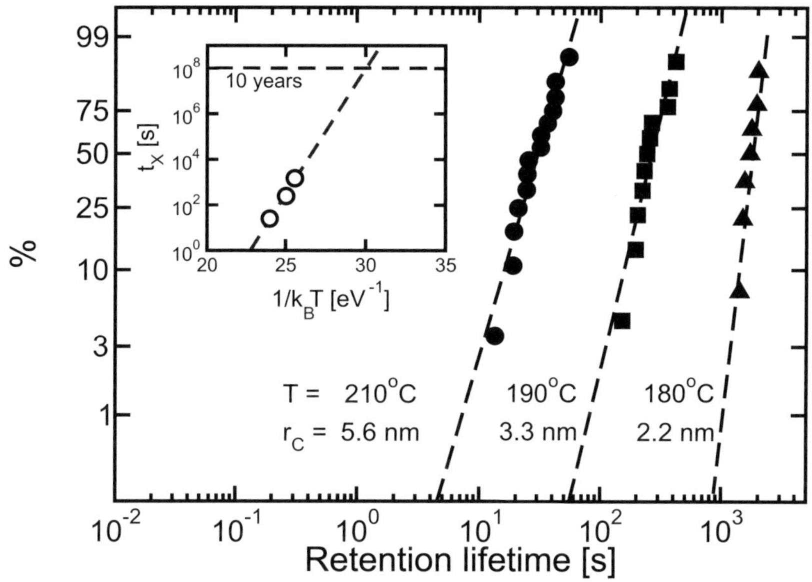

Fig. 14.16. Cumulative distributions of retention lifetime t_x for $T = 180$, 190 and 210°C. The figure also indicates the average sizes of grain size r_C, which is consistent with the experimental Weibull slope according to the percolation theory [14.28]. (© (2007) IEEE, from [14.29]).

The distribution spread increases with temperature, because of the increase of the average size of nanocrystallites as a result of the larger impact of growth against nucleation [14.28].

The inset of the figure shows the average crystallization lifetime t_X, as a function of $1/k_BT$: while the Arrhenius law can account for t_x in the relatively small experimental range of temperatures, t_X results from the detailed T-dependences of N/G, which are known to be non-Arrhenius [14.33-35]. Thus, a careful prediction and extrapolation of the retention time at relatively low temperatures, which are not achievable by simple accelerated experiments, requires the physical modeling of N/G.

To provide an experimental basis for phase change modeling, N/G parameters for the amorphous chalcogenide volume produced by the electrical reset have to be separately determined. To this aim, the average radius r_C of nanocrystallites at the percolation condition can be estimated from the experimental distribution of t_X as a function of temperature. This extraction is possible recalling that the size of crystalline particles relative to the thickness of the amorphous volume controls the statistical spread of the t_X distributions [14.28]. Thus, r_C can be estimated by a) collecting the distributions of t_X obtained from Monte Carlo simulations of nucleation and percolation through the mixed amorphous/crystalline phase [14.30] and b) comparing the calculated statistical slope to the experimental one (see Fig.

14.16). From this comparison, r_C values of 2.2, 3.3, and 5.6 nm were found for T = 180, 190, and 210°C, respectively [14.28, 14.29]. From the extracted r_C and from the measured t_X, the nucleation rate I_N and the growth velocity v_G can be extracted by the simple analytical model shown in the following. The average grain size at percolation condition (time equal to t_X) can be obtained as:

$$r_C = r_N + v_G t_X / 2\,, \tag{14.8}$$

where r_N is the size of the as-nucleated (critical) nucleus and v_G was assumed isotropic and time-independent [14.35–14.37]. The average retention lifetime t_X can be calculated as the time to establish a critical crystalline fraction f_X within the amorphous volume, corresponding to percolation condition, thus yielding:

$$t_X = \frac{f_X}{(1 - f_X/2) I_N \Omega}\,, \tag{14.9}$$

where Ω is the average volume of crystalline grains at percolation condition which can be easily calculated from the average grain radius r_C [14.29]. In Eq. (14.9), the amorphous volume available for nucleation was taken as $V_{TOT}(1 - f_X/2)$, that is the average between the initial amorphous volume V_{TOT} and the final one $V_{TOT}(1 - f_X)$. Results similar to Eqs. (14.8) and (14.9) can be obtained from Monte Carlo simulations of N/G, as shown in Fig. 14.17 for a reference 2D geometry [14.29].

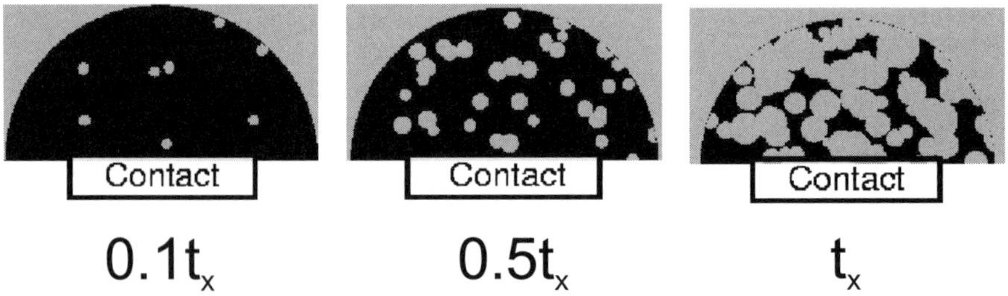

Fig. 14.17. Snapshots from Monte Carlo simulations showing random nucleation and isotropic, time independent growth of crystalline grain size within the amorphous volume in the programmed PCM cell. Snapshots were taken for increasing times corresponding to 10%, 50% and 100% of the crystallization lifetime t_X, from left to right. (© (2007) IEEE, from [14.29]).

Eqs. (14.8) and (14.9) allow to estimate I_N and v_G from experimental values of r_C and t_X. The results are shown in Fig. 14.18 as a function of $1/k_BT$ (f_X = 0.4 was assumed).

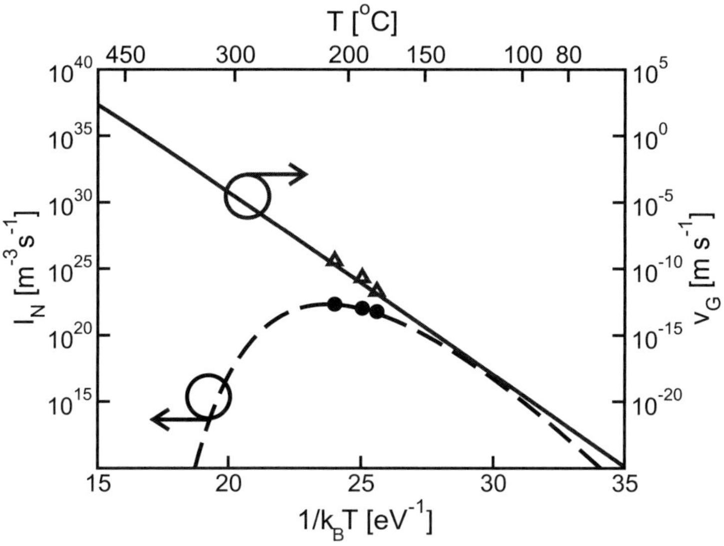

Fig. 14.18. Arrhenius plot of experimentally extracted I_N and v_G and of calculations according to the standard nucleation theory. (© (2007) IEEE, from [14.29]).

For physics-based extrapolations of N/G parameters to lower temperatures, the standard nucleation theory can be used [14.33, 14.34] from which I_N is given by:

$$I_N = I_{N0} \exp\left(-\frac{\Delta G_A + \Delta G^*}{k_B T} \right), \tag{14.10}$$

where I_{N0} is a constant, ΔG_A is the energy barrier for atomic mobility and ΔG^* is the energy barrier for the formation of a critical nucleus with radius r_N. The nucleation energy barrier is obtained as the maximum in the nucleation free energy including bulk and interface energies, thus leading to [14.33]:

$$\Delta G^* = \frac{16\pi\sigma^3}{3\Delta G_V^2}, \tag{14.11}$$

where σ is the interfacial energy per unit area and ΔG_V is the Gibbs free energy difference per unit volume between the amorphous and the crystalline phase. Growth velocity is instead given by

$$v_G = v_{G0} \exp\left(-\frac{\Delta G_A}{k_B T} \right)\left(1 - \exp\left(-\frac{\Delta G_V \omega_P}{k_B T} \right) \right), \tag{14.12}$$

where v_{G0} is a constant and ω_P is the average volume of the particle joining the crystalline grain during growth, which can be estimated by the average atomic volume of about $3\text{x}10^{-23}$ cm^3 [14.29].

Fig. 14.18 shows experimentally extracted I_N and v_G, compared with calculations from Eqs. (14.10) and (14.12). In the calculations, it was assumed σ = 0.12 Jm^{-2}, ΔG_A = 3.2 eV and ΔG_V was given by the Singh-Holz approximation [14.29, 14.34, 14.38]. Note that, while v_G obeys the Arrhenius law in the displayed temperature range, I_N is markedly non-Arrhenius, particularly above 200 °C. In fact the activation energy for I_N is not constant, since I_N is suppressed for increasing temperatures as a result of the decreased driving force for crystallization ΔG_V and the consequent increase of the nucleation barrier ΔG^* [see Eq. (14.11)].

Calculated I_N and v_G in Fig. 14.18 and Eqs. (14.8) and (14.9) allow to predict the retention lifetime t_X for any arbitrary temperature. Fig. 14.19 shows the Arrhenius plot of the measured and calculated t_X.

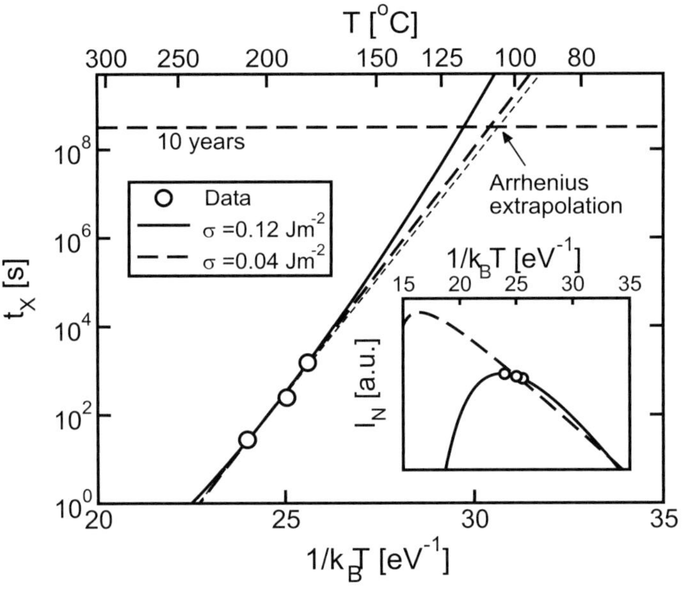

Fig. 14.19. Arrhenius plot of measured and calculated t_X, and calculated I_N (inset). Calculations are shown for two values of the interface energy σ = 0.12 and 0.04 Jm^{-2}. The Arrhenius extrapolations of data are shown for reference. (© (2007) IEEE, from [14.23] and [14.29]).

The Arrhenius extrapolation of data points is also shown for reference: it is clear that the Arrhenius law underestimates the retention lifetime as compared to the standard nucleation theory. This highlights the need for a physics-based extrapolation method for accurate retention prediction. The figure shows simulation results for σ= 0.12 Jm^{-2} [14.34] and 0.04 Jm^{-2} [14.39], while the inset displays the corresponding I_N. These results indicate that retention prediction is strongly sensitive to the value of the interface energy.

Fig. 14.20 shows the local activation energy of t_X, defined as E_X = $d\log(t_X)/d(1/k_BT)$, for σ= 0.12 Jm^{-2}.

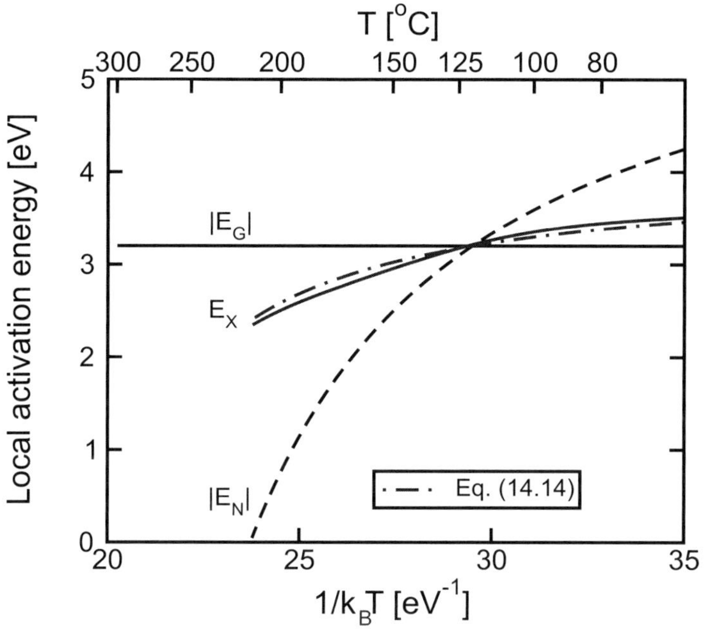

Fig. 14.20. Calculated local activation energies for t_X (E_X) and for N/G parameters I_N and v_G (E_N and E_G). Absolute values of E_N and E_G are shown. E_X calculations according to either Eqs. (14.8-9) or approximation (14.14) are shown. An interface energy for σ = 0.12 Jm^{-2} was assumed. (© (2007) IEEE, from [14.29]).

This is compared to the absolute values of the local activation energies for I_N and v_G, called E_N and E_G respectively, and calculated according to Eqs. (14.10) and (14.12).

For a simplified extraction of the lifetime activation energy, note that, neglecting r_N in Eq. (14.8) and substituting r_C in Eq. (14.9), it can be obtained:

$$t_X^4 \approx \frac{3f_X}{\pi(1-f_X/2)} v_X^{-3} I_N^{-1}, \tag{14.13}$$

which, after calculating the logarithm and derivative with respect to $1/k_BT$, leads to:

$$E_X = -\frac{E_N + 3E_G}{4}, \tag{14.14}$$

which was already obtained for a more general case in [14.36]. From Fig. 14.20, note that $E_G \approx \Delta G_A$ is constant, since Eq. (14.12) can be approximated by a simple Arrhenius for temperatures far from the melting point.

On the other hand, E_N decreases for increasing temperature, as a result of the temperature dependence of ΔG^* and the consequent drop of I_N at high temperature. The non-Arrhenius behavior of t_X in Fig. 14.19 thus results from the non-constant activation energy E_N, predicted by the standard nucleation theory.

The crystallization process can be similarly simulated to predict the set process. Set is achieved by the application of a programming current pulse by which the cell temperature is raised to a high value, yet below the melting point. Set calculations are strongly sensitive to the localization of the current resulting from the switching process and affected by the detailed geometry of the bottom electrode and amorphous volume [14.22]. Indeed, a correlation has been demonstrated between crystallization kinetics in the set process and in annealing experiments below 200 °C, thus suggesting that the same N/G mechanisms control both reliability and set programming in the memory cell [14.40].

14.8 Modeling the Structural Relaxation in the Amorphous Phase

The metastable nature of the amorphous phase of the chalcogenide material manifests itself not only by phase transformation, but also by more subtle structural relaxation processes, which affect the atomic arrangement and the electrical conduction while still maintaining the disordered structure. Structural relaxation (SR) has been observed for a broad set of amorphous materials, including a-Si, a-Ge, metallic glasses, organic glasses, and chalcogenide glasses [14.41-44]. The structural relaxation also affects the amorphous GST in PCM cells, and results in a steady increase (drift) of the electrical resistance even at room temperature [14.45, 14.46]. This is shown in Fig. 14.21, for a PCM cell in either a reset state or a set state: while reset resistance increases according to a power law with an exponent of about 0.11, the set state resistance is practically stable, indicating that only the amorphous phase is affected by the relaxation phenomenon.

This modification of the electrical parameters is the result of a change in the band structure of the chalcogenide glass: To predict the time and temperature dependence of resistance for reliability estimation, the relationship between band structure and electrical parameters have to be carefully assessed. The high temperature annealing is seen to accelerate the relaxation process, which appears to lead to an increase of a) R at low voltage, b) sub-threshold slope STS and c) threshold voltage V_T with annealing time.

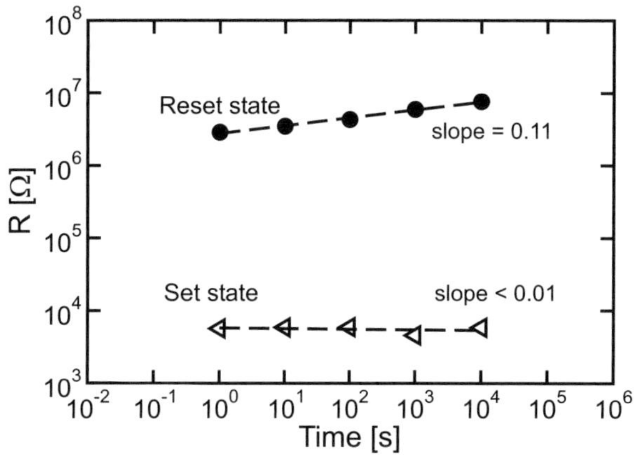

Fig. 14. 21 Measured resistance as a function of time after program, for set and reset states. Reset-state R drifts with time according to a power law, as a result of SR. (© (2007) IEEE, from [14.50]).

For more insight into the physics of structural relaxation, Fig. 14.22 shows *I-V* characteristics measured at room temperature before and after annealing at $T = 60$ or 120°C for 1 day.

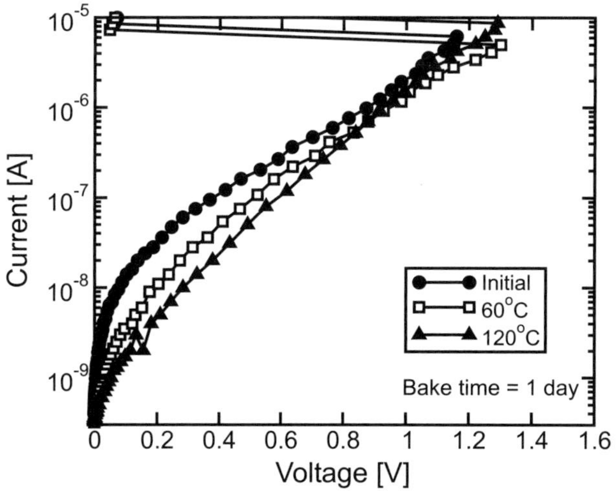

Fig. 14.22. Measured *I-V* curves before and after annealing for one day at T = 60 or 120°C. The *I-V* curves were measured at room temperature. Both low-voltage R and STS increase after bake. (© (2007) IEEE, from [14.50]).

In particular, Eq. (14.5) indicates that, within the PF model for transport in the amorphous phase, the increase of STS can only be explained by an increase in the average distance between Coulombic traps Δz, hence by a defect annihilation process in the amorphous structure.

Further evidence for defect annihilation during relaxation was shown for several amorphous materials by differential scanning calorimetry [14.41, 14.47, 14.48], photoconductivity [14.43], and mechanical relaxation studies [14.49]. The increase of resistance can be explained by the schematic in Fig. 14.23. The overlap of Coulombic potential wells can provide a significant lowering of the mobility edge E_C. Therefore, the annihilation of positively-charged traps can raise the E_C level. The increase of the mobility edge can be similarly explained noting that the reduction of trap density results in a larger trap spacing, hence in a smaller overlap between localized states. Accordingly, the mobility edge, defined as the energy level for which the carrier states become practically delocalized, is shifted to higher energies, to compensate for the larger average distance between the states.

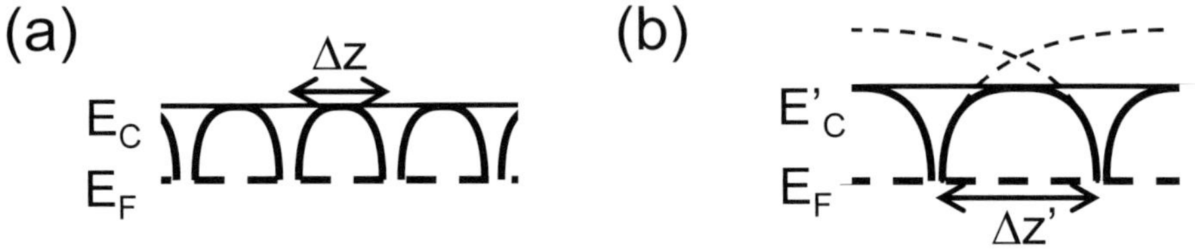

Fig. 14.23. Sketch for the increase of mobility edge in the amorphous chalcogenide upon relaxation. Defect annihilation from (a) to (b) leads to an increase of trap distance from Δz to $\Delta z'$ and, correspondingly, from a mobility edge E_C to E'_C, as a result of the lower interaction between Coulombic wells. (© (2007) IEEE, from [14.50]).

The increase in E_C is consistent with the observed increase of activation energy [14.45] and of V_T [14.46].

Fig. 14.24 shows a schematic of the kinetic model for defect annihilation.

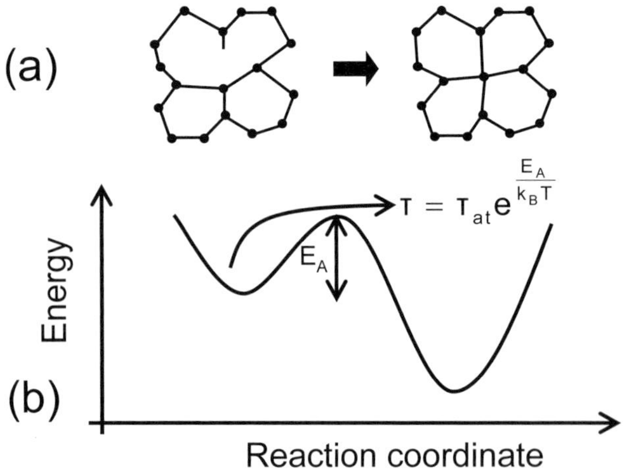

Fig. 14.24. Kinetic model for SR in chalcogenide glasses. Metastable states are healed (a) by thermal excitation over an energy barrier E_A (b). (© (2007) IEEE, from [14.50].

Aim of this model is to predict the time and temperature dependences of the trap concentration N_T, hence of R according to the PF transport model. Defects (e.g. dangling bonds, vacancies, bond distortion, either in length or angle [14.41]) are described as metastable states in the disordered structure (Fig. 14.24a).

Defect annihilation is modeled as a transition from the metastable state to a stable state at lower energy through thermal excitation over an energy barrier E_A. Thermal emission occurs in a characteristic time given by:

$$\tau(E_A) = \tau_{at} \exp\left(\frac{E_A}{k_B T}\right) \tag{14.15}$$

where τ_{at} is an atomic-vibration time describing the attempt to relaxation by thermal excitation (Fig. 14.24b) [14.42]. A mono-molecular kinetic as:

$$\frac{dg_T}{dt} = -\frac{g_T}{\tau(E_A)} \tag{14.16}$$

was assumed, where g_T is the trap density per volume and energy [$\text{cm}^{-3}\text{eV}^{-1}$] [14.41]. Note anyway that the overall time evolution of the total trap concentration N_T does not significantly depend on the order of the kinetics, i.e. whether the process is mono-, bi- or multi-molecular [14.50].

Fig. 14.25a shows calculated trap distribution g_T as a function of the activation energy, for increasing times during an annealing process at room temperature. The initial distribution is assumed to be exponential, to account for the broad spectrum of activation energies observed in differential scanning calorimetry experiments [14.41, 14.47, 14.48]. This exponential distribution of traps is also able to account for a power-law time-dependence of the total trap density $N_T = N_{T0} (t / t_0)^{-\gamma}$, where N_{T0}, t_0 and γ are constants, as observed in [14.43].

According to Eq. (14.15), defects with a low E_A annihilate first, thus the trap distribution after some time appears truncated at low energies due to annealing. The *annealing front*, separating annihilated traps from remaining traps, moves to higher E_A for increasing time, according to Eq. (14.15). From the overall concentration of traps, the PF transport model allows to calculate R as a function of time, as shown in Fig. 14.25b: R increases steadily with time according to a power law, in good agreement with experimental data (also shown).

Note that the activation energies assumed in the relaxation simulations are generally small, as compared to those measured for the crystallization mechanism (see Figs. 14.19 and 14.20). This is, on one hand, consistent with differential scanning calorimetry measurements, where relaxation effects are seen to occur at temperatures well below the crystallization point, e.g. room temperature. On the other hand, the relatively low relaxation energies can be understood considering the sketch in Fig. 14.26, which compares the energy excitation required for SR and crystallization processes. SR requires thermal excitation of few electrons at weakly-bonded atoms in correspondence of a structural defect, which requires the transition from localized states to conduction-band (free) states (Fig. 14.26a).

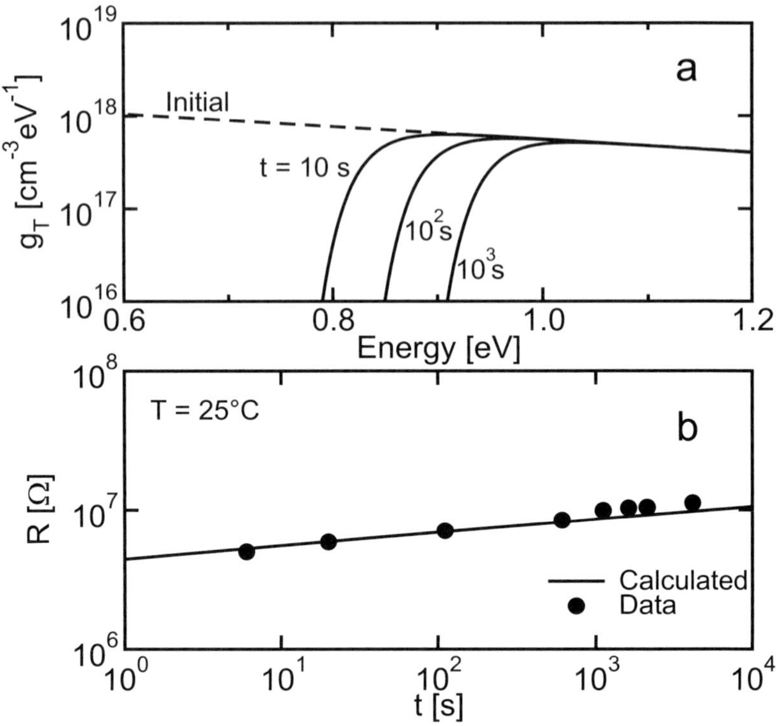

Fig. 14.25. (a) Calculated distribution of traps as a function of activation energies for increasing relaxation times 0 (initial), 10, 10^2 and 10^3 s, at room temperature. (b) Measured and calculated R as a function of relaxation time.

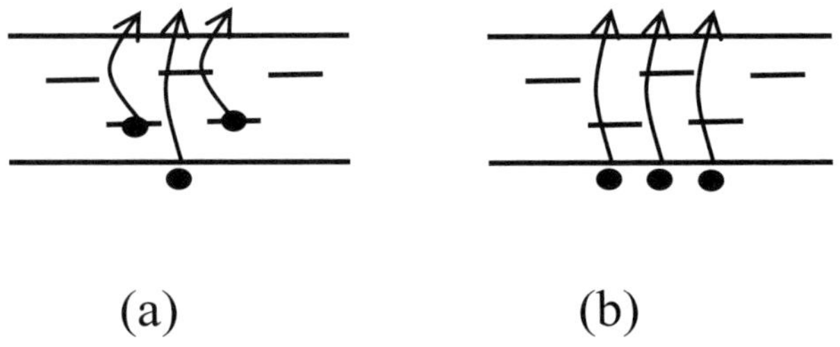

Fig. 14.26. Schematic view of SR (a) and crystallization processes (b), showing the different energies required for the structural transition. SR involves weak bonds i.e. localized states, whereas crystallization requires transition from non-bonding (valence) to bonding (conduction) states. (© (2007) IEEE, from [14.50]).

Crystallization is instead achieved by massive excitation of electrons from bonding (valence) to antibonding (conduction) states, resulting in a higher activation energy (Fig. 14.26b). Note however that exact values of energies within the simulation of Fig. 14.25 could be affected by uncertainties in the Arrhenius pre-exponential factor in Eq. (14.15), which is known to span several orders of magnitude for crystallization and glass transitions [14.42].

To highlight the temperature dependence of relaxation, Fig. 14.27 shows the measured and calculated power-law exponent ν as a function of temperature.

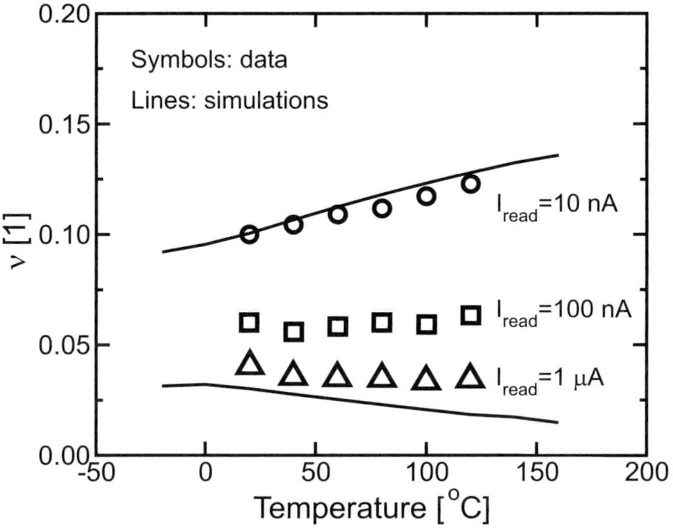

Fig. 14. 27. Measured and calculated ν of resistance $R = V/I_{read}$ evaluated at different read currents I_{read} = 10 nA, 100 nA and 1μA. (© (2007) IEEE, from [14.50]).

In the low-current regime (read current I_{read} = 10 nA), ν increases almost linearly with T, as a result of the temperature acceleration in Eq. (14.15). For higher I_{read}, the I-V curves are less affected by drift: in particular, ν remains below 0.04 in the investigated temperature range for I_{read} = 1 μA, as compared to a typical ν = 0.1 at room temperature and low current.

This can be understood considering the I-V curves in Fig. 14.22: at high I_{read}, the voltage is almost unchanged by the relaxation process, because of the partial compensation between a) the increase of R, leading to a current decrease in the Ohmic region of the I-V curve, and b) the increase of the STS, leading to a current increase in the exponential region of the I-V curve. In addition, reading at high current requires the measurement of relatively high voltage, thus the relative increase of $R = V / I_{read}$ will be smaller. This combination of advantages makes the high-current read particularly beneficial to minimize the impact of relaxation on the cell resistance.

14.9 Summary and Outlook

Modeling of PCM devices requires a detailed understanding of several different aspects of cell physics, ranging from material issues (e.g. hopping transport in amorphous semiconductors, phase transformation and relaxation by atomic rearrangement) to cell- and array-level studies (electrical and thermal transport within the 3D cell structure, heat leakage toward adjacent cells for disturb prediction [14.51]).

Although a significant progress has been achieved in the last 5 years on the modeling of PCM devices, there are several open issues that need to be addressed for a successful development and scaling of PCM in the next technological nodes. According to the ITRS roadmap [14.25], 22-nm technology PCM devices are expected in 2015, featuring a phase-change volume of less than 12000 nm^3 (corresponding to a hemisphere of 17 nm radius), 4 bits per cell (in the multilevel implementation) and 10 years data-retention. To achieve this goal, the physical understanding available to date is probably insufficient. In particular, there are strong concerns that the phase-change properties observed for relatively large devices (e.g. 90 nm technology) cannot be straightforwardly extended to extremely scaled devices due to size-dependent crystallization effects [14.52, 14.53]. To address these issues, dedicated experiments and ab-initio simulations aimed at understanding interface effects in the electrical, thermal and phase-change processes are necessary.

These problems are somehow complicated when one considers that device reliability requires controlled and predictable behaviors for 10^8-10^9 memory cells on the same chip [14.40]. In fact, it should be recalled that, for conventional Flash memories, many reliability issues of the device stem from the failure of only a minor statistical 'tail' of cells in an array. For PCM, accurate statistical modeling of the effects of composition fluctuations, contaminations and interface anomalies is mandatory for a knowledge-based development of reliable devices.

Optimization of the active material may offer a way to improve the reliability, hence the scaling potential of PCM devices. Doping and alloying of the chalcogenide material with, *e.g.*, Si [14.54] and In [14.55] have been shown to be beneficial in terms of the data retention behavior, thus providing a solution to alleviate possible size-dependent crystallization effects for ultra-scaled PCM cell. However, an efficient material engineering in the PCM device requires a detailed understanding and modeling of the relation between chemical composition, crystallographic and electronic structure, electrical/thermal properties and phase change parameters. As for many PCM modeling issues, this challenging task will only be possible through the cooperation of complementary modeling approaches and research fields, from material science to device engineering.

Acknowledgments. The author would like to thank A. L. Lacaita, S. Lavizzari, D. Mantegazza, A. Redaelli, U. Russo, D. Sharma and Y. Zhang who contributed to the research reviewed in this chapter. He also gratefully acknowledges several useful discussions with A. Pirovano, F. Pellizzer, R. Bez, D. Kau and P. Spadini.

References

[14.1] Pellizzer, F., Benvenuti, A., Gleixner, B., Kim, Y., Johnson, B., Magistretti, M., Marangon, T., Pirovano, A., Bez, R., Atwood, G.: A 90nm phase change memory technology for stand-alone non-volatile memory applications. In: Symp. on VLSI Tech. Dig., pp. 122–123. (2006)

[14.2] Ahn, S. J., Song, Y. J., Jeong, C. W., Shin, J. M., Fai, Y., Hwang, Y. N., Lee, S. H., Ryoo, K. C., Lee, S. Y., Park, J. H., Horii, H., Ha, Y. H., Yi, J. H., Kuh, B. J., Koh, G. H., Jeong, G. T., Jeong, H. S., Kim, K., Ryu, B. Y.: Highly manufacturable high density phase change memory of 64 Mb and beyond. In: IEDM Tech. Dig., pp. 907–910. (2004)

[14.3] Pirovano, A., Lacaita, A. L., Benvenuti, A., Pellizzer, F., Bez, R.: Electronic switching in phase-change memories. IEEE Trans. Electron Devices **51**, 452–459 (2004)

[14.4] Ielmini, D., Lacaita, A. L., Pirovano, A., Pellizzer, F., Bez, R.: Analysis of phase distribution in phase-change nonvolatile memories. IEEE Electron Device Lett. **25**, 507–509 (2004)

[14.6] Pellizzer, F., Pirovano, A., Ottogalli, F., Magistretti, M., Scaravaggi, M., Zuliani, P., Tosi, M., Benvenuti, A., Besana, P., Cadeo, S., Marangon, T., Moranti, R., Piva, R., Spandre, A., Zonca, R., Modelli, A., Varesi, E., Lowrey, T., Lacaita, A., Casagrande G., Bez, R.:Novel µtrench phase change memory cell for embedded and stand alone non volatile memory applications. In: Symp. VLSI Tech. Dig., pp. 18-19. (2004)

[14.7] Hindley, N. K.: Random phase model of amorphous semiconductors I. Transport and optical properties. J. Non-Crystalline Solids **5**, 17–30 (1970)

[14.8] Ielmini D., Zhang, Y.: Evidence for trap-limited transport in the sub-threshold conduction regime of chalcogenide glasses. Appl. Phys. Lett. **90**, 192102 (2007)

[14.9] Thomas, C. B.: The temperature dependence of the non-ohmic current and switching characteristics of a chalcogenide glass. J. Phys. D **9**, 2587–2596 (1976)

[14.10] Ielmini, D., Zhang, Y.: Analytical model for subthreshold conduction and threshold switching in chalcogenide-based memory devices. J. Appl. Phys. **102**, 054517 (2007)

[14.11] Jonscher, A. K., Hill, R. M.: Electrical conduction in disordered nonmetallic films. In Hass, G., Francombe, M. H., Hoffman, R. W. (eds.) Physics of thin films, Vol. 8. Academic Press, (1975)

[14.12] Redaelli, A., Pirovano, A., Pellizzer, F., Lacaita, A. L., Ielmini, D., Bez, R.: Electronic switching effect and phase-change transition in chalcogenide materials. IEEE Electron Device Lett. **25**, 684–686 (2004)

[14.13] Ielmini, D., Mantegazza, D., Lacaita, A. L., Pirovano, A., Pellizzer, F.: Parasitic reset in the programming transient of phase change memories. IEEE Electron Device Lett. **26**, 799–801 (2005)

[14.14] Adler, D., Shur, M. S., Silver M., Ovshinsky, S. R.: Threshold switching in chalcogenide-glass thin films. J. Appl. Phys. **51**, 3289–3309 (1980)

[14.15] Prakash, S., Asokan, S., Ghare, D. B.: A guideline for designing chalcogenide-based glasses for threshold switching characteristics. IEEE Electron Device Lett. **18**, 45–47 (1997)

[14.16] Ovshinsky, S. R.: Localized states in the gap of amorphous semiconductors. Phys. Rev. Lett. **36**, 1469–1472 (1976)

[14.17] Adler, D., Henisch, H. K., Mott, N.: The mechanism of threshold switching in amorphous alloys. Rev. Mod. Phys. **50**, 209–220 (1978)

[14.18] Mott, N. F.: Electrons in Glass. In: Lundqvist, S. (ed.) Nobel Lectures, Physics 1971-1980. World Scientific Publishing Co., Singapore (1992)

[14.19] Emin, D.: Current-driven threshold switching of a small polaron semiconductor to a metastable conductor. Phys. Rev. B **74**, 035206 (2006)

[14.20] Jonscher, A. K.: Energy losses in hopping conduction at high electric fields. J. Phys. C **4**, 1331–1340 (1971)

[14.21] Jonscher, A. K., Loh, C. K.: Poole-Frenkel conduction in high alternating electric fields. J. Phys. C **4**, 1341–1347 (1971)
[14.22] Lacaita, A. L., Redaelli, A., Ielmini, D., Pellizzer, F., Pirovano, A., Benvenuti, A., Bez, R.: Electrothermal and phase-change dynamics in chalcogenide-based memories. In: IEDM Tech. Dig., pp. 911-914. (2004)
[14.23] Russo, U., Ielmini, D., Redaelli, A., Lacaita, A. L.: Modeling of programming and read performance in phase-change memories – Part I: cell optimization and scaling. IEEE Trans. Electron Devices **55**, 506-514, (2008)
[14.24] Hwang, Y. N., Lee, S. H., Ahn, S. J., Lee, S. Y., Ryoo, K. C., Hong, H. S., Koo, H. C., Yeung, F., Oh, J. H., Kim, H. J., Jeong, W. C., Park, J. H., Horii, H., Ha, Y. H., Yi, J. H., Koh, G. H., Jeong, G. T., Jeong, H. S., Kim, K.: Writing current reduction for high-density phase-change RAM. In: IEDM Tech. Dig., pp. 893-896. (2003)
[14.25] International Roadmap for Semiconductors—Process Integration, Devices, and Structures. 2007 Release. http://www.itrs.net/ (2007). Accessed 17 April 2008.
[14.26] Pirovano, A., Lacaita, A. L., Benvenuti, A., Pellizzer, F., Hudgens, S., Bez, R.: Scaling analysis of phase change memory technology. In: IEDM Tech. Dig., pp. 699-702. (2003)
[14.27] Pirovano, A., Redaelli, A., Pellizzer, F., Ottogalli, F., Tosi, M., Ielmini, D., Lacaita, A. L., Bez, R.: Reliability study of phase-change non volatile memories. IEEE Trans. Device Mater. Rel. **4,** 422–427 (2004)
[14.28] Redaelli, A., Ielmini, D., Russo, U., Lacaita, A. L.: Intrinsic data retention in nanoscaled phase-change memories – Part II: Statistical analysis and prediction of failure time. IEEE Trans. Electron Devices **53**, 3040–3046 (2006)
[14.29] Russo, U., Ielmini, D., Lacaita, A. L.: Analytical modeling of chalcogenide crystallization for PCM data-retention extrapolations. IEEE Trans. Electron Devices **54**, 2769 – 2777 (2007)
[14.30] Russo, U., Ielmini, D., Redaelli, A., Lacaita, A. L.: Intrinsic data retention in nanoscaled PCMs—Part I: Monte Carlo model for crystallization and percolation. IEEE Trans. Electron Devices **53**, 3032–3039 (2006)
[14.31] Stathis, J. H.: Percolation models for gate oxide breakdown. J. Appl. Phys. **86**, 5757–5766 (1999)
[14.32] Ielmini, D., Spinelli, A. S., Lacaita, A. L., van Duuren, M. J.: Impact of correlated generation of oxide defects on SILC and breakdown distributions. IEEE Trans. Electron Devices **51**, 1281–1287 (2004)
[14.33] Christian, J. W.: The Theory of Transformations in Metals and Alloys. Oxford, U.K. (1975)
[14.34] Senkader, S., Wright, C. D.: Models for phase-change of $Ge_2Sb_2Te_5$ in optical and electrical memory devices. J. Appl. Phys. **95**, 504–511 (2004)
[14.35] Peng, C., Cheng, L., Mansuripur, M.: Experimental and theoretical investigations of laser-induced crystallization and amorphization in phase change optical recording media. J. Appl. Phys. **82**, 4183–4191 (1997)
[14.36] Ruitenberg, G., Petford-Long, A. K., Doole, R. C.: Determination of the isothermal nucleation and growth parameters for the crystallization of thin $Ge_2Sb_2Te_5$ films. J. Appl. Phys. **92**, 3116–3123 (2002)
[14.37] Kalb, J., Spaepen, F., Wuttig, M.: Atomic force microscopy measurements of crystal nucleation and growth rates in thin films of amorphous Te alloys. Appl. Phys. Lett. **84**, 5240–5242 (2004)
[14.38] Singh, H. B., Holz, A.: Stability limit of supercooled liquids. Solid State Commun. **4**, 985–987 (1983)
[14.39] Kalb, J. A., Spaepen, F., Wutting, M.: Kinetics of crystal nucleation in undercooled droplets of Sb- and Te-based alloys used for phase change recording. J. Appl. Phys. **98**, 54910 (2005)

[14.40] Mantegazza, D., Ielmini, D., Pirovano, A., Varesi, E., Lacaita, A. L.: Statistical analysis and modeling of programming and retention in PCM arrays. In: IEDM Tech. Dig. 311–314. (2007)
[14.41] Roorda, S., Sinke, W. C., Poate, J. M., Jacobson, D. C., Dierker, S., Dennis, B. S., Eaglesham, D. J., Spaepen, F., Fuoss, P.: Structural relaxation and defect annihilation in pure amorphous silicon. Phys. Rev. B **44**, 3702–3725 (1991)
[14.42] Khonik, V. A., Kitagawa, K., Morii, H.: On the determination of the crystallization activation energy of metallic glasses. J. Appl. Phys. **87** 8440–8443 (2000)
[14.43] Koughia, K., Shakoor, Z., Kasap, S. O., Marshall, J. M.: Density of localized electronic states in a-Se from electron time-of-flight photocurrent measurements. J. Appl. Phys. **97**, 3706–3716 (2005)
[14.44] Kasap, S. O., Yannacopoulus, S.: Kinetics of structural relaxations in the glassy semiconductor *a*-Se. J. Mater. Res. **4**, 893–905 (1989)
[14.45] Pirovano, A., Lacaita, A. L., Pellizzer, F., Kostylev, S. A., Benvenuti, A., Bez, R.: Low-field amorphous state resistance and threshold voltage drift in chalcogenide materials. IEEE Trans. Electron Devices **51**, 714–719 (2004)
[14.46] Ielmini, D., Lacaita, A. L., Mantegazza, D.: Recovery and drift dynamics of resistance and threshold voltages in phase-change memories. IEEE Trans. Electron Devices **54**, 308–315 (2007)
[14.47] Roorda, S., Doorn, S., Sinke, W. C., Scholte, P. M. L. O., van Loenen, E.: Calorimetric evidence for structural relaxation in amorphous silicon. Phys. Rev. Lett. **62**, 1880–1883 (1989)
[14.48] Kalb, J. A., Wuttig, M., Spaepen, F.: Calorimetric measurements of structural relaxation and glass transition temperatures in sputtered films of amorphous Te alloys used for phase change recording. J. Mater. Res. **22**, 748–754 (2007)
[14.49] Kasap, S. O., Polischuk, B., Aiyah, V., Yannacopoulos, S.: Drift mobility relaxation in a-Se. J. Appl. Phys. **67**, 1918–1922 (1990)
[14.50] Ielmini, D., Lavizzari, S., Sharma D., Lacaita, A. L.: Physical interpretation, modeling and impact on phase change memory (PCM) reliability of resistance drift due to chalcogenide structural relaxation. In: IEDM Tech. Dig., pp. 939–942. (2007)
[14.51] Russo, U., Ielmini, D., Redaelli, A., Lacaita, A. L.: Modeling of programming and read performance in phase-change memories . Part II: Program disturb and mixed scaling approach. IEEE Trans. Electron Devices **55**, 515-522 (2008)
[14.52] Hirasawa, M., Orii, T., Seto, T.: Size-dependent crystallization of Si nanoparticles. Appl. Phys. Lett **88**, 093119 (2006)
[14.53] Raoux, S., Rettner, C. T., Jordan-Sweet, J. L., Deline, V. R., Philipp, J. B., Lung, H.-L., Scaling properties of phase change nanostructures and thin films. In: European Symposium on Phase Change and Ovonic Science. (2006)
[14.54] Qiao, B., Feng, J., Lai, Y., Cai, Y., Lin, Y., Tang, T., Cai B., Chen, B.: Si–Sb–Te films for phase-change random access memory. Semiconductor Science and Technology **21**, 1073-1076 (2006)
[14.55] Morikawa, T., Kurotsuchi, K., Kinoshita, M., Matsuzaki, N., Matsui, Y., Fujisaki, Y., Hanzawa, S., Kotabe, A., Terao, M., Moriya, H., Iwasaki, T., Matsuoka, M., Nitta, F., Moniwa, M., Koga, T., Takaura, N.: Doped In-Ge-Te phase change memory featuring stable operation and good data retention. In: IEDM Tech. Dig., pp. 307–310. (2007)

15. Phase Change Random Access Memory Advanced Prototype Devices and Scaling

Yi-Chou Chen

Abstract This chapter describes the design, fabrication, and testing of advanced prototype Phase Change Random Access Memory (PCRAM) devices that have been used to study the scaling behavior of PCRAM. It compares various PCRAM designs and summarizes what has been learned from the performance of these devices regarding the switching properties and scaling behavior of PCRAM. In addition, specific test equipment and test procedures for the characterization of PCRAM are described.

15.1 Introduction

To explore the limitations of phase change memory, different types of Phase Change Random Access Memory (PCRAM) devices have been fabricated and investigated. In this chapter, we will discuss different advanced prototype devices, compare their advantages and disadvantages, and describe the discoveries and observations that were found using these different devices. Among all the properties, the scaling capability is a major concern. Since scaling is a critical consideration for solid-state devices, especially for solid-state memory, the development of phase change memory also needs to address this difficult challenge. Solid-state devices have obeyed Moore's law for several decays and a new technology will be developed only if it can scale. For example, LOCOS (LOCal Oxidation of Silicon) technology was the dominating isolation technology in the CMOS (Complementary Metal Oxide Semiconductor) process for features as small as half a micron [15.1-3]. However, the bird-beak phenomenon became an un-solvable scaling issue for features smaller than half a micron. LOCOS has now been almost totally replaced by STI (Shallow Trench Isolation) [15.1].

Compared with solid-state devices, scaling plays an even more important role for phase change memory. Phase change memory utilizes melting/quenching (RESET) and crystallization (SET) of the phase change material to switch it between the two different states. The required RESET current to melt the material

S. Raoux, M. Wuttig (eds.), *Phase Change Materials*,
DOI 10.1007/978-0-387-84874-7_15,

and transform it back to the amorphous phase can be rather high. In order to provide enough RESET current, the size of the access device for the memory cell needs to be relatively large and a large cell size directly increases the manufacturing cost. Furthermore, the performance of the chip will be degraded when the writing current of the unit cell is high since the overall writing speed will be reduced. In order to reduce the RESET current, scaling becomes a challenging and important topic for phase change memory. As a result, many advanced prototype devices have been built for evaluating this subject.

In this chapter, we will divide the prototype devices into three major categories. The first category includes devices where scaling is achieved by reducing the contact size between the phase change material and the electrodes. Devices in the second category are scaled by reducing the size of the phase change material itself. In the third category, we will introduce other potential device structures or approaches to scaling and RESET current reduction. In addition to the structures themselves, the required methods and techniques for electrical testing are also discussed in this chapter because there are several special requirements for testing a scaled phase change device, e.g. wide resistance range, fast response speed, and low programming current.

15.2 Device Scaling by Reducing the Electrode Contact Area

In this section, four major structures will be introduced. The first one is the heater structure. It is a widely investigated structure and it is a straightforward approach to scaling. It involves the reduction of the contact size between the bottom electrode and the phase change material by applying techniques to form sub-lithographic bottom contacts. Many innovative fabrication techniques have been developed in the course of trying to reduce the contact area, and these structures will also be presented in this section. The second approach uses so-called edge-contact structures. Edge-contact structures utilize the ability to control film thicknesses at much smaller dimensions than the lithographic process size control, and very small contact sizes were obtained using conventional lithographic processes. A third similar but vertical structure called the "μTrench" (micro-trench) will then be discussed. This structure also uses the superior control of the thickness of thin films to further extend scaling. Lastly, a "ring structure" combines the advantages of sub-lithographic contact size and control of the thickness of thin films. This structure provides not only a reduced contact area but is also characterized by a tight process control on the device properties.

15.2.1 The Heater Structure

One very simple and efficient way to scale the device is to reduce the contact area between the phase change material and the electrode. Lai et al. [15.4] demonstrated a basic concept of this structure as shown in Fig. 15.1.

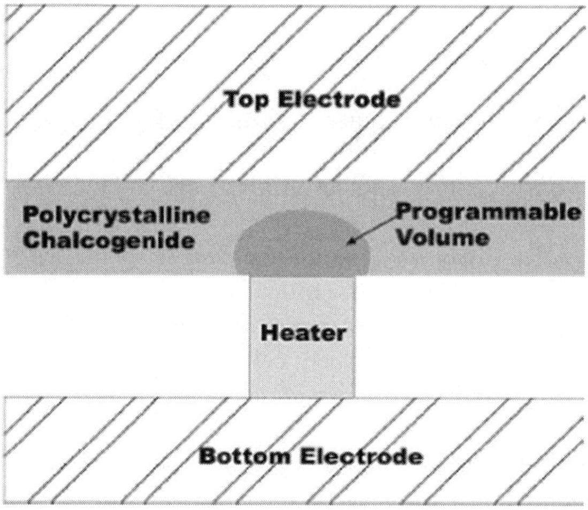

Fig. 15.1. A schematic drawing of the heater structure. (© (2001) IEEE, from [15.4])

This structure contains a bottom contact via conductive electrode called heater beneath the phase change material whose size has been defined by sub-lithographic techniques. After filling the sub-lithographic via with a conducting material, e.g. TiN, the extra TiN is removed by Chemical Mechanical Polishing (CMP), and the wafer is covered by the phase change material. A top electrode is then formed that contacts the phase change material. In this structure, the diameter of the heater determines the power required for changing the resistance state. The relationship between the diameter of the heater and the RESET current is shown in Fig. 15.2 [15.5].

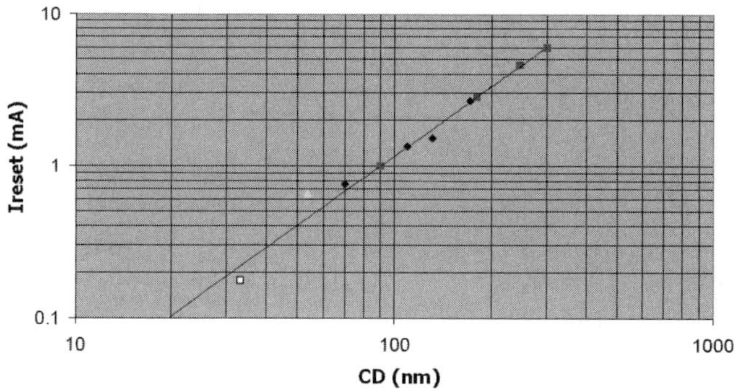

Fig. 15.2. The RESET current as a function of the contact critical dimension (CD). (© (2003) IEEE, from [15.5])

This structure is also known as a mushroom or a matchstick structure due the geometrical shape of the heater and the phase change region. The structure is well suited for scaling studies because the phase change material is deposited on a planar surface, indicating an easy fabrication process with high yield. $Ge_2Sb_2Te_5$ (GST) based material is typically used as the phase change material. GST is a well-established material that has been widely studied and is used in optical storage media (see Chapter 12). The crystallization mechanism for this material is nucleation-dominated, which means there are many crystallization nuclei sites throughout the material when the crystallization process takes place everywhere in the amorphous material instead of predominant crystal growth from the amorphous-crystalline interface. GST was chosen because the material has favorable properties in terms of data retention, crystallization speed, and resistivities in both phases. Prototype test cells have been demonstrated with a writing endurance of more than 10^{12} times [15.4].

The crystallization speed is less than 50 ns and the resistivity is about 0.01 ohm-cm in the crystalline phase and 1000 ohm-cm in the amorphous phase. Simulation studies of the scaling behavior for this structure were performed in 2003 [15.6]. The simulation results indicate that adjacent cells can sustain data for more than 10 years without so-called cross-talk even if the devices have been scaled to less than 65 nm. The sputter deposition process for GST material has been well developed, so thin film fabrication is not very difficult. In addition to the advantages of easy processing, the structure has advantages from the thermal and electrical standpoint. Electrically, the current flows through the relatively large top electrode and phase change material into the confined heater, where the current crowding effect results in a high temperature region just above the heater, thus the size of the phase change (active) region can be controlled. Thermally, the heating area of the device is far away from the top electrode. As a result, the heat lost to the top electrode is reduced. Due to its advantages, many phase change memory studies have been done or derived from this structure, including the investigation of new materials, the influence of the cell geometry on switching properties, and the study of adhesion and interface layers. In addition to the contact size, the structure has been further studied by evaluating the following aspects.

15.2.1.1 Additional Adhesion Layer

It has been reported that the adhesion between a dielectric layer and the phase-change material can be improved by adding an adhesion layer [15.7]. Ti is usually used in the CMOS process and can be considered as an important candidate. The adhesive fracture energy can be improved from ~0.16 J/m^2 to ~0.7 J/m^2. by replacing SiO_2 with Ti. Reports indicate however that Ti can react with Te in the normally used Ge-Sb-Te based phase change material resulting in a degradation of the device properties [15.7, 15.8]. Ta_2O_5 was also proposed as an adhesion layer. The structure is illustrated in Fig. 15.3.

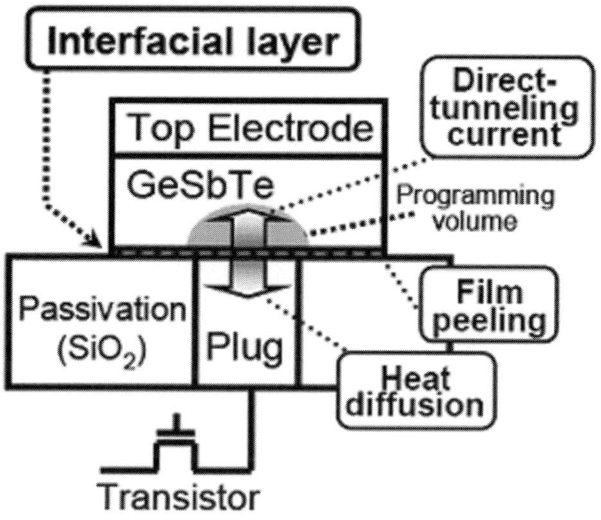

Fig. 15.3. Ta_2O_5 interfacial layer to enhance the adhesion. (© (2006) IEEE, from [15.7])

The adhesion between the dielectric material and the phase change material was significantly improved due to a relatively high adhesive fracture energy (~0.41 J/m^2). In addition, the RESET current was also reduced significantly due to lower heat dissipation [15.7].

15.2.1.2 Size Effect of the Phase Change Material

The size and the thickness of the phase change material were found to influence the RESET current. In the case of a 75 nm diameter heater, the RESET current was reduced from 2.8 mA to 2.0 mA for a square GST volume when the size was reduced from 0.54 μm x 0.54 μm to 0.4 μm x 0.4 μm [15.9]. Hwang et al. explained that the reduction of RESET current is due to current localization [15.10]. Since the size of the GST is much larger than the heater, the shape of the current path may affect the RESET current. In addition to the horizontal size, the thickness of the phase change material is also very important for reducing the RESET current, as shown in Fig. 15.4.

15.2.1.3 Different Phase Change Materials

Many different material studies have been done using the heater structure. It was found that Sn-doped GST possesses a fast operation speed [15.11]. N-doped GST reduces the RESET current by increasing the dynamic resistance [15.9, 15.10].

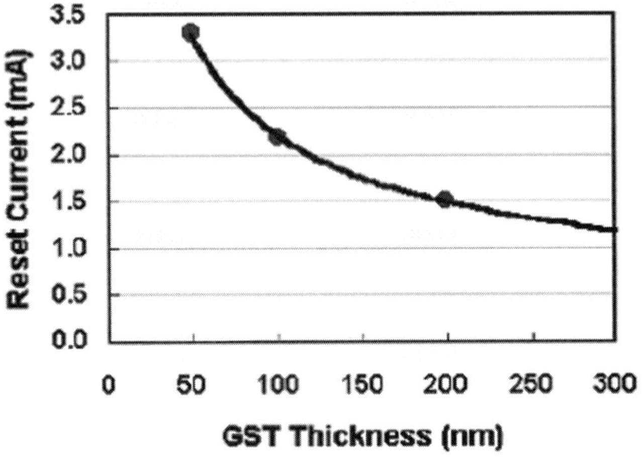

Fig 15.4. RESET current as a function of the GST thickness. (© (2003) IEEE, from [15.10])

This material further possessed a fine crystal size and a high activation energy; both are advantages for scaling. Oxygen doped GST was investigated by Matsuzaki et al. [15.12] and was shown to significantly reduce the RESET current. InGeTe was proposed by Hitachi and it was shown that the data retention is improved compared to GST [15.13].

15.2.1.4 Process Integration Issues for Scaling

There are some special process integration issues in the heater structure. In order to reliably fabricate sub-lithographic contacts, process variation is very critical. We assume that D is the absolute process variation of the feature with size F. In a conventional lithographic process the related process variation is D/F. In a sub-lithographic process however the actual diameter of the heater is a fraction $1/x$ of F. If the absolute value of the process variation stays the same, the related variation of the sub-lithographic feature becomes Dx/F, i.e. the process variation is x times larger than in the conventional process. In addition to difficulties with the process variation, etching damage of the GST was also observed [15.14, 15.15]. Improvements in the etching process conditions and reductions of the exposure area to the etching process can reduce this effect. But some damaged layer at the outer rim of the contact cannot be fully avoided, and the existence of this damaged layer could limit the scaling capabilities of this structure.

15.2.2 The Edge Contact Structure

Introduced by Samsung, the edge contact structure reduces the contact size by using a TiN thin film electrode denoted by BE in Fig. 15.5 [15.16]. Scaling by applying thin films is a very efficient way because the thin film thickness can be much smaller than the dimension of the lithography and can be very well controlled, see Fig. 15.5.

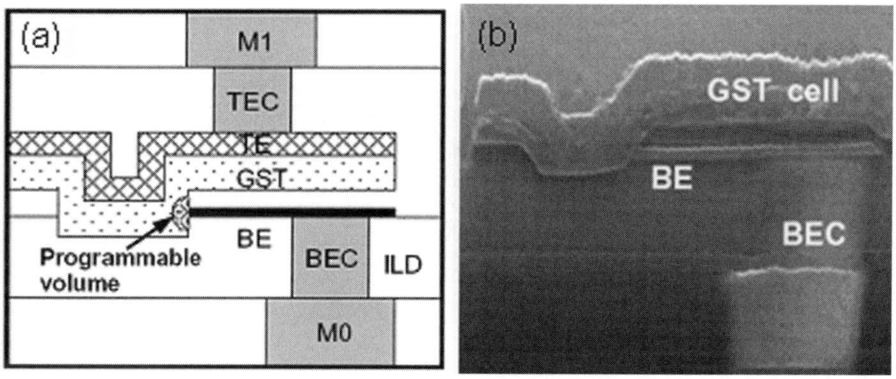

Fig.15.5. The schematic structure (a) and the scanning electron microscope (SEM) image (b) of the edge contact approach. (© (2001) IEEE, from [15.16])

In this work, the contact size was 0.2 μm but the TiN film thickness was only 0.02 μm.

In addition to the reduction of contact size, the film thickness variability can be well controlled (typically within 2% compared to 10% for the lithographic dimension), improving the uniformity over the wafer. The disadvantage of this structure is that the size of the cell is larger than a vertical structure. In the heater process, a $4F^2$ unit cell can be achieved but in the edge contact structure approach the cell size is at least $8F^2$. In addition, the process is more complex; the phase change material needs to be deposited on the sidewall, which may cause reliability issues. The number of masks required to fabricate the structure is also larger than for the heater structure. Simulation results indicate that it is not easy to SET the cell because an over-programming RESET pulse may generate an amorphous shell that is not easily re-crystallized [15.17]. The difficulties in controlling the fabrication process and in programming along with the complexity of the process present problems for this approach.

15.2.3 µTrench Structure

The µTrench approach was introduced by Pellizzer et al. in 2004 [15.18]. It utilizes the advantage of a thin film process to reduce the contact size and it is a vertical device. Similar to the edge contact structure, a TiN thin film is used to scale one dimension of the contact size, although the other dimension (AA' plane in Fig. 15.6) is still defined by conventional lithographic technology. Using a 180 nm process a RESET current of ~ 600 µA was achieved.

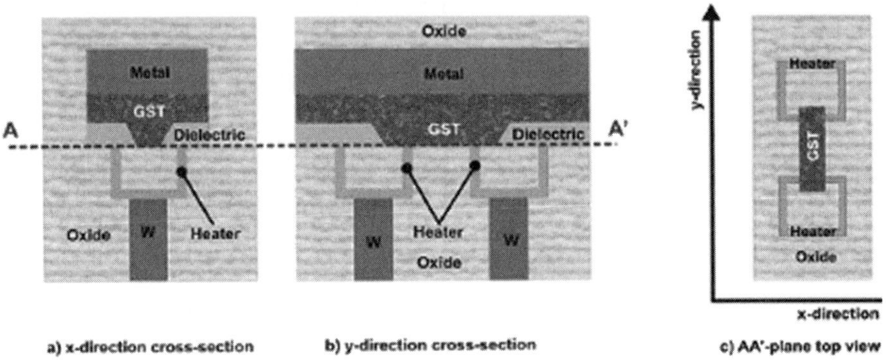

Fig. 15.6. Schematic cross-sectional plot and plain view of the µTrench structure. (© (2004) IEEE, from [15.18])

The GST material is confined in a dielectric-layer-defined µTrench. This dielectric layer is used to determine the current path, because the bottom electrode is a large ring in the shape of a square, and only part of the bottom electrode is electrically connected to the GST material. The structure had very small contact size. Except some considerations regarding the deposition of GST into the µTrench and the patterning of the GST for isolation, the processes used to fabricate this structure are similar to the standard CMOS process. This structure has been highly investigated due to the ease of the fabrication and good electrical performance [15.18-22].

15.2.4 The Ring Structure

Similar to the µTrench structure, the ring structure is a vertical cell, as shown in Fig. 15.7(a) [15.23]. It combines the advantages of a vertical structure, thin film scaling, and a very tight process distribution. The appearance of the ring structure is very similar to the heater structure. Compared to the heater structure, the ring structure uses a hollow cylinder of TiN with a TiN bottom to replace the solid pillar TiN heater. Since the contact area is defined by the shell thickness, it is less dependent on the contact diameter.

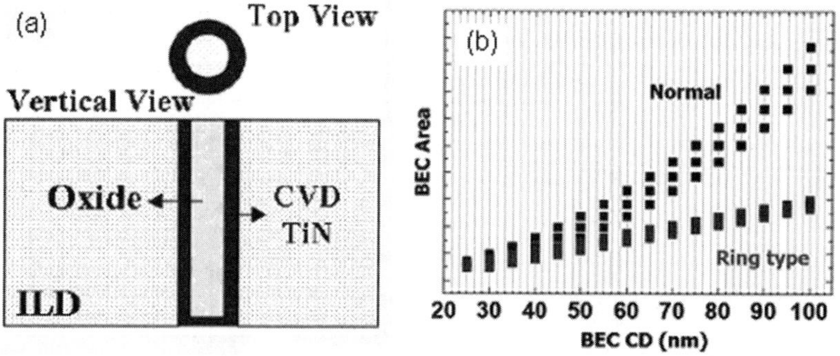

Fig 15.7. (a) Top view and vertical view of the ring structure (© (2005) IEEE, from [15.23]), and (b) contact area variation as a function of the contact CD (critical dimension) (© (2005) IEEE, from [15.24]). BEC is the bottom electric contact, ILD is the inter-layer dielectric.

As a result, process variations are less of a problem for this structure, as shown in Fig. 15.7(b). Simulation results indicate that this structure can reduce the RESET current by about 8 % (from 0.62 mA to 0.57 mA). Though this RESET current reduction is not significant, the tight distribution control can further relax the requirements for current drivability, because in a memory array the cell that needs the highest current determines the writing current. In order to further reduce the process variation, several process improving approaches have been demonstrated, such as 2-step CMP for the ring, encapsulating layer, and heat dissipation structure [15.23-26]. Along with a p-n diode the ring structure device achieved a cell size of 5.8 F^2, which is competitive in cell size with other structures [15.26, 15.27].

15.3 Device Scaling by Reducing the Phase Change Material Volume

In the previous section, the phase change memories are scaled by reducing the contact size. An alternative method is to scale the volume of the phase change material itself. To scale the phase change material, processes should be carefully chosen and fine-tuned. As discussed in the previous section, the etching damage might degrade the electrical properties of the phase change material. When the material size is smaller than two times the depth of the damaged layer there is no undamaged or unmodified material left. An important consideration for these approaches then is how to keep device performance while scaling the material.

15.3.1 The Pillar Structure

In 2006, Happ et al. proposed a structure that scales the phase change material and the contact size [15.28]. Figure 15.8 illustrates the structure and the process flow.

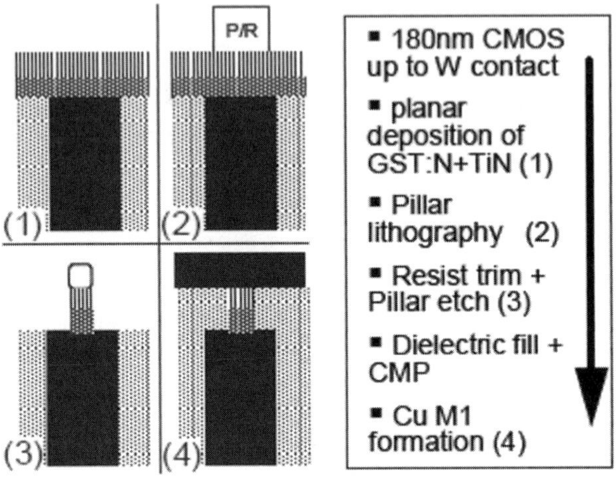

Fig. 15.8. The process flow of the pillar structure. (© (2006) IEEE, from [15.28])

To fabricate the structure a standard 180 nm CMOS process was followed up through the W contact. After planar deposition of N_2 doped GST and TiN thin film, pillar lithography is developed on the TiN layer. Resist trimming and pillar etch is used to construct the pillars. After a dielectric fill and CMP, Cu metal layers are used to make electrical connection. This structure can reduce the RESET current significantly. However, for the pillar cell it is difficult to achieve high aspect ratios (Fig. 15.9), in particular the etching process is very challenging.

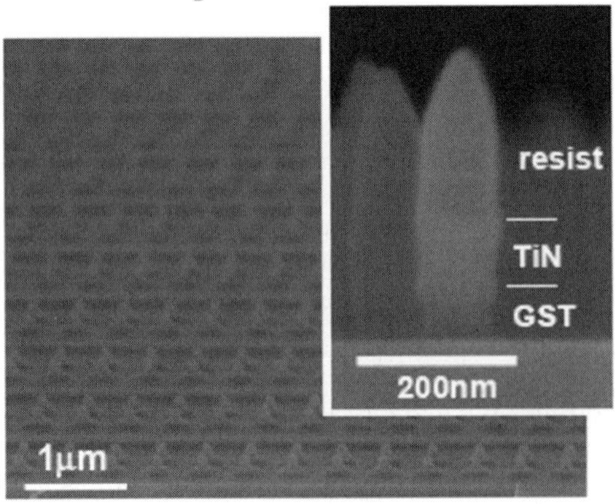

Fig. 15.9. Scanning Electron Microscope (SEM) images of the pillar array after the etching process and a closer view of a pillar (inset). (© (2006) IEEE, from [15.28])

The line structure and the bridge structure introduced in the following sections can reduce the sensitivity to process variations.

15.3.2 The Line Structure

By using a doped SbTe material, researchers from Philips (now NXP) scaled the phase change material by using a lateral "line" structure [15.29]. The structure is shown in the Fig. 15.10.

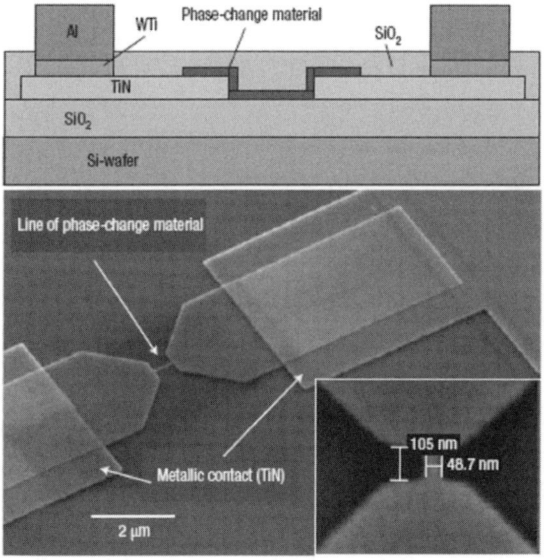

Fig. 15.10. Schematic plot of the line structure (top) and SEM images of the structure after fabrication (bottom). Reprinted by permission from Macmillan Publishers Ltd: Nature Materials [15.29], copyright (2005).

The two electrodes were made by a TiN layer of 50 nm in thickness. The phase change material, doped SbTe, was deposited by sputtering and then patterned by electron beam lithography and RIE (reactive ion etching). The phase change material extends from the two electrodes and then shrinks to a narrow line structure in the middle between the two electrodes. 100 nm of PECVD (plasma enhanced chemical vapor deposition) SiO_2 was subsequently deposited on the wafer to protect the structure. The phase change material was contacted by TiW/Al contact pads for electrical testing. The doped SbTe material is a growth-dominated material. Compared with a conventional nucleation-dominated material (e.g. GST) the growth-dominated material appears to be better in terms of resistance variation [15.21, 15.30], but the material might face problems of data retention [15.13].

When the crystallization speed of the material is v, and the gap between the two electrodes is L, the time for full crystallization of a growth-dominated material is L/v. When we scale the devices to $0.7L$ (one generation), the time for full crystallization becomes $0.7L/v$ so the crystallization speed is a function of device dimension. To ensure both high operation speed and long data retention crystallization speed and the activation energy (E_a) of crystallization of the material should be carefully investigated.

Measurements of the electrical properties of the phase change memory devices are quite different from testing conventional CMOS devices. We will discuss testing procedures of highly scaled structures in Section 15.5.

15.3.3 The Bridge Structure

Another lateral structure called the "bridge" structure which is similar to the "line" structure has been demonstrated by Chen et al. [15.31]. The structure is shown in Fig. 15.11.

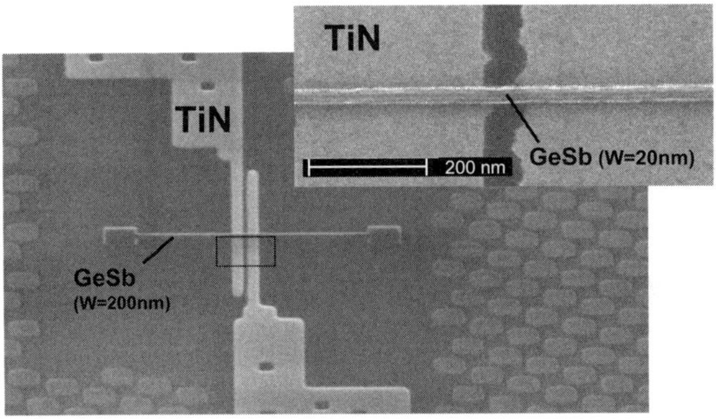

Fig. 15.11. Plane view SEM of a manufactured bridge device. A closer view of the device is shown in the inset. (© (2006) IEEE, from [15.31])

Two TiN electrodes were separated by a dielectric layer and a thin phase change bridge connected the two electrodes by crossing the dielectric layer. The bridge was made from a doped GeSb material. This material is also a growth-dominated material, which has very high crystallization speed. In order to investigate the scaling limit, advanced electron beam (e-beam) lithography was applied to fabricate the smallest feature sizes with good size control.

The TiN bottom electrode structures were fabricated from a thick SiO_2 layer on Si. Using KrF lithography and etch trimming, first two trenches with a very thin oxide wall between them were made. Thin TiN was deposited and then a CMP process, which stopped on the SiO_2 layer, was employed to remove the extra TiN and planarize the sample. After fabrication of the TiN bottom electrodes, a thin (as thin as 3 nm) doped GeSb layer was deposited by PVD (physical vapor deposition) using magnetron sputtering. GeSb is easily oxidized by air exposure, thus a thin SiO_2 layer was deposited in-situ on the GeSb layer to protect the material. To make the bridge cell, e-beam lithography was used to define the phase change bridge itself ranging in width from 20 nm to 200 nm with a large structure on each end to prevent collapse of the photo resist, as shown in Fig. 15.11. The photo resist used was a SiO_x based negative photo resist, so it did not need to be removed after the fabrication process. Ar ion-milling was then used to transfer the pattern into the phase change material, and a thin Al_2O_3 or SiO_2 film was deposited in-situ to prevent oxidation of the phase change material. The structure was then covered by a thick 50 nm layer of SiO_2 for protection, and wiring patterns were made for electrical connection.

The doped GeSb is not a commonly used material system in phase change memory because it is not a chalcogenide. The material was chosen because it shows very good scalability and it has been demonstrated that films as thin as 1.3 nm still show clear crystallization [15.32]. Moreover, the crystallization speed of this material is very fast. One drawback of this material is its relatively low resistivity, so the voltage drop on the serial resistor in the circuit is significant which is not desirable. In order to adjust the resistivity, a dopant was added to the GeSb, and the resistivity (in the crystalline state) was successfully increased to > 1 mohm-cm from about 0.1 mohm-cm for undoped GeSb.

Electrical testing proved that even the smallest devices could be fully operated indicating a further possibility of scaling. The RESET current was 90 μA and the RESET voltage was 1.5 V, with a device of 20 nm x 3 nm x 50 nm in width, thickness, and length, respectively. The testing was performed with a serial resistor of 6 kohm. The details of the testing instruments will be discussed in Section 15.5. The power dissipated on the device was 95 μW with a RESET time less than 60 ns [15.31]. Further investigations revealed that the device can be operated using RESET and SET pulses as short as 40 ns [15.33]. Reliability test results showed a cycle endurance of more than 10^4 times [15.31]. Even though this meets the lowest endurance requirement of flash memory, an endurance of more than 100,000 times can be expected for these devices. Retention tests indicated that the device can sustain a high temperature of up to 220 °C [15.33]. However, more detailed reliability tests need to be performed to fully evaluate the viability of this structure.

Further investigations by Castro et al. [15.34] indicated that there is a Thomson effect in such a structure. As a result the molten zone shifts towards the anode electrode during RESET programming. Therefore it might be possible to reduce the RESET current using an asymmetric cell structure.

15.4 Other Prototype Devices

In addition to device scaling by reducing the contact area or volume of the phase change material, other approaches are also being pursued. By combining the advantages of different strategies, the device performance can be further improved.

In this section, three different device types are described. The first one scales not only the size of the bottom electrode but also the phase change material itself resulting in a very small contact area. The second one uses MLC (multi-level-cell) technology to increase the data storage density. The third one reduces the RESET current by recessing the bottom electrode.

15.4.1 Scaling Both the Material and the Contact

In a structure reported by Chen et al. [15.35] both the electrode and the phase change material were scaled to smaller sizes. This so called cross-spacer structure is shown in the Fig. 15.12.

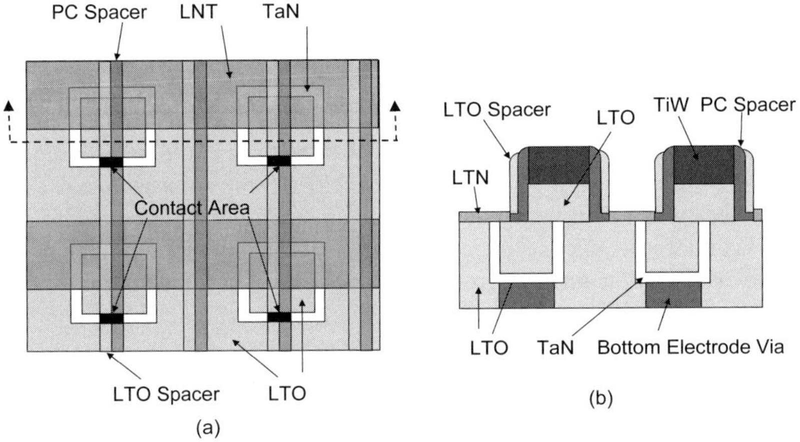

Fig.15.12. The schematic plots of the cross-spacer structure (a) plain view, and (b) cross-sectional view. (© (2007) IEEE, from [15.35])

25 nm of a thin TaN layer was deposited in a via and then filled with oxide. After planarization by back etching the bottom electrode was formed. Half of the bottom electrode was then covered by a thin SiO_2 layer to ensure only one contact between the bottom electrode and the phase change material. After forming of a straight edged LTO (low-temperature SiO_2) and TiW top electrode, 40 nm of GST and 50 nm of SiO_2 were sequentially deposited on the SiO_2/TiW stack, followed by a blanket etch-back to form the spacer. In this way an orthogonal contact between the phase change material and the bottom electrode was made.

In this structure the contact area between the phase change material and the bottom electrode is defined by the thickness of the TaN film and the phase change spacer. The contact area is thus not affected by the lithographic process.

The RESET current was reduced to 80 μA for a cell with a contact area of 500 nm^2. Even though the RESET current was reduced in this structure, the GST film formed on the sidewall may cause reliability issues, especially as GST is the key layer. Since it is not easy to form very smooth and uniform thin films on a vertical sidewall, a CVD process might be a better candidate for the deposition method.

We have introduced several innovative scaling approaches by physically scaling down the phase change material, the contact, or both. There are still some other approaches that can be effective for storing more bits in a unit area described in the next section.

15.4.2 Multi-level Cell

Multi-level-cell (MLC) design is a way to double or multiply the bits stored in the same cell. Since the resistance ratio between the RESET and SET state is very high in phase change memory (usually more than a factor of 100), MLC has been discussed for phase change memory from the very beginning. 16 levels have been demonstrated before 1999 [15.36]. A reasonable multi-level cell needs to fulfill a number of criteria:

- All levels should be separated well from each other even with GB chip sizes, after many cycles, and for years of operation. This is a very strict criterion.
- The read out time should be reasonably short.
- The number of writing steps should be reasonable, typically less than 10.

Nirschl et al. [15.37] have experimentally demonstrated 4 level cells (Fig. 15.13) in a 32 kbit array with < 8 iterative programming steps. Further developments are needed to meet the strict criteria of MLC products.

15.4.3 Confined Structure

Another category is the confined structure studied by several groups including Samsung, ST Microelectronics, IBM, and Intel. Pirovano et al. [15.6] showed by simulations that a structure with a confined GST volume can lead to a RESET current that is reduced by a factor of two. Hwang et al. [15.10] proved the concept experimentally. Multiple experiments have confirmed the concept and all the results indicate that the RESET current can be reduced using a confined cell structure [15.6, 15.10, 15.38-40].

Though the RESET current can be reduced, fabrication process issues have been raised. In this structure the phase change material is deposited into a small contact hole. This causes the step coverage to become an important concern and conformal and dense phase change films are required. A deposition/etch/deposition process was proposed by Cho et al. [15.39] but it is a slow process and the etching damage can be a possible issue.

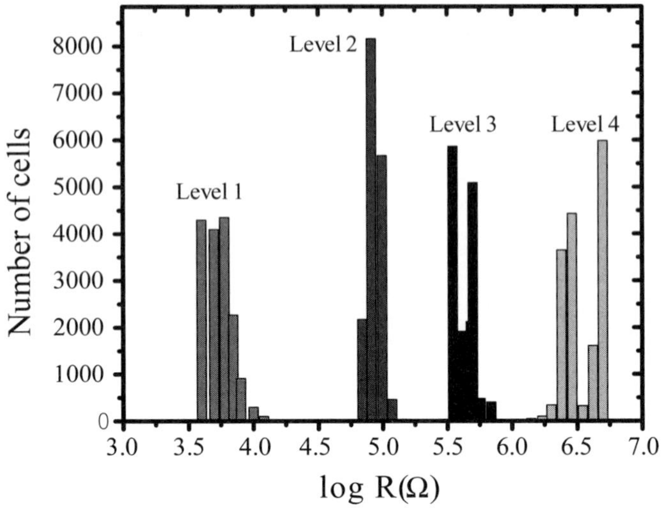

Fig. 15.13. Four level distribution of a 16k physical bit memory page. The memory page is programmed by a discrete-step algorithm, resulting in a 32 kbit phase-change memory. (© (2007) IEEE, from [15.37])

Lee et al. [15.41] demonstrated a confined structure with reduced RESET current (Fig. 15.14) by using CVD and CMP processes. The combination of CVD and CMP solved the difficult issue of filling the phase change material into a very small hole.

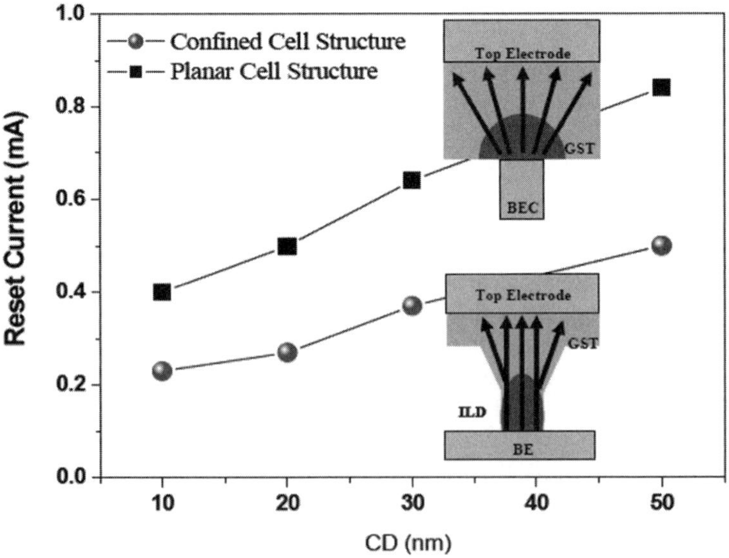

Fig. 15.14. Comparison of RESET current between confined and planar cell structure as a function of the contact diameter. (© (2007) IEEE, from [15.41])

15.5 Advanced Device Testing

In order to measure the transient electrical properties of the phase change devices, several testing circuits have been proposed that need to fulfill specific criteria for testing PCRAM devices:

- The resistance of the device is usually as low as several kohms in the SET state and as high as several Mohms in the RESET state. A system should be able to measure these two very different states accurately.
- The RESET pulse is usually shorter than 100 ns, so the temporal resolution of the testing system should be in the nanosecond range with a sharp trailing edge to enable melt-quenching.
- The RESET current is typically < 100 μA in an advanced device, and the tester should be able to measure the current trace properly.
- Fast phase change devices can be SET in several tens of nanoseconds. The rise time and fall time of the pulses should be as short as possible.

To fulfill the requirement, several approaches have been proposed and the most practical circuits are shown in Fig. 15.15 [15.29, 15.31, 15.42-44].

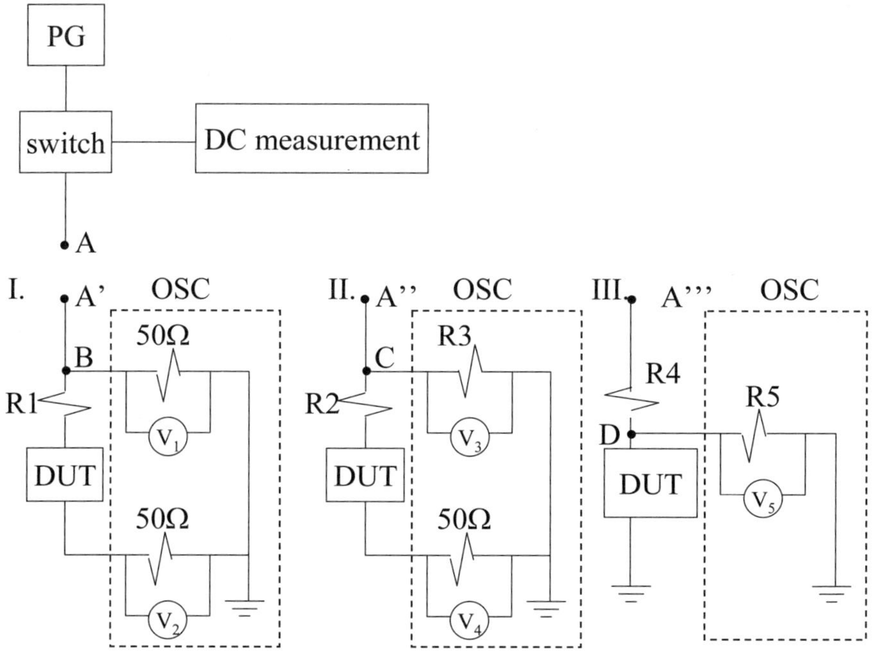

Fig. 15.15. Three typical electrical testing circuits for phase change memory devices.

In Fig. 15.15, the switch is used for separating the pulse generator (PG) and the direct current (DC) measurement tool. It can be a physical switch such as a relay or an electrical switch such as a transistor. The rise time and the leakage current are the major concerns regarding the switch and a switch with suitable bandwidth should be selected to prevent slow rise times and large leakage currents. Three types of circuits (I, II, and III) can be completed by connecting point A to A', A'', or A''', respectively. Using a DC meter to measure the RESET or SET current in the device is not a practical method, especially since these instruments usually provide a current compliance function, and a current/voltage surge in microsecond time scale can be found while measuring a negative resistance device such as phase change memory devices [15.45]. A high speed measuring tool such as an oscilloscope (OSC) is required for measuring RESET or SET currents.

In the circuit of type I, a 50 ohm resistor is connected between point B and the Device Under Testing (DUT), and the voltage at point B, V_1, is very close to the voltage at the device. The major advantage of circuit I is that the equivalent resistance is very close to 50 ohm if the resistance of the DUT is >>50 ohm; as a result there is no significant reflection in the system. A serial resistance $R1$ which matches the resistance of the DUT is connected with the DUT to protect it from the high current surge in the ON state of the device. The resistance of $R1$ is usually several thousand ohms. In spite of the reflection-free pulses, there is no significant RC delay in the current measuring circuit, because the reference resistance of the current measurement is only 50 ohm. It is possible to measure high-speed in-situ current traces from V_2, and the current is equal to $V/R = V_2/50$ A. Though the pulse does not show reflections and the speed can be fast, there are still some flaws in this system:

- The power efficiency is very low, because most of the power is dissipated in the 50 ohm resistor which is connected in parallel to the device. A powerful pulse generator is needed for high voltages and currents.
- DC measurements of the high resistance state are very difficult in this system. The resistance is measured using the difference between the system resistance and 50 ohm. For example, when the DUT and the serial resistance is 1 Mohm, the system resistance is 49.9975 ohm, which is very close to 50 ohm. Even though a precise Ohmic measurement instrument can be used in the system, any contact resistance or variation of the resistors can result in a significant error in the measuring result.
- The current measured in the system is limited. Usually the sensitivity of an oscilloscope is 1 mV/division, therefore one current (I) scale of this system is 20 μA ($I = V/R = 1$ mV/50 ohm = 20 μA). Considering a 10% error, a reasonable lower detecting limit will be 200 μA. To extend the measuring limits, a current amplifier can be used in this system, as well as in the next circuit type.

The second circuit type is completed by connecting A with A'' (Fig. 15.15 II). $R3$ is usually 1Mohm. The current on the device can be measured when $R3$ is very close to the device. In the same manner as the circuit I the current is converted from the voltage V_4 dropping over a 50 ohm resistor. The advantages of the system are:

- The most of the power provided from the pulse generator is used on the DUT, so the power of the pulse generator is efficiently applied.
- The temporal resolution of the current trace is high.
- A precise and accurate DC resistance measurement can be done easily because the DUT is connected in parallel to a 1 Mohm resistor but not a 50 ohm resistor.

The major disadvantage of the circuit is that the mismatch between the device and the pulse generator can introduce a significant reflection issue. The reflection pulse can strongly affect the profile of the electrical pulse. How to reduce or eliminate the reflection pulse is very important for this approach. For example, a pulse generator with low reflection ratio is a good choice. An arbitrary waveform generator can be used to eliminate the reflection also [15.31].

The third approach is shown in Fig. 15.15 III. In this circuit the voltage is measured at point D, *R*5 is usually as high as 1 Mohm. Since the voltage from the pulse generator is known and the value of *R*4 (usually several kohm) is known the current can be identified. The major advantage of this system is that the lower current measuring limit is significantly lower than in the former two circuits. However, there would be a severe *RC* delay when measuring V_5 (*C* is the capacitance). Redaelli et al. [15.44] showed that when the parasitic capacitance of the oscilloscope is 25 pF and *R*4 is 4 kohm, the rise time of the system is about 75 ns which is much too slow for proper device testing. The parasitic capacitance of the system should be carefully eliminated. Kurotsuchi et al. [15.43] used a very low capacitance probe (0.1 pF) and were able to reduce the rise time substantially.

15.6 Summary

In this chapter, several prototype structures for phase change random access memory devices have been described. We started by introducing the most common type for phase change memory scaling - contact scaling devices. Devices using sub-lithographic contact bottom electrodes are called heater structures and there are many variations of this structure. The edge contact structure is a structure using the advantage of thin film scaling, it is a lateral structure. The μTrench structure also uses the thin film for scaling. The structure is not only advantageous in scaling the contact area but also in scaling the cell size because it is a vertical cell. The ring structure uses a hollow cylinder bottom electrode for the device. The distribution of the operating current was very tight compared with other structures. The RESET current was also reduced.

In addition to contact scaling cells, devices which scale the phase change materials were also discussed. The pillar structure is a vertical structure with only one additional mask step for the fabrication, it can reduce the RESET current very efficiently but the fabrication is not easy. The other two lateral structures, line and bridge devices, use lateral thin films to scale down the volume of the phase change region. The cross-sectional area can be effectively reduced but the lateral structure is intrinsically larger than that of the vertical devices.

Three "mixed mode" structures were described in the third section. These structures combine different concepts to improve the cell performance. The cross-spacer structure scales both the phase change material and the contact size. The multilevel devices store multiple bits in one cell to increase the data density. The confined structure contains a recessed bottom electrode to reduce the RESET current and several special process technologies are used to fabricate this structure.

After introducing the structures, we discussed device level testing. The device characterization of PCRAM is different from typical Flash memories, the considerations and test parameters are different, and several typical circuits were described that take into account the testing requirements for characterizing advanced prototype devices with high speed, low RESET current, and high resistance ratio.

Acknowledgements. The author would like to thank the colleagues and managers at Macronix International Co., Ltd.. He thanks Dr. Rich Liu, Dr. Ken Hsieh and Dr. Hsiang-Lan Lung for their support and encouragement. Eric Lee, Shih-Hung Chen, Chieh-Fang Jimmy Chen, Dr. Yen-Hao Shih, Yu-Yu Lin, and Dr. Huai-Yu Michelle Cheng helped to solve difficult questions and experiments. Furthermore, the author would like to thank Dr. Simone Raoux for her mentoring, Dr. Charles Rettner for his enthusiasm for research, and Dr. Geoffrey Burr for many fruitful discussions. Finally, he would like to thank his wife, Lin, for her support.

References

[15.1] Baker, R.J.: CMOS Circuit Design, Layout, and Simulation. 2nd edition, pp.181-182, IEEE Press (2005)

[15.2] Wolf, S.: Silicon Processing for the VLSI Era. Vol. 3: The submicron MOSFET, pp. 307, Lattice Press (1995)

[15.3] Chen, M.L., Leung, C.W., Cochran, W.T., Harney, R., Maury, A. and Hey, H.P.W.: A high performance submicron CMOS process with self-aligned channel-stop and punch through implants (Twin-Tub V), IEDM Tech. Dig., 256-259 (1986)

[15.4] Lai, S., Lowrey, T.: OUM-A 180 nm nonvolatile memory cell element technology for stand alone and embedded applications. IEDM Tech. Dig., 803-806 (2001)

[15.5] Lai, S.: Current status of the phase change memory and its future. IEDM Tech. Dig., 255-258 (2003)

[15.6] Pirovano, A., Lacaita, A.L., Benvennuti, A., Pellizzer, F., Hudgens, S. and Bez, R.: Scaling analysis of phase-change memory technology. IEDM Tech. Dig., 699-702 (2003)

[15.7] Matsui, Y., Kurotsuchi, K., Tonomura, O., Morikawa, T.,Kinoshita, M., Fujisaki, Y., Matsuzaki, N., Hanzawa, S., Terao., M., Takaura, N., Moriya, H., Iwasaki, T., Moniwa, M. and Koga, T.: Ta_2O_5 interfacial layer between GST and W plug enabling low power operation of phase change memories. IEDM Tech. Dig., 769-772 (2006)

[15.8] Cabral, C. Jr., Chen, K.N., Krusin-Elbaum, L. and Deline, V.: Irreversible modification of $Ge_2Sb_2Te_5$ phase change material by nanometer-thin Ti adhesion layers in a device-compatible stack. Appl. Phys. Lett. **90**, 51908-51910 (2007)

[15.9] Hwang, Y.N., Hong, J.S., Lee, S.H., Ahn, S.J, Jeong, G.T., Koh, G.H., Oh, J.H., Kim, H.J., Jeong, W.C., Lee, S.Y., Park, J.H., Ryoo, K.C., Horii, H., Ha, Y.H., Yi, J.H., Cho, W.Y., Kim, Y.T, Lee, K.H., Joo, S.H., Park, S.O., Chung U.I., Jeong, H.S. and Kim, K.: Full integration and reliability evaluation of phase-change RAM based on 0.24 μm-CMOS technologies. Symp. on VLSI Tech. Dig., 173-174 (2003)

[15.10] Hwang, Y.N., Lee,, S.H., Ahn, S.J, Lee, S.Y., Ryoo, K.C., Hong, H.S., Koo, H.C., Yeung, F., Oh, J.H., Kim, H.J., Jeong, W.C., Park, J.H., Horii, H., Ha, Y.H., Yi, J.H.Koh, G.H., Jeong, H.S. and Kim, K: Writing current reduction for high-density phase-change RAM, IEDM Tech Dig., 893-896 (2003)

[15.11] Chen, Y.C., Chen, C.T., Yu, J.Y., Lee, C.Y., Chen, C.F., Lung, S.L. and Liu, R.: 180 nm Sn-Doped Ge2Sb2Te5 chalcogenide phase-change memory device for low power, high speed embedded memory for SoC applications. Proc. of Custom Integrated Circuits Conference (CICC), 395-398 (2003)

[15.12] Matsuzaki, N., Kurotsuchi, K., Matsui, Y., Tonomura, O., Yamamoto, N., Fujisaki, Y., Kitai, N., Takemura, R., Osaka, K., Hanzawa, S., Moriya, H., Iwasaki, T., Kawahara, T., Takaura, N., Terao, M., Matsuoka, M. and Moniwa, M.: Oxygen-doped GeSbTe Phase-change memory cells featuring 1.5V/100 μA standard 0.13-μm CMOS operations. IEDM Tech. Dig., 757-780 (2005)

[15.13] Morikawa, T., Kurotsuchi, K., Kinoshita, M., Matsuzaki, N., Matsui, Y., Fujisaki, Y., Hanzawa, S., Kotabe, A., Terao, M., Moriya, H., Iwasaki, T., Matsuoka, M., Nitta, F., Moniwa, M., Koga, T. and Takaura, N.: Doped In-Ge-Te phase change memory featuring stable operation and good data retention. IEDM Tech. Dig., 307-340 (2007)

[15.14] Ahn, S.J., Song, Y.J., Jeong, C.W., Shin, J.M., Fai, Y., Hwang, Y.N., Lee, S.H., Ryoo, K.C., Lee, S.Y., Psrk, J.H., Horii, H., Ha, Y.H., Yi, J.H., Kuh, B.J., Koh, G.H., Jeong, G.T., Jeong, H.S., Kim, K. and Ryu, B.I.: Highly manufacturable high density phase change memory of 64Mb and beyond. IEDM Tech. Dig., 907-910 (2004)

[15.15] Oh, J.H., Park, J.H. Lim, Y.S., Lim, H.S., Oh, Y.T., Kim, J.S., Shin, J.M., Park, J.H., Song, Y.J., Ryoo, K.C., Lim, D.W., Park, S.S., Kim, J.I., Yu, J., Yeung, F., Jeong, C.W., Kong, J.H., Kang, D.H., Koh, G.H., jeong, G.T., Jeong, H.S. and Kim, K.: Full integration of highly manufacturable 512Mb PRAM based on 90nm technology. IEDM Tech. Dig., 49-52 (2006)

[15.16] Ha, Y.H., Yi, J.H., Horii, H., Park, J.H., Joo, S.H., Park, S.O., Chung, U.-I. and Moon, J.T.: An edge contact type cell for phase change RAM featuring very low power consumption. Symp. on VLSI Tech. Dig., 175-176 (2003)

[15.17] Yi, J.H., Ha, Y.H., Park, J.H., Kuh, B.J., Horii, H., Kim, Y.T., Park, S.O., Hwang, Y.N., Lee, S.H., Ahn, S.J., Lee, S.Y., Hong, J.S., Lee, K.H., Lee, N.I., Kang, H.K., Chung, U.-I, and Moon, J.T.: Novel cell structure of PRAM with thin metal layer inserted GeSbTe. IEDM Tech. Dig. 901-904 (2003)

[15.18] Pellizzer,F., Pirovano, A., Ottogalli, F., Magistretti, M., Scaravaggi ,M., Zuliani, P., Tosi, M., Ben-venuti, A., Besana, P., Cadeo, S., Marangon, T., Morandi, R., Piva, R., Spandre, A., Zonca, R., Mod-elli, A., Varesi, E., Lowrey, T., Lacaita, A., Casagrande, G., Cappelletti, P. and Bez, R.: Novel μ–trench phase-change memory cell for embedded and stand-alone nonvolatile memory applications. Symp. on VLSI Tech. Dig., 18–19 (2004)

[15.19] Bedeschi, F., Bez, R., Boffino, C., Bonizzoni, E., Buda, E.C., Casagrande, G., Costa, L., Ferraro, M., Gastaldi, R., Khouri, O., Ottogalli, F., Pellizer, F., Pirovano, A., Resta, C;, Torelli, G. and Tosi, M.: 4-Mb MOSFET-Selected μ-trench phase change memory experimental chip. IEEE J. Solid-State. Cir. **40**, 1557-1565 (2005)

[15.20] Pellizzer, F., Benvenuti, A., Gleixner, B., Kim, Y., johnson, B., Magistretti, M., Marangon, T., Pirovano, A., Ben, R. and Atwood, G.: A 90 nm phase change memory technology for stand-alone non-volatile memory applications. Symp. on VLSI Tech. Dig., 122-123 (2006)

[15.21] Mantegazza, D., Ielmini, D., Varesi, E., Pirovano, A. and Lacaita, A.L.: Statistical analysis and modeling of programming and retention in PCM arrays. IEDM Tech. Dig., 311-314 (2007)

[15.22] Ielmini, D., Lavizzari, S., Sharma, D. and Lacaita, A.L.: Physical interpretation, modeling and impact on phase change memory (PCM) reliability of resistance drift due to chalcogenide structural relaxation. IEDM Tech. Dig., 939-942 (2007)

[15.23] Ahn, S.J., Hwang, Y.N., Song, Y.J., Lee, S.H., Leee, S.Y., Park, J.H., Jeong, C.W., Ryoo, K.C., Shin, J.M., Park, J.H., Fai, Y., Oh, J.H., Koh, G.H., Jeong, G.T., Joo, S.H., Choi, S.H., Son, Y.H., Shin, J.C., Kim, Y.T., Jeong, H.S. and Kim, K.: Highly reliable 50nm contact cell technology for 256Mb PRAM. Symp. on VLSI Tech. Dig., 98-99 (2005)

[15.24] Jeong, C.W., Ahn, S.J., Hwang, Y.N., Song, Y.J., Oh, J.H., Lee, S.Y., Lee, S.H., Ryoo, K.C., Park, J.H., Shin, J.M., Park, J.H., Yeung, F., Jeong, W.C., Kim, Y.T., Koh., K.H., Jeong, G.T., Jeong, H.S. and Kim, K.N.: Highly reliable ring type contact scheme for high density PRAM. Int. Conf. on Solid State Devices and Mat., 1048-1049 (2005)

[15.25] Song, Y.J., Ryoo, K.C., Hwang, Y.N., Jeong, C.W., Lim, D.W., Park, S.S., Kim, J.I., Kim, J.H., Lee, S.Y., Kong, J.H., Ahn, S.J., Lee, S.H., Park, J.H., Oh., J.H., Oh, Y.T., Kim, J.S., Shin, J.M., Park, J.H., Fai, Y., Koh, G.H., Jeong, G.T., Kim, R.H., Lim, H.S., Park, I.S., Jeong, H.S. and Kim, K: Highly reliable 256Mb PRAM with advanced ring contact technology and novel encapsulating technology. Symp. on VLSI Tech. Dig., 118-119 (2006)

[15.26] Kang, D.H., Kim, J.S., Kim, Y.R., Kim, Y.T., Lee, M.K., Jun, Y.J., Park, J.H., Yeung, F., Jeong, C.W., Yu, J., Kong, J.H., Ha, D.W., Song, S.A., Park, Y.H., Song, Y.J., Eum, C.Y., Ryoo, K.C., Shin, J.M., Lim, D.W., Park, S.S., Kim., J.H., Park, W.I., Sim, K.R., Cheong, J.H., Oh, J.H., Oh, J.H., Park, J.H., Kim, J.I., Oh, Y.T., Lee, K.W., Koh, S.P., Eun, S.H., Kim, N.B., koh, G.H., Jeong, G.T., Jeong, H.S. and Kim, K: Novel heat dissipating cell scheme for improving a reset distribution in a 512Mb phase-change random access memory (PRAM). Symp. on VLSI Tech. Dig., 96-97 (2007)

[15.27] Oh, J.H., Park, J.H., Lim, Y.S., Lim, H.S., Oh, Y.T., Kim, J.S., Shin, J.M., Oark, J.H., Song, Y.J., Ryoo, K.C., Lim, D.W., Park, S.S., Kim, J.I., Kim, J.H., Yu, J., Yeung, F., Jeong, C.W., Kong, J.H., Kang, D.H., Koh, G.H., Jeong, G.T., Jeong, H.S. and Kim, K.: Full integration of highly manufacturable 512Mb PRAM based on 90nm technology. IEDM Tech. Dig., 49-52 (2006)

[15.28] Happ, T., Breitwisch, M., Schrott, A., Philipp, J., Lee, M., Cheek, R., Nirschl, T., Lamorey, M., Ho, C., Chen, S., Chen, C., Joseph, E., Zaidi, S., Burr, G., Yee, B., Chen, Y.C., Raoux, S., Lung, H., Bergmann, R. and Lam, C.: Novel one-mask self-heating pillar phase change memory. Symp. on VLSI Tech. Dig., 120-121 (2006)

[15.29] Lankhorst, M.H.R, Ketelaars, B.W.S.M.M. and Wolters, R.A.M.: Low-cost and nanoscale non-volatile memory concept for future silicon chips. Nature Mater., **4**, 347-352 (2005)

[15.30] Redaelli, A., Ielmini, D., Lacaita, A.L., Pellizzer, F. and Ben, R.: Impact of crystallization statistics on data retention for phase change memories. IEDM Tech. Dig., 761-764 (2005)

[15.31] Chen, Y.C., Rettner, C.T., Raoux, S., Burr, G.W., Chen, S.H., Shelby, R.M., Salinga, M., Risk, W.P., Happ, T.D., McCleland, G.M., Breitwisch, M., Schrott, A., Philipp, J.B., Lee, M.H., Cheek, R., Nirschl, T., Lamorey, M., Checn, C.F., Joseph, E., Zaidi, S., Yee, B., Lung, H.L., Bergmann, R. and Lam, C.: Ultra-thin phase-change bridge memory device using GeSb. IEDM Tech. Dig., 777-780 (2006)

[15.32] Raoux, S., Rettner, C.T., Jordan-Sweet, J.L., Deline, V.R., Philipp, J.B. and Lung, H.L.: Scaling properties of phase change nanostructures and thin films. European Phase Change and Ovonics Symposium (2006), http://www.epcos.org/papers/pdf_2006/pdf_Invited/Raoux.pdf, Accessed 20 Feb. 2008

[15.33] Raoux, S., Rettner, C.T., Chen, C.T., Jordan-Sweet, J., Zhang, Y., Caldwell, M., Wong, H.-S.P., Milliron, D. and Cha, J.: Scaling properties of phase change materials. Non-Volatile Memory Tech. Symp. 2007, 30-35 (2007)

[15.34] Castro, D.T., Goux, L., Hurkx, G.A.M., Attenborough, K., Delhouhne, R., Lisoni, J., Jedema, F.J., in't Zandt, M.A.A., Wolters, R.A.M., Gravesteijn, D.J., Verheijen, M.A., Kaiser, M., Weemaes, R.G.R. and Wouters., D.J.: Evidence of the thermo-electric Thomson effect and influence on the program conditions and cell optimization in phase-change memory cells. IEDM Tech. Dig., 315-318 (2007)

[15.35] Chen, W.S., Lee, C.M., Chao, D.S., Chen, Y.C., Chen, F., Chen, C.W.,Yen, P.H., Chen, M.J., Wang, W.H., Hsiao, T.C., Yeh, J.T., Chiou, S.H., Liu, M.Y., Wang, T.C., Chein, L.L., Huang, C.M., Shih, N.T., Tu, L.S., Huang, D., Yu, T.H., Kao, M.J. and Tsai, M.-J.: A novel cross-spacer phase change memory with ultra-small lithography independent contact area. IEDM Tech. Dig., 319-322 (2007)

[15.36] Ovonic Unified Memory, http://ovonyx.com/technology/technology.pdf (1999). Accessed 20 Feb. 2008

[15.37] Nirschl, T., Philipp, J.B., Happ, T.D., Burr, G.W., Rajendran, B., Lee, M.-H., Schrott, A., Yang, M., Breitwisch, M., Chen, C.-F., Joseph, E., Lamorey, M., Cheek, R., Chen, S.-H., Zaidi, S., Raoux, S., Chen, Y.C., Zhu, Y., Bergmann, R., Lung, H.-L. and Lam, C.: Write Strategies for 2 and 4-bit multi-level phase-change memory. IEDM Tech. Dig., 461-464 (2007)

[15.38] M. Breitwisch, M., Nirschl, T., Chen, C.F., Zhu, Y., Lee, M.H., Lamorey, M., Burr, G.W., Joseph, E., Schrott, A., Philipp, J.B., Cheek, R., Happ, T.D., Chen, S.H., Zaidi, S., Flaitz, P., Bruley, J., Dasaka, R., Rajendran, B., Rossnagel, S., Yang, M., Chen, Y.C., Bergmann, R., Lung, H.L. and Lam., C.: Novel lithography-independent pore phase change memory. Symp. on VLSI Tech Dig., 100-101(2007)

[15.39] Cho, S.L., Yi, J.H., Ha, Y.H., Kuh, B.J., Lee, C.M., Park, J.H., Nam, S.D., Horii, H., Cho, B.O., Ryoo, K.C., Park, S.O., Kim, H.S., Chung, U.I., Moon, J.T. and Ryu, B.I.: Highly scalable on-axis confined cell structure for high density PRAM beyond 256Mb. Symp. on VLSI Tech. Dig., 96-97 (2005)

[15.40] Kencke, D.L., Karpov, I.V., Johnon, B.G., Lee, S.J., Kau, D., Hudgens S.J., Reifenberg, J.P., Savransky, S.D., Giles, M.D. and Spadini, G.: The role of interfaces in Damascene phase-change memory. IEDM Tech. Dig., 323-326 (2007)

[15.41] Lee, J.I., Park, H., Ch, S.L., Park, Y.L., Bae, B.J., Park, J.H., Park, J.S., An.H.G., Bae, J.S., Ahn, D.H., Kim, Y.T., Horii, H., Song, S.A., Shin, J.C., Park, S.O., Kim., H.S., Chung, U.I., Moon, J.T., Ryu, B.I.: Highly Scalable Phase Change Memory with CVD GeSbTe for Sub 50 nm Generation, Symp. on VLSI Tech., 102-103 (2007)

[15.42] Pirovano, a., Lacaita, A.L., Pellizzer, F., Kostylev, S.A., Benvenuti, A. and Bez, R.: Low-field amorphous state resistance and threshold voltage drift in chalcogenide materials. IEEE Trans. on Electronic Dev., 51, 714-719 (2004)

[15.43] Kurotsuchi, K., Takaura, N., Matsuzaki, N., Matsui, Y., Tonomura, O., Fujisaki, Y., Kitai, N., Takemura, R., Osada, K., Hanzawa, S., Moriya, H., Iwasaki, T., Kawahara, T., Terao, M., Matsuoka, M. and Moniwa, M.: Measurement method for transient programming current of 1T1R phase-change memory. IEEE Int. Conf. on Microelectronic Test Structures, 43-46 (2006)

[15.44] Redaelli, A., Lacaita, A.L., Benvenuti, A. and Pirovano, A.: Comprehensive numerical model for phase-change memory simulations. Proceeding of IEEE International Conference on Simulation of Semiconductor Devices and Processes (SISPAD), 280-282 (2005)

[15.45] Shaw, M.P. and Gastman, I.J.: Circuit controlled current instabilities in "S-shaped" negative differential conductivity elements. Appl. Phys. Lett. **19**, 243-245 (1971)

16. Phase Change Memory Cell Concepts and Designs

Roberto Bez, Robert J. Gleixner, Fabio Pellizzer, Agostino Pirovano and Greg Atwood

Abstract Phase-Change Memories are the most promising candidates for next-generation semiconductor non-volatile memories, having the potential of improved performance compared to Flash as well as scalability to the deca-nanometer range. The development of this technology requires a deep understanding of the cell concept and a specific design of the cell architecture, as well as an extensive characterization of the electrical behavior and reliability of the cells. To this aim, in this chapter the most important topics of the phase change memory concept and designs will be reviewed. An up-dated technology overview will be presented, together with the methodology to electrically characterize the cell element; finally an extensive assessment of the cell reliability will be reported.

16.1 Introduction

Flash memories have been able to scale for more than 15 years, boosting the recent impressive growth of the portable equipment market and becoming the mainstream Non-Volatile Memory (NVM) technology. Projecting into the next decade, though, there are several fundamental issues that must be solved to push the floating-gate concept beyond the 32 nm technology node. The increasing complexity of floating gate scaling leaves room for the investigation of alternative NVM concepts that promise better scalability, improved performances, and competitive cost with Flash. Phase-Change Memories (PCM) are the most promising candidates for next-generation NVM, having the potential of improved performance compared to Flash. This includes faster random access time, read throughput, and write throughput, as well as other features including direct write, bit granularity, high endurance and scalability to the deca-nanometer range.

In order to be able to validate the concept and to design the correct cell architecture there are three fundamental areas that must be fully understood:

1. a robust technology architecture, able to minimize the reset current (and to guarantee a clear scaling path with dimensional scaling) together with very tight process control; this will enable tight bit distributions considering large

S. Raoux, M. Wuttig (eds.), *Phase Change Materials*,
DOI 10.1007/978-0-387-84874-7_16, © Springer Science+Business Media, LLC 2009

cell arrays for very high density memory (the reset current is the current required to switch the cell from the crystalline, low resistivity phase to the amorphous, high resistivity phase by melt-quenching);
2. an extensive cell characterization and analysis in order to extract the functional mode and the critical design parameters;
3. a detailed study of the reliability of the cell with a deep understanding of the potential failure mode and the related physical mechanisms with the focus on retention and endurance.

This chapter will review each of these areas. In Section 16.2 we will present an up-date of the technology that can be used to minimize the reset current while maximizing the process controllability; in Section 16.3 we will analyze the methodology to characterize the cell and identify the critical cell parameters that must be taken into account; finally in Section 16.4 an extensive assessment of the cell reliability will be reported.

16.2 Technology Overview

Conceptually the structure of a PCM cell is simple, consisting of a memory layer and a heater (Fig. 16.1) which form the so-called storage element. The most commonly used memory material is the chalcogenide alloy $Ge_2Sb_2Te_5$ (GST).

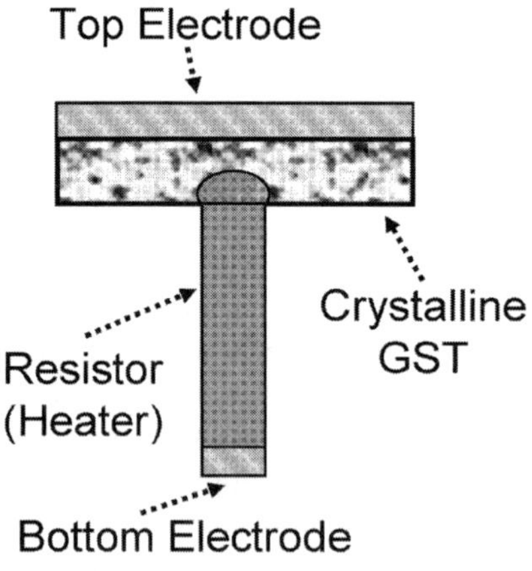

Fig. 16.1. Schematic depiction of the PCM storage element.

Since this storage element is a programmable resistor it requires a selecting device in order to be decoded inside a regular array. Two primary solutions (Fig. 16.2) have been investigated for high-volume manufacturing:

- Vertical Bipolar Junction Transistor (BJT)
- Planar metal-oxide-semiconductor field effect transistor (MOSFET).

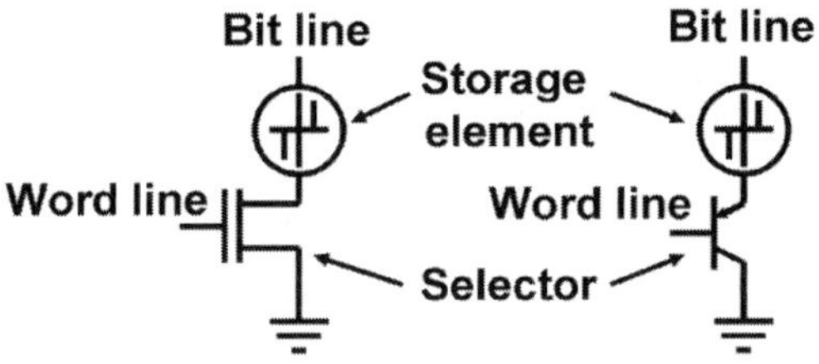

Fig. 16.2. Schematic depiction of the PCM cell structure with MOSFET and BJT selecting devices.

Both solutions have pros and cons. The vertical BJT gives a smaller cell size (6 – 8 F^2, where F is the minimum half-pitch that defines the technology) still being able to deliver the required reset current to the storage element with a small mask overhead with respect to a standard complementary-metal-oxide-semiconductor (CMOS) process. The planar MOSFET is already available in baseline CMOS and does not require any additional mask, but its width must be sufficient to provide the reset current and results in a larger cell area (15 – 25 F^2).

The aim of process integration is to build a compact and efficient PCM storage element coupled with its selector. In the following description it has been assumed that the selector is a vertical BJT, built with dedicated steps in the front-end of the process of a 90 nm CMOS technology [16.4]. The largest efforts are being dedicated to the integration of a compact PCM cell structure with the chalcogenide compound to achieve full compatibility with an advanced CMOS technology and to reduce the programming current without degrading the compelling features of the PCM technology. Several approaches have been proposed to achieve low reset currents, none of them being completely satisfactory. The so called lateral cell approach has demonstrated currents as low as 200 μA [16.5], but its integration in multi-megabit arrays has not yet been proven. Nitrogen doping [16.3] is another effective technique to reduce the programming current, but with the drawback of significantly increasing the set resistance, thus degrading the capability for a fast random read of the memory state (the set state is the crystalline, low resistivity state reached by heating the material above the crystallization temperature for a long enough time). The in-line memory cell [16.6, 16.7] with the self-heating concept promises enhanced performance in terms of programming currents (less than 100 μA), but the cell size is larger due to the planar structure of the storage element. Moreover the reported results are not well understood and its potential at this stage cannot be clearly assessed.

Among these alternatives, the μTrench PCM cell architecture represents an optimization of all the features of the PCM technology, having been demonstrated to simultaneously achieve low programming currents, small cell size, good dimensional control, and proven multi-megabit manufacturability [16.2]. Moreover, the fine tuning capabilities of the PCM μTrench cell and its potential to obtain very low programming currents have been largely demonstrated, with an optimized μTrench cell that achieved a programming current of 450 μA using 180 nm lithography technology [16.8]. Despite the good results so far achieved and the capability to scale the μTrench PCM cell to the 90 nm technology with good electrical characteristics [16.4], the more stringent lithographic alignment requirements of this architecture with respect to other options could impact the integration in multi-megabit array and the final manufacturing yield of PCM products.

With this in mind, the so called Self-Aligned (SA) μTrench is a cost-effective PCM solution capable of simplifying the integration process at 90 nm technology and beyond by reducing the lithographic requirements, still preserving the attractive features of the standard (STD) μTrench PCM concept. This novel architecture, called Self-Aligned (SA) μTrench, is characterized by much less demanding requirements for the lithographic steps employed in the cell fabrication while still providing far superior electrical performances in terms of low programming currents with good reproducibility. The successful integration of a SA-μTrench 128 Mb chip [16.9] demonstrates its suitability for the production of high-density PCM arrays at the 90 nm technology and beyond, allowing also the projection of reset current reduction with new technology nodes (Fig. 16.3).

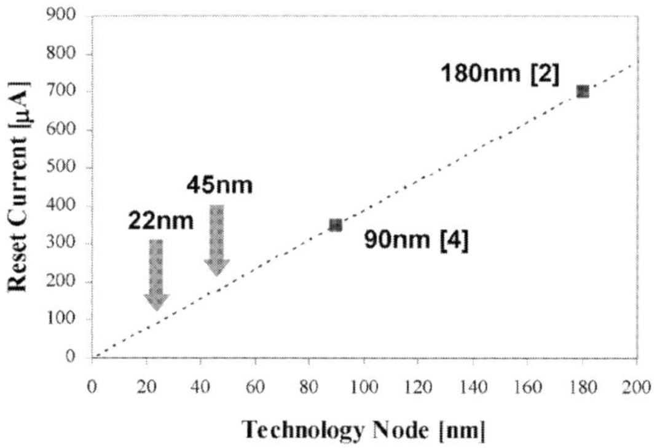

Fig. 16.3. Projected reset current reduction with technology scaling. Squares are experimental points obtained for the 90 nm [16.4] and 180 nm [16.2] technologies; dotted line is the interpolation line. The arrows show the extrapolated reset current at the 45 nm and 22 nm technology node [16.10].

The PCM architecture was originally developed considering the small cell size requirements, the process cost, and the high performance characteristics, in particular in terms of fast random access-time typical of NOR Flash applications [16.2, 16.4, 16.8]. In the STD μTrench storage element proposed for the 90 nm platform, the active storage region is achieved at the intersection between a vertical thin-film metallic layer, called the heater and deposited inside an opening on a tungsten plug, and the thin layer of chalcogenide material - actually the compound $Ge_2Sb_2Te_5$ - capped with a TiN barrier and deposited inside a sub-lithographic trench, the so called μTrench. Although this architecture has been demonstrated to achieve superior results in terms of programming current and dimensional control of the sub-lithographic features, the integration in an array architecture demands additional lithographic requirements, in particular the alignment tolerance with the word-line plugs, to avoid short-circuits in the array. Moreover, these constraints are expected to become increasingly important with the scaling of the technological node, thus becoming a potential yield detractor in next-generation PCM products. In order to reduce the lithographic constraints and simplify the μTrench cell fabrication, a self-aligned approach has been adopted [16.9].

Fig. 16.4 shows cross-sections along both directions depicting the fabrication steps for both the STD and the SA μTrench architecture.

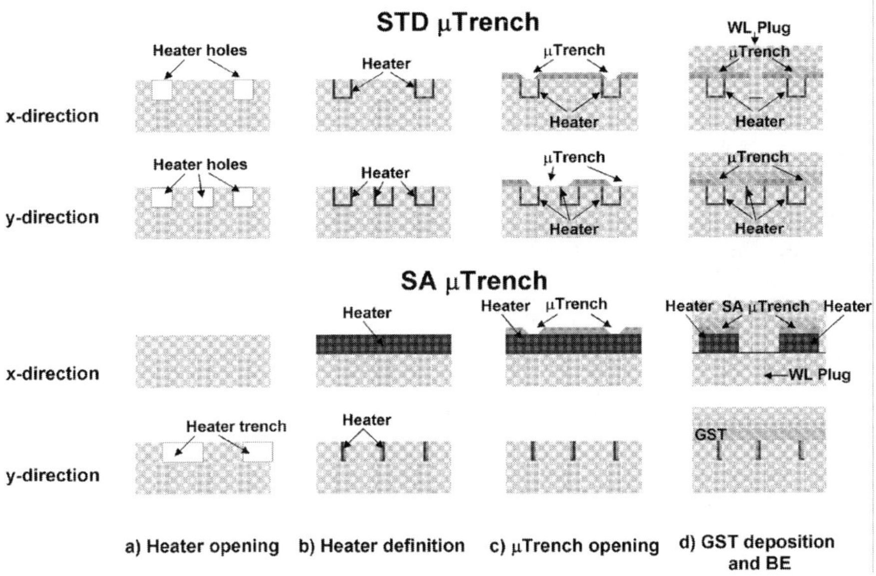

Fig. 16.4. Schematic description of STD and SA μTrench fabrication steps (from [16.9], © (2007) IEEE).

In the first step reported (Fig. 16.4a), called the heater opening, it can be clearly appreciated that in the SA μTrench architecture larger trenches are opened to host the heater, while in the STD approach almost minimum size holes must be opened in the dielectric. After the heater thin-film deposition, in the SA approach a deposition-etch step is performed to electrically isolate heaters belonging to different cells, thus obtaining vertical metallic layers along the x-directions that, at this stage, are in contact with plugs along the same word-line (WL) direction (as clearly seen in the x-direction cross section in Fig. 16.4c). In the STD μTrench approach, the heater thin film is deposited in small openings, followed by a dielectric filling and a final planarization.

Fig. 16.4c reports the μTrench opening. For this process step, the SA approach allows the definition of a continuous trench along the entire bit-line, while in the STD μTrench small elongated rectangles with a sub-lithographic dimension must be patterned to separate cells belonging to the same bit-line. Finally, the GST layer is deposited and the cell structure is finished (Fig. 16.4d). For the STD μTrench the final etching procedure is required to define the bit-line, and the word-line plugs are subsequently opened and filled. On the other hand, in the SA approach the heater along the word-line is not yet defined. The SA approach thus consists of a single etching step that defines the GST bit-line and segments at the same time the heater along the word-line. The word-line plugs are then defined with a direct alignment over the SA bit-line architecture.

Fig. 16.5 shows a scanning electron microscope (SEM) picture of the SA μTrench array after the GST etching step. With this approach, the word-line plugs are directly aligned on the storage elements, thus intrinsically avoiding short circuits between the WL plug and the heater element.

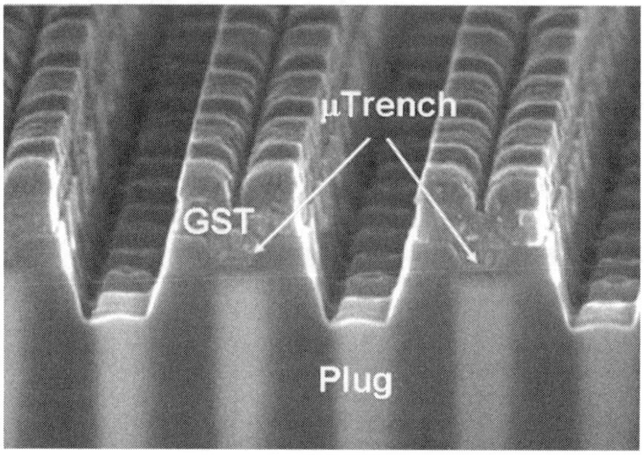

Fig. 16.5. SEM picture of SA μTrench array along the word-line direction after the GST self-aligned etching step (from [16.9], © (2007) IEEE).

Moreover, as reported in the Transmission Electron Microscope (TEM) cross-sectional image along the y-direction of Fig. 16.6, the heater formation is achieved through the opening of trenches of width $2F$, where F is the technology node, thus saving a critical mask step with respect to the STD μTrench approach, where the heater element is patterned at the minimum lithographic size.

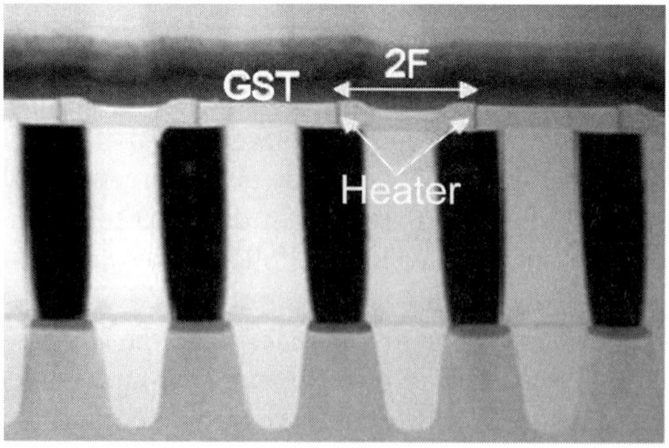

Fig. 16.6. TEM cross-section (along y-direction) of the SA μTrench array (from [16.9], © (2007) IEEE).

The low programming current and the good dimensional control of the sub-lithographic features achieved with the μTrench structure, combined with a self-aligned patterning strategy that simplifies the integration process in terms of lithographic requirements and fabrication costs, seem to give the best compromise for high-volume manufacturing of PCM memory.

16.3 Phase Change Memory Cell Electrical Characterization

In order to exploit the PCM technology for high-performance applications, fast write and read times are mandatory while preserving good data retention capabilities. One of the main concerns to speed up the writing operation is the trade-off between fast crystallization and good non-volatility properties. In order to assure 10 years data retention capabilities at temperatures in the range of 110°C -120°C, the $Ge_2Sb_2Te_5$ compound represents the preferred choice. However, better retention capabilities imply longer programming pulses to crystallize the GST.

Fig. 16.7 shows the programming curves of a MOSFET-selected PCM cell for several programming pulse widths [16.11].

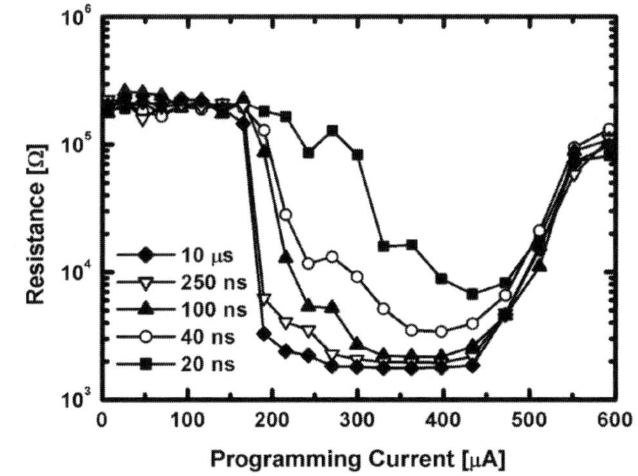

Fig. 16.7. Programming curves of a MOSFET-selected PCM cell for several programming pulse widths (from [16.11], © (2004) IEEE).

For very long pulses (10 μs), a complete crystallization is easily achieved with a resistance change of two orders of magnitude between the set (crystalline phase, low resistance) and the reset state (amorphous phase, high resistance). However, this programming time is unacceptable for high-performance products. By reducing the pulse width, the GST is not able to fully crystallize, resulting in a higher set resistance. However, for pulses as short as 20 ns, a factor 10 in the resistance change between the two programmed states is still achieved. Despite the fact that the read margin is reduced for very short pulses, Fig. 16.7 clearly demonstrates a suitable working window for read out operations with a 20 ns set time.

Fig. 16.8 reports the programmed amorphous resistance as a function of the reset pulse widths. The three curves correspond to different pulse amplitudes, i.e., programmed reset states ranging from 10^5 to 10^6 Ω.

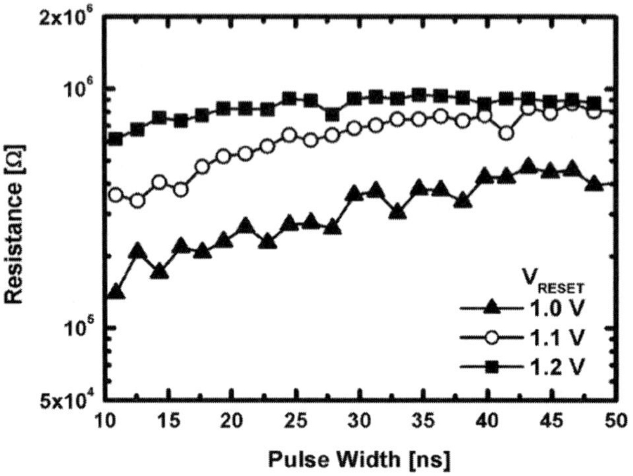

Fig. 16.8. Reset state resistance as a function of the programming pulse width for several voltage amplitudes (from [16.11], © (2004) IEEE).

Even if the pulse width is reduced to as short as 10 ns, only a small decrease in the programmed resistance is seen. This effect could be related to the existence of a delay to reach the thermal steady state condition in the heated volume, mainly due to the high thermal resistivity of the GST film. In any case, the read margin is still sufficient and programming pulses of 20 ns and 10 ns for the set and reset operations, respectively, are suitable for a 10 x resistance change of the PCM cell. These programming/erase capabilities, combined with read access times shorter than 50 ns in a multi-megabit demonstrator, clearly confirm that the PCM technology is suitable for high performance embedded non-volatile memory products.

An advantage related to the reduction of the reset pulse width is the corresponding improvement of the PCM cell endurance. Fig. 16.9 reports the cycle life measured on different cells as a function of the reset pulse width, *W*.

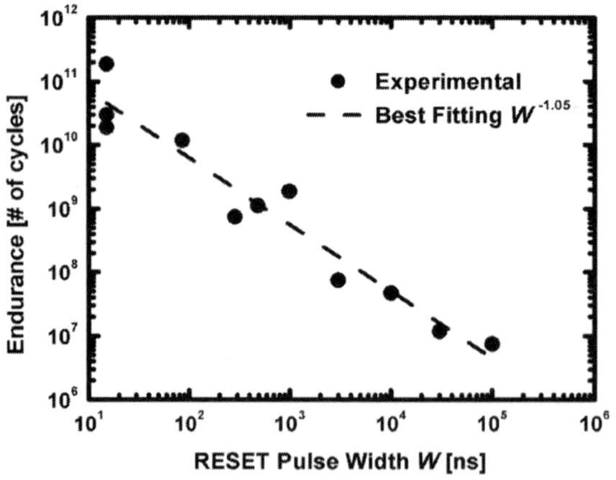

Fig. 16.9. PCM cell endurance as a function of the reset pulse width (from [16.11], © (2004) IEEE).

The experimental data have been fitted with the power law $W^{-1.05}$. This dependence on the reset pulse width demonstrates that the cell endurance depends on the overall time elapsed by the cell at the higher temperature (reset operation). Analogous measurements performed with variable set and fixed reset pulse widths do not show any cycle life dependence on the programming strategy. This result demonstrates that the actual limit for PCM cell endurance is related to the total energy dissipated inside the device, now being in the order of few Joules.

Despite the large potential of the PCM concept, many conceptual and practical challenges have to be addressed. In particular, as discussed in the previous Section, great efforts are being dedicated to reduce the programming current without degrading the features of the PCM technology.

The μTrench PCM cell, as described in Section 16.2, is able to simultaneously achieve low programming currents, small cell size, good dimensional control, and proven multi-megabit manufacturability. In fact, since the μTrench can be defined by sub-litho techniques and the heater thickness by film deposition, the cell performance can be optimised by tuning the resulting contact area while maintaining good dimensional control. The main geometrical tuning parameters exploited to optimise the cell programming current and the overall performance are: 1) the heater thickness, 2) the heater height, and 3) the GST thickness. All these parameters can be determined in terms of layer thickness that allows far better dimensional control than lithographic patterning.

Table 16.1 qualitatively reports the expected trends (higher ↑, lower ↓, almost equal ~) for the main electrical parameters of the PCM storage element.

Table 16.1: Expected trends (higher ↑, lower ↓, almost equal ~) for the main electrical parameters as a function of the geometrical variations of some cell structure parameters. I_{prog}, V_{prog}, and R_{set} are the programming current and voltage and the resistance after set operation, respectively (from [16.8], © (2005) IEEE).

Parameter	Variation	I_{prog}	V_{prog}	R_{set}
Heater thickness	↑	↑	~	↓
Heater height	↑	↓	↑	↑
GST thickness	↑	↓	↑	↑

For example, by increasing the heater thickness, the contact area increases, thus leading to a higher programming current with a lower set resistance. On the other side, the heater resistance decreases at the same rate, resulting in an almost constant voltage drop across the device.

While these trends provide rough guidelines for the PCM cell optimisation, a more detailed investigation based on numerical simulations is required to quantitatively assess the benefits. To clarify the trade-offs, Fig. 16.10 reports the reset programming current, the corresponding programming voltage, and the dissipated power in the PCM cell as a function of the heater height.

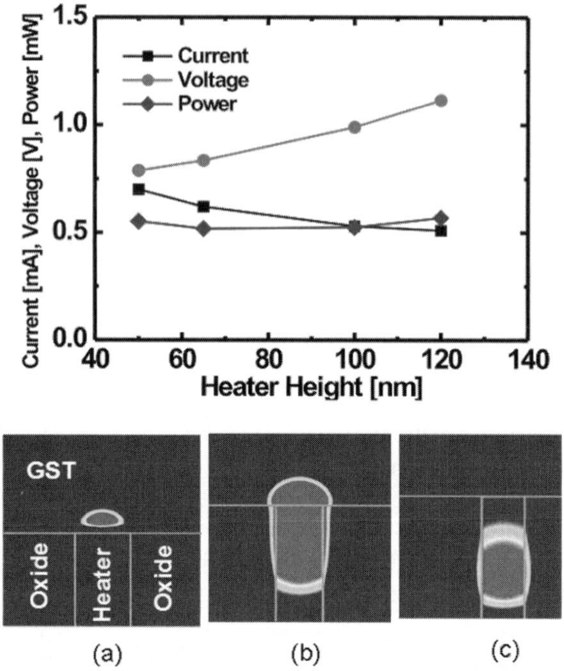

Fig. 16.10. Programming current, programming voltage and dissipated power as a function of the heater height (from [16.8], © (2005) IEEE).

The reported values were obtained with a 3D self-consistent electro-thermal simulation of the μTrench architecture, including the phase-change dynamics [16.12]. Here it is found that the programming power has a minimum that corresponds to the optimum placement of the temperature hot-spot at the heater-GST interface (see the inset (b) of Fig. 16.10).

This condition is a trade-off between the current-consuming small heater (inset a), where the poor thermal insulation provided by the heater moves the hot-spot into the chalcogenide, and the tall heater condition (inset c), where the better thermal confinement leads to lower programming currents, but the increased heater resistance requires a higher programming voltage.

From this point of view, the optimal PCM cell designed for total power minimization does not correspond to the requirement for the minimum programming current. Fig. 16.11 reports a comparison between the power consumption for the PCM storage element, already shown in Fig. 16.10, and the simulated total dissipated power for a cell inside an array (thus including the selecting element and the parasitic line resistances). In this case, the additional voltage drops across the selecting element (in this case a BJT device) and on the line resistance significantly change the dependence of the power consumption on the heater height. In fact, for the taller heater, the benefit of the programming current reduction also implies a lower voltage drop on the programming element and on the parasitic resistances. This effect largely compensates the increased programming voltage on the storage element.

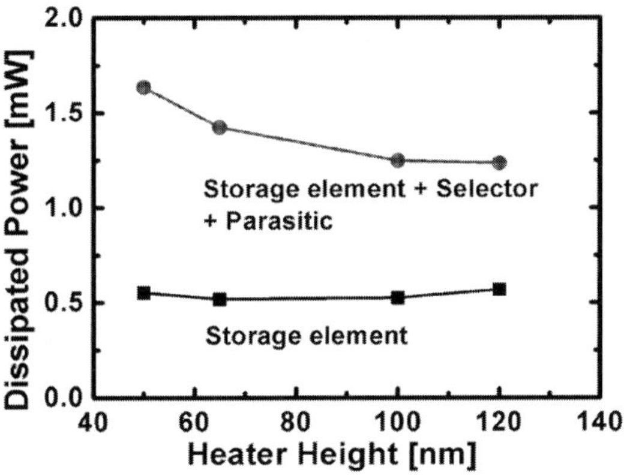

Fig. 16.11. Comparison of the dissipated power for the storage element alone and for the PCM cell + selector + parasitic resistances, as a function of the heater height (from [16.8], © (2005) IEEE).

In terms of overall system power consumption, the optimisation goal for the heater height variation is thus the programming current minimization. Moreover, a lower programming current corresponds to less demanding current driving capabilities for the selecting element, thus simplifying the selector design and dimensional shrinking.

The dependences of the cell parameters have been experimentally investigated employing μTrench PCM cells integrated in a standard CMOS technology [16.2]. Fig. 16.12 shows the TEM pictures of two cells with a heater thickness of 25 nm and 10 nm, respectively, while Fig. 16.13 reports the programming currents and voltages and the set resistances for PCM cells with different heater thickness.

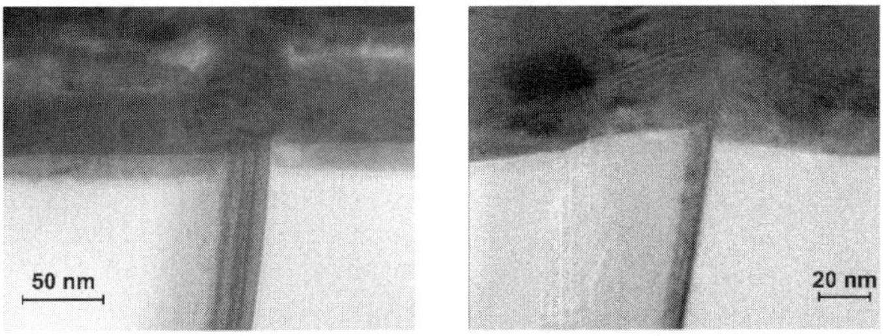

Fig. 16.12: TEM pictures of two cells with different heater thicknesses of 25 nm and 10 nm. Note also the GST crystal grown in the cell's active region (from [16.8], © (2005) IEEE).

As expected, there is a strong sensitivity of the programming current and resistance on this parameter. On the other side, the voltage drop slightly changes. This effect can be easily understood considering that the heater thickness directly affects the dissipated power. Roughly speaking, the power dissipated at the heater-GST interface, P, is given by

$$P = V_h I + \eta R_{heater} I^2, \qquad (16.1)$$

where V_h is the voltage drop inside the chalcogenide, I is the programming current, η is the heater efficiency, and R_{heater} is the heater resistance.

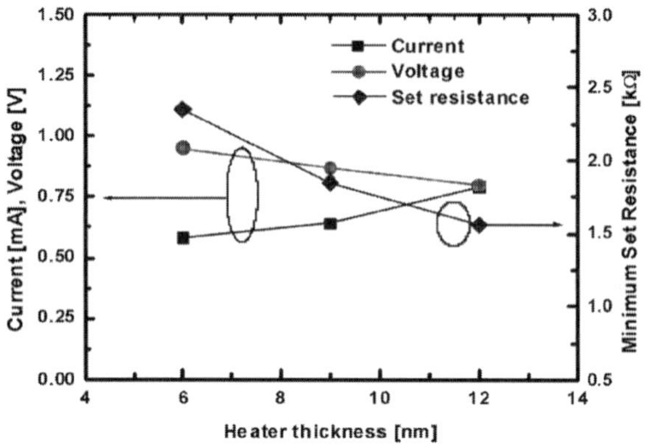

Fig. 16.13. Dependence of the PCM cell electrical properties on the heater thickness (from [16.8], © (2005) IEEE).

Decreasing the heater thickness by a factor k ($k<1$), the electrical resistance R_{heater} and also the overall thermal resistance, R_{th}, go up. Since to melt the GST a temperature variation $\Delta T = R_{th}P$ of 630 °C must be provided, the programming currents must scale by the same factor k. Correspondingly the overall voltage drop across the storage element $V=V_h+R_{heater}I$ remains unchanged, as reported in Fig. 16.13.

The dependence of the PCM cell performance on the heater height is shown in Fig. 16.14.

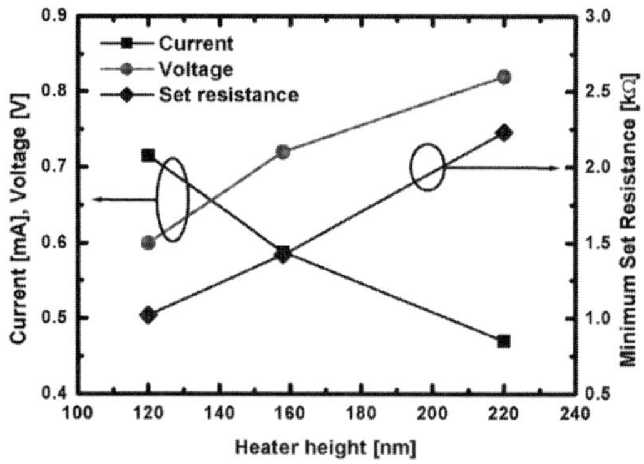

Fig. 16.14. Dependence of the PCM cell electrical properties on the heater height (from [16.8], © (2005) IEEE).

In this case, a taller heater provides a better thermal insulation, with an increased thermal resistance, and a higher electrical resistance. However, a large part of this resistance is placed far from the heater-GST interface, thus not contributing to the GST heating. In practice, the heater efficiency η is lower for taller heaters. Correspondingly, the programming current scales less than linearly with k, while the electrical resistance grows as k. As reported in Fig. 16.14, a slightly higher programming voltage is thus reached.

Finally, Fig. 16.15 reports the dependence of the cell electrical properties on the GST thickness.

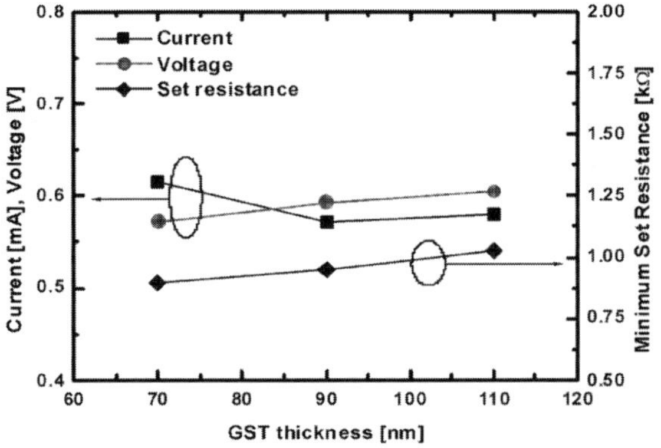

Fig. 16.15. Dependence of the PCM cell electrical properties on the GST thickness (from [16.8], © (2005) IEEE).

Since the μTrench cell structure is not able to confine the current flowing in the GST in regions far from the interface, at some tenths of nanometers from the GST-heater interface the current density is reduced dramatically and the impact of the material properties in this region is almost negligible. Fig. 16.15 shows a slight programming current reduction, consistent with the numerical estimates and resulting from the increased thermal resistance.

Despite the high thermal resistivity of GST, the current spreads into a very large area and the contribution to the overall thermal resistance is very small. As in the case of increasing heater height, the voltage drop far from the interface cannot be completely offset by the decreasing current, thus resulting in a slight increase in programming voltage. This result clearly demonstrates that the optimization rules reported in Tab. 16.1 can be significantly affected by the overall cell structure and by the material parameters. From this point of view, it is key to underline the importance of placing the temperature hotspot close to the heater/GST interface. This goal can be achieved by tuning different cell parameters and the effectiveness of optimizing each of them largely depends on the device architecture. Only by relying on a physically-based numerical model and on well-calibrated material parameters the PCM can cell be effectively optimized.

Based on the previous results and on accurate estimates for the best trade-off among the different requirements, an optimized μTrench cell structure able to achieve very low programming current was designed. The main purpose of the cell engineering was to increase the thermal confinement and the power dissipation in the heating region with a well-balanced tuning of the thermal resistance of both the heater and the GST region. Figs. 16.16 and 16.17 report the key electrical characteristics of the optimized cell, showing that a programming current of 450 μA with a voltage drop of 1.5 V can be achieved. With respect to other approaches, the fine tuning capability of the μTrench cell approach has enabled the smallest programming current so far reported for a PCM cell with demonstrated multi-Mbit capabilities [16.2, 16.4].

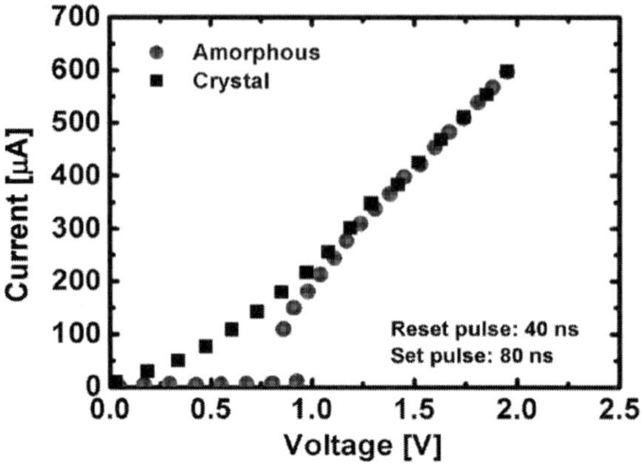

Fig. 16.16. Current-voltage characteristics of the optimized μTrench PCM cell with a programming current of 450 μA at 1.5 V (from [16.8], © (2005) IEEE).

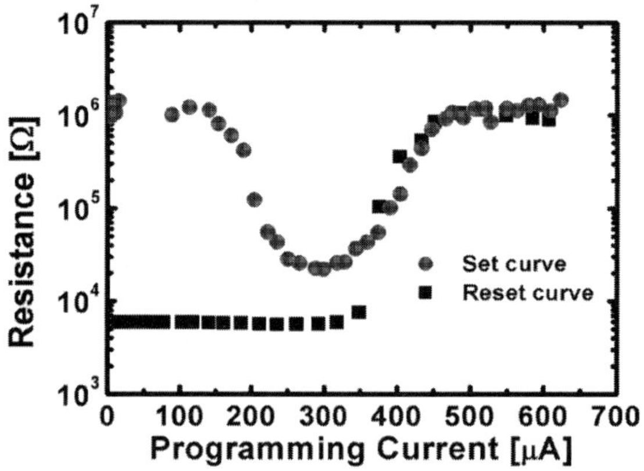

Fig. 16.17. Programming curve of the optimized μTrench PCM cell. The GST interface starts to melt at 350 μA, achieving at 450 μA a programming window with a minimum set resistance of 5 kΩ and a reset of 1 MΩ (from [16.8], © (2005) IEEE).

Moreover, this result has been obtained while preserving a set resistance of 5 kΩ, demonstrating that there is still room for current optimization without degrading the fast sensing capabilities of this technology.

16.4 Phase Change Memory Cell Reliability

To support large NVM arrays, PCM cells must be capable of meeting the requirements for data retention and re-writes (cycling endurance) to very low defect levels (ideally < 1 part-per-billion). In this section, we discuss the types of reliability failures that can exist for PCM, particularly those that relate to the storage element, as extensively described in [16.13].

16.4.1 Data Retention Characterization

Among the failure mechanisms generally seen in non-volatile memories, retention loss of amorphous (reset) cells is the most fundamental to PCM. This results from the fact that amorphous GST is unstable with respect to crystalline (set) GST, limited only by the kinetics of the crystallization process. Few papers on data retention of reset cells show that data retention of GST is much longer than 10 years at 85 °C, and therefore adequate for typical non-volatile memory applications. [16.14, 16.15]. These studies, however, are generally limited in two ways. First, the data are generally collected at very high temperatures (> 180 °C) to accelerate data collection, which requires a significant extrapolation to product use temperatures. And second, data are collected on a very small number of cells, which would not expose possible defect failure modes that may be observed on a large array product. The study reported in [16.13] examines data retention of reset cells at the part-per-million (ppm) level, across a broad range of temperatures (85 °C to 160 °C) and at test times as long as one year. These experiments have exposed a failure mode that is not seen in single cell testing and occurs earlier than the single-cell data would predict. It has been demonstrated that this failure mode can be removed through either process or write algorithm optimization. Moreover it has been shown that the intrinsic retention time-to-failure and acceleration with temperature are not impacted by scaling from the 180 nm to the 90 nm lithography node.

To measure the reset retention behavior of arrays, accelerated bake testing on PCM cells and arrays having the μTrench architecture [16.2, 16.4] have been performed. When subject to elevated temperatures, the resistance of a reset PCM cell evolves with time as shown in Fig. 16.18.

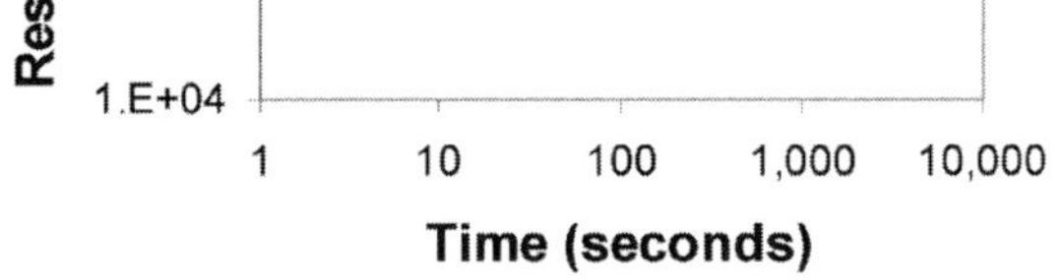

Fig. 16.18. Resistance vs. time behaviour of a reset PCM cell at 180 °C (from [16.13], © (2007) IEEE).

Here the cell is initially reset to about 1 MΩ and the resistance is monitored at a constant 180 °C. Initially the resistance increases due to drift, a phenomenon seen in amorphous chalcogenides and related to the dynamics of the intrinsic traps [16.16]. Eventually crystallization in the amorphous phase results in a drop in resistance and thereby loss of data in the cell. This process can be modeled by crystallite nucleation, growth, and the eventual formation of a percolation path across the amorphous region [16.17]. Since set cells do not drift and are already crystallized, their resistance is stable with time. Data loss in this state is therefore not a significant risk.

When measured at the array level, a similar behavior but a broad distribution of data retention failure times has been seen. Figure 16.19 contains resistance distributions for 512 kb of reset cells that were exposed to successive high temperature bake steps.

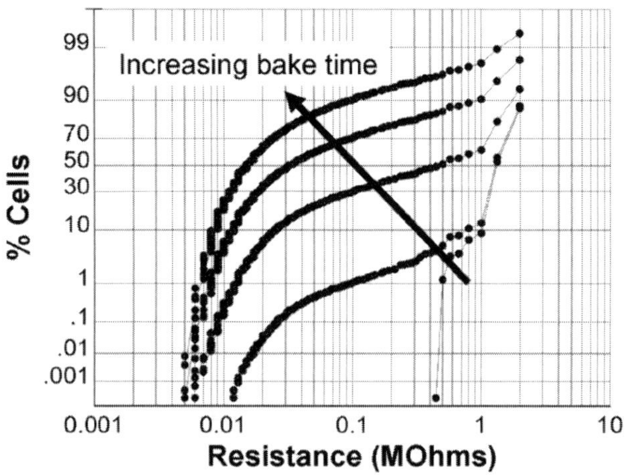

Fig. 16.19. Resistance distribution of initially reset PCM cells with increasing bake time at 180 °C (from [16.13], © (2007) IEEE).

The drift component is difficult to observe in this case due to the loss of measurement resolution above 1 MΩ. Here there is a significant variation in crystallization time across the distribution of cells: after the final bake, cells range from essentially set (< 10 kΩ) to fully reset.

To estimate failure rates at product use conditions, an acceleration model for data loss as a function of bake temperature has been established. The experimental procedure consisted of resetting arrays of cells and baking them at elevated temperatures until retention loss was observed. Readouts were performed at room temperature and at a constant voltage (~ 0.4 V). Cells were considered fails once the resistance dropped below 100 kΩ. This procedure was repeated at 160 °C, 140 °C, and 125 °C on the same cells. Data were then fit to the Arrhenius equation, where time t is accelerated with temperature T according to the form:

$$t \propto \exp\left(\frac{E_a}{k_B T}\right) \tag{16.2}$$

where E_a is the activation energy for crystallization and k_B the Boltzman constant. Very complex models have been developed to describe the crystallization process [16.17], but also the simple Arrhenius model is able to describe the failure process in the range of temperatures considered with an E_a of 2.5 ± 0.1 eV (Fig. 16.20).

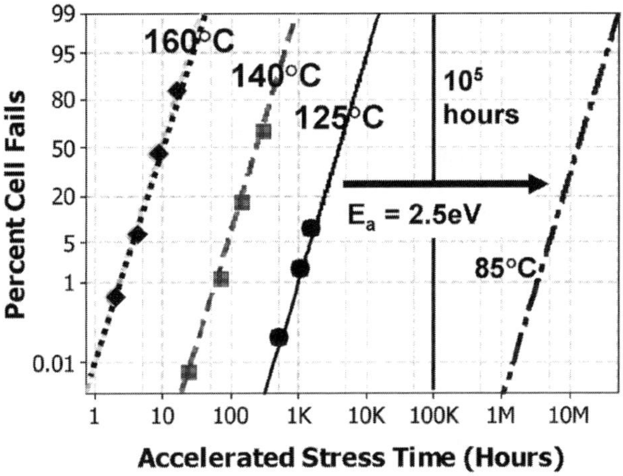

Fig. 16.20. Data retention failure rate as a function of time at multiple temperatures. Data are fit to an Arrhenius equation with $E_a = 2.5$ eV (from [16.13], © (2007) IEEE).

Additionally, a comparison of the median times-to-failure from Fig. 16.20 to single cell data collected from 180 °C to 210 °C [16.18] shows Arrhenius behavior across a much wider range of temperatures (Fig. 16.21).

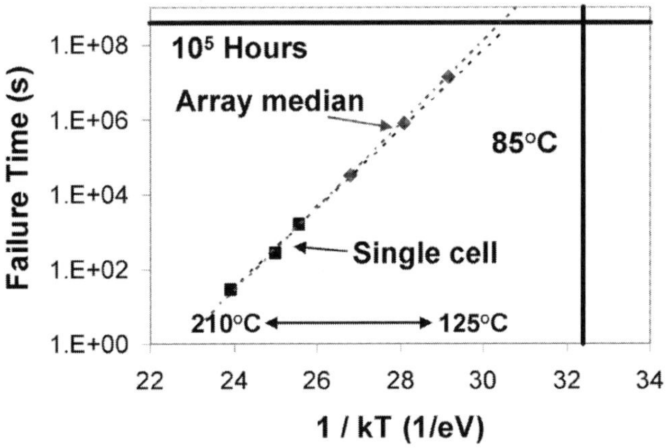

Fig. 16.21. Arrhenius plot of data retention failure time vs. temperature, including both array and single cell data (from [16.13], © (2007) IEEE).

Finally, it is important to note that the array-level failure distribution is log-normal in time, suggesting that the variation is driven by process variability as opposed to a defect population. Scaling these data to 85 °C gives a cell failure rate of much less than 1 ppb (parts-per-billion) at 100,000 hours, adequate for high density array applications.

While the previous data indicate that retention performance of PCM cells, based on the variability of the intrinsic population, is more than adequate, the risk of cells that do not extrapolate according to the prior model must be considered. Such failures may not be observed in the data shown in Fig. 16.20 as the bake times and temperatures of the first readout generally show significant numbers of failing cells and therefore the time of the first cell failure is uncertain. The risk of these cells is that they could fail earlier than the lognormal fit would predict, not extrapolate with the same Arrhenius model, or both.

To understand the risk of the early failing cells, bake retention studies at lower temperatures than the prior experiment (specifically 125 °C, 115 °C, and 105 °C) have been performed. These data (contained in Fig. 16.22) expose early failing cells and are best fit by a Weibull distribution.

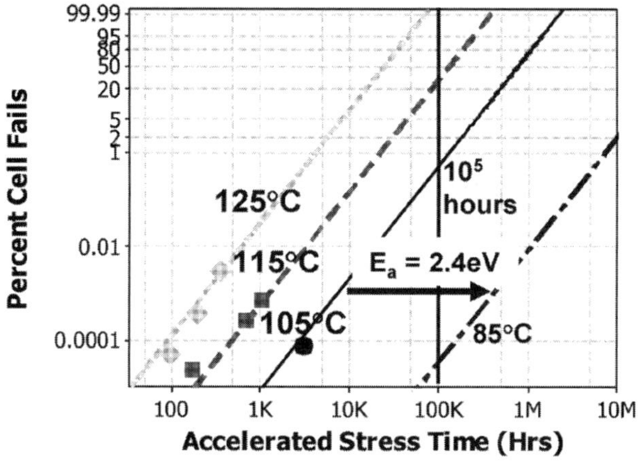

Fig. 16.22. Early retention failure behaviour as a function of bake time and temperature (from [16.13], © (2007) IEEE).

This distribution is consistent with a "defect" or weak-link failure mechanism. Examining the shift in the distribution with temperature gives an activation energy of 2.4 ± 0.1 eV, similar to the 2.5 eV measured in the case of intrinsic failures. This suggests that the crystallization mechanism is not fundamentally altered between intrinsic and early fails. Extrapolating these data to 85 °C indicates that the time to reach 1 ppb of fails is 38,000 hours, less than the 100,000 hour goal. Suppression of these failures is therefore desired. To further increase the level of confidence in the early fail model, devices were baked at 85 °C. These studies lasted for over one year with no retention loss (only drift was observed). While it would take many more years at 85 °C to validate the Arrhenius model, these data provide some level of confidence that additional failure mechanisms are not present.

To further understand the mechanism of early failing cells, the initial resistance of these cells as compared to the intrinsic distribution has been examined. The results are shown in Fig. 16.23; while there is a biasing of the early fails toward a lower initial resistance, the distribution curves show significant overlap.

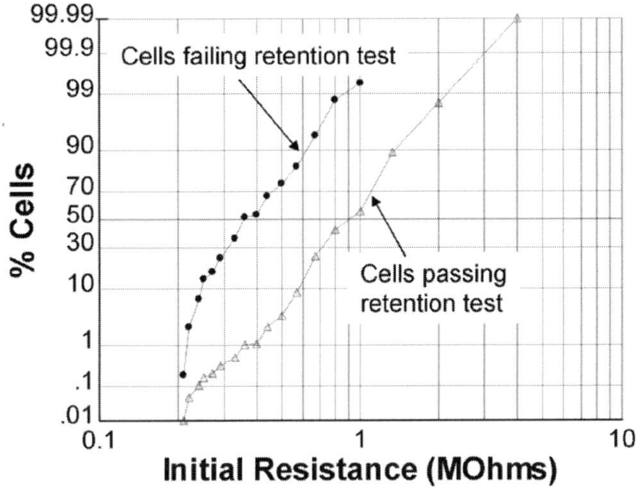

Fig. 16.23. Initial resistance distributions of cells that either passed or failed data (from [16.13], © (2007) IEEE).

This indicates that the initial resistance of the early fails is not clearly distinguishable from the intrinsic cells, and that early retention failures are able to form an amorphous region that is electrically similar to the passing cells.

Since the early failing cells can be reset to a similar level as the intrinsic cells, it must be understood whether the failure mechanism results from defects that are "permanent" to the cell or whether they are related to a weakness in the amorphous region that is unique to the immediate prior reset operation (so-called "soft" defects). This has implications on the solution path, as permanent defects could be removed via a screening operation in manufacturing while soft defects would require suppression of the failure mode. To examine this behavior, an array of cells was repeatedly reset and baked to investigate whether the same cells failed each time. The first observation was that the total number of failing cells was consistent through the repeated bakes (to within 10%). However, when examining the individual cell failures, it was found that different cells had failed each time through subsequent bakes. In fact, less than 1% of cells failed repeatedly. Figure 16.24 shows examples of this "erratic" cell behavior after multiple reset – read – bake – read operations. In this plot, cells were specifically chosen that had failed only the second bake test (denoted Bake 2), but passed the first, third, and fourth bakes. The data in Fig. 16.24 show that the cells could be reset very well prior to each of the bakes and consistently achieve the same reset level each time they were written. In addition, while the cells all failed after the second bake (again, they were selected by this criterion), they showed behavior ranging from drift (higher resistance) to modest loss of resistance after the other three bakes. This suggests that the behavior is unique to the particular reset operation immediately prior to the bake stress, and that any defect must be associated with the newly formed amorphous region. From a product reliability perspective, it also suggests that a simple bake screen would not effectively remove early-failing cells.

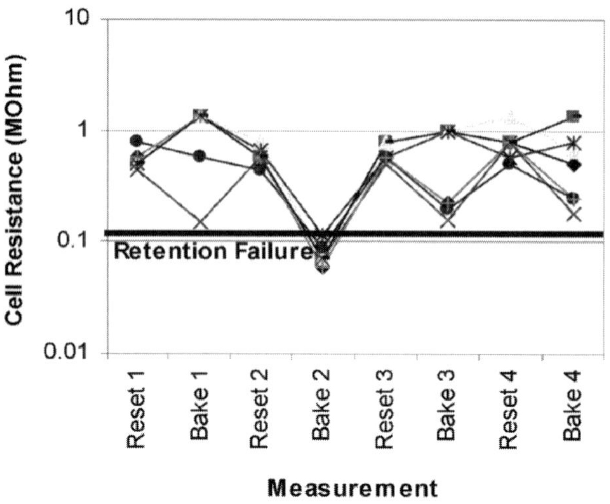

Fig. 16.24. Examples of cells showing erratic retention behaviour. Cells were chosen that failed bake 2 while passing other bakes (from [16.13], © (2007) IEEE).

To understand the impact of write cycles on early retention failures, data retention testing on devices, that had been written (reset) either 10 or 100,000 times, was performed. In this case different reset conditions were chosen (sampled equally) such to measure failures rates from several ppm (parts-per-million) to almost 1% of cells after the bake. The results are shown in Fig. 16.25, where it is shown that the overall failure rate is very stable with cycling and may be showing a slight improvement.

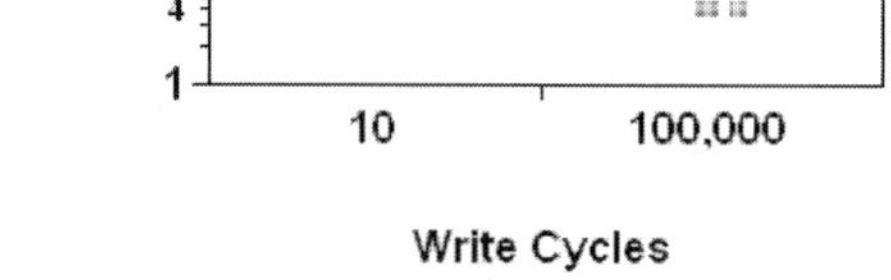

Fig. 16.25. Retention failure rate, in terms of defect-per-million (DPM) as a function of write cycles (sampling cells at different reset conditions) (from [16.13], © (2007) IEEE).

The prior data suggest that early retention fails are not the result of a hard "defect" in the cell, but instead are related to a unique property of the amorphous region immediately after a cell is reset. The activation energy study suggests that the mechanism of crystallization is not changed. And since the initial reset resistance of early failing cells is very high, a small or otherwise defective amorphous zone is not the likely reason. Referring to the crystallite growth model for retention loss described earlier, the observed behavior is consistent with a condition in which pre-existing nucleation sites are arranged in a configuration such that when the thermal energy is applied, very little growth is required before a resistance decrease is observed. A possible model for this behavior is shown in Fig. 16.26.

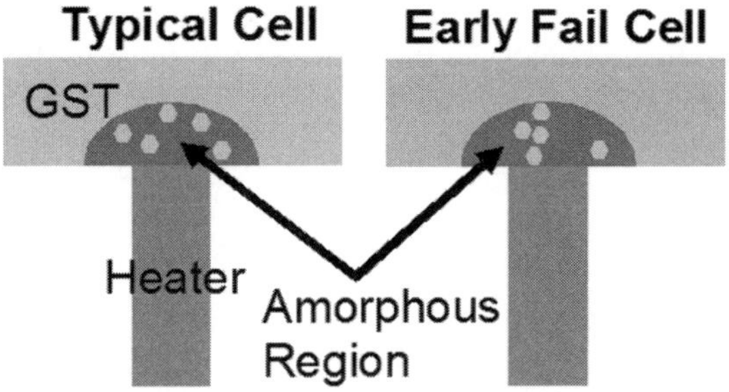

Fig. 16.26. Depiction of the failure model for early retention failures. Here the early failure has a configuration of pre-existing nucleation sites that leads to rapid development of a crystalline path and a cell failure (from [16.13], © (2007) IEEE).

Here it has been assumed that nucleation sites exist in all cells immediately after reset, and retention is determined by the time required for them to grow and join to form a crystalline percolation path across the amorphous region. Each time a cell is melted and solidified during the reset process, the configuration of these nuclei changes. If the number of nuclei is fairly constant over the population of cells, this model can explain the observations that the early fails are both erratic and show similar initial (reset) resistance as compared to the typical cells.

Given the above model, reduction in early retention fails would require improving the quality of the amorphous region. This could be accomplished either during manufacturing by changes in the cell process or during operation by optimizing the write scheme. Fig. 16.27 shows the result of changes in the process architecture to lower the rate of early retention fails, where essentially a complete removal of early failures can be obtained with an optimized process, contrasting with as many as 10 ppm of cell failures in the original process.

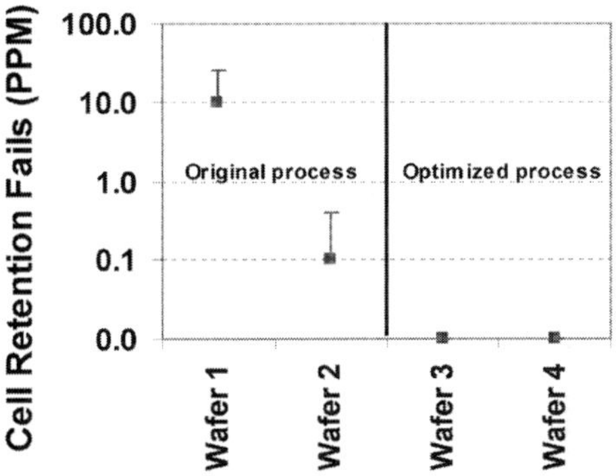

Fig. 16.27. Early fail removal by cell process optimization (from [16.13], © (2007) IEEE).

This optimization not only suppressed the tails, but also improved the intrinsic data retention. The importance of optimized programming on achieving a good reset was highlighted in a recent paper [16.19]. While that work did not address early data retention failures, it showed that subtle differences in the reset operation can significantly impact the reset state of the cell. Similar to that work, an optimized programming technique to improve data retention performance have been identified in [16.13].

The results are contained in Fig. 16.28, where after repeated bake retention tests (similar to that done in the erratic cell analysis) the cells are subjected to the optimized programming technique.

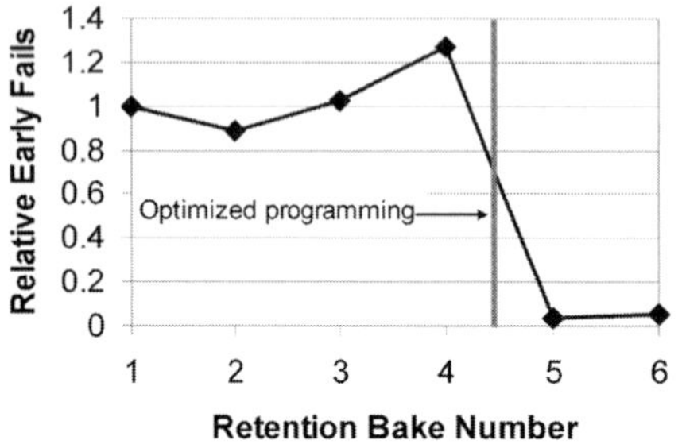

Fig. 16.28. Reduction in early fails with optimized programming (reset operation) (from [16.13], © (2007) IEEE).

Unlike the result described in [16.19], this technique does not increase the cell resistance. However, a 100 times reduction in early fails once it is implemented has been observed.

16.4.1 Retention Behavior with Device Scaling

As PCM devices are lithographically scaled, reduction in the volume of the amorphized region raises concerns regarding the data retention capability. To assess this risk, cell level retention data were compared across both the 180 nm and 90 nm technology nodes. The results of this comparison are shown in Fig. 16.29, where the retention behavior is similar between the two technologies, suggesting that at these particular technology nodes the intrinsic data retention performance meets the 85°C / 100,000 hour specification.

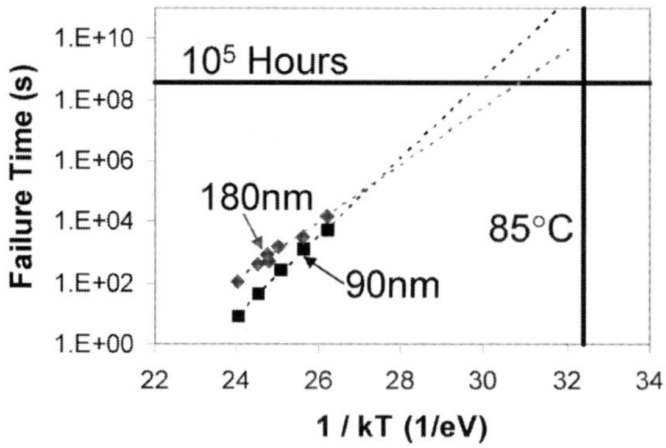

Fig. 16.29. Comparison of cell-level data retention between 180 nm and 90 nm devices (from [16.13], © (2007) IEEE).

16.4.3 Cycling Endurance

One of the expected advantages of PCM over floating-gate NVM technologies is an increase in the maximum write cycles from 10^5 to well over 10^6 writes. At a fundamental level it appears that there is no intrinsic "wear-out" of the chalcogenide material itself, with single devices showing stable behavior up to 10^{13} write operations (Fig. 16.30) [16.19].

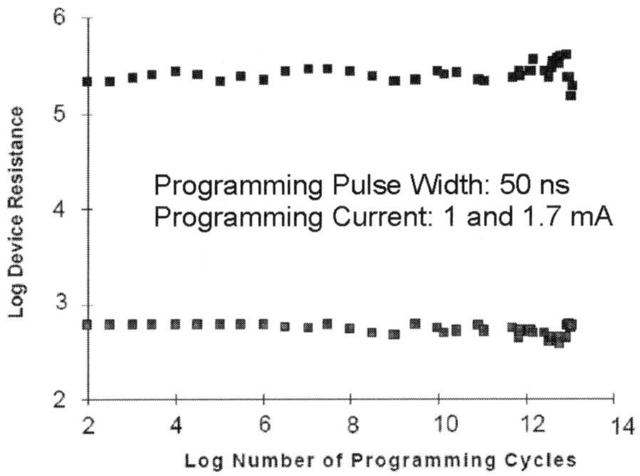

Fig. 16.30. Single cell PCM device showing high endurance capability, reprinted with permission from Ovonyx Inc., from [16.20].

Array data with fully integrated cells [16.14], however, generally shows somewhat reduced cycling endurance of 10^8 – 10^{10}, still a significant improvement over floating gate technologies. Examples of both failures to set and reset are seen in Fig. 16.31.

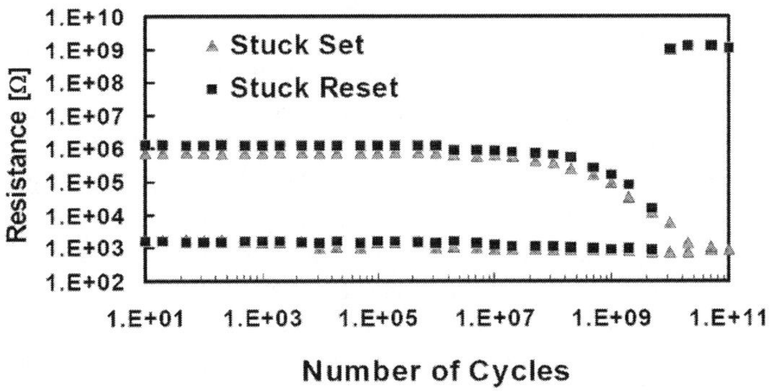

Fig. 16.31. Cycling failures (both set and reset) in a PCM array after ~10^{10} write operations (from [16.14], © (2005) IEEE).

In the case of the stuck reset (open cell), the observed mechanism is voiding or delamination at the heater/chalcogenide interface. Stuck set failures, though, are believed to be produced by compositional changes in the active volume of GST, perhaps through interaction with the surrounding electrode contact materials. In addition, the endurance of PCM cells is shown to vary inversely with the reset energy (Fig. 16.32).

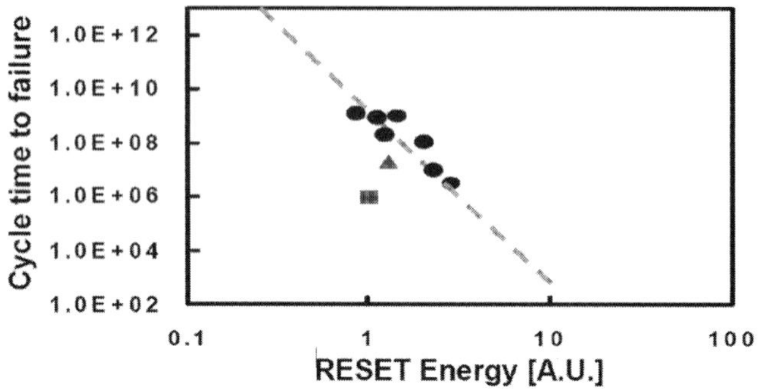

Fig. 16.32. Cycling endurance varies inversely with the applied write energy (from [16.14, © (2005) IEEE).

Since the current required to reset the cells in an array can vary significantly, typically the reset current is much higher than that required for most of the cells. As a result, many cells are over-reset during operation, which reduces there cycling endurance from the ideal case. Such issues can be addressed with reduced reset current variability and through the use of product-level cell placement algorithms.

16. 5 Summary and Outlook

In this chapter we have presented and reviewed the key topics to validate the phase change memory concept and to design the phase change memory cell. We have shown that the PCM technology development has been focused on the definition of robust cell architecture, able to minimize the reset current and to guarantee a clear scaling path with dimensional scaling. Thanks to the cell characterization methodology and to the understanding of the basic physical mechanism that have been developed so-far, it is possible to describe and to analyze in detail the cell behavior. This allows to extract the fundamental information of the cell structure and to engineer the critical cell parameters in such a way as to optimize the memory functionality. The detailed studies of the cell reliability with a deep understanding of the potential failure modes and the related physical mechanisms were presented with attention to retention and endurance. Data retention of the amorphous state in PCM was characterized at the statistical level showing that the intrinsic variation in retention lifetime predicts a failure rate more than adequate for large array application.

Considering all these aspects, we can summarize that phase change memory based on chalcogenide alloy has demonstrated its potential as non-volatile memory, with fast read and write, good retention and high write cycles. The challenges for the near future for the PCM technology is to enter into the high volume manufacturing phase with the demonstration to be able to satisfy all the typical manufacturing requirements. Another fundamental topic, that will be matter of continuous work, is the experimental validation of the scalability. This definitely will be strictly related to the consistent scalability of the cell structure and of the electrical parameters, but in particular of the reset current, without any relaxation of all the reliability aspects. From this point of view the process integration improvements, as well as the chalcogenide and active materials engineering and optimization will be of fundamental importance and they will remain the focus of constant development effort.

References

[16.1] Bez, R., Atwood, G.: Chalcogenide phase change memory: Scalable NVM for the next decade?. Non-Volatile Silicon Memory Workshop '06, pp. 12-15 (2006)

[16.2] Pellizzer, F., Pirovano, A., Ottogalli, F., Magistretti, M., Scaravaggi, M., Zuliani, P., Tosi, M., Benvenuti, A., Besana, P., Cadeo, S., Marangon, T., Morandi, R., Piva, R., Spandre, A., Zonca, R., Modelli, A., Varesi, E., Lowrey, T., Lacaita, A., Casagrande, G., Cappelletti, P., Bez, R.: Novel μTrench phase change memory cell for embedded and stand alone non volatile memory applications. Symp. on VLSI Tech. Dig., pp. 18-19, (2004)

[16.3] Ahn, S. J. , Song, Y.J., Jeong, C.W., Shin, J.M., Fai, Y., Hwang, Y.N. , Lee, S.H., Ryoo, K.C., Lee, S.Y. , Park, J.H. , Horii, H., Ha, Y.H., Yi, J.H., Kuh, B.J., Koh, G.H. , Jeong, G.T., Jeong H.S., Kim, K.: Highly manufacturable high density phase change memory of 64 Mb and beyond. IEDM Tech. Dig., pp. 907–910 (2004)

[16.4] Pellizzer F., Benvenuti A., Gleixner B., Kim Y., Johnson B., Magistretti M., Marangon T, Pirovano A., Bez R., Atwood G.: A 90nm phase change memory technology for stand-alone non-volatile memory applications. Symp. on VLSI Tech. Dig., pp. 122–123, (2006)

[16.5] Ha, Y. H., Yi, J. H., Horii, H., Park, J.H., Joo, S.H., Park, S.O., Chung, U-In, Moon, J.T.: An edge contact type cell for phase change RAM featuring very low power consumption. Symp. on VLSI Tech. Dig., pp. 175-176 (2003)

[16.6] Merget, F, Kim D.-H., Hadam, B., Haring Bolivar, P., Kurz, H.: Novel lateral cell design for low current phase change RAM memories. Non-Volatile Silicon Memory Workshop '04, pp. 30-31 (2004)

[16.7] Lankhorst, M. H. R., Ketelaars, B. W. S. M. M., Wolters, R. A. M.: Low-cost and nano-scale non-volatile memory concept for future silicon chips. Nature Materials **4**, 347-352 (2005)

[16.8] Pirovano, A., Pellizzer, F., Redaelli, A., Tortorelli, I., Varesi, E., Ottogalli ,F., Tosi, M., Besana, P., Cecchini, R., Piva, R., Magistretti, M., Scaravaggi, M., Mazzone, G., Petruzza, P., Bedeschi, F., Marangon, T., Modelli, A., Ielmini, D., Lacaita, A. L., Bez, R.: μTrench phase-change memory cell engineering and optimization. Proc. European Solid-State Device Research Conference '05, pp. 313-316 (2005)

[16.9] Pirovano, A., Pellizzer, F., Tortorelli, I., Harrigan, R., Magistretti, M., Petruzza, P., Varesi, E., Erbetta, D., Marangon, T., Bedeschi, F., Fackenthal, R., Atwood, G., Bez, R.: Self-Aligned μTrench phase-change memory cell architecture for 90nm technology and beyond. Proc. European Solid-State Device Research Conference '07, pp. 222-225 (2007)

[16.10] Bez, R., Pellizzer, F.: Progress and perspective of phase-change memory. European\Phase-Change and Ovonic Symposium '07, http://www.epcos.org/library/papers/pdf_2007/paper03_RobertoBez.pdf (2007)

[16.11] Ottogalli, F., Pirovano, A., Pellizzer, F., Tosi, M., Zuliani, P., Bonetalli, P., Bez, R.: Phase-change memory technology for embedded applications". Proc. European Solid-State Device Research Conference '04, pp. 293-296 (2004).

[16.12] Lacaita, A. L., Redaelli, A., Ielmini, D., Pellizzer, F., Pirovano, A., Benvenuti, A., Bez, R.: Electrothermal and phase-change dynamics in chalcogenide based memories. International Electron Device Meeting Tech. Dig., pp. 911-914 (2004)

[16.13] Gleixner, B., Pirovano, A., Sarkar, J., Ottogalli, F., Tortorelli, I., Tosi, M., Bez, R.: Data retention characterization of phase-change memory arrays. Proc. International Reliability Physics Symposium '07, pp. 542-546 (2007)

[16.14] Kim, K., Ahn, S.-J.: Reliability investigations for manufacturable high density PRAM. Proc. International Reliability Physics Symposium '05, pp. 157-162 (2005)

[16.15] Pirovano, A., Redaelli, A., Pellizzer, F., Ottogalli, F., Tosi, M., Ielmini, D.: Reliability study of phase-change nonvolatile memories. IEEE Trans. On Device and Materials Reliability **4**, n. 3, 422-427 (2004)

[16.16] Pirovano, A., Lacaita, A. L., Pellizzer, F., Kostylev, S. A., Benvenuti, A., Bez, R.: Low-field amorphous state resistance and threshold voltage drift in chalcogenide materials. IEEE Trans. Electron Devices **51**, 714-719 (2004)

[16.17] Redaelli, A., Ielmini D., Lacaita, A. L., Pellizzer, F.: Pirovano, A., Bez, R.: Impact of crystallization statistics on data retention for phase change memories. International Electron Device Meeting Tech. Dig., pp. 742-745 (2005)

[16.18] Redaelli, A., Pirovano, A., Ielmini, D., Lacaita, A.: Intrinsic data retention in nanoscaled phase-change memories—part II: statistical analysis and prediction of failure time. IEEE Trans. Electron Devices **53**, 3040-3046 (2006)

[16.19] Mantegazza, D., Ielmini, D., Pirovano, A., Gleixner, B., Lacaita, A. L., Varesi, E., Pellizzer, F., Bez R.: Electrical characterization of anomalous cells in phase change memory arrays. International Electron Device Meeting Tech. Dig., pp. 1-4 (2006)

[16.20] Ovonic Unified Memory presentation, http://ovonyx.com/technology/technical-presentation.html. Accessed August 22, 2008

17. Phase Change Random Access Memory Integration

Matthew J. Breitwisch

Abstract This chapter reviews the basic process integration and structural design issues regarding the phase change random access memory cell. Basic memory cell design, phase change device characteristics and access device requirements will be reviewed, and then a detailed discussion of the phase change memory device design follows. Various cell designs, including the mushroom cell, pillar cell, μTrench cell and pore cell will be evaluated in terms of RESET current (the current to switch the cell to the amorphous state by melting and rapidly quenching) and RESET current variability minimization. Finally, multi-level phase change random access memory will be discussed.

17.1 Introduction

The ultimate integration goal for Phase Change Random Access Memory (PCRAM) is to demonstrate a low cost set of processing steps which reproducibly and reliably produce large arrays (Megabits to Gigabits) of phase change memory cells, which can be seamlessly incorporated into an existing semiconductor technology. In order to be adapted by the semiconductor industry for replacing existing and enabling new types of memory technologies, the most important issues are the cost of manufacturing, the functionality of the memory, and the scalability of the memory for future technologies.

The cost of manufacturing is dependent on the physical cell size and on the complexity of the processing steps required to build the devices. The smaller the cell size, the larger the number of memory chips that can fit onto a Si wafer (i.e., distributing the wafer processing cost among many chips). Furthermore, the smaller the memory chip and the fewer the number of required processing steps, the more likely it will be on a given wafer to produce a chip without any defects. A smaller memory chip built with fewer processing steps will result in a larger number of functional chips per wafer, and hence be more cost efficient.

The functionality of a memory cell can be described in terms of the speed at which the memory can be programmed and read (speed), the number of times the memory can be programmed and read (endurance), the duration over which the

S. Raoux, M. Wuttig (eds.), *Phase Change Materials*,
DOI 10.1007/978-0-387-84874-7_17,

memory can retain its information (retention), and the memory's ability to withstand harsh environmental conditions (e.g., radiation, shock, high temperatures, etc.).

The scalability of a memory cell can be described in terms of the physical difficulties associated with shrinking the dimensions of the memory cell and in terms of the scalability of the device characteristics as a function of the dimensions of the memory cell. As technology generations scale to smaller and smaller dimensions, the memory cell manufacturability and functionality need to remain robust.

When designing the physical structure of the phase change memory cell, all of these considerations must be kept in mind. These considerations translate into specific design goals for the phase change memory cell. This chapter will review the practical aspects of realizing these goals for phase change memory technology.

17.2 Phase Change Random Access Memory Design Basics

What is the principle on which this phase change non-volatile random access memory device operates? The phase change memory cell is comprised of a variable resistor in series with an access device (Fig. 17.1).

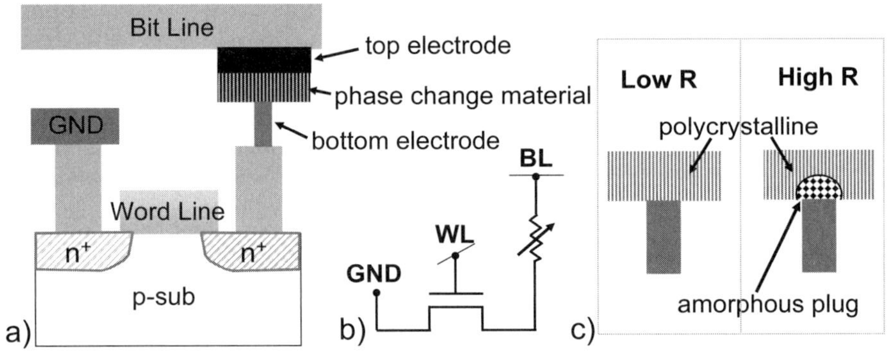

Fig. 17.1. The phase change memory cell is comprised of a variable resistor in series with an access device. (a) Illustration of the mushroom cell showing an FET (Field Effect Transistor) access device comprised of the n-doped regions and the p-doped substrate in series with a phase change memory element. (b) Schematic of cell. GND – ground, WL – word line, BL – bit line. (c) The states of the memory element are defined by the resistance of the cell. The low resistance (Low R) state is obtained when the phase change material directly above the bottom electrode is polycrystalline. The high resistance (High R) state is obtained when an amorphous "mushroom-like" plug of phase change material covers the bottom electrode, thereby restricting the current flow.

The state of the memory cell is determined by the resistance of the phase change variable resistor. This phase change resistor element is programmed using Joule heating to melt (> 630 °C) and then quench in nanoseconds to produce the highly resistive amorphous phase, and Joule heating to crystallize in tens to hundreds of nanoseconds to produce the highly conductive crystalline phase. The phase change material in the amorphous phase undergoes threshold switching above a critical electric field at the applied threshold voltage V_{th} which enables low voltage amorphous-to-crystalline programming. The state of the cell is read using low enough currents so as not to modify the state of the cell. Retention of the memory element is influenced by the ability to avoid unintentional re-crystallization of the amorphous state. Endurance of the memory element is influenced by the stability of the materials and interfaces (at temperatures > 650 °C) and the programming methodology.

The purpose of the access device is to isolate a given memory element during programming and reading of the cell. The memory cell can be a three terminal device with the access device being a field effect transistor (FET) or a bipolar junction transistor (BJT), or a two terminal device with the access device being a diode (further discussion of the access device can be found in Section 17.3). Figure 17.2 shows the schematic of an array of three terminal FET access device memory cells.

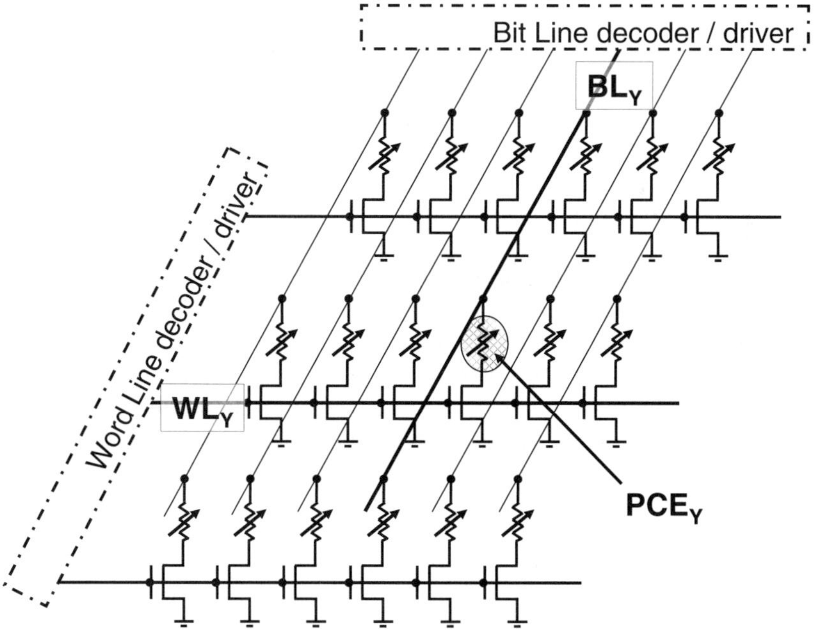

Fig. 17.2. The phase change memory array is arranged in bit lines and word lines, where the word lines are connected to the gates of the FETs and the bit lines are connected to the phase change elements. Any given cell can be accesses (programmed or read) by selecting the appropriate word line and bit line, and de-selecting all other word lines and bit lines.

The array is wired by connecting the gates of the FETs along word lines (WLs) and by connecting the termination of the variable resistors along bit lines (BLs). Any given memory element can be accessed (programmed or read) without disturbing the other cells using the bit line decoder and the word line decoder.

In order to program or read phase change element Y, PCE_Y, (of Fig. 17.2) without disturbing the other cells, all WLs and BLs are grounded except WL_Y and BL_Y, which are biased appropriately to either read the cell, or to program the cell into a specified resistance level. During the read of PCE_Y, the voltages on BL_Y and WL_Y must be chosen so that the current passing through PCE_Y does not heat the phase change material enough to alter the phase of the material. Typically, the selected BL voltage is approximately 0.2 V to 0.4 V during read.

During a SET operation (amorphous to crystalline phase change), BL_Y is held above V_{th} and the current through the device is controlled with the FET gate bias (WL_Y) so as to generate enough Joule heating to raise the temperature of the phase change material above the crystallization temperature (but below the melting temperature) for a duration sufficient to fully crystallize the material. Typically, the SET pulse width is in the range of 30 – 150 ns.

During a RESET operation (crystalline to amorphous phase change), BL_Y and WL_Y are biased to deliver enough current to generate enough Joule heating to raise the temperature of the phase change material above the melting temperature of the phase change material. Once the phase change material is melted (i.e., in the molten state), BL_Y is then shut off quickly so as to quench the melted region into the amorphous phase. Typically, the RESET pulse is in the range of 10 - 50 ns, with a pulse fall time of a few nanoseconds.

It should be noted that phase change memory is a direct write technology, meaning that either the SET or RESET state of the cell can be directly written, regardless of the previous state of the cell; no erase is required. However, for efficient multi-level programming in which intermediate resistance values are targeted, some write schemes first bring the cell into a known state (either fully crystallized, fully amorphous, or melted) and then attempt to bring the cell to the desired intermediate resistance value. Multi-level aspects of phase change memory will be discussed further in Section 17.6.

In order to read cell Y, BL_Y is held at a voltage well below V_{th} (see Fig. 17.3) so as to not to disturb the cell. The current that flows during this read condition, I_{BL}, is then a direct function of the applied voltage V_{BL}, the resistance of the memory element R_{PCM}, plus the resistance of the access device $R_{access\ device}$, and the resistance of the wiring R_{wiring}, given by the equation

$$I_{BL} = V_{BL} / (R_{PCM} + R_{access\ device} + R_{wiring}). \tag{17.1}$$

Typically, $R_{access\ device}$ during the read condition is < 1 kOhm, R_{PCM} ranges from 1 - 50 kOhm when in the SET state and 1-10 MOhm when in the RESET state, and R_{wiring} is a few tens of Ohms. With a V_{BL} of 0.4 V, the read current then ranges from 200 μA to 0.4 μA. In a single-bit-per-cell memory array a reference current (for example 4 μA corresponding to 100 kOhm) will be compared to the read current to determine the state of the cell.

Figure 17.4(a) shows the temperature of the phase change memory element versus time obtained from applying a RESET current programming pulse and a SET current programming pulse.

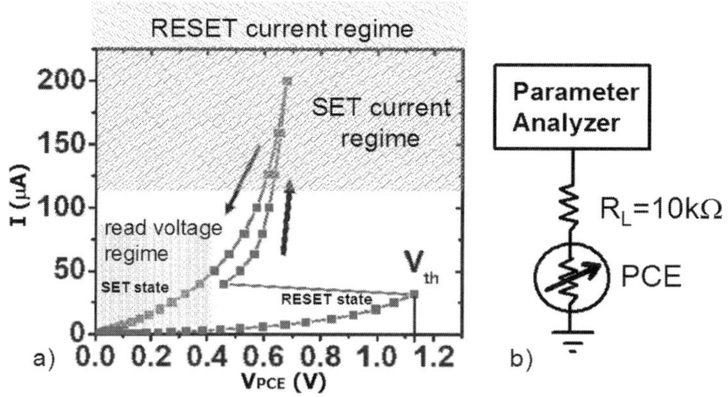

Fig. 17.3. Current – voltage characteristics of the phase change element (PCE). (a) The cell begins in the amorphous phase. Current is applied (schematic shown in (b)) and the voltage across the phase change element is measured. (Actually, the voltage drop across the PCE plus the load resistor R_L is measured, and V_{PCE} is calculated by subtracting off the voltage drop across the known load resistor ($I \cdot R_L$).) At V_{th} the device snaps back to a lower voltage. Further increasing the current brings the device to the crystalline phase. Then, reducing the current to zero brings the device back along the SET state *I-V* curve. The cell finishes in the crystalline phase.

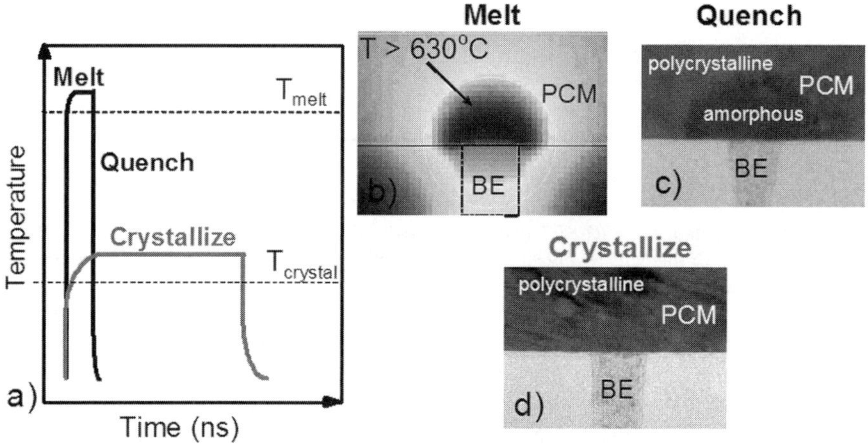

Fig. 17.4. Programming the cell. (a) Temperature vs. time obtained within the cell during RESET (melt then quench) and SET (crystallize) current pulses. T_{melt} and $T_{crystal}$ are the melting point and the crystallization temperature of the phase change material, respectively. (b) Temperature profile within the cell during the melt part of the RESET programming. BE is the bottom electrode, PCM the phase change material. (c) Transmission Electron Microscope (TEM) image of cell in the high resistance RESET state. (d) TEM image of cell in the low resistance SET state.

The RESET pulse brings the temperature of the phase change element above the melting temperature (~ 630 °C) and actually forms a molten region (Fig. 17.4(b)), and is then shut off in a few nanoseconds so as to quench (Fig. 17.4(c)) the phase change material into the amorphous phase. The SET pulse brings the temperature of the phase change element above the crystallization temperature for a duration sufficient to crystallize the phase change material (Fig. 17.4(d)).

17.3 Review of Desired Phase Change Memory Cell Characteristics

Figure 17.5 shows a typical histogram displaying the RESET and SET resistance distributions for a large array of cells (100,000s of bits).

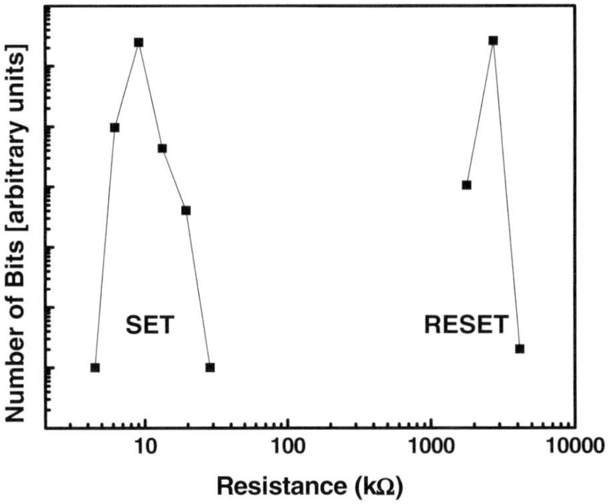

Fig. 17.5. SET resistance and RESET resistance distributions for a collection of cells. To obtain this graph, first, all of the cells receive a SET programming current pulse, and the SET resistance of each cell is read. Then, all of the cells receive a RESET programming current pulse, and the RESET resistance of each cell is read. Even though all of the cells are intended to be identical by design, inherent process variations result in the cells having a distribution of characteristics.

These resistance distributions were obtained by programming all of the cells with a single SET programming pulse and then reading the resistance of each cell (SET resistances), followed by programming all of the cells with a single RESET programming pulse and then reading the resistance of each cell (RESET resistances). For a given programming pulse, with a certain current pulse amplitude, width, and fall time, cell-to-cell variability results in a range of resultant resistance values (instead of all of the cells being exactly at 10 kOhm for the SET resistance and all of the cells being exactly at 2 MOhm for the RESET resistance). The critical device characteristics for a PCRAM cell can be understood in terms of these RESET and SET resistance distributions.

The first and most obvious desirable characteristic is that the SET and RESET resistance distributions have a large margin between them so that the states of the cells can be easily distinguished. This means that out of all the memory elements in a given memory array the resistance of the cell with the highest SET resistance must be much lower than the resistance of the cell with the lowest RESET resistance. This can be achieved in a number of ways. Minimizing the cell variability is paramount. Small variations in the cell structure can strongly affect the way in which the cell responds to a given programming pulse. The smaller the cell-to-cell variability, the sharper the resistance distributions will be. The way in which the cells are programmed will also affect the resistance distributions. Iterative write schemes, where the cell is programmed, read and programmed again, repeatedly, until the resistance value achieved is acceptably high (for RESET) or low (for SET), can also be utilized to widen the margin between the resistance distributions. The duration of the SET pulse (or pulses) can also influence the SET resistance distribution (Fig. 17.6) [17.1].

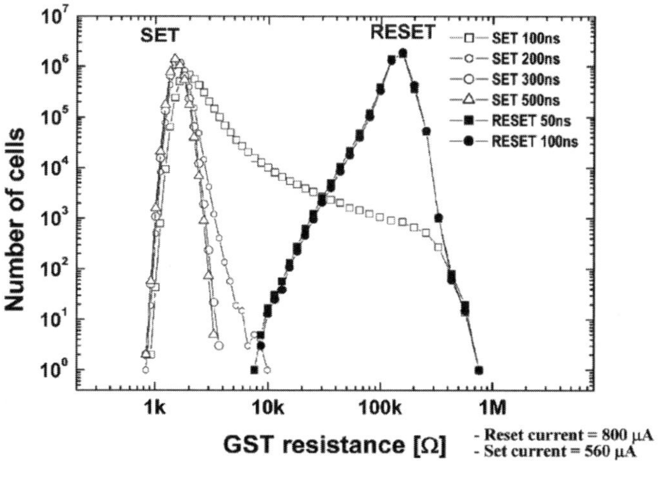

Fig. 17.6. SET resistance and RESET resistance distributions as a function of the programming pulse width. In this example, 100 ns is sufficient to SET most of the cells to below 2 kOhm, however, the 100 ns SET pulse results in many cells having a resistance of greater than 10 kOhm. Extending the 50 ns RESET pulse to 100 ns has no noticeable effect on increasing the resistance of the RESET tail. GST refers to the phase change material used in this experiment, $Ge_2Sb_2Te_5$ (© (2007) IEEE, from [17.1]).

A longer SET programming pulse will allow the slowest SETing bits to more fully crystallize. These programming methods then become a tradeoff in the required programming time with the read noise margin.

It should be noted that in addition to the cell-to-cell variability, there is also an intra-cell variability. Upon repeated application of a specific programming pulse to a given cell, first initialized to a given state, there will be a distribution of final resistance values; programming a phase change memory device is not a uniquely deterministic process. Furthermore, applying an identical programming condition to a given cell beginning in a slightly different initial state (for example starting from a SET state of resistance 3.0 kOhm as compared to 3.8 kOhm) may result in slightly different final states. (It is a complicated situation, indeed!)

Another requirement of the PCRAM cell is that it must possess the ability to switch between the resistance levels with accessible electrical pulses. The access device must be able to shut the current off quickly enough so as to be able to quench the melted region of phase change material into the amorphous phase. Likewise, the local thermal conductivity environment surrounding the volume of phase change material that undergoes phase transformations must allow for fast thermal dissipation during quench. Most importantly, the access device must be able to supply enough current to melt the phase change material. This last requirement has driven a great deal of work as it directly influences the smallest achievable footprint of the PCRAM cell.

The required phase change memory element RESET power determines the access device current requirements. The larger the needed RESET power, the larger the required size of the access device. The access device must be able to deliver enough current to melt the phase change element. Hence, the footprint of the PCRAM cell can be minimized both by maximizing the current per area of the access device, and by reducing the required RESET power of the phase change element.

There are many factors which influence the RESET current of the cell (and they will be discussed in more detail in a subsequent section), but most basically, they are the cell geometry and the thermal and electrical properties of the materials that comprise the cell. The higher the current density through the cell, the larger the Joule heating will be. Hence, part of the critical cell design must include designing a highly sub-lithographically sized feature through which the current flows. The electrical and thermal properties of the phase change material and the electrode materials, and the thermal properties of the encapsulating materials will determine how much Joule heating is generated and how that heat is distributed within the cell.

There are tradeoffs between the electrical resistivities of the materials and the lowest SET resistance achievable, which in turn influences the achievable read speed. Higher electrical resistivities of the phase change material and the electrodes will result in more Joule heating for a given current density, but will also result in an increase in the SET resistance. Furthermore, there are tradeoffs between the thermal resistivities of all of the materials and the cells ability to quench into the amorphous phase. If the heat generated during a RESET programming is unable to quickly dissipate once the current is shut off, the cell will risk cooling at a rate slow enough to crystallize the once-melted phase change region. On the other hand, if the thermal insulation inadequately confines the heat generated during the RESET programming, a larger amount of heat generation will be required to achieve a temperature profile within the phase change material needed to melt the phase change material. All of these components provide an extremely complex set of interactions, so complex, that although some rough qualitative understanding can be obtained using analytical analysis (trying to write out all of the physics equations to see what will happen), ultimately, experiments and finite element simulations (whose parameters are tuned with experimental data) are needed.

The RESET programming of a cell typically can occur in tens of nanoseconds. Once the RESET current is flowing through the cell, thermal equilibrium is reached very quickly (in a few nanoseconds). The melted volume of phase change material is formed in less than a few nanoseconds and then the current just needs to be turned off quickly so that the material quenches into the amorphous phase. The speed limiter for PCRAM is the SET programming, i.e., the time it takes to crystallize the amorphous phase change region. The crystallization speed is mainly a function of the details of the composition of the phase change material itself. For a given base phase change material, doping (or alloying) the phase change material typically slows down the crystallization speed, but also has the affect of increasing the crystallization temperature, which typically improves the retention. Different base phase change materials will have different crystallization speeds. Those materials in which the dominant crystallization mechanism is the growth-front mechanism (as opposed to the nucleation-dominated mechanism) typically crystallize faster (see Chapter 5.3 for crystallization mechanisms). Since there is a trade-off between SET speed and retention, the phase change material may be specific to the specific intended application. For example, if the intended application would be for an automotive application (where the normal operation temperatures are much higher than those experienced in other applications like in a cell phone) the retention requirement would be more stringent. If the intended application were a pseudo-DRAM replacement, programming speed would be more important. In any case, fast SET speeds and long retention are both desirable characteristics of the PCRAM cell.

The PCRAM cell must not only possess the ability to switch between the resistance levels with accessible electrical pulses, but also must be able to switch between these resistance levels many, many times. That means that the phase change material must undergo repeated melting and quenching without the constituents of the phase change material separating or redistributing too substantially. It also means that the cell must undergo the repeated volume expansions and contractions associated with the volume changes of the amorphous vs. the crystalline phases without breaking, cracking, or becoming disconnected along the current path.

Since current flow is the means by which the cell is programmed, and current flow is also required to read or sense the state of the cell, care must be taken to avoid altering the state of the cell during the read. The read current must be kept low enough so as not to disturb the cell when in the amorphous phase. If the read current becomes too high, enough heat may be generated to partially re-crystallize the phase change material. Furthermore, during the read, the bit line voltage must be well below the threshold voltage of the RESET state of the cell so as to avoid transforming the amorphous state into the highly conductive state.

The states of the cell must not only be resilient to change caused by the read conditions, but also by other external influences. A typical retention specification would be that the cell be able to maintain its state at 85 °C for 10 years. Depending on the details of the phase change material properties and the state of the cell (fully vs. partially amorphous) the resistance may either decrease with time (crystallization) or increase with time (relaxation of conduction traps) at elevated temperatures. Both mechanisms will be present, and the resulting change in resistivity depends on which mechanism dominates, which strongly depends on the temperature of the cell. In addition to the possibility of a given state of a particular cell drifting over time, another type of drift is associated with the cycling of the cell. Repeated cycling of the cell between the amorphous and crystalline phases may cause the cell to change in such a way as to be effected differently by a given current pulse [17.2, 17.3]. A cell that has been cycled only 100 times may react differently to a given RESET current pulse than a cell that has been cycled 10,000 times. Cell drift in all of its forms must be minimized or at least be predictable.

These device criteria have been described in terms of a cell with two distinct resistance levels. Ideally, there would be as many distinct resistance levels as possible squeezed in between the fully amorphous and fully crystalline states, thus enabling multi-bit per cell PCRAM. Further discussion of multi-level PCRAM will follow.

17.4 The Access Device

The purpose of the access device is to allow programming currents and read currents intended for a particular memory element to interact with, and only with, that particular memory element, i.e., without disturbing neighboring cells. The access device must be able to provide enough current to RESET any given memory element, and must also be designed such that during this programming nearby cells remain isolated from this programming current. Typically, bipolar junction transistors and diodes can supply more current per unit area, but field effect transistors provide better isolation between cells. The issue with the diode scheme is that there can be bipolar turn-on effects of nearest neighbor cells during programming [17.4], which can lead to write disturb failures. However, these effects can be minimized by careful device design.

As mentioned previously, the size of the memory cell directly relates to the cost of the memory. The size of the memory is often quoted as the number of bits per feature squared (#bits / F^2), where one feature (1 F) is the minimum lithographically printable feature size for a given technology node (e.g., 1 F for the 90 nm node technology is 90 nm; 1 F^2 = (90nm)^2). The smallest cross-point-like device would then be 1 F for the structure width plus 1 F for the space between the structures along one dimension, and likewise for the perpendicular direction, which adds up to 4 F^2 for the total structure footprint. 4 F^2 for the PCRAM structure is extremely challenging to achieve, mainly because of the RESET current requirements of the memory element.

For a given cell size, or footprint, a bipolar transistor or diode access device can typically deliver more current than a conventional CMOS (Complementary Metal–Oxide–Semiconductor) FET device (at least down to the 16 nm node [17.5]). However, depending on the intended applications, a slightly larger CMOSFET + PCM cell may have a cost advantage over other memory technology alternatives, as the CMOSFET device is a well developed device that comes "for free" with a given technology. For example, if the targeted application is embedded NOR flash (embedded in the sense that the memory array is embedded on the same chip as the logic circuitry) a 25 F^2 PCRAM cell would be competitive. On the other hand, if the intended application is stand-alone NAND flash, a 25 F^2 PCRAM cell is much too large (approximately a factor of 5 times too large). NAND and NOR refer to the architecture of the arrays.

A 5.8 F^2 PCRAM diode cell has been demonstrated using a 90 nm technology in which the diode was able to supply 1.8 mA at 1.8 V [17.4]. In comparison, a 90 nm 10 F^2 tri-gate FET could only supply approximately half of the current [17.4]. Figure 17.7 - 9 show examples of an FET based cell (Fig. 17.7), a BJT based cell (Fig. 17.8), and a diode based cell (Fig. 17.9).

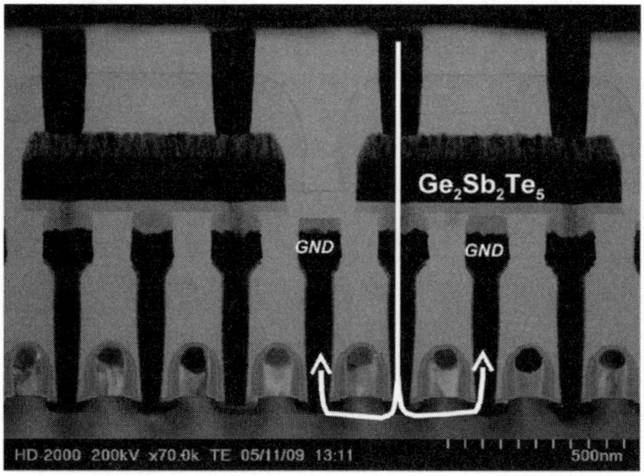

Fig. 17.7. TEM cross-section of FET based PCRAM cell. In this example, the current for each phase change element is supplied by two FETs in parallel. 256 Mb PRAM using 100 nm technology with a 0.166 μm^2 (16.6 F^2) cell size. 16.6 F^2 tri-gate nMOS current: 1.1 mA at 3 V ((© (2007) IEEE, from [17.1]).

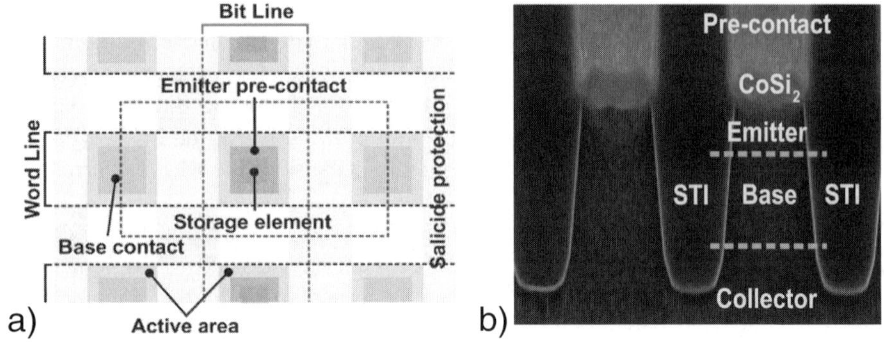

Fig. 17.8. BJT based PCRAM cell, (a) layout, and (b) TEM cross-section. Multi-Mb PCRAM using 90 nm technology with a 0.0968 μm^2 (12 F^2) cell size. 12 F^2 pnp-BJT current: ~ 1.7 mA at 1.8 V. STI – shallow trench isolation (© (2006) IEEE, reprinted with permission from [17.6]).

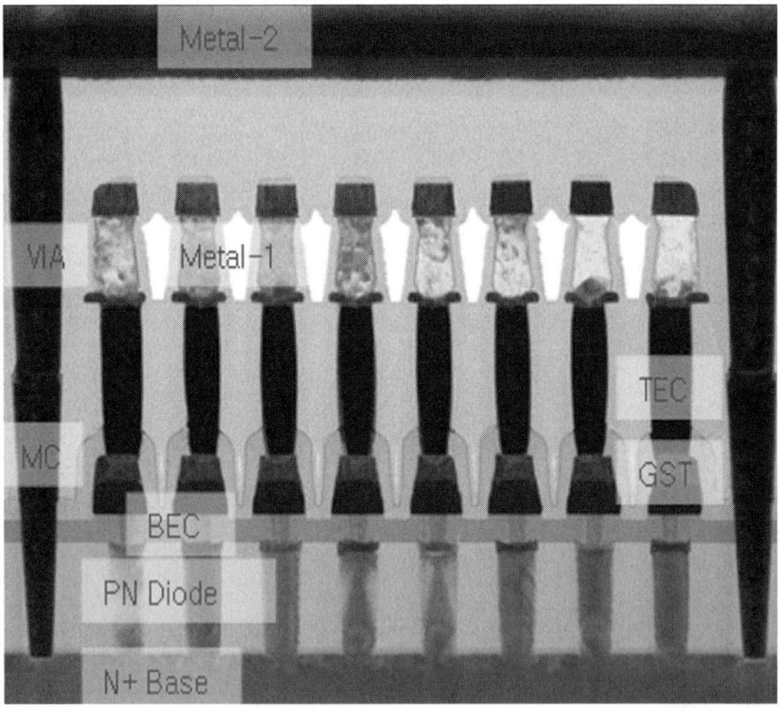

Fig. 17.9. TEM cross-section of diode based PCRAM cell. 512 Mb PRAM using 90 nm technology with a 0.047 μm^2 (5.8 F^2) cell size. 5.8 F^2 pn-diode current: 1.8 mA at 1.8 V (© (2006) IEEE, reprinted with permission from [17.7]).

17.5 Device Design Considerations

The basic strategy for designing the PCM element is to construct the device such that the current passes through a very small aperture. As this aperture dimension shrinks, so does the volume of phase change material that must be melted and quenched into an amorphous plug during the RESET programming. The power (and thus current) requirements decrease as the aperture size decreases. The key then is to be able to define a highly sub-lithographically sized feature, which is very controllable, through which the current flows. Maintaining tight control of the aperture size while scaling is paramount. Variations of the aperture size translate into variations in cell resistance which has the following consequences: a narrowing of the margin between the RESET and SET resistance distributions, an increase in the required worst-case RESET pulse amplitude, a reduction of endurance by exposing some cells to unnecessarily high RESET pulses, and a complication of the design of a SET pulse that can successfully switch any cell in the array [17.8].

There have been many different structures attempted with varying degrees of success to meet this challenge. Most notable are the mushroom cell [17.9], the ring bottom electrode mushroom cell [17.10-13], the pillar cell [17.14], the bridge cell [17.15], the μTrench cell [17.16] and the pore cell [17.17-20].

The following discussion will review basic structural and material design considerations beginning with the commonly studied mushroom cell, and from that starting point move towards enhancements aimed at both reducing the RESET current and increasing the controllability and manufacturability of the structure.

17.5.1 The Mushroom Cell without or with Bottom Ring Electrode

The mushroom cell uses a highly scaled cylindrical bottom electrode in contact with a planar phase change material. Figure 17.10 shows TEM images of an example of the mushroom cell in the SET state (a) and the RESET state (b). In the RESET state the amorphous plug is clearly visible above the bottom electrode and has the resemblance of a mushroom (hence the name). The phase change material is capped with titanium nitride which is then contacted on the top with a via which then connects to the bit line.

The mushroom cell has many advantages in terms of fabrication. The bottom electrode can be formed by etching a sub-lithographically sized via into the dielectric material, filling the via with CVD (chemical vapor deposition) titanium nitride, and polishing the titanium nitride back down to the surface of the dielectric, thereby exposing the top of the bottom electrode.

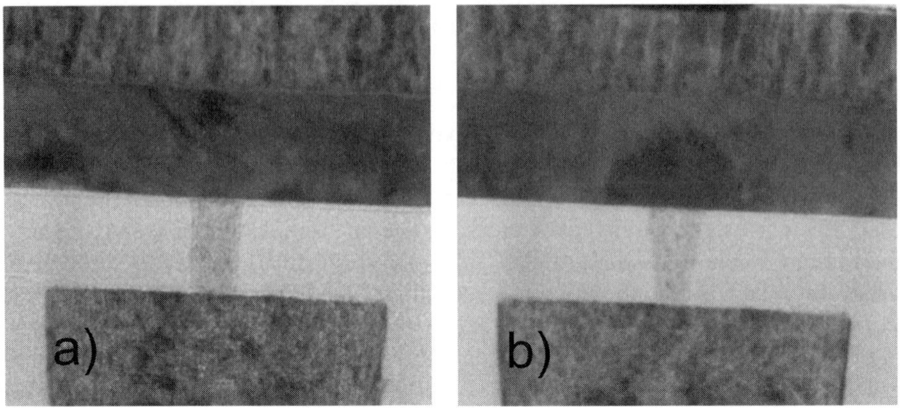

Fig. 17.10. TEM cross-sections of a mushroom cell phase change memory element in the (a) SET state and (b) RESET state. In the SET state the phase change material is polycrystalline throughout. In the RESET state, a mushroom of amorphous phase change material restricts the current flow through the bottom electrode.

The main advantage of the mushroom cell is that since the phase change material is located on top of a flat surface, the phase change material can be deposited using a standard PVD (physical vapor deposition) sputtering system. The desired phase change material composition is simply made into a large sputtering target, and then the phase change material is sputtered from the sputtering target onto the surface of the wafer. (In a PVD sputtering system, argon gas is ionized by applying a direct current (DC) bias to the sputtering target. The ionized argon atoms smash into the sputtering target which then sputters the material of the sputtering target in all directions, including the direction of the wafer.) After the phase change material is deposited, a capping layer of (PVD) titanium nitride is deposited without breaking vacuum (i.e., without exposing the wafer to air), so as to avoid any oxidation of the phase change material. The titanium nitride / phase change material stack is patterned (to isolate the phase change memory elements) using standard photo-lithography and RIE (reactive ion etching) techniques. These titanium nitride / phase change material islands are then encapsulated with silicon nitride, which has been found to help avoid oxidation penetration which can oxidize the interface between the bottom electrode and the phase change material [17.12].

The choice of the materials which comprise the bottom and top electrodes are extremely important. Specifically, the thermal conductivity of the electrodes must be low enough so as to provide thermal insulation during the programming operations, while the electrical conductivity of the electrodes must be large enough so as not to add unnecessarily high additional series resistance. Furthermore, the electrode materials must be able to withstand the high temperatures (> 650 °C) associated with the RESET programming condition. Lastly, the electrode materials must not interact chemically with the phase change material itself. The RESET current may be reduced by increasing the resistivity of the bottom electrode (see Fig. 17.11) [17.21, 17.22].

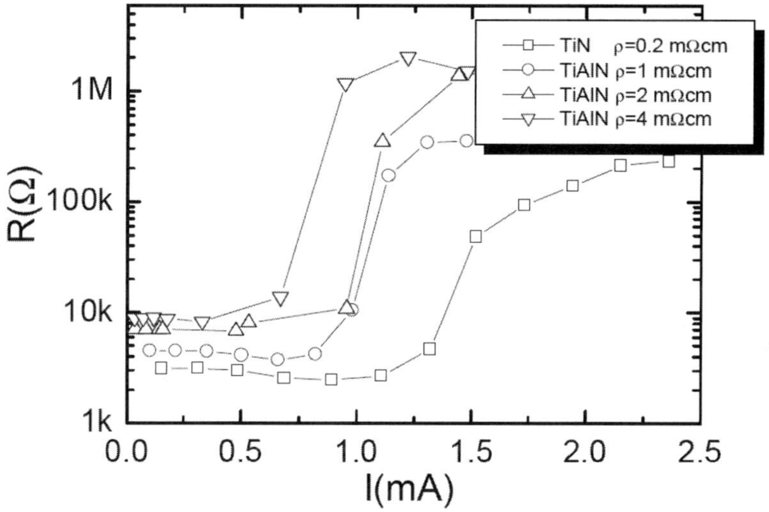

Fig. 17.11. Resistance vs. applied programming current as a function of mushroom cell bottom electrode material. Increasing the resistivity of the bottom electrode material decreases the RESET current while increasing the SET resistance. Reprinted with permission from the MRS, Warrendale, PA (2005), from [17.22].

However, both the RESET current and the RESET power must be considered when optimizing the cell design. Although continually increasing the resistivity of the bottom material will continually reduce the RESET current, the RESET power will also rise to unacceptable levels at some point.

The RESET current of the mushroom cell is a complex function of the bottom electrode dimensions and electrical and thermal properties, the thermal properties of the material which surrounds the bottom electrode and lies beneath the phase change material, and the phase change material thickness and electrical and thermal properties. The more efficient the cell structure is at utilizing the current flowing through the cell for heating the phase change material which is in the direct vicinity of the bottom electrode, the lower the required RESET current will be.

Hence, an important design point would be to thermally insulate the structure as well as possible (without, of course, insulating the cell so well that it cannot quench from the melted state into the amorphous phase). The best thermal insulation would be achieved if it were possible to completely surround the structure by vacuum. Silicon dioxide and silicon nitride are both standard back-end-of-line

(BEOL) materials which have been extensively characterized, silicon nitride being slightly less thermally insulating, and are both stable at the elevated programming temperatures associated with the RESET operation.

The phase change material itself is a good thermal insulator. The region of phase change material surrounding the switching volume can insulate that switching volume from heat loss escaping through the top electrode and laterally as well. In order to achieve the high current densities required during the RESET programming (~10^6 A/cm^2) a highly scaled electrode is needed. One method of creating a sub-lithographically sized electrode is to begin with a lithographically defined hole and use a collar process to shrink the size of the hole (Fig. 17.12(a)) [17.10].

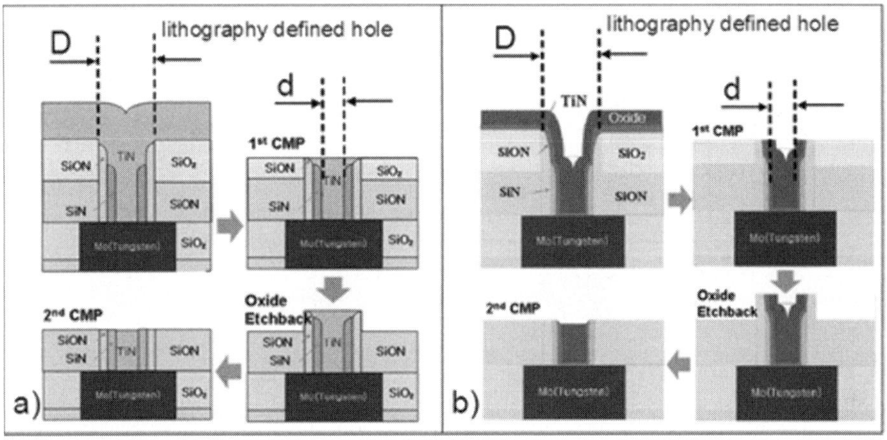

Fig. 17.12. (a) A collar process is used to create a sub-lithographically sized TiN bottom electrode. First, a lithographically defined hole of diameter *D* is etched into an SiON/SiN stack. A first collar is formed by depositing a conformal SiON layer followed by a collar RIE step. A second collar is formed in the same manner. Next, the CVD TiN is deposited to fill the hole. Finally, a series of CMP (Chemical-Mechanical Polishing) and oxide etchback processes are performed resulting in a cylindrical TiN bottom electrode. (b) A TiN ring electrode is constructed in a similar manner except that only a thin layer of CVD TiN is deposited into the hole, and then the center of the hole is filled with oxide (© (2005) IEEE, from [17.10]).

This construction has the disadvantage of directly transferring the full lithographic size variability down to the final electrode size. For example, if the initial lithographically defined hole diameter *D* is nominally 200 nm with a variation of ΔD = +/- 20 nm, and a 50 nm thick collar is used to shrink the nominally sized 200 nm down to 100 nm, the final electrode diameter, *d*, will be 100 nm +/- 20 nm! The contact area of the bottom electrode is given by

$$\textit{Contact area of electrode} = \pi/4\ (d + \Delta D)^2. \tag{17.2}$$

In a similar manner, a ring electrode can be formed (Fig. 17.12(b)). Here the contact area depends only linearly on ΔD, and is given by (where t is the ring width)

$$\text{Contact area of ring electrode} = \pi\,(d + \Delta D - t)\,t. \tag{17.3}$$

In addition to having the advantage of a reduced coupling between the lithographic size variability and the contact area, the ring electrode also has a slight advantage in terms of the RESET current. Compared to the full bottom electrode mushroom cell, the ring electrode mushroom cell results in an approximately 10% reduction in RESET current [17.10].

17.5.2 The Pillar Cell

In order to achieve a sub-lithographic aperture, there are two main categories of structures: contact-minimized [17.9-13, 17.16, 17.23] and volume-minimized [17.7, 17.14, 17.15, 17.17-19, 17.23-25]. A contact-minimized structure is one in which the cross-sectional area of an electrode in contact with the phase change material is highly scaled; the mushroom cell is an example of a contact-minimized cell structure. A volume-minimized structure is one in which the size of the phase change material itself is minimized somewhere within the cell. Compared to the contact-minimized structure, for a given critical dimension (*CD*), the volume-minimized structure has the advantage of offering a potentially lower RESET current [17.14, 17.23, 17.24]. This makes the idea of a volume-minimized structure quite attractive. However, achieving such a structure can be a processing challenge; the phase change material has to be etched into a small structure (a pillar cell, see Fig. 17.13), deposited into a small structure (a pore cell), or grown as a small structure (a nano-wire phase change cell).

Figure 17.14 shows the processing sequence used to fabricate a sub-lithographic pillar of phase change material. The TiN layer, deposited in step (a) of Fig. 17.14, serves as both a hard mask for the RIE step, and also as the top electrode contact. There are three critical steps to this process. First, uniform, well controlled islands of photo resist must be patterned. The semiconductor industry has a great deal of experience patterning lines of photo resist used to form the gates of FETs, and patterning holes into photo resist used to form vias, however, patterning islands of photo resist is a less mature art (which is not to say that it is in principle more or less inherently difficult). Second, the photo resist is trimmed using RIE, and is then transferred down through the TiN and phase change material. The trimming step may help even out the initial non-uniformity of the sizes of the photo resist islands as a larger island of photo resist will have a larger surface area which when exposed to a reactive etching may etch faster.

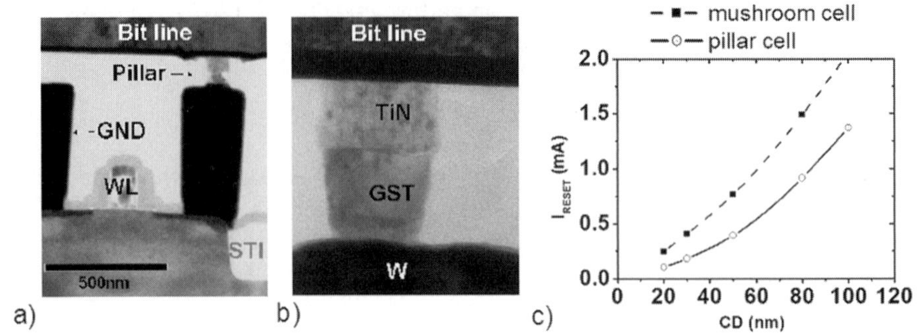

Fig. 17.13. (a) TEM cross-section of a pillar cell with an FET access device. (b) Close-up TEM cross-section of GST/TiN pillar. (c) Simulated RESET current vs. *CD* for a mushroom cell and a pillar cell showing the advantage of the volume-confined structure (© (2006) IEEE, from [17.14]).

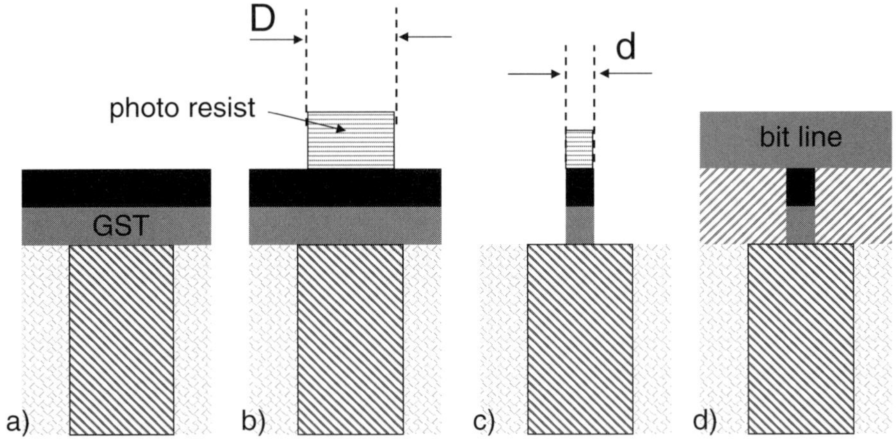

Fig. 17.14. Pillar cell processing sequence. (a) Deposit PVD GST followed by TiN. (b) Expose islands of photo resist. (c) Trim the photo resist and transfer the pattern down using RIE to make pillars of TiN/GST. (d) Encapsulate with a dielectric material, CMP down to expose top of TiN pillar, form Cu bit line wiring.

If α represents this trim uniformity factor, then the pillar cross-section area is given as follows:

$$\textit{Cross-section area of pillar} = \pi/4\ (d + \alpha\Delta D)^2. \tag{17.4}$$

It should be noted that it has yet to be shown experimentally that α is substantially different from unity. If α is not substantially different from unity, the variability of the size of the pillar will be directly inherited from the lithographic size variability. The RIE of the GST also presents a challenging process. It has been found that compared to TiN, GST has an extremely fast etch rate, so a well controlled TiN/GST etch process is required to avoid undercutting the GST. Furthermore, the RIE of the GST can form a damaged layer approximately 10 nm thick on the surface of the GST in which antimony is depleted and germanium oxide is formed [17.26].

After the TiN/GST pillar is formed, the next step is to encapsulate with a dielectric material and planarize with CMP exposing the top of the TiN electrode. The bit line can then directly make contact to this TiN electrode.

The relatively few processing steps required to fabricate the pillar cell and the potential for low RESET current for a given *CD* make it an attractive design. However, the difficulty of controlling of the *CD* of the pillar and the challenges of etching a highly scaled pillar of TiN/GST are formidable.

17.5.3 The μTrench Cell

The μTrench cell (Fig. 15.6) is a combined approach between a contact-minimized structure and a volume-minimized structure. The PCM element is formed at the intersection of a sidewall deposited CVD TiN bottom electrode and a trench of phase change material. The thickness of the sidewall film, *t*, can be well controlled even below a thickness of 10 nm. A sub-lithographic sized trench can be formed using a collar process to shrink a lithographically defined trench. (See the previous chapter for further details regarding the μTrench cell fabrication and characteristics.) Only one dimension of the μTrench cell inherits lithographic size variability. The cross-sectional area of contact is given by the following equation:

$$\textit{Cross-section area of } \mu\textit{Trench} = (d + \Delta D)\, t. \tag{17.5}$$

If a PVD phase change material deposition is used, the aspect ratio of the trench is limited. If a CVD phase change material deposition is available, the aspect ratio of the trench can be increased further to lower the required RESET current.

17.5.4 The Pore Cell

The pore cell is a volume-confined cell. In this design, a sub-lithographically sized aperture is etched into an insulating layer, exposing a lower conductive electrode. The aperture is then filled with phase change material. The aspect ratio of the pore is limited by the filling capabilities of the phase change material deposition available. The pore must be filled without producing an unwanted seam or void within the phase change material. Figure 17.15(a) shows a TEM cross-section of a pore cell in which a PVD GST deposition was used to fill a 1 : 1.5 (height : width) aspect ratio pore having a sidewall slope of approximately 60° [17.19]. Figure 17.15(b) shows simulations of the pore cell RESET current as a function of *CD* for both a 60° and 90° sidewall slope.

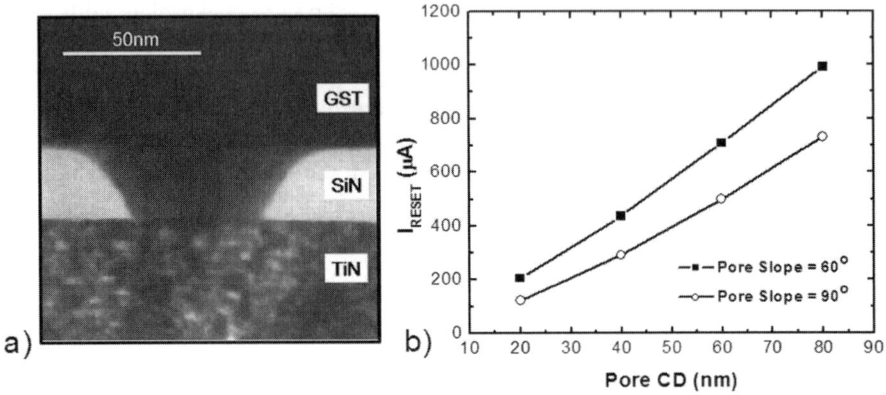

Fig. 17.15. (a) TEM cross-section of a 45 nm bottom *CD* low aspect ratio pore cell filled with a PVD GST process. (b) Simulated RESET current vs. *CD* for a pore slope of 60° and 90° (with the SiN thickness fixed as shown in (a)), (© (2007) IEEE, from [17.19]).

Compared to the pore cell with a 60° sidewall slope, a pore cell with a 90° sidewall slope has a substantially lower RESET current for a given *CD*. However, the steeper the sidewall, the more difficult it is to fill the pore with a PVD process. To extend the filling capabilities of a PVD process, it can be combined with an etchback process whereby there is a first PVD deposition followed by an *in situ* etchback of the material (which will have the effect of preferentially etching the corners of the phase change material due to a field enhancement at the corners, thereby opening up the top of the partially filled pore) followed by a second PVD deposition. This scheme has been demonstrated to be capable of filling a 50 nm bottom *CD* 2:1 aspect-ratio pore (Fig. 17.16) [17.18].

There is currently a great deal of effort being put forth to develop CVD processes for depositing phase change materials. A first demonstration of a CVD GST filled high aspect ratio pore was given in 2007 by J.I. Lee et. al. [17.20] (Fig. 17.17). In this demonstration, a sub-lithographically sized pore was formed using a collar process, the GST was deposited using a CVD process, the GST was polished back and then capped with a top electrode material. This fill-and-polish scheme has the additional advantage of avoiding possible damage of the phase change material caused by the RIE process [17.4, 17.26]. In order to isolate neighboring devices along adjacent bit lines, a RIE process is normally utilized (see Fig. 17.9 for example).

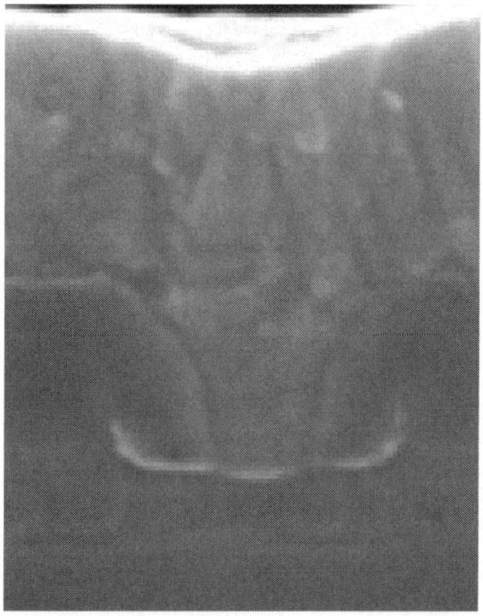

Fig. 17.16. TEM cross-section of a 50 nm bottom *CD* 2:1 aspect-ratio pore filled with a deposition/etch/deposition PVD process. (© (2005) IEEE, from [17.18]).

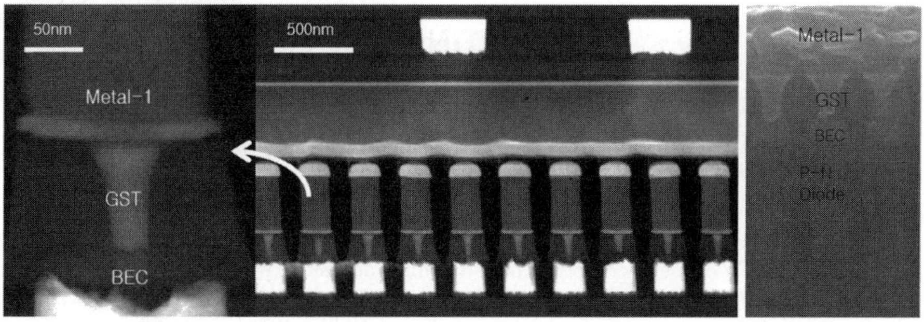

Fig. 17.17. TEM cross-section of a 40 nm bottom *CD* high aspect-ratio pore filled with a CVD GST. The left image shows a close-up of a single pore. The central image shows an array of pores. The right image shows the pore integrated with a pn-diode access device (© (2007) IEEE, from [17.20]).

As with the collar-defined bottom electrode for the mushroom cell, if the sub-lithographically sized pore is fabricated using a collar process to shrink a lithographically defined hole, the pore size will directly inherit the full variability associated with the lithographically define hole (ΔD):

$$\textit{Cross-section area of collar-defined pore} = \pi/4\ (d + \Delta D)^2. \qquad (17.6)$$

To eliminate this problem, an integration scheme which completely decouples the final aperture size from the lithographic dimension, and hence, decouples the final aperture size from any variability associated with lithography can be utilized.

Figure 17.18(a) illustrates the sequential processing steps of the keyhole-transfer process [17.26] which creates a final aperture whose size is independent from the initial lithographically defined hole size. This keyhole-transfer process combines an accurate etch-back with a conformal deposition into a lithographically-defined hole to create a keyhole whose size is independent of the lithography-defined hole size. First, a lithographically-defined hole is etched into a SiN-SiO_2-SiN stack stopping on the bottom SiN (Fig. 17.18(a)-1).

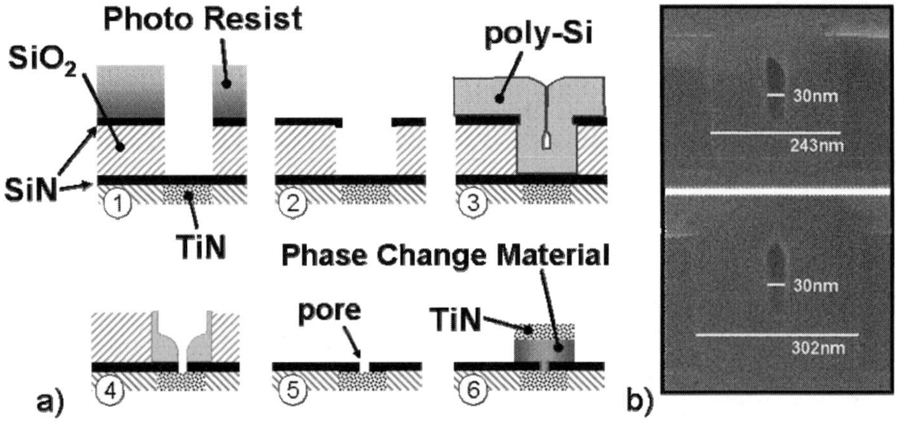

Fig. 17.18. (a) A sub-lithographic size lithography-independent feature is fabricated using the keyhole-transfer process: 1) A lithographically-defined hole is etched, and 2) the middle SiO_2 layer is recessed. 3) A highly conformal poly-Si film is deposited, producing a sub-lithographic keyhole whose diameter is equal to the recess of the SiO_2 layer. 4) The keyhole is transferred into the underlying SiN layer to define a pore, followed by 5) removal of the SiO_2 and poly-Si. 6) The phase change and top electrode (TiN) materials are then deposited and the cell is patterned for isolation. (b) An SEM cross-section corresponding to step 3), showing keyholes for two different sized lithographically-defined holes. Since the keyhole size does not depend on lithography, the phase change *CD* is decoupled from any lithographic variability. Reprinted with permission from [17.8].

Next, a selective wet etch is used to recess the SiO_2 layer with respect to the SiN layer, resulting in an overhang of the SiN layer (Fig. 17.18(a)-2).

An example wet etch chemistry which can be used to obtain a highly selective etch of the SiO_2 layer with respect to SiN is a buffered oxide etch (BOE) which is a mixture of ammonium fluoride and hydrofluoric acid. The etch rate can be controlled with the dilution of water into the BOE. After the overhang is formed, a highly conformal poly-Si film is deposited, producing a sub-lithographic keyhole in the poly-Si whose diameter is equal to the recess of the SiO_2 layer (Fig. 17.18(a)-3). Figure 17.18(b) demonstrates the keyhole size independence from the initial hole size. Next, this keyhole is transferred into the underlying SiN layer using a RIE process to define a pore in the SiN layer (Fig. 17.18(a)-4). The SiO_2 layer and poly-Si are removed with selective wet etches (Fig. 17.18(a)-5). Examples of selective wet etches that can be used for this step are a BOE to remove the oxide layer and potassium hydroxide etch to remove the poly-Si. After the pore is formed the phase change and top electrode (TiN) materials are deposited, and finally the cell is patterned for isolation (Fig. 17.18(a)-6).

Combining this keyhole-transfer process, forming a high aspect ratio pore, with a CVD phase change material deposition process would create a very efficient (low required RESET power) and controllable phase change memory element. The pore size variability would be completely decoupled from lithographic size variability:

$$\textit{Cross-section area of keyhole-defined pore} = \pi/4\, d^2. \quad (17.7)$$

17.6 Multi-Level Phase Change Random Access Memory

Once the cell size (cell area / F^2) has been reduced as far as possible, there are two options for further increasing the memory density. The first option would be to pursue a three-dimensional (3-D) integration scheme where on top of the first layer of access devices plus memory elements would be additional layers of access devices plus memory elements. The main challenge of this approach is to identify a suitable 3-D stackable access device which can be fabricated within the BEOL temperature budget constraints (typically < 400 °C) and still be capable of delivering a high current density and high on-off ratio.

The second option would be to store more than one bit per memory cell, thus enabling multi-level PCRAM. The resistance distributions shown in Fig. 17.5 display the two distinct resistance levels, representing "0" and "1", required for storing one bit per cell. Four distinct levels are required for storing two bits ("00", "01", "10", or "11"). 2^N distinct levels are required for storing N bits. The main issues to address for multi-level PCRAM are the programming methodology and the stability and retention of the levels of intermediate resistance.

In order to achieve an intermediate resistance state, there are a number of programming techniques which can be utilized. One method [17.27] is to first fully RESET the cell with a first current pulse, and then apply nanosecond current pulses of sequentially higher amplitude, thereby crystallizing the cell more and more with each current pulse. The resistance is read in between each current pulse and this iterative write-verify cycle is stopped once the desired resistance is achieved. Another programming method is to use a single current pulse with amplitude large enough to melt the phase change material, and then control the final resistance of the cell by controlling the fall time of the current pulse (see Fig. 17.19) [17.28].

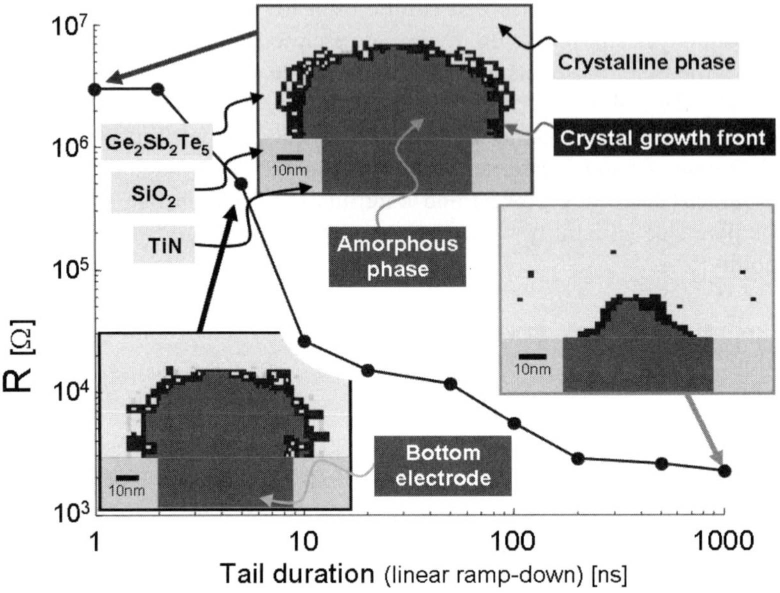

Fig. 17.19. Simulation of different multi-level states as a function of pulse-tail duration (linear ramp-down at pulse-end). The shape and size of the amorphous plug at the top of the pillar-heater influences the resistance of the phase change element, allowing intermediate resistance states (© (2007) IEEE, from [17.28]).

Figure 17.20 illustrates the need for an iterative write-verify scheme for this programming method as well. For a collection of cells, a current pulse with a given fall time intended to program the cells into the intermediate resistance state of level 3 (of Fig. 17.20) results in a wide range of final resistance values. This cell-to-cell programming variability can be minimized by reducing the process induced cell variability, but to achieve a maximum number of distinguishable resistance levels a write-verify scheme must be utilized.

Figure 17.21 demonstrates 16 levels per cell (4 bits per cell) using an average of approximately three write-verify iterations. Hence, with a few write-verify iterations, each phase change memory element can be programmed into a well defined resistance level. However, reliability of these intermediate resistance levels will be a challenge. Even so, techniques involving redundancy and error correction coding can be used to enhance the reliability of multi-level phase change memory.

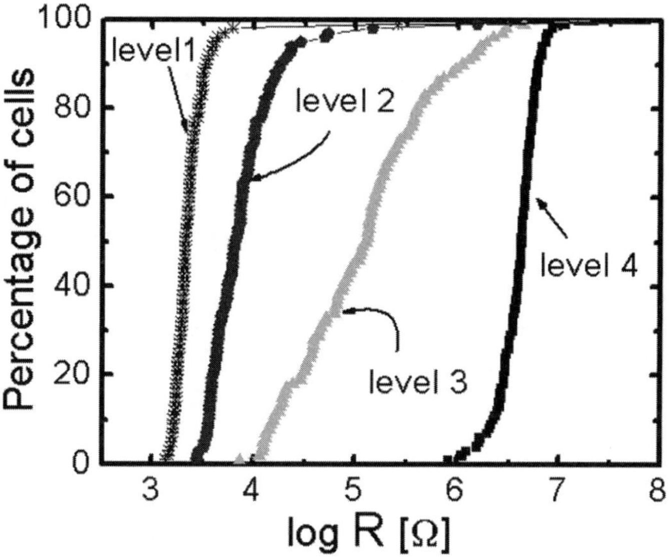

Fig. 17.20. Resistance distribution of a four-level cell using single pulse programming. Process induced variations cause distributions to overlap because the same applied voltage pulse leads to different temperatures in different cells (© (2007) IEEE, from [17.28]).

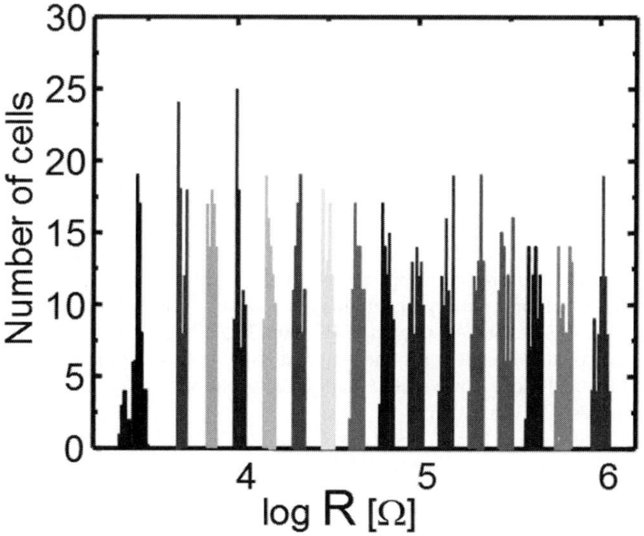

Fig. 17.21. 10x10 array test structure programmed into 16 levels. Tight, well controlled distributions allow 4 bit/cell. Adjustment of pulse slopes depending on resistance is essential to achieve narrow distributions (© (2007) IEEE, from [17.28]).

17.7 Concluding Remarks

This chapter has reviewed the main process integration challenges associated with the design and fabrication of the phase change memory cell. Careful design of the structural geometry and choice of materials are required for minimizing the RESET programming current, which, in combination with the access device capabilities, ultimately determine the achievable memory cell size. Furthermore, controlling the geometrical variability of the highly scaled dimensions within the phase change memory element is key for achieving robust operability of large memory arrays. And for further increasing the achievable memory density, phase change memory's wide resistance parameter space is well suited for a multi-bit per cell non-volatile memory technology.

References

[17.1] Kang, S., Cho, W. Y., Cho, B. H., Lee, K. J., Lee, C. S., Oh, H. R., Choi, B. G., Wang, Q., Kim, H. J., Park, M. H., Ro, Y. H., Kim, S., Ha, C. D., Kim, K. S., Kim, Y. R., Kim, D. E., Kwak, C. K., Byun, H. G., Jeong, G., Jeong, H., Kim, K. and Shin, Y.: A 0.1-μm 1.8-V 256-Mb phase change random access memory (PRAM) with 66-Mhz synchronous burst-read operation. IEEE Journal of Solid-State Circuits **42**, 210-218 (2007)

[17.2] Rajendran, B., Lee, M. H., Breitwisch, M., Burr, G. W., Shih, Y. H., Cheek, R., Schrott, A., Chen, C. F., Lamorey, M., Joseph, E., Zhu, Y., Dasaka, R., Flaitz, P. L., Baumann, F. H., Lung, H. L., and Lam, C.: On the dynamic resistance and reliability of phase change memory. VLSI (2008)

[17.3] Pirovano, A., Redaelli, A., Pellizzer, F., Ottogalli, F., Tosi, M., Ielmini, D., Lacaita, A. L., and Bez, R.: Reliability study of phase change nonvolatile memories. IEEE Transactions on Device and Materials Reliability, 4(3):422-427 (2004)

[17.4] Oh, J. H., Park, J. H., Lim, Y. S., Lim, H. S., Oh, Y. T., Kim, J. S., Shin, J. M., Park, J. H., Song, Y. J., Ryoo, K. C., Lim, D. W., Park, S. S., Kim, J. I., Kim, J. H., Yu, J., Yeung, F., Jeong, C. W., Kong, J. H., Kang, D. H., Koh, G. H., Jeong, G. T., Jeong, H. S., and Kim, K.: Full integration of highly manufacturable 512Mb PRAM based on 90nm technology. IEDM, San Francisco, CA, p. 49-53 (2006)

[17.5] Lung, H. L, Breitwisch, M., Happ, T., and Lam, C.: Phase change memory – present and future. Int. Conf. on Memory Technology and Design, Giens, France (2007)

[17.6] Pellizzer, F., Benvenuti, A., Gleixner, B., Kim, Y., Johnson, B., Magistretti, M., Marangon, M., Pirovano, A., Bez, R., and Atwood, G.: A 90nm phase change memory technology for stand-alone non-volatile memory applications. In Symposium on VLSI Technology, pages 122-123 (2006)

[17.7] Oh, J. H., Park, J. H., Lim, Y. S., Lim, H. S., Oh, Y. T., Kim, J. S., Shin, J. M., Park, J. H., Song, Y. J., Ryoo, K. C., Lim, D. W, Park, S. S., Kim, J. I., Kim, J. H., Yu, J., Yeung, F., Jeong, C. W., Kong, J. H., Kang, D. H., Koh, G. H., Jeong, G. T., Jeong, H. S., and Kim, K.: Full integration of highly manufacturable 512Mb PRAM based on 90nm technology. In IEDM Technical Digest, page 2.6 (2006)

[17.8] Raoux, S., Burr, G. W., Breitwisch, M. J., Rettner, C. T., Chen, Y. C., Shelby, R. M., Salinga, M., Krebs, D., Chen, S. H., Lung, H. L., and Lam, C. H.: Phase change random access memory — a scalable technology. IBM J. Res. & Dev. **52**, 465-480 (2008)

[17.9] Lai, S. and Lowrey, T.: OUM - a 180 nm nonvolatile memory cell element technology for stand alone and embedded applications: International Electron Devices Meeting. Technical Digest Washington, DC, USA 2-5 Dec. 2001, p. 3651-3654 (2001)

[17.10] Ahn, S. J., Hwang, Y. N., Song, Y. J., Lee, S. H., Lee, S. Y., Park, J. H., Jeong, C. W., Ryoo, K. C., Shin, J. M., Park, J. H., Fai, Y., Oh, J. H., Koh, G. H., Jeong, G. T., Joo, S. H., Choi, S. H., Son, Y. H., Shin, J. C., Kim, Y. T., Jeong, H. S., and Kim, K.: Highly reliable 50nm contact cell technology for 256Mb PRAM. Symposium on VLSI Technology Digest of Technical Papers, p. 98-99 (2005)

[17.11] Jeong, C. W., Ahn, S. J., Hwang, Y. N., Song, Y. J, Oh, J. H., Lee, S. Y., Lee, S. H., Ryoo, K. C., Park, J. H., Park, J. H., Shin, J. M., Yeung, F., Jeong, W. C., Kim, J. I., Koh, G. H., Jeong, G. T., Jeong, H. S., and Kim, K.: Highly reliable ring-type contact for high-density phase change memory. Jap. J. Appl. Phys. 45, 3233-3237 (2006)

[17.12] Song, Y. J., Ryoo, K. C., Hwang, Y. N., Jeong, C. W., Lim, D. W., Park, S. S., Kim, J. I., Kim, J. H., Lee, S. Y., Kong, J. H., Ahn, S. J., Lee, S. H., Park, J. H., Oh, J. H., Oh, Y. T, Kim, J. S., Shin, J. M., Park, J. H., Fai, Y., Koh, G. H., Jeong, G. T., Kim, R. H., Lim, H. S., Park, I. S., Jeong, H. S., and Kim, K.: Highly reliable 256Mb PRAM with advanced ring contact technology and novel encapsulating technology: Symposium on VLSI Technology, p. 15-16 (2006)

[17.13] Ryoo, K. C., Song, Y. J., Shin, J. M., Park, S. S., Lim, D. W., Kim, J. H., Park, W. I., Sim, K. R., Jeong, J. H., Kang, D. H., Kong, J. H., Jeong, C. W., Oh, J. H., Park, J. H., Kim, J. I., Oh, Y. T., Kim, J. S., Eun, S. H., Lee, K. W., Koh, S. P., Fai, Y., Koh, G. H., Jeong, G. T., Jeong, H. S., and Kim, K.: Ring contact electrode process for high density phase change random access memory. Jap. J. Appl. Phys. 46, p. 2001-2005 (2007)

[17.14] Happ, T. D., Breitwisch, M., Schrott, A., Philipp, J. B., Lee, M. H., Cheek, R., Nirschl, T., Lamorey, M., Ho, C. H., Chen, S. H., Chen, C. F., Joseph, E., Zaidi, S., Burr, G. W., Yee, B., Chen, Y. C., Raoux, S., Lung, H. L, Bergmann, R., and Lam, C.: Novel one-mask self-heating pillar phase change memory. Symp. on VLSI Technology, Honolulu, Hawaii, p. 15-16 (2006)

[17.15] Chen, Y. C., Rettner, C. T., Raoux, S., Burr, G. W., Chen, S. H., Shelby, R. M., Salinga, M., Risk, W. P., Happ, T. D., McClelland, G. M., Breitwisch, M., Schrott, A., Philipp, J. B., Lee, M. H., Cheek, R., Nirschl, T., Lamorey, M., Chen, C. F., Joseph, E., Zaidi, S., Yee, B., Lung, H. L., Bergmann, R., and Lam, C.: Ultra-thin phase-change bridge memory device using GeSb. IEDM Technical Digest, San Francisco, CA, p. S30P3 (2006)

[17.16] Pellizzer, F., Pirovano, A., Ottogallic, F., Magistretti, M., Scaravaggi, M., Zuliani, P., Tosi, M., Benvenuti, A., Besana, P., Cadeo, S., Marangon, T., Morandi, R., Piva, R., Spandre, A., Zonca, R., Modelli, A., Varesi, E., Lowrey, T., Lacaita, A., Casagrande, G., Cappelletti, P., and Bez, R.: Novel μ-trench phase-change memory cell for embedded and stand-alone non-volatile memory applications. Symposium on VLSI Technology, p. 18-19 (2004)

[17.17] Tyson, S., Wicker, G., Lowrey, T., Hudgens, S., and Hunt, K.: Nonvolatile, high density, high performance phase change memory. IEEE Aerospace Conference Proceedings, Vol. 5, Big Sky, MT, USA, 18-25 March 2000, p. 385-390 (2000)

[17.18] Cho, S. L., Yi, J. H., Ha, Y. H., Kuh, B. J., Lee, C. M., Park, J. H., Nam, S. D., Horii, H., Cho, B. O., Ryoo, K. C., Park, S. O., Kim, H. S., Chung, U. I., Moon, J. T., and Ryu, B. I.: Highly scalable on-axis confined cell structure for high density PRAM beyond 256Mb. Symp. on VLSI Technology Digest of Technical Papers, p. 96-97 (2005)

[17.19] Breitwisch, M., Nirschl, T., Chen, C. F., Zhu, Y., Lee, M. H., Lamorey, M., Burr, G. W., Joseph, E., Schrott, A., Philipp, J. B., Cheek, R., Happ, T. D., Chen, S. H., Zaidi, S., Flaitz, P., Bruley, J., Dasaka, R., Rajendran, B., Rossnagel, S., Yang, M., Chen, Y. C., Bergmann, R., Lung, H. L., and Lam, C.: Novel lithography-independent pore phase change memory. Symposium on VLSI Technology, p. 6B-3 (2007)

[17.20] Lee, J. I., Park, H., Cho, S. L., Park, Y. L., Bae, B. J., Park, J. H., Park, J. S., An, H. G., Bae, J. S., Ahn, D. H., Kim, Y. T., Horii, H., Song, S. A., Shin, J. C., Park, S. O., Kim, H. S., Chung, U. I., Moon, J. T., and Ryu, B. I.: Highly scalable phase change memory with CVD GeSbTe for sub 50nm generation. Symp. on VLSI Technology, Kyoto, Japan, p. 102-103 (2007)

[17.21] Czubatyj, W., Lowrey, T., Kostylev, S., and Asano, I.: Current reduction in Ovonic memory devices. Proc. Europ. Phase Change and Ovonic Science Symp., Grenoble, France, p. 143-152 (2006)

[17.22] Lee, S., Song, Y. J., Hwang, Y. N., Lee, S. H., Park, J. H., Ryoo, K. C., Ahn, S. J., Jeong, C. W., Oh, J. H., Shin, J. M., Yeung, F., Jeong, W. C., Kim, Y. T., Park, J. B., Koh, K. H., Jeong, G. T., Jeong, H. S., and Kim, K.: Effect of the bottom electrode contact (BEC) on the phase transformation of N_2 doped $Ge_2Sb_2Te_5$ (N-GST) in a phase-change random access memory. In MRS Proceedings, volume 830, page D7.9.1 (2005)

[17.23] Pirovano, A., Lacaita, A. L., Benvenuti, A., Pellizzer, F., Hudgens, S., and Bez, R.: Scaling analysis of phase-change memory technology. IEDM Tech. Dig. p. 699-703 (2003)

[17.24] Kim, Y, T., Hwang, Y. N., Lee, K. H., Lee, S. H., Jeong, C. W., Ahn, S. J, Yeung, F., Koh, G. H., Jeong, H. S., Chung, W. Y., Kim, T. K., Park, Y. K., Kim, K. N., and Kong, J. T.: Programming characteristics of phase change random access memory using phase change simulations. Jap. J. Appl. Phys. 44, 2701-2705 (2005)

[17.25] Lankhors, M. H. R., Ketelaars, B. W. S. M. M., and Wolters, R. A. M.: Low-cost and nanoscale non-volatile memory concept for future silicon chips.: Nature Mater. 4, 347-352 (2005)

[17.26] Joseph, E. A., Happ, T. D., Chen, S. H., Raoux, S., Totir, G., Pyzyna, A., Chen, C. F., Breitwisch, M., Schrott, A. G., Zaidi, S., Dasaka, R., Yee, B., Zhu, Y., Bergmann, R., Lung, H. L., and Lam, C.: Patterning of $N{:}Ge_2Sb_2Te_5$ films and the characterization of etch induced modification for non-volatile phase change memory applications. VLSI TSA (2008)

[17.27] Bedeschi, F., Fackenthal, R., Resta, C., Donze, E., Jagasivamani, M., Buda, E., Pellizzer, F., Chow, D., Fantini, A., Calibrini, A., Calvi, G., Faravelli, R., Torelli, G., Mills, D., Gastaldi, R., and Casagrande, G.: A multi-level-cell bipolar-selected phase-change memory. In ISSCC Technical Digest, volume paper 23.5 (2008)

[17.28] Nirschl, T., Philipp, J. B., Happ, T. D., Burr, G. W., Rajendran, B., Lee, M. H., Schrott, A., Yang, M., Breitwisch, M., Chen, C. F., Joseph, E., Lamorey, M., Cheek, R., Chen, S. H., Zaidi, S., Raoux, S., Chen, Y. C., Zhu, Y., Bergmann, R., Lung, H. L., and Lam, C.: write strategies for 2 and 4-bit multi-level phase-change memory. Int. Electron Devices Meeting, Washington, DC (2007)

18. Reconfigurable Logic

James Lyke

Abstract. With the soaring price of custom electronics at advanced technology nodes, reconfigurable electronics are becoming even more important. In reconfigurable electronics, versatile architectures allow customized functions to be bound after fabrication. This paradigm of software-definable hardware offers intriguing benefits that include flexibility through in situ reprogrammation, fault tolerance through circumlocution, and rapid response solutions to be formed by configuring pre-built components. The penultimate example in current practice is the field programmable gate array (FPGA), which has evolved from being a curiosity to a multi-billion dollar disruptive technology. Phase change materials have not traditionally played a role in reconfigurable electronics, but offer powerful advantages by virtue of their non-volatility, durability, and the possibility of multi-state configurations. These concepts lead potentially to more compact computation architectures based on threshold logic, or more ambitiously dense arrays of artificial neurons and other novel hybrid (digital plus analog) signal processing architectures.

18.1 Introduction

There is a rich body of work on configurable digital systems, such as processors, memory, and field programmable gate arrays (FPGAs) of increasing sophistication, while configurable analog concepts remain a relatively fragmented body of work. This chapter explores configurable digital and analog systems from the standpoint of extending the former *into* the latter. The approach chosen for this exploration begins with digital concepts, their extension to configurable systems, and proceeds then to consider concepts based on multi-valued resistive elements (enabled by summing junctions) and analog-based reconfigurable architectures. The presentation is decidedly eclectic, even though many of the basic tenets of configurable analog are not new.

S. Raoux, M. Wuttig (eds.), *Phase Change Materials*,
DOI 10.1007/978-0-387-84874-7_18,

18.2 Digital System Basics

Digital systems are simply those components or networks that interoperate on a set of two states or values drawn from the set {0,1}. Physically, these values are formed in a number of ways, such as voltages, currents, charge patterns, or other discernable physical phenomena that can effectively be resolved into two different states, one called "0" (zero) or "1" (one). The most prevalent implementations of digital systems are based on CMOS (complementary metal-oxide semiconductor) integrated circuits. A canonical digital circuit, the invertor ("NOT" gate, or more descriptively, an inverting thresholded comparator), is shown in Fig. 18.1.

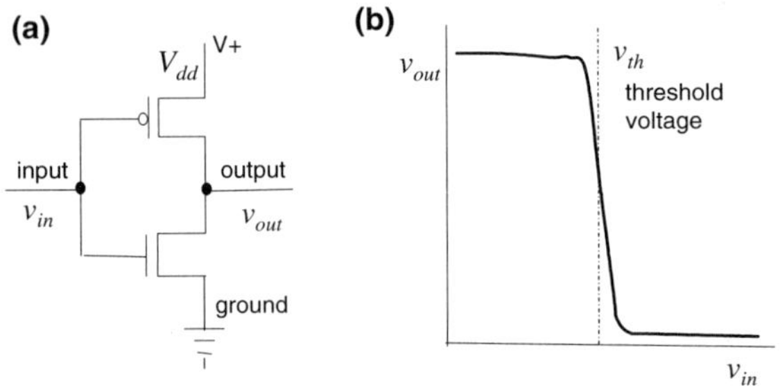

Fig. 18.1. Complementary metal-oxide semiconductor (CMOS) invertor. (a) Schematic representation. (b) Transfer characteristic.

The CMOS inverter circuit (Fig. 18.1a) ideally implements a very simple "digital" function, one that receives an input voltage v_{in} element of $0 < v < v_{dd}$ (v_{dd} is the positive supply voltage) and produces an output voltage v_{out} as shown in the transfer curve (Fig. 18.1b). In this idealized case, we would expect the transfer function to resolve inputs into one of two possible states defined as

$$v_{out} = \begin{cases} 0, & v_{in} < v_{th} \\ v_{dd} & v_{in} \geq v_{th} \end{cases}, \tag{18.1}$$

where 0 corresponds to logical zero, and v_{dd} corresponds to logic one.

The resolution of voltages through digital functions obviously destroys information, collapsing two partitions demarcated by threshold voltage (v_{th}) into two single values. This resolution of subranges into values gives reliability and robustness to digital systems. In real systems, physical voltages are often corruptible due to noise and variations in processing and operating environment. Minor corruptions are easily mitigated through (in this case) an inverting threshold comparison.

The invertor is an extremely simple logic gate, which performs a simple "computation" on a single signal variable v_{in}, with mapping function over $\{0,1\} \rightarrow \{0,1\}$ to produce a simple output variable v_{out} that is the logical complement of the input. More useful computations require the introduction of Boolean functions that map a larger Boolean input space $\{0,1\}^k$ to single or multiple Boolean outputs, in general $\{0,1\}^k \rightarrow \{0,1\}^m$. These include of course the common Boolean functions, such as AND and OR which operate as logical conjunction or disjunction, respectively (simply put, conjunction outputs a "1" when all inputs are "1", and disjunction outputs "1" when any input is "1"), and other functions according to the rules of Boolean logic. Simple gates, if not constructed directly in the transistors as circuit layouts, are built as cascades of simpler building blocks. Much discussion of the very many styles of digital logic circuitry useful in these constructions is not undertaken here; the reader is referred to [18.1] as a useful reference. Suffice to say that all Boolean functions are expressable by some (non-unique) compositions of gates drawn from complete basis sets of Boolean functions (such as {AND, NOT}, {OR, NOT}, etc.).

Boolean networks based on cascades of simpler functions are sometimes called *combinational logic*. Combinational logic networks compute "statelessly", meaning that for any static input combination, final output values are produced (after accounting for finite signal propagations through the gate-networks) that never change so long as the inputs to the network are frozen, but lose these values when the inputs are removed. In order to hold the outputs computed by Boolean networks so that they do not continue to change, a mechanism to store Boolean state is required. The addition of storage elements and feedback dramatically expand the utility of digital systems. Storage provides a *registration* mechanism, conveying the notion of states that, once locked in, do not change even as inputs continue to vary. Feedback provides the ability to modify future states even when inputs do not vary. It is possible to capture all of these concepts in the generalization of a clocked-domain digital system, shown in Fig. 18.2. Here, a combinational Boolean circuit computes outputs (y_c and y_d) from inputs (x_a and x_b). A flip-flop array (one deep) registers all outputs. Through the inclusion of feedback, we achieve history-dependent behavior. This history-dependent behavior, being an advancement of the circuit behavior from one state to another (the description is sometimes called the *state transition matrix*), is captured by dedicating some of the inputs and outputs to the computation of state using feedback. The output states of the flip-flop array latch the outputs of the combinational Boolean circuit in synchronization with a clock, freezing both state and outputs between clock pulses. Fig. 18.2 is one hardware interpretation of a *finite state machine (FSM)*.

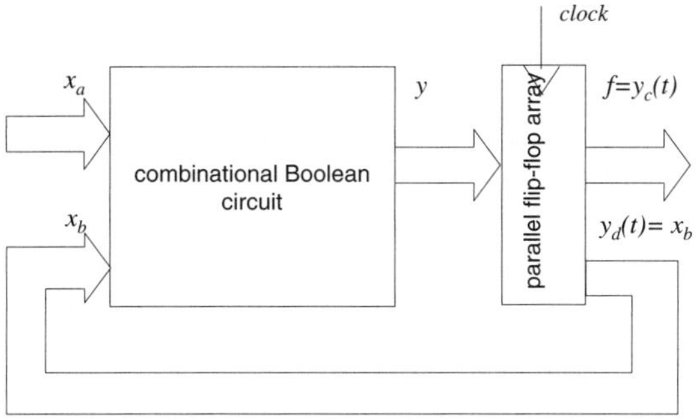

Fig. 18.2. Generalized representation of clocked-mode sequential digital circuit (with single clock-domain).

The discipline of clocking digital systems emerged as a means of coping with the complexities of digital systems with feedback in the presence of non-uniform delays. Not all paths in classical real-world combinational circuits are isochronal, leading to some parts of a complex calculation of an input taking longer than others. In circuits with feedback, *race* conditions can result, leading to erroneous calculations or circuit states that are difficult to control and in fact become dependent on circuit layout, temperature, and other factors. In effect, the introduction of clocking produces a temporal firewall, a discipline that eliminates race conditions by construction. Obviously, the clocking interval is typically chosen so that enough time has elapsed for the worst-case (slowest) timing path to complete its calculation. Clocked-domain sequential design, the practice of design circuits with clocks, is the dominant form of digital design today, and most synthesis tools expect this style of design. Disciplines for *clockless* (asynchronous) and *wave-pipelined* systems are much rarer and have evolved separately. These arcane approaches achieve performance through self-timed circuitry. In clockless circuitry, output states change only in response to specific changes in input. Wave pipelining exploits the latency of signal propagation in deeply-staged combinational logic to "launch" multiple computations within an overall clocking cycle (the reader is referred to [18.2-4] for excellent discussions of these topics). Clockless approaches are promising but less comfortable to designers due to their additional complexity and limited support in electronic design automation today.

Hence, sequential digital design involves memory. Memory plays at least three major roles in Boolean systems:

- ***Finite state machines (FSMs)***, in which memory cells (flip-flops / registers) are dispersed throughout combinational logic, which computes the values of these memory cells. The design objective in FSMs is to maximize performance of state machines, for some combination of power, density, and latency (the latter dictating the operating speed of calculations).
- ***Bulk storage***, where large amounts of state information in the form of user data are to be preserved and retrieved. In this case, combinational logic is used primarily to access the stored data for reading or writing. The design objective in bulk storage is to maximize the density of memory bits available to a user.
- ***Configuration storage***, where state information is used to customize the functions of circuits. The design objective is to minimize the overhead of circuit added to manage configuration.

In each of these three cases, it is possible to consider both volatile and non-volatile versions of storage. It is also possible to consider one-time-programmable (OTP) versions of bulk storage (e.g., read-only memories) and configuration storage (used in some anti-fuse field programmable gate arrays), except for FSMs, which are trivially reduced to Boolean constant values.

For the scattered bits of storage in FSM implementations, the use of single-bit memories (flip-flops) are convenient, but the implementation of larger amounts of storage is easier with a managed array of densely arranged storage elements. Two rudimentary concepts are shown in Fig. 18.3.

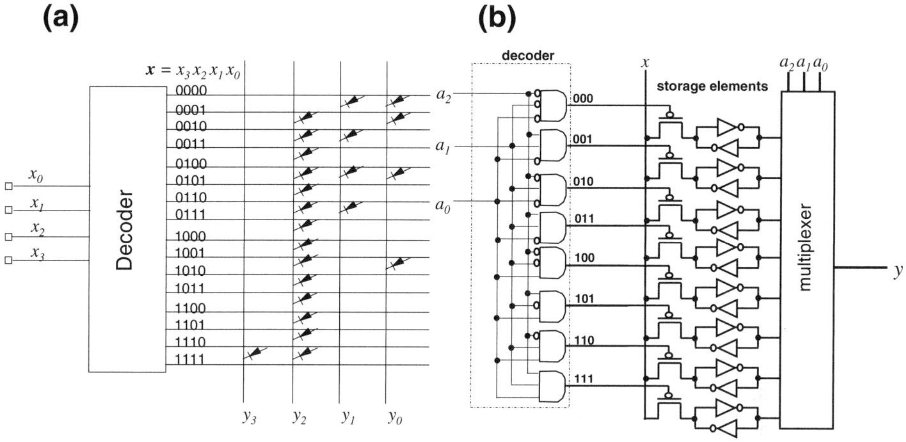

Fig. 18.3. Example read-only memory (ROM) and write-able random access memory (RAM) circuit implementations. (a) 16 x 4 ROM using a decoder and diodes. (b) 4 x 1 RAM based on flip-flops.

They take advantage of special combinational circuit blocks called *decoders* and *multiplexers*. A decoder converts a n-bit input variable space into 2^n output variables, of which only one output is "asserted" as logical state "1" for any setting of input space, and all unique settings of input space are typically mapped to individual outputs in lexicographical order of the input settings. Fig. 18.3a depicts a simple diode memory based on this principle in which a read-only memory (ROM) is organized as rows and columns, the rows being decoder outputs, and the columns being a 2^n-bit memory formed by including or omitting diodes to create the memory pattern. Implementing a writeable memory (Fig. 18.3b) can be implemented with a similar address decoding circuit, but requires replacing diodes with alterable bit storage structures. In this case, storage is created through an array of cross-coupled invertors, which form storage cells as bistable elements whose state can be overridden with a pass-connected transistor (shown left in each of the eight invertor pairs). The pass-connected transistor (or "pass transistor" or "transmission gate") provides a high-impedance isolation unless the gate is driven, in which case a current path is opened between each side (i.e., the source and drain are connected). Each output decoder row drives one pass transistor gate, connecting input x to a particular storage cell. Hence, the "write" side of the memory is formed. The "read side" of this simple memory is formed with a *multiplexer* (another combinational circuit) which simply conveys as its output the state present on a single one of its 2^n data inputs (with n being 3 in this example), each connected to single bistable element, as uniquely specified by the n-bit address on its control input port. The address lines (the a_i inputs to the multiplexer) need not be connected to the address lines of the decoder; separating them produces a rudimentary two-port memory in which one port is write-only and the second read-only. Additional circuitry can produce a memory that reads and writes using the same port.

Memory architectures, such as these pedagogical examples, can be used to implement logic functions directly (as an alternate to using logic gate networks). Logic functions can be represented by truth tables, as shown in the simple example of a majority function in Fig. 18.4.

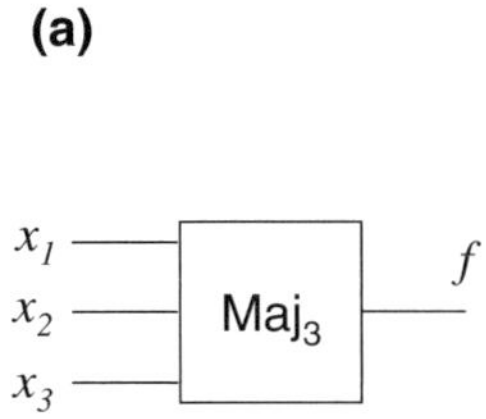

(b)

x_1	x_2	x_3	f
0	0	0	0
0	0	1	0
0	1	0	0
0	1	1	1
1	0	0	0
1	0	1	1
1	1	0	1
1	1	1	1

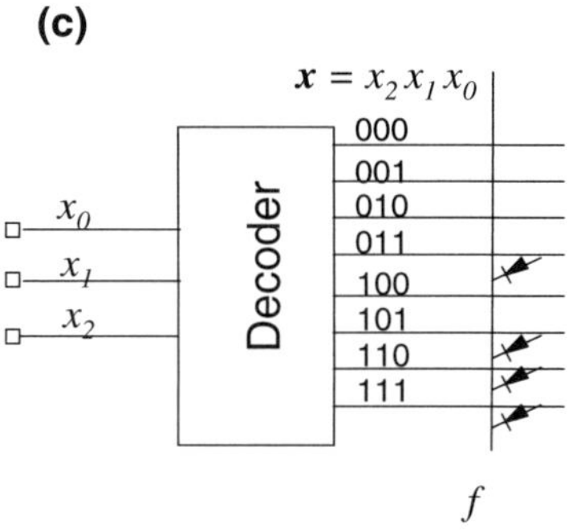

Fig. 18.4. Implementation of Boolean function in memory. (a) Majority gate function. (b) Truth table. (c) Mapping into three-input (8x1) memory.

The m-bit majority function is defined as "1" only when $\lfloor m/2 \rfloor$ of inputs are "1", as reflected in its truth table (Fig. 18.4). The implementation of this (or any) Boolean function in a memory circuit is achieved simply by translating the entries in the output column of the truth table (Fig. 18.4b) as a pattern in a corresponding three-input (8-bit) memory. In the case of a diode memory (Fig. 18.4c), diodes are placed on the rows having a true ("1") assignment. When a memory is used this way, it is referred to as a look-up table (LUT), in this example a 3-input LUT or "3-LUT". In general, an m-LUT requires 2^m memory bits, which form 2^{2^m} unique patterns, each being a possible Boolean function of m inputs. LUTs are extremely important in the formation of programmable or configurable logic systems, most notably the field programmable gate array (FPGA).

18.3 Simple Configurable Digital Systems

This section introduces some simple pedagogical examples of mythical field programmable gate arrays (FPGAs), which at one level can be thought of as pre-built digital chips whose function is like a blank slate that is effectively user-configurable after fabrication. The LUT is the key building block, which will be used in these examples, which will be called the "SU" (somewhat useless) series of FPGAs. While remarkably unsophisticated, these examples will expose a number of key principles underlying the most complex reconfigurable systems.

The simplest possible FPGA would probably be based on a single LUT. The "FPGA" embodying this notion is called the "Somewhat Useless" (SU-1) FPGA, shown in Fig. 18.5.

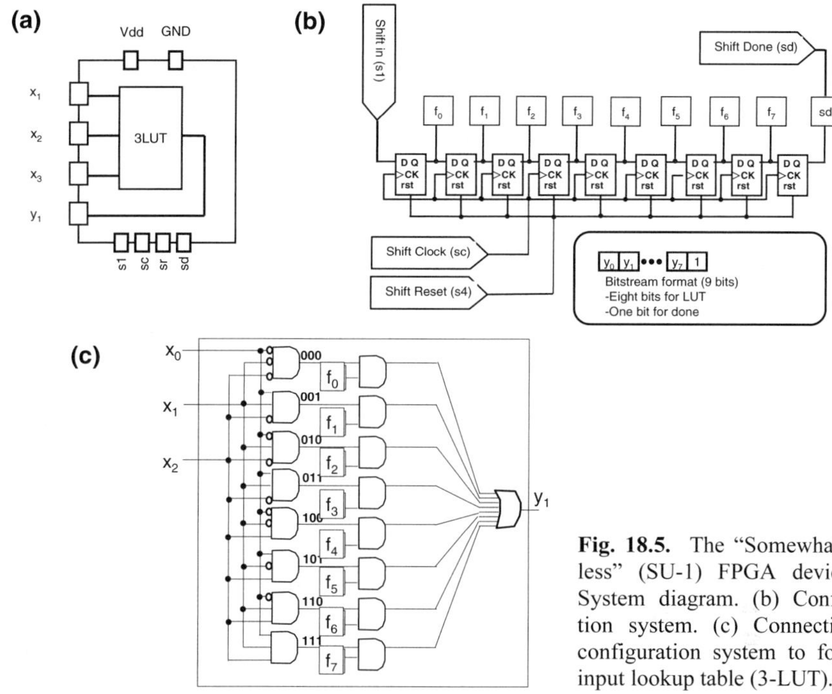

Fig. 18.5. The "Somewhat Useless" (SU-1) FPGA device (a) System diagram. (b) Configuration system. (c) Connection of configuration system to form 3-input lookup table (3-LUT).

This minimal FPGA provides pins for three user inputs and a single user output, with the remaining pins necessary to power and configure the FPGA. In Fig. 18.4, a 3-LUT was shown based on a diode memory. Since the diode memory in Fig. 18.4. is not configurable after fabrication, it is more useful to consider writeable memories as a basis for the LUTs to be described in this series. For the SU-1, the configuration "system" for the 3-LUT employs a shift register memory. The f_i implement the column of the truth table (identically the memory pattern needing to implement the 3-LUT function) by applying the *bitstream* pattern as a linear stream on a configuration "shift-in" (s1) pin, in synchronization with a shift clock (sc), which advances the shifted pattern through the shift register, which is terminated with an extra dummy position to indicate completion of the loading of the bitstream. A shift register reset (s4) is convenient in initializing or reconfiguring the FPGA. The shift register is embedded within additional logic to form the overall 3-LUT (Fig. 18.5c).

Since the *m*LUT is an effective mechanism for implementing simple functions, why not simply build FPGAs as *m*LUTs for larger values of *m*, which would enable the implementation of arbitrarily large Boolean functions? Obviously, the exponential growth of the memory with *m* would serve as a significant barrier. Even a simple circuit, such as a 16-bit adder, would require a 2^{32} = 4 Gb (32 inputs being needed to accommodate each addend) for every bit of the output result, which would result in a 68 Gbit memory! Fortunately, most digital functions can be expressed as a relatively compact composition of simpler digital functions, and powerful synthesis approaches have been developed [18.5] to find those representations.

In this spirit, the "SU-2" has added a second LUT. While seemingly a straightforward extension of the SU-1 (Fig. 18.6a), adding a second LUT requires the addition of circuitry to manage the connections of signals between the LUTs and the external user pins of the FPGA device (Fig. 18.6b). The circles in Fig. 18.6b represent switches that are configurably set to short the associated row and column together using a pass transistor backed by a configuration memory cell. In this manner, a programmable wiring system is created to route signals between user inputs, outputs and the inputs and outputs of the LUTs. The logical appearance of the shift register chain (Fig. 18.6c) is very similar to that used in SU-1, except that it must be longer to accommodate the addition LUT and the settings for the routing switch manifold. The bitstream logical position must, of course, be correlated to the physical location of particular switches, as suggested notionally in Fig. 18.6d.

The SU-1 and SU-2 "FPGAs" are at most capable of only implementing a few simple combinational circuits. To implement FSMs, it is necessary to somehow integrate storage cells that are accessible to the user (as opposed to the many storage cells inaccessible to the user, but necessary to configure the FPGA device). In the next progression of the "SU family", the SU-3 FPGA shown in Fig. 18.7 provides the integration of user storage by adding a bypassable flip-flop in the LUT. In fact, the six configurable logic blocks (CLBs) in Fig. 18.7a represent a more generalized structure in FPGA architectures, often containing a number of simpler LUTs, bypassable flip-flops, and other features. The CLB shown in Fig. 18.7b is very simple, containing only a 3LUT and a bypassable flip-flop. The flip-flop, when "engaged", provides a registration structure, useful for building FSMs. A single global clock is distributed directly to the clock signal of each flip-flop in SU-3, whereas in real-world FPGA devices, the clocking signals would themselves be separable and configurable, exploiting enhanced routing structures to maintain high-performance.

The routing system for SU-3, shown in Fig. 18.7c, is progressively more sophisticated. The configuration system is not shown, but follows the same general approach used in SU-1 and SU-2, with additional shift register positions in the bitstream being dedicated to configure the bypass settings of each CLB.

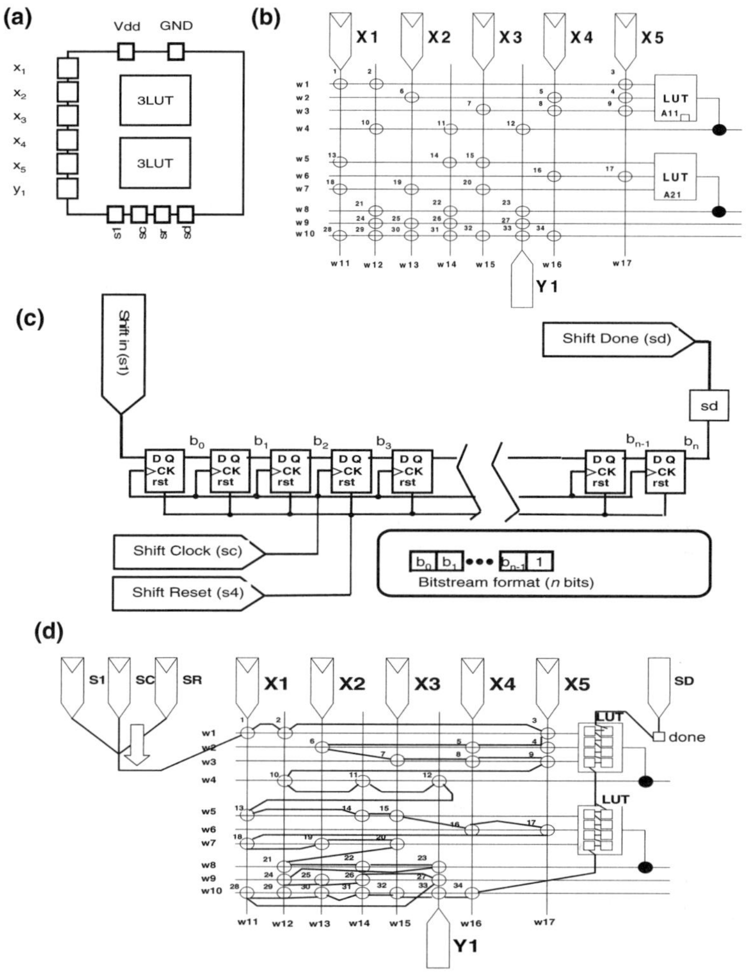

Fig. 18.6. SU-2 FPGA. (a) System block diagram. (b) Internal detail of routing network. (c) Configuration bitstream circuitry. (d) Notional distribution of shift register circuitry.

All digital systems are comprised of logic, interconnect, and memory. FPGAs, which aim to emulate these systems, at a very basic level are formed by creating a sufficiently expressive fabric of wiring together with large numbers of CLBs. In the very simple FPGA examples presented in this section, key concepts that ultimately resulted in a number of CLBs are interconnected through a configurable wiring fabric to form circuits. Key to the success of FPGA architectures are the choices in the composition of the CLB and the LUTs they might contain, the number of CLBs and the arrangement of them necessary to route the nettist (set of defined connections between CLBs and the input/output of the FPGA component).

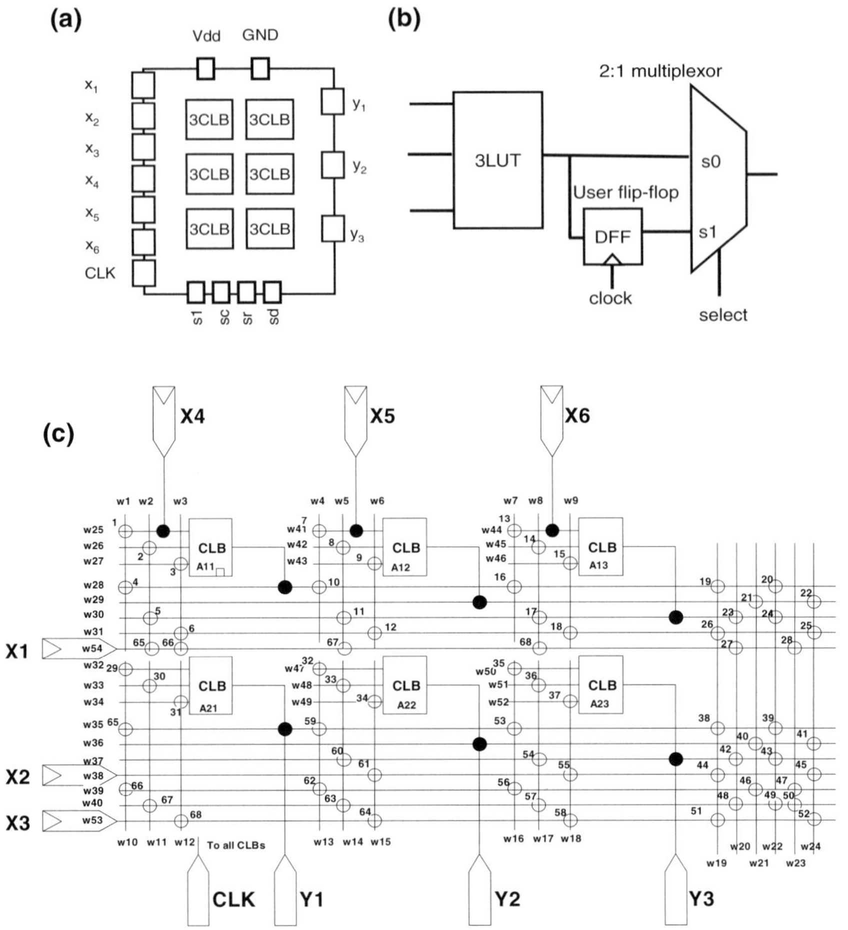

Fig. 18.7. SU-3 FPGA architecture. (a) System diagram. (b) Configuration logic block (CLB) details. (c) Detailed diagram (clock signal CLK is distribution to all CLBs).

Using the configuration machinery, all settings of logic, memory and interconnects are conveyed as a series of 0-1 programming decisions in the form of a configuration memory bitstream. To be sure, modern FPGAs are vastly more sophisticated than these simple examples. Over the years, modern FPGA architectures have supplemented the simple building blocks with a variety of modular intellectual property blocks that are "sprinkled" through the architecture to include block memories (means contiguous regions of storage memory, usually > 1kb), multiply accumulate blocks, configurable input/output (I/O) blocks, dedicated hardcore processors, clock management modules (for multiple clock domains), and dedicated encryption cores. Each of these modules can be thought of as black boxes whose termini connect to the internal interconnection manifold and whose functions are shaped through internal configurations joined to the overall configuration bitstream. This concept of digitally defined systems is fundamentally important, as it provides us insights that are useful to the extension of digital systems to configurable systems of any description. In this context any configurable system can be viewed as an ensemble of configurable components, an idea that might be true at several levels of hierarchy.

18.4 Considerations in Computation Architectures

Computation with digital circuits tends toward two extremes. At the one end lie those architectures that are optimized for temporal computation, the so-called von Neumann (stored program) architectures, which are in effect the real-world implementations of Turing machines [18.6]. In computer science, all Turing machines are basically equivalent. Though the microprocessor in a digital watch seems much different than a high-end processor core in a supercomputer, each is fundamentally capable of computing exactly the same set of algorithms. In other words, *anything* that can be computed by one, can be computed by the other (one may just require a lot more time than the other). The other extreme of computation is spatially-oriented, in which individual computations are processed through combinational logic. These computations are "circuitize-able", meaning that they can be mapped into stateless logic. We tend to believe that such computers are "faster" since we do not have to sequence computations through the complex state machinery of a Turing machine, which focuses on a single stage of computation at a time, and we can exploit parallelism (in space) to achieve more work in a unit of time. Unfortunately, circuitize-able computation is severely limited in that many calculations that can be performed very compactly with simple Turing machines require circuits that grow exponentially in the length of the inputs that are fed into the computation. As such, most computation architectures are a complex compromise of temporal and spatial computation styles. Modern Turing machines implement many exotic techniques (caching, pipelining, speculative execution) to reduce temporal latencies in calculation (the reader will find [18.7] an excellent reference discussing these techniques).

An important consideration in the temporal-spatial trades made in computation architecture is the nature of the networks required to perform computation and those induced by the structure of computation architectures. The origins of understanding the nature of such networks has evolved eclectically. For example, E. F. Rent studied the relationship between the number of pins and modules in circuitry blocks, establishing the well-known "Rent's rule" [18.8]:

$$T = A \cdot G^p , \tag{18.2}$$

where T is the number of terminals, G is the number of logic gates, A is the average number of pins per gate, and $0 < p < 1$ is Rent's exponent. Researchers have found that complex architectures are characterized by a Rent's exponent range of $0.5 < p < 0.8$. Rent's exponent is low for systems with regular structure, such as memories, and is highest for complex Application Specific Integrated Circuits (ASICs). Random circuitry has no Rent's rule (i.e., $p = 1$). Others have commented extensively on the existence and prevalence of power-law relationships such as Rent's rule throughout other man-made systems (most notably, Albert Lazlio-Barabasi [18.9]), and they are used to define the interconnection manifold densities for FPGA architectures [18.10]. In spatial computation, matching the wiring supply of circuit networks to the wiring demands of computations embedded within them is an important objective, since strong imbalances (marked by interconnection starvation (or overabundance) lead to resource inefficiencies. In temporal computation, an analogous problem occurs in the form of memory bandwidth and the structure of the memory hierarchy in computation. The nature of computation (such as the principle that most computation is spatially localized in memory) has resulted in the creation of multi-level caching architectures to compensate for the time it takes to access larger stores of memory.

18. 5 Multi-valued Systems

The physical implementation of logic gates as building blocks requires the ability to generate (output) and unambiguously resolve (input) measurable quantities into at least two distinct levels. Noise content in electrical signal ultimately limits the ability to extend digital two-value logic into multi-valued logic (including logic that operates on continuum value ranges). The variations in signal due to noise must not cause the partitions in voltage range defining states to overlap. It is a necessary practice to define gaps between those ranges. Those gaps are referred to as noise margins, as they represent the separation of voltage ranges defining state values. When noise causes signal to blur outside those partitions, erroneous operation results. The idealized transfer function in Fig. 18.1 is always present with a number of degradations, and the robustness of a two-value system of logic simplifies the problems of effectively cascading simple elements into complex arrangement of many of them to form a system. First, the value of threshold voltage (v_{th}) is not stable but depends on process, operating temperature, and aging conditions such as exposure to radiation. The value of v_{th} may differ between transistors even in the same monolithic integrated circuit, and it will certainly vary in transistors in separate components. The ideal transfer function would be a crisp step function. The transfer function of real world circuits, as depicted in Fig.18.1, cannot have infinite vertical slope. As such, there are values of v_{in} which will provide any v_{out} between the two clearly definite logic levels. Usually, the measured and generated signals are conductively-coupled voltages, but they could be conducted currents, magnetic fields, liquid flow, reactivity rates or other observable phenomena.

When the process of generation in logic is compatible with the process of measurement, then the logic system is cascade-able and can support non-trivial circuit depth. If the logic system is infinitely cascadeable, then the logic system is *regenerative*. This ability to chain the output of one logic circuit to another is obviously fundamental to construct complex Boolean circuits, since without it, additional, separate structures are required to query and drive the circuits. If these structures are small enough, they can be considered part of the logic circuit and need not be distinguished as a separate structure. If they are not small enough, then the circuit density of the overall scheme is limited by the density of the supporting structure. CMOS became CMOS, in part because the more minimal nMOS and pMOS technologies (using only n-type or p-type transistors, respectively) that preceded them were imbalanced, each capable of "pulling" signals only toward a particular "rail" (a "rail" being either electrical ground or the positive voltage power source). CMOS, in effect, combined both approaches into dual-rail system in a single circuit.

The requirements on logic circuit input and output (I/O) level compatibility is now made more precise for the case involving electrical voltages, as is the case in traditional saturating digital logic (e.g. CMOS, transistor-transistor logic (TTL) and its higher speed variants). Given a single-ended supply (a positive "rail" voltage v_{DD} and ground voltage *GND*, usually assumed as zero potential), it is usually the goal of a logic drive design to produce an output signal corresponding to a logical "1" very close to this supply rail, whereas logical "0" would ideally be identical to the ground potential (*GND*). In practice, of course, this ideality is never achieved. Instead the drive specification of a logic system defines a minimum voltage for a logical "1" ($v_{OH,min}$) and maximum voltage for a logical zero ($v_{OL,max}$) guaranteed over a certain set of usage conditions (such as temperature and fan-out). The typical logic input circuit would ideally be designed to resolve all signals between ground and v_{DD}, with exactly one ambiguous voltage level at a design threshold $v_{DD}/2= v_{IN,thresh}$. All signals measured above $v_{IN,thresh}$ are resolved as logical "1", whereas all signals below $v_{IN,thresh}$ are resolved as logical "0". Rather than an infinitesimal range, however, the ambiguity zone is necessarily made wider to account for variations in process and usage conditions. The size and placement of this window is defined by two parameters: $v_{IL,max}$, the highest voltage "safely" resolved as a logical "0", and $v_{IH,min}$ the minimum voltage safely defined as a logical "1". Clearly, the relationship of these parameters for any realizable logic implementation must be:

$$GND < v_{OL,max} < v_{IL,max} < v_{IH,min} < v_{OH,min} < v_{DD} \qquad (18.3)$$

The magnitude $| v_{IL,max} - v_{OL,max}|$ is sometimes referred to as the *low* or Δ0 noise margin, while $| v_{IL,max} - v_{OL,max}|$ is termed the *high* or Δ1 noise margin.

Regenerative logic structures must have some type of signal amplification or gain, to restore degraded input signal levels. Sometimes in a complex VLSI (very-large-scale integration) design, complex circuits drawing from a pool of different structural types are combined. For example, a series of CMOS transmission gates, which are cascade-able but not regenerative, are often combined with a traditional CMOS inverter on the back end of the chain to provide signal restoration. This concept is illustrated in Fig. 18.8, in which a simple two-stage binary select network of nMOS transistors configured as pass gates, followed by a CMOS inverter, which provides the signal restoration.

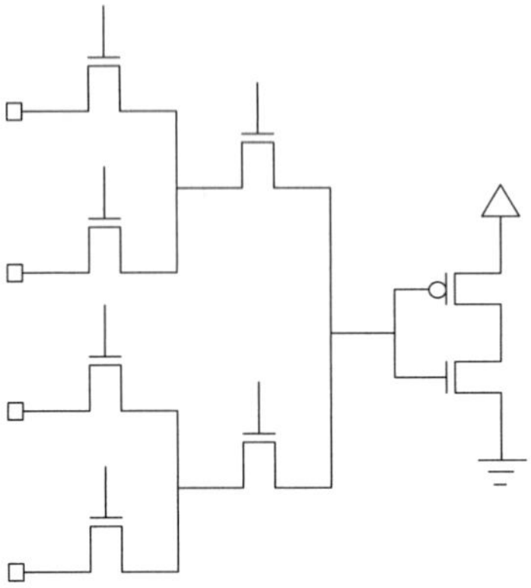

Fig. 18.8. A combination of non-regenerative logic and regenerative logic. The first two transistor stages (from left to right) are nMOS pass transistors which do not provide signal gain, connected to a CMOS invertor (see Fig. 18.1) which restores the signal to one of two logic levels.

Multi-valued systems are complicated by the need to extend the Δ0, Δ1 concepts to an exponentially growing number of $\Delta_{i\text{-}j}$ margins, one of many reasons that multi-valued logic has not found popular use.

18.6 Threshold Logic

A perhaps more manageable intermediate between pure Boolean and multi-valued logic systems involves systems that have digital inputs and digital outputs but employ multi-valued concepts to compute a Boolean output. In particular, threshold logic employs an analog summing junction that is digitally resolved with a thresholding element. The thresholding element can be as simple as a CMOS invertor. An example of a configurable threshold circuit based on the single-input case is shown in Fig. 18.9.

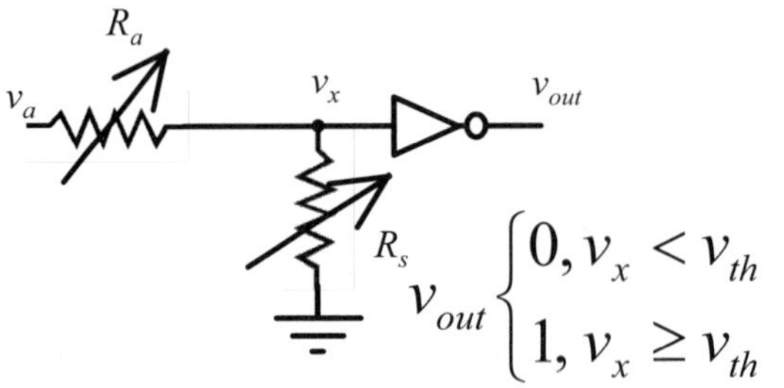

Fig. 18.9. A nearly trivial thresholded logic circuit employing two programmable resistors.

In this case, the idealized Fig. 18.1 invertor is preceded by a voltage ladder in which the resistors R_a and R_s can be configured. From elementary circuit theory, $v_x = v_a R_s / (R_s+R_a)$, and v_{out} resolves into a Boolean value. For the case where $v_{a,max} / v_{th} = 0.5$ (where $v_{a,max}$ corresponds to "1", $v_a = 0$ corresponding to "0") it is necessary to have $R_s / (R_s+R_a) \geq 0.5$ in order to have a non-constant output. This particular circuit is of limited utility, however. Of the four possible Boolean functions of one variable (two constant functions: "0" and "1", "TRUE", and "FALSE") it is only possible to configure R_s and R_a to produce two of them ("1", NOT).

More useful thresholded circuits involve a greater number of inputs, and a great class of functions can be implemented with a single configurable resistance (i.e, R_s). Threshold logic has been widely studied [18.11] but has not gained acceptance owing to the success of digital CMOS logic and the difficulty in implementing a competitive version of threshold logic in MOS. Phase change materials offer new hope in reviving threshold logic, by establishing a possible basis for reconfigurable logic that exploits configurable resistances. To illustrate the density advantage of threshold logic, Fig. 18.10 depicts the relative size of an nMOS implementation of a four-bit parity function with a prospective chalcogenide version.

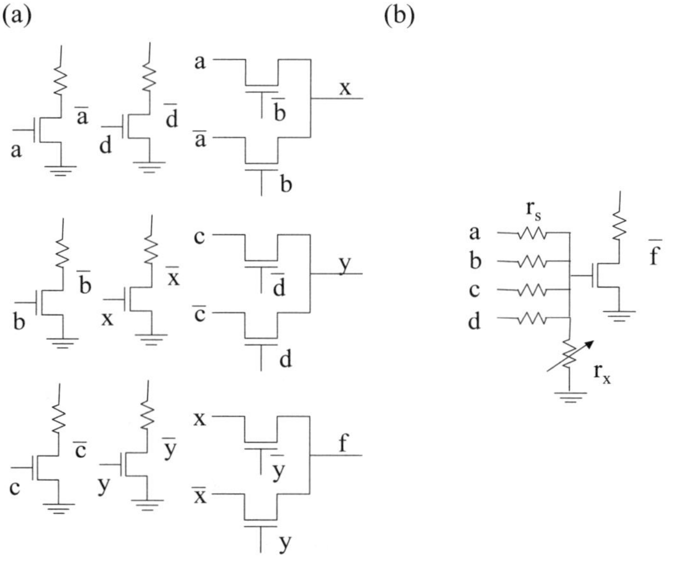

Fig. 18.10. n-Channel MOS (nMOS) implementation and threshold logic implementation of four-input parity function. (a) nMOS representation requires 18 devices. (b) threshold circuit implements complement of parity using three devices. The in-line series resistances (r_s) are due to parasitic interconnects and are not included in device count but are necessary for proper operation.

Clearly, the nMOS input voltage is decided by the simple voltage ladder formed through the combination of series input resistances with a shunt chalcogenide element:

$$V_{gate}(V_i, \alpha) = \frac{\sum_{i=1}^{n} V_i}{(n+\alpha)}, \tag{18.4}$$

where α is the ratio of series resistance r_s to chalcogenide resistance r_x. Through proper design of the ratio between threshold voltage and gain of the MOS device, it is possible to create a compact threshold gate, capable of implementing any symmetric Boolean function. Symmetric Boolean functions are those that depend only on the number of inputs, which include the class of parity, OR, AND, and majority gate functions.

Circuits realized with threshold logic gates have special significance in circuit complexity theory [18.12]. It is well known, for example, that AC, the class of Boolean circuits with unbounded fan-in (number of inputs), are contained within TC, the class of Boolean circuits that contain threshold circuits. As such, circuits based on threshold gates can be more compact than circuits based only on the standard set of Boolean primitive functions (e.g., AND, OR, NOT). In fact, many open questions remain regarding the expressive capacity of threshold digital circuits, including the possibility that NP (the class of non-deterministic polynomial time computation functions) might be contained in the class of constant-depth thresholded circuits with unbounded fan-in (TC^0) [18.11].

The ability to engineer programmable resistance may lead to much more powerful FPGA architectures. If a device can be engineered according to a template suggested in Fig. 18.11, then it is possible that a planar array of these gates can be used to form a scalable digital fabric with density and performance comparable, perhaps even superior, to those that define contemporary FPGAs.

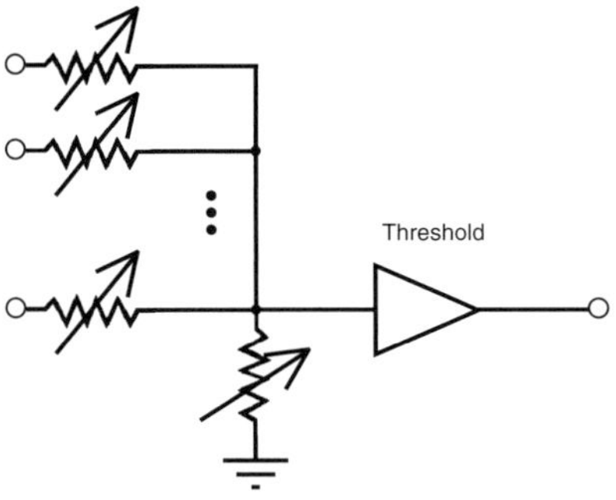

Fig. 18.11. Generic template for a thresholded lookup table (TLUT). Each rectangle is a programmable resistor formed from chalcogenide. Different valuations of resistance express different equivalent Boolean functions.

A properly defined threshold gate structure can be reconfigured to directly implement any of symmetric functions by programming the resistance state of chalcogenide, replacing the look-up-table (LUT) by a far more compact structure. Such an architecture does not supplant but complements the traditional base of structures provided by CMOS for standard digital computation. It is conceivable that (if advantageous to do so), future FPGA components could intermingle standard logic with threshold logic sub-regions. The degree of improvement possible, as measured in increased performance due to reduced logic depth or as measured in effective functional density, is itself an open research question.

18.7 Artificial Neural Networks

Thresholded circuits of the type described in the last section are sometimes called *Boolean linear thresholded functions* or *Boolean neurons* [18.13], denoted hard-thresholded functions. Even the invertor in Fig. 18.1 does not provide a hard threshold. An obvious extension of chalcogenide to analog applications, which Ovshinsky himself may have contemplated in his early work on chalcogenides (in [18.14], he compares the principles of chalcogenide to the human brain, see also the foreword), is in directly implementing simple artificial neural networks (ANNs). The connection between a common building block in ANNs, the perceptron, and an approximating chalcogenide based circuit is straightforward, as shown in Fig. 18.12.

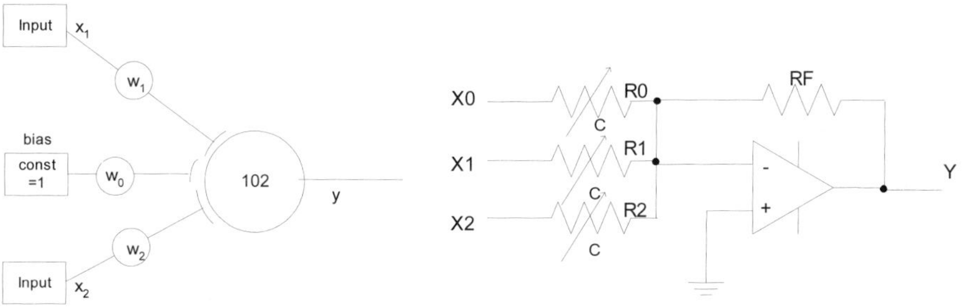

$$y = \sum_{i=0}^{m=2} x_i w_i \qquad\qquad y \approx \sum_{i=0}^{m=2} x_i \frac{R_f}{R_i}$$

Fig. 18.12. Perceptron and approximate implementation using chalcogenide.

Perceptrons using linear combiners [18.15] are ideally represented as a simple summing junction. The activation function, which is usually a non-linear function, is shown as an ideal operational amplifier (which, with high gain, becomes saturated and therefore non-linear), but in fact could be replaced with a single MOS device. The weights, which are the adjustable parameters in neural networks, are replaced with programmable resistors.

When the weights within perceptrons are adjusted as part of a training process, ANNs are considered adaptive. One framework for training ANNs, including the well-known back-propagation algorithm [18.16], is represented in Fig. 18.13. Although a number of silicon implementations of ANNs have been attempted, they are for the most part digital simulations of neural networks. Programmable resistors with weights that are tunable *in situ* offer unique opportunities for direct implementation of not only the ANN, but also for implementing the training system directly in MOS circuitry.

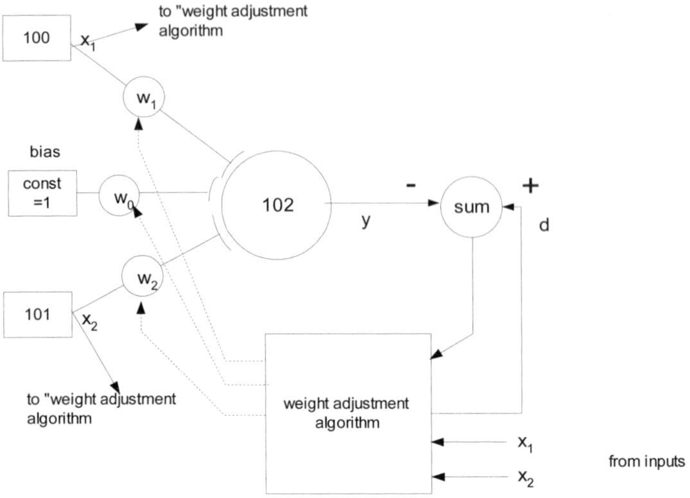

Fig. 18.13. Training "machinery" to adjust weights in an adaptive neural network.

Such an ANN implementation would be able to adapt on-line (during operation), preserving learned weight values non-volatilely. Since chalcogenide resistance values are not continuously variable, but are the by-products of controlled re-heating, then the training mechanism is more comparable to a sampled-analog feedback system. Convergence of such a system can be very rapid, limited only by the timescale under which the chalcogenide material is adjusted.

18.8 Other Analog-domain Programmable Systems

The concept of analog field programmable array architectures is not new, as a number of researchers and developers have been exploring ideas for some time [18.17-21]. Many of these earlier concepts were interesting but not commercially successful. More recent attempts [18.23] have built on the success of the FPGA. The simplest approach to integrate reconfigurable analog functionality is through the introduction of analog conversion elements. Other architectures have explored programmable transistor arrays [18.24] and switched capacitor elements [18.25] to build analog functions, but none so far have employed configurable resistance elements.

One near-term prospective application for configurable resistance is in creation of configurable, direct solvers for differential equations based on finite difference method (FDM) implementations. The finite difference method uses a discrete approximation of differential equations to reduce them to a system of algebraic equations. For example, it is possible to reduce the solution of Poisson's equation ($\nabla^2 v = \rho$ where v is the potential and ρ the charge density) over a 2-D space to an equation at each of many discrete points on a grid formed onto this space, where the voltage at any interior grid point x, y is given in its simplest form as :

$$v(x, y) = 1/4\ \{v(x + dx, y) + v(x-dx, y) + v(x, y + dy) + v(x, y-dy)\} \quad (18.5)$$

An electrical analog of this discretization can be realized by using a grid network of resistors, as shown in Fig. 18.14. Boundary conditions are simulated by impressing voltages on particular nodes.

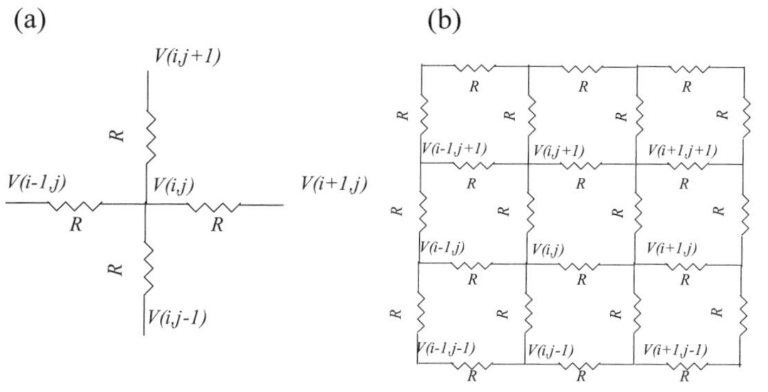

Fig. 18.14. Implementation of Laplace equation using finite difference approximation resistance network. (a) Neighborhood about a single grid point. (b) Section of 2-D grid.

These sources correspond to Dirichlet boundary conditions. Of course, Poisson's equation reduces to Laplace's equation in source-free regions ($\nabla^2 v = 0$). The parasitic resistance intrinsic in CMOS elements (without configurability) can form a partial differential equation (PDE) solver, if the values of these resistances can be tightly controlled (held to constant value with good precision). The solver is an analog-domain equation solver, which can converge much faster than a corresponding digital implementation of the finite difference method. Fig. 18.14a illustrates the implementation of a single grid point using resistances, and Fig. 18.14b depicts an extension of this cell to a larger subregion.

It is possible to make a self-contained circuit, in which a number of digital-to-analog converters (DACs) and analog-to-digital converters (ADCs) are added to circuits of the form of Fig. 18.14 produce a hybrid computer. A number of obvious design details are omitted in this discussion, such as: (1) the bussing arrangements and enable signals for the different peripheral blocks, (2) that the operating windows of the ADC/DAC components must be aligned with the effective operating voltage range of the integrated circuit elements involved, (3) the DAC outputs must be themselves isolated (using possibly a transmission gate switch) when they are not used actively in a particular problem formulation. The primary function of the DACs is to inject boundary conditions, while the function of the ADCs is to read analog voltages corresponding to nodal solutions. Therefore the number of nodal boundary conditions are limited by the number of DAC circuits.

The number of ADCs may be as few as one. If the ADC employs an infinite impedance front end, then it is conceivable that an ADC can be tied to a single node of the resistive grid. It is then possible, one by one, to close switches between the measurement node (defined as the single pin connected to an analog-to-digital converter) and particular nodes in an active equation under solution for the purposes of measurement. This measurement might be undertaken by using FPGA-like routing to form a probe infrastructure as suggested in Fig. 18.15.

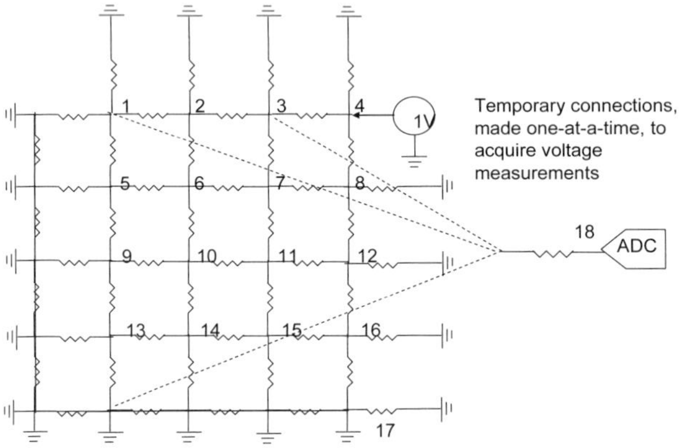

Fig. 18.15. A possible probe network for a hybrid PDE solver employing dynamically route-able probes and an analog-to-digital convertor (ADC). Connection at point 4 is an impressed voltage, implementing a Dirichlet boundary condition.

When the input impedance of the ADC is infinite, then the parasitic resistance formed by the temporarily switched measurement connections may not contribute to measurement errors with careful design. In the case of finite resistance, a systematic (correctable) error occurs due to the corresponding voltage ladder effects.

The infrastructure for the PDE solver described can benefit from the inclusion of programmable resistance elements in a number of ways. First, within a resistive grid, the inclusion of programmable resistance allows the modeling of changes in continuous media being simulated by the PDE solver. A region of fixed resistances implements the simulation of continuous media, such as a particular section of dielectric in an electromagnetic simulation. Altering the resistance from one grid point to another would, in this case, implement a dielectric boundary of (for example) a material having a different electric permittivity. A second benefit is derived in a multi-epoch simulation, in which only particular fractions of a large problem must be simulated. In this case, a programmable resistance grid can be rapidly reconfigured to implement each portion of a problem, advancing the solution in a piecewise fashion, with a single cycle evaluation of all pieces constituting one simulation "epoch".

18.9 Conclusions

In this chapter, a number of important concepts in configurable systems has been introduced. The foundations of basic digital systems were reviewed, and the role of memory in the creation of configurable systems was examined. Several themes of how programmable resistance elements (such as those possible in phase change materials) could be introduced in microelectronics to form configurable systems were explored, from summing junctions to programmable equation solvers.

Configurable resistance elements offer a promising technology for extending the utility of field programmable gate arrays (FPGAs). Through approaches such as threshold logic, for example, they may provide improvements in performance. As newer hybrid FPGA concepts, such as those described in the last section, become more prevalent, configurable resistance elements will provide more powerful techniques for blending continuous circuitry with digital elements.

References

[18.1] Bakoglu, H.B. Circuits, Interconnections, and Packaging for VLSI, Addison Wesley, Reading, MA (1990)

[18.2] Martin, A. J., Alain J., Nystrom, M.: Asynchronous techniques for system-on-chip design. Proc. of the IEEE **94**, 1089 – 1120 (2006)

[18.3] Gray, C.T., Liu, W. and Cavin, R.K., Wave Pipelining: Theory and CMOS Implementation, Kluwer Academic Publications (1993)

[18.4] Burleson, W.P., Ciesielski, M., Klass and F. Liu, W.: Wave-pipelining: A tutorial and research survey. IEEE Transactions On Very Large Scale Integration (VLSI) Systems **6**, 464-474 (1998)

[18.5] Brayton, R.K, Hachtel, G.D. and Sangiovanni-Vincentelli, A.L.: Multilevel logic synthesis. Proceedings of the IEEE **78**, 254-300 (1990)

[18.6] Moret, B. M.: The Theory of Computation. Addison-Wesley, Reading, MA (1998)

[18.7] Hennessy, J. L., and Patterson, D. A.: Computer Architecture: A Quantitative Approach (Third Edition). Morgan Kaufmann Publishers, San Francisco, CA (2003)

[18.8] Donath, W.: Placement and average interconnection lengths of computer logic, IEEE Trans. Of Circuits and Systems, **CAS-26**(4) (April 1979)

[18.9] Barabási, A.-L. and Réka, A.: Emergence of scaling in random networks. Science **286**, 509-512 (1999)

[18.10] De Hon, A.: Reconfigurable Architectures for General-Purpose Computing. AI Technical Report 1586, MIT Artificial Intelligence Laboratory, Cambridge, MA, (1996)

[18.11] Maciel, A. and Therien, D.: Threshold circuits of small majority-depth. Information and Computation, **146**,55-82 (1998)

[18.12] Vollmer, H.:. Introduction to Circuit Complexity. Springer, New York (1999)

[18.13] Parberry, I.: Circuit Complexity and Neural Networks, MIT Press, Cambridge, MA (1994)

[18.14] Brown, K.: Inventors at Work : Interviews With 16 Notable American Inventors, Microsoft Press (1988)

[18.15] Minsky, M. L. and Papert, S. A.: Perceptrons (expanded edition), MIT Press, Cambridge MA (1988).

[18.16] Haykin, S.: Neural Networks: A Comprehensive Foundation. Prentice Hall, Upper Saddle River, NJ (1999)
[18.17] Sackinger, E., Lee, E. K. F. and Gulak, P.G.: A CMOS Field-programmable analog array. IEEE Solid-State Circuits Conference. (ISSCC), 186-314 (1991)
[18.18] Gulak, P.G.: Field programmable analog arrays: past, present and future perspectives. Proceedings of International Conference on Microelectronics. pp. 123-126 (6 November 1995).
[18.19] Schweber, B.: Programmable analog ICs: Designer's delight or dilemma?. Electronic Design News. 72-84 (13 April 2000)
[18.20] Bindra, A.: Field programmability pervades analog devices. Electronic Design, 56-62 (9 July 2001)
[18.21] Edwards, R.T., Strohbehn, K. and Jaskulek, S.E.: A field-programmable mixed-signal array architecture using antifuse interconnects. Proceedings from the 2000 IEEE International Symposium on Circuits and Systems, Geneva, 319-322 (2000)
[18.22] Lee, E.K.F. and Gulak, P.G.: Field programmable analogue array based on MOSFET transconductors. Electronics Letters **28**, 292-293 (January 1992).
[18.23] Actel Fusion Programmable System Chips, Revision 1.2 product announcement, Actel Corporation (http:/www.actel.com) (2007).
[18.24] Stoica, A., Keymeulen, D., Zebulum, R. S., Mojarradi, M., Kat-koori, S., Daud, T.: Adaptive and Evolvable Analog Electronics for Space Applications. Proceedings of the 7th International Conference on Evolvable Systems (ICES), 379-390, (2007)
[18.25] Pierzchala, E., Gulak, G. and Chua, L. (editors): Field Programmable Analog Arrays, Springer, New York (1994)
[18.26] Green, H.E.: The numerical solution of some important transmission-line problems. IEEE Transactions on Microwave Theory and Techniques. MTT-**13**, 676-692 (May 1965)

Author Bios

Dr. Simone Raoux, IBM Almaden Research Center, 650 Harry Road, San Jose, CA 95120, USA. She is Research Staff Member at the IBM Almaden Research Center. She received her MS degree in 1984 and Ph.D. degree in physics in 1988 from Humboldt University, Berlin, Germany. Before joining IBM she worked at the Institute for Electron Physics, Berlin, and Lawrence Berkeley National Laboratory, Berkeley, CA. Her current research interests include the physics and materials science of phase change materials.

Prof. Matthias Wuttig, 1. Physikalisches Institut (1A), RWTH Aachen University, 52056 Aachen, Germany. He received his diploma from Cologne (1986) and Ph.D. from Aachen University (1988). He is a Full Professor of Physics at RWTH Aachen since 1997 and is presently Dean of the Faculty of Mathematics, Computer Science and the Natural Sciences. As a visiting scientist he has worked at NIRIM Tsukuba (Japan), Bell Labs (USA), $CRMC^2$ Marseille (France), Zhejiang University (China), IBM Research Center and UC Berkeley (USA). His main interest is the development of novel materials.

Dr. Stanford R. Ovshinsky, Ovshinsky Innovation LLC, 1050 East Square Lake Road, Bloomfield Hills, MI 48304, USA. He is the co-founder of Energy Conversion Devices, Inc. (ECD) and is the primary inventor of ECD's technology including the Ovonic Universal Memory (OUM), the Ovonic optical memory and various other chalcogenide devices, Ovonic thin-film photovoltaic technology and its continuous web multi-junction roll-to-roll machine, the Ovonic NiMH battery which enabled the electric and hybrid automotive industry, as well as the Ovonic solid hydrogen storage technology. He started the field of amorphous and disordered materials in the 1950s by atomically designing devices that have unique electronic, chemical and structural mechanisms. In 2007 he formed an independent new company called *Ovshinsky Innovation LLC.*

Dr. Chung Lam, T. J. Watson Research Center, Yorktown Heights, NY 10598, USA. He received his B.Sc. in Electrical Engineering at Polytechnic University of New York in 1978, and joined IBM General Technology Division in Burlington in 1978 as a memory circuit designer. In 1984, he was awarded the IBM Resident Study Fellowship and received his M.Sc. and Ph.D., both in Electrical Engineering, at Rensselaer Polytechnic Institute in 1987 and 1988 respectively. In 2003, Dr. Lam transferred to IBM Research Division, and was named Distinguished Engineer in 2007. He has managed the Phase-change memory Research Project at the IBM Research Division in T.J. Watson Research Center since 2003.

Dr. Wojciech Welnic, Laboratoire des Solides Irradies, Ecole Polytechnique, 91128 Palaiseau, France. He received his Ph.D. in 2006 from the Technical University Aachen working on phase change materials. He spent time at the Forschungszentrum, Juelich, Germany and the Ecole Polytechnique in Paris, France to study computational methods in solid state physics. He is currently a postdoctoral fellow at the Ecole Polytechnique and the ESRF in Grenoble working on electronic and structural properties of phase change materials.

Prof. Punit Boolchand, University of Cincinnati, 820 Rhodes Hall, Cincinnati, OH 45221-0030, USA. He received his Ph.D. in Physics from Case-Western Reserve University in 1969. He is currently a Professor of Electrical and Computer Engineering, and Physics at University of Cincinnati. He has been interested in nature of glass transition and molecular structure of network glasses probed by thermal, optical, nuclear and electrical methods. He has held visiting positions at Stanford (USA), Univ. of Paris VI (France), Leuven University (Belgium). He is a Fellow of the American Physical Society.

Prof. Matthieu Micoulaut, Laboratoire de Physique Théorique de la Matière Condensée, CNRS UMR 7600 Université Pierre et Marie Curie, Boite 121, 4 place Jussieu 75252 Paris Cedex 05, France. He received his Ph.D. in Theoretical Physics from University Paris VI (1993). He is currently Maitre de Conférences (Associate Professor) in theory of condensed matter (CNRS and UPMC-Paris VI). His interests include theoretical aspects of glass transition, glasses and amorphous solids, and topology based approaches and molecular simulations to investigate structure and dynamics of supercooled liquids.

Ping Chen, Department of Electrical and Computer Engineering, University of Cincinnati, Cincinnati, OH 45221-0030, USA. He received his M.S. degree in Microelectronics from Nanjing University (2002). He is currently graduate student in the Department of Electrical and Computer Engineering at University of Cincinnati and working towards his Ph.D. degree. He is interested in molecular structure, reversibility window and aging in alkali-germanate and As-chalcogenide glasses.

Prof. Stephen Elliott, Department of Chemistry, University of Cambridge, Lensfield Road, Cambridge CB2 1EW, UK. He is Professor of Chemical Physics in the Department of Chemistry, and a Fellow of Trinity College, at the University of Cambridge. His research interests centre on studies of disordered materials, particularly chalcogenide glasses, from both experimental and theoretical/computer-simulation points of view. He was the recipient, in 2001, of the first Ovshinsky Prize for research on chalcogenide materials.

Dr. Liesbeth van Pieterson, Philips Research, High Tech Campus 4 (Box WAG 12), 5656 AE Eindhoven, The Netherlands. She received her PhD degree in chemistry of condensed matter from Utrecht University in 2001. She is a senior scientist at Philips Research Laboratories in The Netherlands. Her research interest is in the area of solid-state chemistry with a current focus on the physics and chemistry of materials for optical recording.

Dr. Johannes Kalb, Intel Corporation, Robert Noyce Building, Mail Stop RNB-3-01, 2200 Mission College Blvd., Santa Clara, CA 95054, USA. He received his MS (2002) and PhD degree (2006) from the Technical University of Aachen, Germany. As a graduate student, he spent two years at the School of Engineering and Applied Sciences at Harvard University, working on stresses and crystallization kinetics in phase change materials. In 2006 he joined the Department of Materials Science and Engineering at MIT as a Postdoctoral Researcher. Since 2007 he has been with Intel Corporation, continuing research on phase change materials.

Dr. Paul Fons, Advanced Institute of Industrial Science and Technology, Tsukuba, Japan. He received a masters in physics and a Ph.D. in materials science from the University of Illinois, graduating in 1990. He was visiting research fellow at the Applied Physics Department of the University of Tsukuba in Japan. He became a permanent staff member of the Optoelectronics division of the Electrotechnical Laboratory of AIST in 1993. In 2003, he moved to the Center for Applied Near-Field Optics Research to work on materials characterization of optical disk materials where he is now head of the Nano Optics research group.

Dr. Bong-Sub Lee, University of Illinois at Urbana-Champaign, 1-110 Engineering Sciences Building, 1101 W. Springfield Ave., Urbana, IL 61801, USA. He is a post-doctoral research associate at the University of Illinois at Urbana-Champaign, where he received his Ph. D. in Materials Science and Engineering in 2006. He received his M.S. and B.S. (cum Laude) at Seoul National University, Korea. He is a winner of the E*PCOS 07 Presentation Award, European Phase Change and Ovonics Symposium (2007), as well as other awards and scholarships.

Prof. Stephen Bishop, University of Illinois at Urbana-Champaign, 153 Everitt Laboratory, 1406 West Green St, MC-702, Urbana, IL 61801, USA. He is Professor of Electrical and Computer Engineering and Physics at the University of Illinois at Urbana-Champaign (UIUC). Before joining the UIUC in 1989, he was a research physicist for 23 years at the Naval Research Laboratory (NRL) in Washington, D.C. He received his BA from Gettysburg College and a Ph.D. in physics from Brown University.

Dr. Noboru Yamada, AV Core Technology Development Center, Matsushita Electric Industrial Co., Ltd. (Panasonic), 3-1-1 Yagumo-Nakamachi, Moriguchi, Osaka, Japan. He graduated in Electronic Science and Engineering in 1974 and obtained a Ph.D. in Engineering in 2001, both from Kyoto University. He joined Matsushita Electric Ind. Co., Ltd. in 1974. Since then he has been working in the Corporate R&D division. He was the first to propose GeTe-Sb_2Te_3 phase-change materials in 1987 and he pioneered several optical disc media such as DVD-RAMs and Dual layer Blu-ray discs using this material. Currently, he is General Manager of the Storage Media Group, AV Core Technology Development Center.

Dr. Delia J. Milliron, The Molecular Foundry, Lawrence Berkeley National Laboratory, 1 Cyclotron Road, Berkeley, CA 94720, USA. She is the Facility Director for Inorganic Nanostructures at the Molecular Foundry, and was previously a Research Staff Member at IBM. She obtained her A.B. from Princeton University in 1999 and her Ph.D. from the University of California, Berkeley in 2004, both in Chemistry. Her research interests include chemical synthesis and materials integration of colloidal nanoparticles and solution deposition of metal chalcogenide thin films.

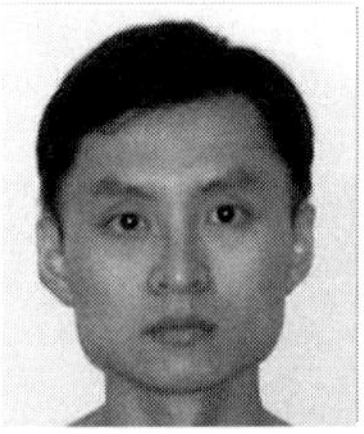

Dr. Qiang Huang, IBM T. J. Watson Research Center, P. O Box 218, Yorktown Heights, NY 10598, USA. He is Research Staff Member at the IBM TJ Watson Research Center. He obtained his B.E. from Zhejiang University in 1997, and his PhD from Louisiana State University in 2004, both in Chemical Engineering. His research interest includes Cu interconnects, electrodeposition of semiconductor materials, magnetic alloys, nanomaterials and nanodevices.

Dr. Yu Zhu, IBM T. J. Watson Research Center, P.O.Box 218, Yorktown Heights, NY 10598, USA. He is advisory engineer at IBM TJ Watson Research Center. He obtained his Ph.D. from College of Nanoscale Science and Engineering of University at Albany – SUNY. His research interests include atomic layer deposition and chemical vapor deposition of thin films for electronic device application.

Dr. Luping Shi, Data Storage Institute, DSI Building, 5 Engineering Drive 1, Singapore 117608. He received a master degree in solid physics from Shan Dong University, P.R. China in 1988, and doctor of science degree from Cologne University, Germany, in 1992. He joined the Data Storage Institute (DSI), Singapore, in 1996 as a Senior Engineer. Currently, he is senior Research Scientist, division manager of the Optical Materials & System division. He is in charge of optical storage, solid state random access memory (PCRAM) and artificial cognitive sensor and memory researches at DSI.

Dr. Junji Tominaga, Center for Applied Near-Field Optics Research, CAN-FOR, National Institute of Advanced Industrial Science & Technology, Tsukuba Central #4, 1-1-1 Higashi, Tsukuba, 305-8562, Japan. He received his Ph. D. from Cranfield Univ. UK in 1991. After research on rewritable optical phase-change disks at TDK corporation, he moved to the National Institute of Advanced Industrial Science & Technology (AIST) in 1997. He is currently director of the Center for Applied Near-Field Optics Research and has been the leader of the super-RENS ultra-high density optical storage project.

Prof. Daniele Ielmini, Dipartimento di Elettronica e Informazione, Politecnico di Milano, piazza Leonardo da Vinci 32, 20133 - Milano (MI), Italy. He received the Laurea (cum laude) and the Ph. D. degrees from the Politecnico di Milano, Italy in 1995 and 1999, respectively. He has been an Assistant Professor at Dipartimento di Elettronica e Informazione, Politecnico di Milano, since 2002. His main research interest is characterization and modelling of non volatile memories, including Flash, discrete-trap, phase-change and resistive-switching memories.

Dr. Yi-Chou Chen, Emerging Central Lab, Macronix Int. Co., Ltd., 16 Li-Hsin Road, Science Park, Hsinchu, Taiwan. He received his B.S. degree and Ph.D. degree in chemical engineering in 1995 and 2000, respectively, both from National Taiwan University. In 2000, he joined Macronix and worked on technology development of lithography, moving to the Emerging Central Lab in 2001 where he started working on phase change memory. From 2005 to 2007, he was on assignment at the IBM Almaden Research Center. His current research interests include both phase change material and memory devices.

Dr. Roberto Bez, STMicroelectronics M6 s.r.l. (at the time of publication with Numonyx), Via C.Olivetti 2, 20041, Agrate Brianza, Milan, Italy. He received the doctor degree in physics from the University of Milan, Italy, in 1985. In 1987 he joined STMicroelectronics and since then has worked on the Non-Volatile Memory technology development in the R&D department. He has been director of the Phase Change Memory, in the NVM Technology Development. He has authored many papers, conference contributions and patents on topics related to NVM and been lecturer on Non-Volatile Memory Devices at many Italian universities.

Dr. Robert J. Gleixner, Intel Corp. (at the time of publication with Numonyx), 2200 Mission College Blvd., Santa Clara, California. He received his Ph.D. degree in materials science from Stanford University in 1998. He joined Intel's Corporate Quality and Reliability group in 1998 and has worked on microprocessor, microdisplay, and non-volatile memory technologies and products. Since 2004, his work has focused on developing and productizing advanced Phase Change Memory technologies.

Dr. Fabio Pellizzer, STMicroelectronics M6 s.r.l. (at the time of publication with Numonyx), Via C.Olivetti 2, 20041, Agrate Brianza, Milan, Italy. He received the doctor degree in electronic engineering in 1996 from the University of Padova, Italy, with a thesis on characterization and reliability of thin gate oxides. In 1998 he joined the Central R&D department of STMicroelectronics in Agrate Brianza (Italy). After 2002 he has been in charge of process development for phase-change memories based on chalcogenide materials. He has authored many papers, conference contributions, and patents on phase-change memories.

Dr. Agostino Pirovano, STMicroelectronics M6 s.r.l. (at the time of publication with Numonyx), Via C.Olivetti 2, 20041, Agrate Brianza (Milan), Italy. He received the Laurea degree in electrical engineering from the Politecnico di Milano, Italy, in 1997, and the Ph.D. degree at the Department of Electrical Engineering, Politecnico di Milano, Italy, in 2000. Since 2001 he has worked on the electrical characterization and modeling of phase-change memories. In 2003 he joined the Non-Volatile Memory Technology Development Group of STMicroelectronics and he has been involved in the investigation of PCM and of emerging NVM technologies.

Dr. Greg Atwood, Intel Corp. (at the time of publication with Numonyx), 2200 Mission College Blvd., Santa Clara, California. He received the master degree in physics from Purdue University in 1979 at which time he joined Intel working initially in Micro-Processor technology development and latter in Non-Volatile Memory technology development, achieving the appointment of Intel Fellow in 1996. Since 2000, his primary focus has been on Phase Change Memory Technology and its application to electrical Non-Volatile Memories.

Dr. Matthew Breitwisch, IBM T. J. Watson Research Center, P. O Box 218, Yorktown Heights, NY 10598, USA. He is Research Staff Member at the IBM T.J. Watson Research Center in Yorktown Heights, NY. He received his B.S. degree in physics, mathematics, and astrophysics from the University of Wisconsin at Madison in 1994, and a Ph.D. degree in physics from Iowa State University in 1999. He joined IBM at the Microelectronics Center in Essex Junction, VT, and since 2005 has worked in the IBM Research exploratory memory group focusing on integration schemes for phase change memory devices.

Dr. James Lyke Air Force Research Laboratory (AFRL/RVSE), 3550 Aberdeen Ave SE, KAFB, NM 87117-5776, USA. He has a BS from the University of Tennessee, Knoxville, an MS from the Air Force Institute of Technology, and a PhD through the University of New Mexico, all in electrical engineering. He serves as technical advisor to the Space Electronics Branch of the Air Force Research Laboratory's Space Vehicles Directorate (AFRL/VSSE) at Kirtland Air Force Base New Mexico. His primary pursuits are the development of novel space systems architectures, especially those exploiting reconfigurable / plug-and-play approaches. He also maintains a keen interest in 3-D packaging and molecularly-scalable computational approaches.

Index

C